# Student Study Guide/Solutions Manual

to accompany

# General, Organic, and Biological Chemistry

## Second Edition

**Erin Smith Berk**

and

**Janice Gorzynski Smith**
*University of Hawai'i at Mānoa*

Student Study Guide/Solutions Manual to accompany
GENERAL, ORGANIC, AND BIOLOGICAL CHEMISTRY, SECOND EDITION
ERIN SMITH BERK AND JANICE GORZYNSKI SMITH

Published by McGraw-Hill Higher Education, an imprint of The McGraw-Hill Companies, Inc., 1221 Avenue of the Americas, New York, NY 10020. Copyright © 2013, 2010 by The McGraw-Hill Companies, Inc. All rights reserved. Printed in the United States of America.

✪ This book is printed on recycled, acid-free paper containing 10% post consumer waste.

2 3 4 5 6 7 8 9 0 QVR/QVR 1 0 9 8 7 6 5 4 3

ISBN: 978-0-07-733230-3
MHID: 0-07-733230-X

# Contents

# Preface

This Student Study Guide and Solutions Manual accompanies *General, Organic, and Biological Chemistry, Second Edition* by Janice Gorzynski Smith. Each chapter correlates with a chapter in the text and gives additional learning aids for increased understanding of the material. Each chapter consists of the following parts.

- **Chapter Review** — The Chapter Review uses the Key Concepts portion of the text as a framework for the most important topics in a chapter, with examples to illustrate key points.

- **Problem Solving** — The Problem Solving section consists of a series of worked out examples that parallel the sample problems in the text. Each example consists of a problem (different from the sample problems within the text), analysis, and step-by-step solution to illustrate how to solve problems in the chapter. These new problems provide additional help in how to solve the most commonly asked types of questions within a chapter.

- **Self-Test** — The Self-Test consists of 24–30 short answer questions that test an understanding of definitions, reactions, and other material encountered within a chapter. Answers to each question are provided at the end of the section.

- **Solutions** — The Solutions section consists of worked-out solutions—not merely short answers—to *all* in-chapter problems and *all* odd-numbered end-of-chapter problems. This section often contains references to particular examples in Problem Solving, so that students can see how similar problems are tackled.

The authors hope that the supplementary material contained in this study guide provides a valuable learning tool that will help the student better understand and appreciate the world of chemistry.

Erin Smith Berk
Janice Gorzynski Smith

# Chapter 1 Matter and Measurement

## Chapter Review

**[1] Describe the three states of matter. (1.1, 1.2)**
- Matter is anything that has mass and takes up volume. Matter has three common states:
  - The **solid state** is composed of highly organized particles that lie close together. A solid has a definite shape and volume.
  - The **liquid state** is composed of particles that lie close together but are less organized than the solid state. A liquid has a definite volume but not a definite shape.
  - The **gas state** is composed of highly disorganized particles that lie far apart. A gas has no definite shape or volume.

**[2] How is matter classified? (1.3)**
- Matter is classified in one of two categories:
  - A pure substance is composed of a single component with a constant composition. A pure substance is either an element, which cannot be broken down into simpler substances by a chemical reaction, or a compound, which is formed by combining two or more elements.
  - A mixture is composed of more than one component and its composition can vary depending on the sample.

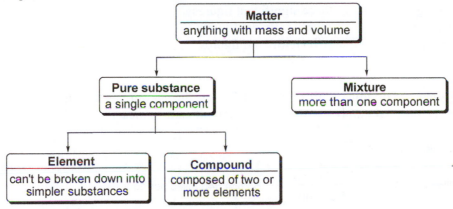

**[3] What are the key features of the metric system of measurement? (1.4)**
- The metric system is a system of measurement in which each type of measurement has a base unit and all other units are related to the base unit by a prefix that indicates if the unit is larger or smaller than the base unit.
- The base units are meter (m) for length, gram (g) for mass, liter (L) for volume, and second (s) for time.

### Common Prefixes Used for Metric Units

| Prefix | Symbol | Meaning | Numerical Value | Scientific Notation |
|--------|--------|---------|-----------------|---------------------|
| mega- | M | million | 1,000,000. | $10^6$ |
| kilo- | k | thousand | 1,000. | $10^3$ |
| deci- | d | tenth | 0.1 | $10^{-1}$ |
| centi- | c | hundredth | 0.01 | $10^{-2}$ |
| milli- | m | thousandth | 0.001 | $10^{-3}$ |
| micro- | μ | millionth | 0.000 001 | $10^{-6}$ |
| nano- | n | billionth | 0.000 000 001 | $10^{-9}$ |

## [4] What are significant figures and how are they used in calculations? (1.5)

- Significant figures are all digits in a measured number including one estimated digit. All nonzero digits are significant. A zero is significant only if it occurs between two nonzero digits, or at the end of a number with a decimal point. A trailing zero in a number without a decimal point is not considered significant.

- All nonzero digits are always significant.

| | |
|---|---|
| 65.2 g | three significant figures |
| 1,265 m | four significant figures |
| 25 μL | two significant figures |
| 255.345 g | six significant figures |

- In multiplying and dividing with significant figures, the answer has the same number of significant figures as the original number with the fewest significant figures.

four significant figures

$$\text{miles per hour} = \frac{351.2 \text{ miles}}{5.5 \text{ hours}} = 63.854\ 545 \text{ miles per hour}$$

two significant figures

The answer must contain only **two** significant figures.

$$= 64$$

- In adding or subtracting with significant figures, the answer has the same number of decimal places as the original number with the fewest decimal places.

weight at one year = 10.11 kg

weight at birth = 3.6 kg

      10.11 kg    two digits after the decimal point

     − 3.6 kg    one digit after the decimal point

weight gain = 6.51 kg

last significant digit

- The answer can have only **one** digit after the decimal point.
- Round 6.51 to 6.5.
- The baby gained 6.5 kg during his first year of life.

**[5] What is scientific notation? (1.6)**

- Scientific notation is a method of writing a number as $y \times 10^x$, where $y$ is a number between 1 and 10, and $x$ is a positive or negative exponent.
- To convert a standard number to a number in scientific notation, move the decimal point to give a number between 1 and 10. Multiply the result by $10^x$, where $x$ is the number of places the decimal point was moved. When the decimal point is moved to the left, $x$ is positive. When the decimal point is moved to the right, $x$ is negative.
- To convert a number in scientific notation to a standard number, move the decimal point to the right for positive exponents, and to the left for negative exponents.

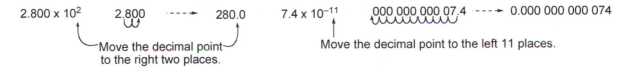

$$2.800 \times 10^2 \qquad 2.800 \dashrightarrow 280.0 \qquad 7.4 \times 10^{-11} \qquad 000\,000\,000\,07.4 \dashrightarrow 0.000\,000\,000\,074$$

Move the decimal point to the right two places.

Move the decimal point to the left 11 places.

**[6] How are conversion factors used to convert one unit to another? (1.7, 1.8)**

- A conversion factor is a term that converts a quantity in one unit to a quantity in another unit. To use conversion factors to solve a problem, set up the problem with any unwanted unit in the numerator of one term and the denominator of another term, so that unwanted units cancel.

conversion factor

$$130\,\cancel{lb} \quad \times \quad \frac{1\ kg}{2.20\,\cancel{lb}} \quad = \quad \boxed{59\ kg} \quad \text{answer in kilograms}$$

Pounds (lb) must be in the denominator to cancel the unwanted unit (lb) in the original quantity.

**[7] What is temperature and how are the three temperature scales related? (1.9)**

- Temperature is a measure of how hot or cold an object is. The Fahrenheit and Celsius temperature scales are divided into degrees. Both the size of the degree and the zero point of these scales differ. The Kelvin scale is divided into kelvins, and one kelvin is the same size as one degree Celsius.

To convert from Celsius to Fahrenheit:

$$°F = 1.8(°C) + 32$$

To convert from Fahrenheit to Celsius:

$$°C = \frac{°F - 32}{1.8}$$

To convert from Celsius to Kelvin:

$$K = °C + 273$$

To convert from Kelvin to Celsius:

$$°C = K - 273$$

**[8] What are density and specific gravity? (1.10)**

- **Density** is a physical property reported in g/mL or g/cc that relates the mass of an object to its volume. A less dense substance floats on top of a more dense liquid.
- **Specific gravity** is a unitless quantity that relates the density of a substance to the density of water. Since the density of water is 1.00 g/mL at common temperatures, the specific gravity of a substance equals its density but it contains no units.

$$\text{specific gravity} \ = \ \frac{\text{density of a substance (g/mL)}}{\text{density of water (g/mL)}}$$

## Problem Solving

### [1] Significant Figures (1.5)

**Example 1.1** How many significant figures does each number contain?

a. 0.309          b. 72,000          c. 52.00          d. 94.03

**Analysis**
All nonzero digits are significant. A zero is significant only if it occurs between two nonzero digits, or at the end of a number with a decimal point.

**Solution**
Significant figures are shown in **bold.**

a. 0.**309** (three)          b. **72**,000 (two)          c. **52.00** (four)          d. **94.03** (four)

---

**Example 1.2** Round off each number to two significant figures.

a. 2.7983          b. 0.010 023 9          c. 42,980.0

**Analysis**
If the answer is to have *two* significant figures, look at the *third* number from the left. If this number is 4 or less, drop it and all remaining numbers to the right. If the third number from the left is 5 or greater, round the number up by adding one to the second digit.

**Solution**
a. 2.8          b. 0.010          c. 43,000          (Omit the decimal point after the 0. The number 43,000. has five significant figures.)

---

**Example 1.3** Carry out each calculation and give the answer using the proper number of significant figures.

a. 4.29 × 0.023          b. 3,482 ÷ 52.7

**Analysis**
Since these calculations involve multiplication and division, the answer must have the same number of significant figures as the original number with the fewest number of significant figures.

**Solution**

a. $4.29 \times 0.023 = 0.098\ 67$

- Since 0.023 has only two significant figures, round the answer to give it two significant figures.

0.098 67　　Since this number is 6 (5 or greater), round the 8 to its left up by one.

- **Answer:** 0.099

---

b. $3{,}482 \div 52.7 = 66.072\ 106$

- Since 52.7 has three significant figures, round the answer to give it three significant figures.

66.072 106　　Since this number is 7 (5 or greater), round the 0 to its left up by one.

- **Answer:** 66.1

---

**Example 1.4** On a trip to the market, you bought three packages of meat weighing 1.17 lb, 2.4 lb, and 0.97 lb. How much meat did you buy in all?

**Analysis**

Add up the amount of weight in each package to get the total weight purchased. When adding, the answer has the same number of decimal places as the original number with the fewest decimal places.

**Solution**

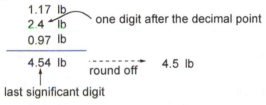

```
  1.17  lb
  2.4   lb      one digit after the decimal point
  0.97  lb
 _____
  4.54  lb   ----------> 4.5  lb
     ↑        round off
last significant digit
```

- Since 2.4 lb has only one digit after the decimal point, the answer can have only one digit after the decimal point.
- Round 4.54 to 4.5.
- Total meat purchased: 4.5 lb

---

## [2] Scientific Notation (1.6)

**Example 1.5** Write the numbers in scientific notation.

　　a. 62,000　　　　b. 0.000 07

**Analysis**

Move the decimal point to give a number between 1 and 10. Multiply the number by $10^x$, where $x$ is the number of places the decimal point was moved. The exponent $x$ is (+) when the decimal point moves to the left and (−) when the decimal point moves to the right.

## Solution

a. $62000 = 6.2 \times 10^4$    the number of places the decimal point was moved to the left

   ↑ Move the decimal point four places to the left.

- Write the coefficient as 6.2 (two significant figures), since 62,000 contains two significant figures.

---

b. $0.000\,07 = 7 \times 10^{-5}$    the number of places the decimal point was moved to the right

   ↑ Move the decimal point five places to the right.

- Write the coefficient as 7 (one significant figure), since 0.000 07 contains one significant figure.

---

**Example 1.6** Convert $1.83 \times 10^4$ to a standard number.

### Analysis

The exponent in $10^x$ tells how many places to move the decimal point in the coefficient to generate a standard number. The decimal point goes to the right when $x$ is positive and to the left when $x$ is negative.

### Solution

**Answer:**

$1.83 \times 10^4$     $1.8300$  ---→   18,300

   ↑ Move the decimal point to the right four places.

---

## [3] Conversion Factors (1.7, 1.8)

**Example 1.7** Write two conversion factors for each pair of units.

      a. kilograms and pounds        b. liters and milliliters

### Analysis

Use the equalities in Tables 1.3 and 1.4 to write a fraction that shows the relationship between the two units.

### Solution

a. Conversion factors for kilograms and pounds:

$$\frac{2.20\ \text{lb}}{1\ \text{kg}} \quad \text{or} \quad \frac{1\ \text{kg}}{2.20\ \text{lb}}$$

b. Conversion factors for liters and milliliters:

$$\frac{1000\ \text{mL}}{1\ \text{L}} \quad \text{or} \quad \frac{1\ \text{L}}{1000\ \text{mL}}$$

---

**Example 1.8** How many liters are in one gallon (1.0) of milk?

### Analysis and Solution

**[1] Identify the original quantity and the desired quantity.**

1.0 gal                    ? L

original quantity        desired quantity

---

**[2] Write out the conversion factors.**

- We have no conversion factor that directly relates gallons to liters. We do, however, know conversions for gallons to quarts, and quarts to liters.

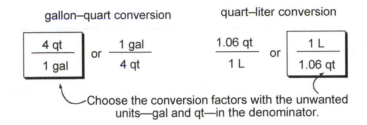

Choose the conversion factors with the unwanted units—gal and qt—in the denominator.

---

**[3] Solve the problem.**

- To set up the problem so that unwanted units cancel, arrange each term so that the units in the numerator of one term cancel the units of the denominator of the adjacent term. In this problem we need to cancel both gallons and quarts to get liters.
- The single desired unit, liters, must be in the **numerator** of one term.

Liters do not cancel.

$$1.0 \text{ gal} \times \frac{4 \text{ qt}}{1 \text{ gal}} \times \frac{1 \text{ L}}{1.06 \text{ qt}} = 3.8 \text{ L}$$

Gallons cancel.    Quarts cancel.

---

**[4] Check**

- Since there are four quarts in a gallon and a quart is about the same size as a liter, one gallon should be about four liters. The answer, 3.8, is just about 4.
- Write the answer with two significant figures since the original quantity, 1.0 gal, has two significant figures.

---

## [4] Temperature (1.9)

**Example 1.9** An elderly patient who was brought in from his cold home in winter had a temperature of 92 °F. Convert this temperature to both °C and K.

**Analysis**

First convert the Fahrenheit temperature to degrees Celsius using the equation °C = (°F – 32)/1.8. Then convert the Celsius temperature to kelvins by adding 273.

**Solution**

[1] Convert °F to °C:

$$°C = \frac{°F - 32}{1.8}$$

$$= \frac{92 - 32}{1.8} = 33\,°C$$

[2] Convert °C to K:

$$K = °C + 273$$

$$= 33 + 273 = 306\ K$$

## [5] Density (1.10)

**Example 1.10** Calculate the mass in grams of 20.0 mL of an antibiotic solution that has a density of 1.21 g/mL.

**Analysis**

Use density (g/mL) to interconvert the mass and volume of a liquid.

**Solution**

$$20.0\ \text{mL} \quad \times \quad \frac{1.21\ g \ \ (density)}{1\ \text{mL}} \quad = \quad 24.2\ g\ \text{of antibiotic solution}$$

Milliliters cancel.

The answer, 24.2 g, has three significant figures to match the number of significant figures in both factors in the problem.

## Self-Test

**[1] Fill in the blank with one of the terms listed below.**

| | | |
|---|---|---|
| Chemical properties (1.2) | Liquid (1.2) | Physical properties (1.2) |
| Element (1.3) | Mass (1.4) | Solid (1.2) |
| Gas (1.2) | Matter (1.1) | Temperature (1.9) |
| Inexact number (1.5) | Mixture (1.3) | Weight (1.4) |

1.   _____ is the force that matter feels due to gravity.
2.   An _____ is a pure substance that cannot be broken down into simpler substances by a chemical reaction.
3.   _____ is anything that has mass and takes up volume.
4.   A _____ has no definite shape or volume.  The particles move randomly and are separated by a distance much larger than their size.
5.   An _____ results from a measurement or observation and contains some uncertainty.
6.   _____ are those properties that can be observed or measured without changing the composition of the material.
7.   A _____ is composed of more than one component.  The composition can vary depending on the sample.

8. _____ is a measure of the amount of matter in an object.
9. A _____ has a definite volume, but takes on the shape of the container it occupies. The particles are close together but they can randomly move around.
10. _____ is a measurement of how hot or cold an object is.
11. A _____ has a definite shape and volume. The particles lie close together, and are arranged in a regular three-dimensional array.
12. _____ are those properties that determine how a substance can be converted to another substance by a chemical reaction.

**[2] Fill in the blank with one of the terms listed below.**

     a. Kelvin               b. Celsius               c. Fahrenheit

13. The zero point on the _____ scale is called absolute zero.
14. On the _____ scale, water freezes at 32 ° and boils at 212 °.
15. On the _____ scale, water freezes at 0 ° and boils at 100 °.

**[3] Pick the appropriate unit of measurement for each object.**

     a. meter               c. centimeter
     b. milliliter            d. kilogram

16. The weight of a newborn infant
17. The amount of vaccine in a syringe
18. The length of a football field
19. The diameter of a pencil

**[4] Match the standard number and the scientific notation.**

     a. $1.04 \times 10^{-8}$          b. $1.04 \times 10^{4}$          c. $1.04 \times 10^{-2}$

20. 10,400
21. 0.0104
22. 0.000 000 010 4

**[5] How many significant figures are in each of the numbers below?**

23. 54,800
24. 0.067 00
25. 119,000.0

## Answers to Self-Test

| | | | | |
|---|---|---|---|---|
| 1. Weight | 6. Physical properties | 11. solid | 16. d | 21. c |
| 2. element | 7. mixture | 12. Chemical properties | 17. b | 22. a |
| 3. Matter | 8. Mass | 13. a | 18. a | 23. 3 |
| 4. gas | 9. liquid | 14. c | 19. c | 24. 4 |
| 5. inexact number | 10. Temperature | 15. b | 20. b | 25. 7 |

## Solutions to In-Chapter Problems

**1.1**   *Natural* materials are isolated from natural sources, whereas *synthetic* materials are produced by chemists in the laboratory.  Natural: water, rock, tree.  Synthetic: tennis ball, CD, plastic bag.

**1.2**   Naturally occurring: ice, blood.  Synthetic: gloves, mask, plastic syringe, stainless steel needle.

**1.3**   *Physical properties* can be observed or measured without changing the composition of the material (a and d).  *Chemical properties* determine how a substance can be converted to another substance by chemical reactions (b, c, and e).

**1.4**   This represents a chemical change because the "particles" on the left are different from the particles on the right.  For example, on the left side there are particles containing only two red balls, while on the right there are none of these.

**1.5**   Representation (a) is a pure substance since each particle contains one red and two gray spheres.  Representation (b) is a mixture since some of the particles are only red, and some are red and black.

**1.6**   A *pure substance* is composed of a single component and has a constant composition regardless of the sample size (d).  A *mixture* is composed of more than one component.  The composition of a mixture can vary depending on the sample (a, b, and c).

**1.7**   An *element* is a pure substance that cannot be broken down into simpler substances by a chemical reaction (a).  A *compound* is a pure substance formed by combining two or more elements together (b, c, and d).

**1.8**   Use Table 1.2 to determine the prefix for each unit.
      a. a million liters = megaliter          c. a hundredth of a gram = centigram
      b. a thousandth of a second = millisecond     d. a tenth of a liter = deciliter

**1.9**   One nanometer = 0.000 000 001 m (one billionth of a meter); therefore, 1 m = 1,000,000,000 nm.

**1.10**  One microgram = 0.000 001 g (one millionth of a gram); therefore, 1 g = 1,000,000 µg.

**1.11**  Use Table 1.2 to determine which quantity is larger.
     a. 3 cL                           c. 5 km
     b. 1 µg                          d. 2 mL

**1.12**  All nonzero digits are significant.  A zero is significant only if it occurs between two nonzero digits, or at the end of a number with a decimal point.  The significant figures are in **bold.**

a. **23.45**
   4 significant figures

c. **23**0
   2 significant figures

e. 0.**202**
   3 significant figures

g. **1,245,006**
   7 significant figures

b. **23.057**
   5 significant figures

d. **231.0**
   4 significant figures

f. 0.003 **60**
   3 significant figures

h. **1,2**00,000
   2 significant figures

**1.13** All nonzero digits are significant. A zero is significant only if it occurs between two nonzero digits, or at the end of a number with a decimal point. The significant figures are in **bold.**

a. **10,040**
  4 significant figures

c. **1,004.00**
  6 significant figures

e. **1.0040**
  5 significant figures

g. 0.001 **004**
  4 significant figures

b. **10,040.**
  5 significant figures

d. **1.004**
  4 significant figures

f. 0.**1004**
  4 significant figures

h. 0.**010 040 0**
  6 significant figures

**1.14** A zero is significant only if it occurs between two nonzero digits, or at the end of a number with a decimal point.

a. 0.003 04    b. 26,045    c. 1,000,034    d. 0.304 00

  No  Yes      Yes        Yes         No  Yes

**1.15** When the number to be rounded off is 4 or less, it and all other digits to the right are dropped. When the number is 5 or greater, 1 is added to the digit to its left.

a. 1.2735   Since this number is 7 (5 or greater), round the 2 to its left up by one.

1.3

c. 3,836.9   Since this number is 3 (4 or less), drop it and all numbers to its right.

3,800

b. 0.002 536 22   Since this number is 3 (4 or less), drop it and all numbers to its right.

0.0025

**1.16** The answers must have the same number of significant figures as the original number with the fewest number of significant figures.

a. $10.70 \times 3.5 = 37.45$
Since 3.5 has only 2 significant figures, round the answer to give it two significant figures.
**37**

b. $0.206 \div 25,993 = $ **0.000 007 93**
Since 0.206 has 3 significant figures, the answer has the appropriate number of significant figures.

c. $1,300 \div 41.2 = 31.553\ 398$
Since 1,300 has only 2 significant figures, round the answer to give it 2 significant figures.
**32**

d. $120.5 \times 26 = 3,133$
Since 26 has only 2 significant figures, round the answer to give it 2 significant figures.
**3,100**

**1.17** The answers must have the same number of decimal places as the original number with the fewest decimal places.

a. $27.8 \text{ cm} + 0.246 \text{ cm} = 28.046 \text{ cm}$
Since 27.8 has one digit after the decimal point, round the answer to one digit after the decimal point.
**28.0 cm**

b. $102.66 \text{ mL} + 0.857 \text{ mL} + 24.0 \text{ mL} = 127.517 \text{ mL}$
Since 24.0 has one digit after the decimal point, round the answer to one digit after the decimal point.
**127.5 mL**

c. 54.6 mg – 25 mg = 29.6 mg
Since 25 has zero digits after the decimal point, round the answer to the nearest whole number.
**30. mg**

d. 2.35 s – 0.266 s = 2.084 s
Since 2.35 has 2 digits after the decimal point, round the answer to 2 digits after the decimal point.
**2.08 s**

**1.18** To write a number in scientific notation:
[1] Move the decimal point to give a number between 1 and 10.
[2] Multiply the result by $10^x$, where $x$ is the number of places the decimal point was moved.

0.000 098 g/dL $=$ 9.8 x $10^{-5}$ g/dL ⌐the number of places the decimal point was moved to the right

Move the decimal point five places to the right.

**1.19** To write a number in scientific notation:
[1] Move the decimal point to give a number between 1 and 10.
[2] Multiply the result by $10^x$, where $x$ is the number of places the decimal point was moved.

a. 93,200 = $9.32 \times 10^4$
The decimal point was moved 4 places to the left.

c. 6,780,000 = $6.78 \times 10^6$
The decimal point was moved 6 places to the left.

e. 4,520,000,000,000 = $4.52 \times 10^{12}$
The decimal point was moved 12 places to the left.

b. 0.000 725 = $7.25 \times 10^{-4}$
The decimal point was moved 4 places to the right.

d. 0.000 030 = $3.0 \times 10^{-5}$
The decimal point was moved 5 places to the right.

f. 0.000 000 000 028 = $2.8 \times 10^{-11}$
The decimal point was moved 11 places to the right.

**1.20** The exponent in $10^x$ tells how many places to move the decimal point in the coefficient to generate a standard number. The decimal point goes to the right when $x$ is positive and to the left when $x$ is negative.

**Answer:**

6.02 x $10^{21}$      6.020000000000000000000      ----→      6,020,000,000,000,000,000,000

Move the decimal point to the right 21 places.

**1.21** The exponent in $10^x$ tells how many places to move the decimal point in the coefficient to generate a standard number. The decimal point goes to the right when $x$ is positive and to the left when $x$ is negative.

a. $6.5 \times 10^3$ = 6,500
The decimal point was moved 3 places to the right.

c. $3.780 \times 10^{-2}$ = 0.037 80
The decimal point was moved 2 places to the left.

e. $2.221 \times 10^6$ = 2,221,000
The decimal point was moved 6 placed to the right.

b. $3.26 \times 10^{-5}$ = 0.000 032 6
The decimal point was moved 5 places to the left.

d. $1.04 \times 10^8$ = 104,000,000
The decimal point was moved 8 places to the right.

f. $4.5 \times 10^{-10}$ = 0.000 000 000 45
The decimal point was moved 10 places to the left.

**1.22** Use the equalities in Tables 1.3 and 1.4 to write a fraction that shows the relationship between the two units.

a. $\dfrac{0.621\ mi}{1\ km}$ $\qquad$ $\dfrac{1\ km}{0.621\ mi}$ $\qquad$ c. $\dfrac{454\ g}{1\ lb}$ $\qquad$ $\dfrac{1\ lb}{454\ g}$

b. $\dfrac{1000\ mm}{1\ m}$ $\qquad$ $\dfrac{1\ m}{1000\ mm}$ $\qquad$ d. $\dfrac{1000\ \mu g}{1\ mg}$ $\qquad$ $\dfrac{1\ mg}{1000\ \mu g}$

**1.23** To convert 4,120 km to miles:
[1] Identify the original quantity and the desired quantity, including units.
[2] Write out the conversion factor(s) needed to solve the problem.
[3] Set up and solve the problem.
[4] Write the answer using the correct number of significant figures and check by estimation.

[1] $\qquad$ 4,120 km $\qquad\qquad$ ? mi
$\qquad$ original quantity $\qquad\qquad$ desired quantity

[2] Two possible conversion factors: $\qquad \dfrac{1\ km}{0.621\ mi}$ or $\boxed{\dfrac{0.621\ mi}{1\ km}}$ $\qquad$ Choose this factor to cancel the unwanted unit, km.

[3] $\qquad\qquad$ **conversion factor**

$\qquad$ 4,120 km $\quad$ x $\quad \dfrac{0.621\ mi}{1\ km}$ $\quad$ = $\quad$ 2,558.52 mi
$\qquad$ original quantity $\qquad\qquad\qquad$ desired quantity

The number of km (unwanted unit) cancels.

[4] The initial number has three significant figures, so the final answer is rounded to 2,560 mi.

**1.24** Use conversion factors to solve the problems.

a. $\quad$ 25 L $\quad$ x $\quad \dfrac{10\ dL}{1\ L}$ $\quad$ = $\quad$ 250 dL

b. $\quad$ 40.0 oz $\quad$ x $\quad \dfrac{28.3\ g}{1\ oz}$ $\quad$ = $\quad$ 1,132 g = 1,130 g rounded to 3 significant figures

c. $\quad$ 32 in. $\quad$ x $\quad \dfrac{2.54\ cm}{1\ in.}$ $\quad$ = $\quad$ 81.28 cm = 81 cm rounded to 2 significant figures

d. $\quad$ 10 cm $\quad$ x $\quad \dfrac{10\ mm}{1\ cm}$ $\quad$ = $\quad$ 100 mm

**1.25** Use conversion factors to solve the problems.

$\qquad\qquad\qquad\qquad\qquad$ Kilometers do not cancel.

a. $\quad$ 6,250 ft $\quad$ x $\quad \dfrac{1\ mi}{5,280\ ft}$ $\quad$ x $\quad \dfrac{1\ km}{0.621\ mi}$ $\quad$ = $\quad$ 1.91 km

$\qquad$ Feet cancel. $\qquad\qquad$ Miles cancel.

Liters do not cancel.

b. 3 cups $\times$ $\dfrac{1\ qt}{4\ cups}$ $\times$ $\dfrac{1\ L}{1.06\ qt}$ = 0.7 L

Cups cancel.  Quarts cancel.

Centimeters do not cancel.

c. 4.5 ft $\times$ $\dfrac{12\ in.}{1\ ft}$ $\times$ $\dfrac{2.54\ cm}{1\ in.}$ = 140 cm

Feet cancel.  Inches cancel.

**1.26**  Convert mL to tsp.

7.5 mL $\times$ $\dfrac{1\ tsp}{5.0\ mL}$ = 1.5 tsp

**1.27**

0.100 mg $\times$ $\dfrac{1000\ \mu g}{1\ mg}$ $\times$ $\dfrac{1\ tablet}{25\ \mu g}$ = 4 tablets

**1.28**

160 mg $\times$ $\dfrac{5\ mL}{100\ mg}$ = 8 mL Children's Motrin

**1.29**  Convert from °C to °F and K using the formulas listed in Section 1.9.

°F = 1.8(°C) + 32          K = °C + 273

   = 1.8(28.5) + 32 = 83.3 °F          = 28.5 + 273 = 302 K

**1.30**

a. °F = 1.8(°C) + 32          c. °C = K − 273

   = 1.8(20.) + 32 = 68 °F          = 298 − 273 = 25 °C

                                °F = 1.8(°C) + 32

                                   = 1.8(25) + 32 = 77 °F

b. °C = $\dfrac{°F\ -\ 32}{1.8}$          d. K = °C + 273

                                   = 75 + 273 = 348 K

   = $\dfrac{150.\ -\ 32}{1.8}$ = 66 °C

**1.31**  To convert volume (mL) to mass (g), multiply the volume by the density (g/mL).

10.0 mL $\times$ $\dfrac{0.713\ g}{mL}$ = 7.13 g

Milliliters cancel.

**1.32**

a. $100. \text{g} \times \dfrac{1 \text{ mL}}{0.92 \text{ g}} = 110 \text{ mL}$   b. $110 \text{ mL} \times \dfrac{1 \text{ L}}{1000 \text{ mL}} = 0.11 \text{ L}$

Grams cancel.

**1.33**  a. Given equal volumes of **A** and **B**, the mass of **A** will be larger.
b. Given equal masses of **A** and **B**, **B** will have the larger volume.

**1.34**

a.  $\text{specific gravity} = \dfrac{\text{density of a substance (g/mL)}}{\text{density of water (g/mL)}} = \dfrac{0.80 \text{ g/mL}}{1 \text{ g/mL}} = 0.80$

b.  $2.3 = \dfrac{\text{density of a substance (g/mL)}}{1 \text{ g/mL}}$   density = 2.3 g/mL

## Solutions to Odd-Numbered End-of-Chapter Problems

**1.35**  Representation (a) is a pure element since each particle contains only gray spheres.  Representation (b) is a pure compound since each particle contains gray and black spheres.  Representation (c) is a mixture, because some of the particles are only gray, and some are gray and black. Representation (d) is a mixture, because some of the particles are gray, and some are blue.

**1.37**  An element is a pure substance that cannot be broken down into simpler substances by a chemical reaction.  A compound is a pure substance formed by combining two or more elements.

**1.39**

| Phase | a. Volume | b. Shape | c. Organization | d. Particle Proximity |
|-------|-----------|----------|-----------------|-----------------------|
| Solid | Definite | Definite | Very organized | Very close |
| Liquid | Definite | Assumes shape of container | Less organized | Close |
| Gas | Not fixed | None | Disorganized | Far apart |

**1.41**  A *chemical change* converts one substance to another substance by a chemical reaction (b).
A *physical change* can be observed or measured without changing the composition of the material (a and c).

**1.43**  This is a physical change since the compound $CO_2$ is unchanged in this transition.  The same "particles" exist at the beginning and end of the process.

**1.45**  a and b.  The temperature on the Fahrenheit thermometer is 76.5 °F, which has three significant figures.

c.  $°C = \dfrac{°F - 32}{1.8} = \dfrac{76.5 - 32}{1.8} = 24.7 \, °C$

**1.47** An exact number results from counting objects or is part of a definition, such as having 20 people in a class. An inexact number results from a measurement or observation and contains some uncertainty, such as the distance from the earth to the sun, $9.3 \times 10^7$ miles.

**1.49** Compare the measurements using Table 1.2. (< means *less than*; > means *greater than*.)

a. 5 mL < **5 dL**    b. **10 mg** > 10 µg    c. **5 cm** > 5 mm    d. **10 Ms** > 10 ms

**1.51** All nonzero digits are significant. A zero is significant only if it occurs between two nonzero digits, or at the end of a number with a decimal point. The significant figures are in **bold.**

a. **16.00**
  4 significant figures

b. **16**0
  2 significant figures

c. 0.001 **60**
  3 significant figures

d. **16**00,000
  2 significant figures

e. **1.06**
  3 significant figures

f. 0.**1600**
  4 significant figures

g. **1.060** $\times 10^{10}$
  4 significant figures

h. **1.6** $\times 10^{-6}$
  2 significant figures

**1.53** When the number to be rounded off is 4 or less, it and all other digits to the right are dropped. When the number is 5 or greater, 1 is added to the digit to its left.

a. 25,401 = 25,400
b. 1,248,486 = 1,250,000

c. 0.001 265 982 = 0.001 27
d. 0.123 456 = 0.123

e. 195.371 = 195
f. 196.814 = 197

**1.55** The answers in problems with multiplication and division must have the same number of decimal places as the original number with the fewest decimal places. The answers in problems with addition and subtraction must have the same number of digits after the decimal point as the original number with the fewest digits after the decimal point.

a. $53.6 \times 0.41 = 21.976$
Since 0.41 has two significant figures, round the answer to two significant figures.
**22**

b. $25.825 - 3.86 = 21.965$
Since 3.86 has two digits after the decimal point, round the answer to two digits after the decimal point.
**21.97**

c. $65.2 \div 12 = 5.43333$
Since 12 has two significant figures, round the answer to two significant figures.
**5.4**

d. $41.0 + 9.135 = 50.135$
Since 41.0 has one digit after the decimal point, round the answer to one digit after the decimal point.
**50.1**

e. $694.2 \times 0.2 = 138.84$
Since 0.2 has one significant figure, round the answer to one significant figure.
**100**

f. $1,045 - 1.26 = 1,043.74$
Since 1,045 has no digits after the decimal point, round the answer to the closest whole number.
**1,044**

**1.57** To write a number in scientific notation:
[1] Move the decimal point to give a number between 1 and 10.
[2] Multiply the result by $10^x$, where $x$ is the number of places the decimal point was moved.

a. $1{,}234 \text{ g} = 1.234 \times 10^3 \text{ g}$
The decimal point was moved 3 places to the left.
b. $0.000\ 016\ 2 \text{ m} = 1.62 \times 10^{-5} \text{ m}$
The decimal point was moved 5 places to the right.
c. $5{,}244{,}000 \text{ L} = 5.244 \times 10^6 \text{ L}$
The decimal point was moved 6 places to the left.

d. $0.005\ 62 \text{ g} = 5.62 \times 10^{-3} \text{ g}$
The decimal point was moved 3 places to the right.
e. $44{,}000 \text{ km} = 4.4 \times 10^4 \text{ km}$
The decimal point was moved 4 places to the left.

**1.59** The exponent in $10^x$ tells how many places to move the decimal point in the coefficient to generate a standard number. The decimal point goes to the right when $x$ is positive and to the left when $x$ is negative.

a. $3.4 \times 10^8 = 340{,}000{,}000$
The decimal point was moved 8 places to the right.
b. $5.822 \times 10^{-5} = 0.000\ 058\ 22$
The decimal point was moved 5 places to the left.

c. $3 \times 10^2 = 300$
The decimal point was moved 2 places to the right.
d. $6.86 \times 10^{-8} = 0.000\ 000\ 068\ 6$
The decimal point was moved 8 places to the left.

**1.61** Compare the two numbers. The number in **bold** is larger.

a. $\mathbf{4.44 \times 10^3}$ or $4.8 \times 10^2$
b. $5.6 \times 10^{-6}$ or $\mathbf{5.6 \times 10^{-5}}$

c. $\mathbf{1.3 \times 10^8}$ or $52{,}300{,}000$
d. $\mathbf{9.8 \times 10^{-4}}$ or $0.000\ 089$

**1.63** Write the number in scientific notation.

a. $0.000\ 400$ g of folate $= 4.00 \times 10^{-4}$ g
The decimal point was moved 4 places to the right.
b. $0.002$ g of copper $= 2 \times 10^{-3}$ g
The decimal point was moved 3 places to the right.

c. $0.000\ 080$ g of vitamin K $= 8.0 \times 10^{-5}$ g
The decimal point was moved 5 places to the right.
d. $3{,}400$ mg of chloride $= 3.4 \times 10^3$ mg
The decimal point was moved 3 places to the left.

**1.65** Use conversion factors to solve the problems.

a. $1.5 \text{ kg} \times \dfrac{1000 \text{ g}}{1 \text{ kg}} = 1{,}500 \text{ g}$

b. $1.5 \text{ kg} \times \dfrac{2.20 \text{ lb}}{1 \text{ kg}} = 3.3 \text{ lb}$

c. $1{,}500 \text{ g} \times \dfrac{1 \text{ oz}}{28.3 \text{ g}} = 53 \text{ oz}$

**1.67**   Use conversion factors to solve the problems.

a.   $300 \not g \times \dfrac{1000 \text{ mg}}{1 \not g} = 300{,}000 \text{ mg}$

b.   $2 \not L \times \dfrac{1{,}000{,}000 \text{ μL}}{1 \not L} = 2{,}000{,}000 \text{ μL}$

c.   $5.0 \not{\text{cm}} \times \dfrac{1 \text{ m}}{100 \not{\text{cm}}} = 0.050 \text{ m}$

d.   $300 \not g \times \dfrac{1 \text{ oz}}{28.3 \not g} = 10.60 \text{ oz} = \textbf{10 oz}$ rounded
to one significant figure

e.   $2 \not{\text{ft}} \times \dfrac{12 \not{\text{in.}}}{1 \not{\text{ft}}} \times \dfrac{1 \text{ m}}{39.4 \not{\text{in.}}} = \begin{array}{l} 0.6091 \text{ m} = \textbf{0.6 m} \text{ rounded} \\ \text{to one significant figure} \end{array}$

f.   $3.5 \not{\text{yd}} \times \dfrac{3 \not{\text{ft}}}{1 \not{\text{yd}}} \times \dfrac{12 \not{\text{in.}}}{1 \not{\text{ft}}} \times \dfrac{1 \text{ m}}{39.4 \not{\text{in.}}} = \begin{array}{l} 3.198 \text{ m} = \textbf{3.2 m} \text{ rounded} \\ \text{to two significant figures} \end{array}$

**1.69**   Use conversion factors to solve the problems.

a.   $234 \not{\text{lb}} \times \dfrac{1 \text{ kg}}{2.20 \not{\text{lb}}} = \begin{array}{l} 106.36 \text{ kg} = \textbf{106 kg} \text{ rounded} \\ \text{to three significant figures} \end{array}$

b.   $50. \not{\text{in.}} \times \dfrac{2.54 \text{ cm}}{1 \not{\text{in.}}} = \begin{array}{l} 127 \text{ cm} = \textbf{130 cm} \text{ rounded} \\ \text{to two significant figures} \end{array}$

c.   $3.0 \not{\text{pints}} \times \dfrac{1 \not{\text{qt}}}{2 \not{\text{pints}}} \times \dfrac{1 \text{ L}}{1.06 \not{\text{qt}}} = \begin{array}{l} 1.415 \text{ L} = \textbf{1.4 L} \text{ rounded} \\ \text{to two significant figures} \end{array}$

d.   °F =   1.8(°C) + 32

   =   1.8(37.7) + 32 = 99.9 °F

**1.71**   Use conversion factors to solve the problems.

a.   $1 \not{\text{qt}} \times \dfrac{946 \text{ mL}}{1 \not{\text{qt}}} = \textbf{946 mL}$ rounded to three significant figures

b.   $1 \not L \times \dfrac{1.06 \not{\text{qt}}}{1 \not L} \times \dfrac{32 \text{ fl oz}}{1 \not{\text{qt}}} = 33.92 \text{ fl oz} = \textbf{33.9 fl oz}$ rounded to three significant figures

**1.73**   Convert from °C to °F and K using the formulas listed in Section 1.9.

a.   °F =   1.8(°C) + 32          K  =   °C + 273

   =   1.8(53) + 32  =  127 °F          =   53 + 273  =  326 K

b.  $°C = \dfrac{°F - 32}{1.8}$  $K = °C + 273$

$= 177 + 273 = 450.\ K$

$= \dfrac{350. - 32}{1.8} = 177\ °C$

**1.75**  Convert the temperatures to a common unit to compare.

a.  $°C = \dfrac{°F - 32}{1.8}$  b.  $°C = \dfrac{°F - 32}{1.8}$

higher temperature

$= \dfrac{10 - 32}{1.8} = -12\ °C < -10\ °C$  $= \dfrac{-50 - 32}{1.8} = -45\ °C > -50\ °C$

↑ higher ↑
10 °F temperature −50 °F

**1.77**  Density is the mass per unit volume, usually reported in g/mL or g/cc.  Specific gravity is the ratio of the density of a substance to the density of water and has no units.

**1.79**

$\dfrac{122\ g}{121\ mL} = 1.01\ g/mL$

**1.81**

$1\ qt \times \dfrac{946\ mL}{1\ qt} \times \dfrac{1.03\ g}{1\ mL} \times \dfrac{1\ kg}{1000\ g} = 0.974\ 38\ kg = \mathbf{0.974\ kg}$

**1.83**  The density of a substance determines whether it floats or sinks in a liquid.  The less dense liquid is the upper layer.  The density of water is 1.0 g/mL.

a. heptane  
(0.684 g/mL < 1.0 g/mL)

c. water  
(1.0 g/mL < 1.49 g/mL)

b. olive oil  
(0.92 g/mL < 1.0 g/mL)

d. water  
(1.0 g/mL < 1.59 g/mL)

**1.85**

a.  specific gravity $= \dfrac{\text{density of mercury (g/mL)}}{\text{density of water (g/mL)}} = \dfrac{13.6\ g/mL}{1\ g/mL} = 13.6$

b.  $0.789 = \dfrac{\text{density of ethanol (g/mL)}}{1\ g/mL}$  density = 0.789 g/mL

**1.87**  Use conversion factors to solve the problems.

a.  $\dfrac{186\ mg}{dL} \times \dfrac{1\ g}{1000\ mg} = 0.186\ g/dL$  b.  $\dfrac{186\ mg}{dL} \times \dfrac{10\ dL}{1\ L} = 1{,}860\ mg/L$

**1.89**

$$1.5 \text{ g} \times \frac{1000 \text{ mg}}{1 \text{ g}} \times \frac{1 \text{ tablet}}{500 \text{ mg}} = 3 \text{ tablets}$$

**1.91**

$$3.5 \text{ g} \times \frac{1000 \text{ mg}}{1 \text{ g}} \times \frac{1 \text{ banana}}{451 \text{ mg}} = 7.8 \text{ bananas (8 bananas)}$$

**1.93**

a. $\dfrac{20 \text{ mL}}{1 \text{ dose}} \times \dfrac{\$10.00}{300. \text{ mL}} = \dfrac{\$0.67}{\text{dose}}$    b.  2 tablespoons = 30. mL

$$\frac{30 \text{ mL}}{1 \text{ dose}} \times \frac{\$10.00}{300. \text{ mL}} = \frac{\$1.00}{\text{dose}}$$

**1.95**

2 tablets x 325 mg/tablet = 650. mg

$$0.510 \text{ kg} \times \frac{1000 \text{ g}}{1 \text{ kg}} \times \frac{1000 \text{ mg}}{1 \text{ g}} \times \frac{1 \text{ dose}}{650. \text{ mg}} = 784.6 = 784 \text{ full doses}$$

**1.97**

$$4 \text{ times} \times \frac{2.0 \text{ g}}{\text{time}} \times \frac{1000 \text{ mg}}{1 \text{ g}} \times \frac{1 \text{ tablet}}{500. \text{ mg}} = 16 \text{ tablets}$$

**1.99**

$$\frac{2.0 \text{ mg}}{1 \text{ kg}} \times \frac{1 \text{ kg}}{2.20 \text{ lb}} \times 110 \text{ lb} = 1.0 \times 10^2 \text{ mg}$$

**1.101**

$$42 \text{ lb} \times \frac{1 \text{ kg}}{2.20 \text{ lb}} \times \frac{10 \text{ mg}}{1 \text{ kg}} \times \frac{1 \text{ tablet}}{80 \text{ mg}} = 2.4 = \textbf{2 tablets}$$

**1.103**

$$1.5 \text{ tsp} \times \frac{5.0 \text{ mL}}{1 \text{ tsp}} \times \frac{100. \text{ mg}}{5 \text{ mL}} \times \frac{1 \text{ g}}{1000 \text{ mg}} = 0.15 \text{ g}$$

# Chapter 2 Atoms and the Periodic Table

## Chapter Review

**[1] How is the name of an element abbreviated and how does the periodic table help to classify it as a metal, nonmetal, or metalloid? (2.1)**
- An element is abbreviated by a one- or two-letter symbol. The periodic table contains a stepped line from boron to astatine. All metals are located to the left of the line. All nonmetals except hydrogen are located to the right of the line. The seven elements located along the line are metalloids.

**[2] What are the basic components of an atom? (2.2)**
- An atom is composed of two parts: a dense nucleus containing positively charged protons and neutral neutrons, and an electron cloud containing negatively charged electrons. Most of the mass of an atom resides in the nucleus, while the electron cloud contains most of its volume.
- The **atomic number** of a neutral atom tells the number of protons and the number of electrons.
- The **mass number** is the sum of the number of protons and the number of neutrons.

**[3] What are isotopes and how are they related to the atomic weight? (2.3)**
- **Isotopes** are atoms that have the same number of protons but a different number of neutrons. The **atomic weight** is the weighted average of the mass of the naturally occurring isotopes of a particular element.

**[4] What are the basic features of the periodic table? (2.4)**
- The periodic table is a schematic of all known elements, arranged in rows **(periods)** and columns **(groups)**, organized so that elements with similar properties are grouped together.
- The vertical columns are assigned group numbers using two different numbering schemes—1–8 plus the letters A or B; or 1–18.
- The periodic table is divided into the main group elements (groups 1A–8A), the transition metals (groups 1B–8B), and the inner transition metals located in two rows below the main table.

**[5] How are electrons arranged around an atom? (2.5)**
- Electrons occupy discrete energy levels, organized into shells (numbered 1, 2, 3, and so on), subshells ($s$, $p$, $d$, and $f$), and orbitals.
- Each orbital holds two electrons.

**Distribution of electrons in the first four shells**

| Shell | Number of electrons in a shell |
|---|---|
| 4 | 32 |
| 3 | 18 |
| 2 | 8 |
| lowest energy → 1 | 2 |

Increasing energy →
Increasing number of electrons →

**[6] What rules determine the electronic configuration of an atom? (2.6)**

- To write the ground state electronic configuration of an atom, electrons are added to the lowest energy orbitals, giving each orbital two electrons. When two orbitals are equal in energy, one electron is added to each orbital until the orbitals are half-filled.
- Orbital diagrams that use boxes for orbitals and arrows for electrons indicate electronic configuration. Electron configuration can also be shown using superscripts to indicate how many electrons an orbital contains. For example, the electron configuration of the six electrons in a carbon atom is $1s^2 2s^2 2p^2$.

Orbital diagram for C:  ⇅  ⇅  ↑ ↑

       1s   2s   2p

**[7] How is the location of an element in the periodic table related to its electronic configuration? (2.6, 2.7)**

- The periodic table is divided into four regions—the $s$ block, $p$ block, $d$ block, and $f$ block—based on the subshells that are filled with electrons last.
- Elements in the same group have the same number of valence electrons and similar electronic configurations.

**[8] What is an electron-dot symbol? (2.7)**

- An electron-dot symbol uses a dot to represent each valence electron around the symbol for an element.

| | H | C | O | Cl |
|---|---|---|---|---|
| Number of valence electrons: | 1 | 4 | 6 | 7 |
| Electron-dot symbol: | H· | ·Ċ· | ·Ö· | ·Cl: |

**[9] How are atomic size and ionization energy related to location in the periodic table? (2.8)**

- The size of an atom decreases across a row and increases down a column.

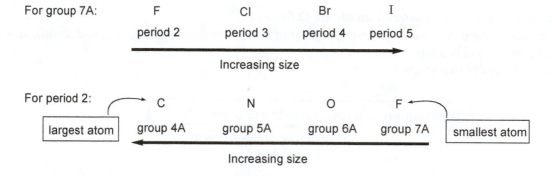

- **Ionization energy**—the energy needed to remove an electron from an atom—increases across a row and decreases down a column.

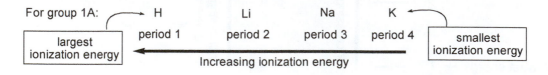

For period 2:    C         N         O        F

             group 4A    group 5A    group 6A    group 7A

Increasing ionization energy

## Problem Solving

## [1] Structure of the Atom (2.2)

**Example 2.1** How many protons, neutrons, and electrons are contained in an atom of sodium, which has an atomic number of 11 and a mass number of 23?

**Analysis**
- In a neutral atom, the atomic number = the number of protons = the number of electrons.
- The mass number = the number of protons + the number of neutrons.

**Solution**
The atomic number of 11 means that sodium has 11 protons and 11 electrons. To find the number of neutrons, subtract the atomic number ($Z$) from the mass number ($A$).

$$\text{number of neutrons} = \text{mass number} - \text{atomic number}$$
$$= 23 - 11$$
$$= 12 \text{ neutrons}$$

## [2] Isotopes (2.3)

**Example 2.2** For each atom, give the following information: [1] the atomic number; [2] the mass number; [3] the number of protons; [4] the number of neutrons; [5] the number of electrons.

    a. $^{109}_{47}\text{Ag}$        b. $^{81}_{35}\text{Br}$

**Analysis**
- The superscript gives the mass number and the subscript gives the atomic number for each element.
- The atomic number = the number of protons = the number of electrons.
- The mass number = the number of protons + the number of neutrons.

**Solution**

| | Atomic Number | Mass Number | Number of Protons | Number of Neutrons | Number of Electrons |
|---|---|---|---|---|---|
| a. $^{109}_{47}\text{Ag}$ | 47 | 109 | 47 | $109 - 47 = 62$ | 47 |
| b. $^{81}_{35}\text{Br}$ | 35 | 81 | 35 | $81 - 35 = 46$ | 35 |

**Example 2.3** Determine the number of neutrons in each isotope.

        a. sulfur-36        b. chlorine-37

**Analysis**
- The identity of the element tells us the atomic number.
- The number of neutrons = mass number – atomic number.

**Solution**
    a. Sulfur's atomic number ($Z$) is 16. Sulfur-36 has a mass number ($A$) of 36.

$$\text{number of neutrons} = A - Z$$
$$= 36 - 16 = 20 \text{ neutrons}$$

    b. Chlorine's atomic number is 17 and the mass number of the given isotope is 37.

$$\text{number of neutrons} = A - Z$$
$$= 37 - 17 = 20 \text{ neutrons}$$

---

## [3] The Periodic Table (2.4)

**Example 2.4** Give the period and group number for each element.
        a. lead        b. calcium

**Analysis**
Use the element symbol to locate an element in the periodic table. Count down the rows of elements to determine the period. The group number is located at the top of each column.

**Solution**
a. Lead (Pb) is located in the sixth row (period 6), and has group number 4A (or 14).
b. Calcium (Ca) is located in the fourth row (period 4), and has group number 2A (or 2).

---

## [4] Electronic Configuration (2.6)

**Example 2.5** Give the orbital diagram for the ground state electronic configuration of the element fluorine. Then, convert this orbital diagram to noble gas notation.

**Analysis**
- Use the atomic number to determine the number of electrons.
- Locate the element in the periodic table, and use Figure 2.8 to determine the order of orbitals filled with electrons. Read the table from left to right, row-by-row, beginning at the upper left corner and ending at the element in question. To fill orbitals of the same energy, place electrons one at a time in the orbitals until they are half-filled.
- To convert an orbital diagram to noble gas notation, replace the electronic configuration corresponding to the noble gas in the preceding row by the element symbol for the noble gas in brackets.

**Solution**

The atomic number of fluorine is 9, so nine electrons must be placed in orbitals. Four electrons are added in pairs to the $1s$ and $2s$ orbitals. The remaining five electrons are then added to the three $2p$ orbitals to give two pairs of electrons and one unpaired electron.

F
fluorine
9 electrons

$1s$ $2s$ $2p$

Since fluorine is in the second period, use the noble gas helium in the preceding row to write the electronic configuration in noble gas notation. **Substitute [He] for the electrons in the first shell.**

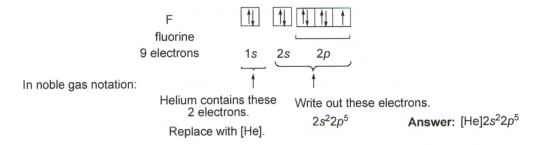

F
fluorine
9 electrons

$1s$ $2s$ $2p$

In noble gas notation:

Helium contains these 2 electrons.
Replace with [He].

Write out these electrons.
$2s^22p^5$

**Answer:** [He]$2s^22p^5$

---

**Example 2.6** Give the ground state electronic configuration of the element phosphorus. Convert the electronic configuration to noble gas notation.

**Analysis**

- Use the atomic number to determine the number of electrons.
- Locate the element in the periodic table and use Figure 2.8 to determine the order of orbitals filled with electrons.
- To convert the electronic configuration to noble gas notation, replace the electronic configuration corresponding to the noble gas in the preceding row by the element symbol for the noble gas in brackets.

**Solution**

The atomic number of phosphorus is 15, so 15 electrons must be placed in orbitals. Twelve electrons are added in pairs to the $1s$, $2s$, three $2p$, and $3s$ orbitals. The remaining three electrons are added to the $3p$ orbitals. Since phosphorus is an element in period 3, use the noble gas neon in period 2 to write the noble gas configuration.

Electronic configuration for P   =   $1s^22s^22p^63s^23p^3$   =   [Ne]$3s^23p^3$

(15 electrons)

The noble gas neon contains these 10 electrons.
Replace with [Ne].

noble gas notation

## [5] Valence Electrons (2.7)

**Example 2.7** Determine the number of valence electrons and give the electronic configuration of the valence electrons of each element.

     a. oxygen        b. beryllium

**Analysis**

The group number of a main group element = the number of valence electrons. Use the general electronic configurations in Table 2.6 to write the configuration of the valence electrons.

**Solution**

a. Oxygen is located in group 6A, so it has six valence electrons. Since oxygen is a second period element, its valence electronic configuration is $2s^2 2p^4$.

b. Beryllium is located in group 2A, so it has two valence electrons. Since beryllium is a second period element, its valence electronic configuration is $2s^2$.

---

**Example 2.8** Write an electron-dot symbol for each element.

     a. calcium        b. oxygen

**Analysis**

Write the element symbol for each element and use the group number to determine the number of valence electrons for a main group element. Represent each valence electron with a dot.

**Solution**

a. The element symbol for calcium is Ca. Ca is in group 2A and has two valence electrons. Electron-dot symbol:

           Ca·

b. The element symbol for oxygen is O. O is in group 6A and has six valence electrons. Electron-dot symbol:

           ·Ö·

---

## Self-Test

### [1] Fill in the blank with one of the terms listed below.

| | | | |
|---|---|---|---|
| Atomic number (2.2) | Ground state (2.6) | Metal (2.1) | Period (2.4) |
| Atomic weight (2.3) | Group (2.4) | Metalloid (2.1) | |
| Chemical formula (2.1) | Isotope (2.3) | Nucleus (2.2) | |
| Electron cloud (2.2) | Mass number (2.2) | Orbital (2.5) | |

1. The _____ is the weighted average of the mass of the naturally occurring isotopes of a particular element reported in atomic mass units.

2. A _____ is a shiny material that is a good conductor of heat and electricity. All are solids except for mercury, which is a liquid.

3. The _____ is a dense core that contains protons and neutrons. Most of the mass of an atom resides here.

4. A row in the periodic table is called a _____. Elements in the same row are similar in size.

5. _____ are atoms of the same element having a different number of neutrons.

6. An _____ is a region of space where the probability of finding an electron is high. Each of these regions can hold *two* electrons.

7. The lowest energy arrangement of electrons is called the _____.

8. The _____ is the number of protons in the nucleus of an atom.

9. A _____ uses element symbols to show the identity of elements forming a compound and subscripts to show the ratio of atoms (the building blocks of matter) contained in the compound.

10. The _____ is the number of protons plus the number of neutrons.

11. A column in the periodic table is called a _____. Elements in the same column have similar electronic and chemical properties.

12. The _____ , composed of electrons that move rapidly in the almost empty space surrounding the nucleus, comprises most of the volume of an atom.

13. A _____ has properties intermediate between metals and nonmetals. These include boron (B), silicon (Si), germanium (Ge), arsenic (As), antimony (Sb), tellurium (Te), and astatine (At).

**[2] Fill in the blank with one of the terms listed below. You may use a term more than once.**

        a. alkali metals         b. halogens         c. noble gases

14. The _____ are especially stable as atoms, so they rarely combine with other elements to form compounds.

15. The _____ are located on the far left side of the periodic table.

16. The _____ and _____ are located on the far right side of the periodic table.

**[3] Pick the statement that describes the type of orbital.**

        a. *s* orbital         b. *p* orbital

17. An _____ has a sphere of electron density. It is lower in energy than other orbitals in the same shell because electrons are kept closer to the positively charged nucleus.

18. A _____ is an orbital that has a dumbbell shape.

**[4] Fill in the blank with one of the terms listed below.**

        a. *s* block     b. *p* block     c. *d* block     d. *f* block

19. The _____ consists of groups 3A–8A (except helium).

20. The _____ consists of the 10 columns of transition metals.

21. The _____ consists of groups 1A and 2A and the element helium.

22. The _____ consists of the two groups of 14 inner transition metals.

**[5] Match the element symbol with the element name.**

        a. As         b. Al         c. Ar

23. Argon
24. Aluminum
25. Arsenic

## Answers to Self-Test

| | | | | |
|---|---|---|---|---|
| 1. atomic weight | 6. orbital | 11. group | 16. b and c | 21. a |
| 2. metal | 7. ground state | 12. electron cloud | 17. a | 22. d |
| 3. nucleus | 8. atomic number | 13. metalloid | 18. b | 23. c |
| 4. period | 9. chemical formula | 14. c | 19. b | 24. b |
| 5. Isotopes | 10. mass number | 15. a | 20. c | 25. a |

## Solutions to In-Chapter Problems

2.1   Each element is identified by a one- or two-letter symbol.  Use the periodic table to find the symbol for each element.

  a.  Ca          b.  Rn          c.  N          d.  Au

2.2   Use the periodic table to find the symbol for each element.

  a.  Cu and Zn          b.  Cu and Sn          c.  Sn, Sb, and Pb

2.3   Use the periodic table to find the element corresponding to each symbol.

  a.  neon      b.  sulfur      c.  iodine      d.  silicon      e.  boron      f.  mercury

2.4   *Metals* are shiny materials that are good conductors of heat and electricity. *Nonmetals* do not have a shiny appearance, and they are generally poor conductors of heat and electricity. *Metalloids* have properties intermediate between metals and nonmetals.

  a, d, f, h: metals          b, c, g:  nonmetals          e:  metalloid

2.5   Use Figure 2.1 and the definitions in Answer 2.4 to determine if the micronutrients are metals, nonmetals, or metalloids.

  As, B, Si: metalloids     Cr, Co, Cu, Fe, Mn, Mo, Ni, Zn: metals          F, I, Se: nonmetals

2.6   Use Figure 2.3 to determine which elements are represented in the molecular art.

  a. 4 hydrogens, 1 carbon      b.  3 hydrogens, 1 nitrogen      c. 6 hydrogens, 2 carbons, 1 oxygen

**2.7** The subscript tells how many atoms of a given element are in each chemical formula.

    a. NaCN (sodium cyanide) = 1 sodium, 1 carbon, 1 nitrogen

    b. $H_2S$ (hydrogen sulfide) = 2 hydrogens, 1 sulfur

    c. $C_2H_6$ (ethane) = 2 carbons, 6 hydrogens

    d. $SnF_2$ (stannous fluoride) = 1 tin, 2 fluorines

    e. CO (carbon monoxide) = 1 carbon, 1 oxygen

    f. $C_3H_8O_3$ (glycerol) = 3 carbons, 8 hydrogens, 3 oxygens

**2.8** Use Figure 2.3 to determine which elements are represented in the molecular art.

Halothane contains 2 carbons, 1 hydrogen, 3 fluorines, 1 bromine, and 1 chlorine atom.

**2.9**

    a. In a neutral atom, the number of protons and electrons is equal; 9 protons = 9 electrons.
    b. The atomic number = the number of protons = 9.
    c. This element is fluorine.

**2.10** The atomic number is unique to an element and tells the number of protons in the nucleus and the number of electrons in the electron cloud.

| | Atomic Number | Element | Protons | Electrons |
|---|---|---|---|---|
| a. | 2 | Helium | 2 | 2 |
| b. | 11 | Sodium | 11 | 11 |
| c. | 20 | Calcium | 20 | 20 |
| d. | 47 | Silver | 47 | 47 |
| e. | 78 | Platinum | 78 | 78 |

**2.11** Answer the question as in Sample Problem 2.4.
    a. There are 4 protons and 5 neutrons.
    b. The atomic number = the number of protons = 4.
        The mass number = the number of protons + the number of neutrons = 4 + 5 = 9.
    c. The element is beryllium.

**2.12** In a neutral atom, the atomic number ($Z$) = the number of protons = the number of electrons. The mass number ($A$) = the number of protons + the number of neutrons.

| | Protons | Neutrons ($A - Z$) | Electrons |
|---|---|---|---|
| a. | 17 | 18 (35 – 17) | 17 |
| b. | 14 | 14 (28 – 14) | 14 |
| c. | 92 | 146 (238 – 92) | 92 |

**2.13** The mass number ($A$) = the number of protons + the number of neutrons.

    a. 42 protons, 42 electrons, 53 neutrons
       42 + 53 = **95**

    b. 24 protons, 24 electrons, 28 neutrons
       24 + 28 = **52**

**2.14**   The superscript gives the mass number and the subscript gives the atomic number for each element. The atomic number = the number of protons = the number of electrons in a neutral atom. The mass number = the number of protons + the number of neutrons.

|   | Atomic Number | Mass Number | Protons | Neutrons | Electrons |
|---|---|---|---|---|---|
| a. $^{13}_{6}C$ | 6 | 13 | 6 | 7 | 6 |
| b. $^{121}_{51}Sb$ | 51 | 121 | 51 | 70 | 51 |

**2.15**   The identity of the element tells us the atomic number. The mass number = the number of protons + the number of neutrons.

|   |   | Protons | Electrons | Atomic Number | Mass Number |
|---|---|---|---|---|---|
| With 12 neutrons: | $^{24}_{12}Mg$ | 12 | 12 | 12 | 12 + 12 = 24 |
| With 13 neutrons: | $^{25}_{12}Mg$ | 12 | 12 | 12 | 12 + 13 = 25 |
| With 14 neutrons: | $^{26}_{12}Mg$ | 12 | 12 | 12 | 12 + 14 = 26 |

**2.16**   Multiply the isotopic abundance by the mass of each isotope, and add up the products to give the atomic weight for the element.

a. Magnesium

| Mass due to Mg-24: | $0.7899 \times 23.99$ amu | = | 18.9497 amu |
|---|---|---|---|
| Mass due to Mg-25: | $0.1000 \times 24.99$ amu | = | 2.499  amu |
| Mass due to Mg-26: | $0.1101 \times 25.98$ amu | = | 2.8604 amu |
|  | Atomic weight | = | 24.3091 amu rounded to 24.31 amu |

**Answer**

b. Vanadium

| Mass due to V-50: | $0.00250 \times 49.95$ amu | = | 0.12488 amu |
|---|---|---|---|
| Mass due to V-51: | $0.99750 \times 50.94$ amu | = | 50.8127  amu |
|  | Atomic weight | = | 50.93758 amu rounded to 50.94 amu |

**Answer**

**2.17**   Use the element symbol to locate an element in the periodic table. Count down the rows of elements to determine the period. The group number is located at the top of each column.

| Element | Period | Group |
|---|---|---|
| a. Oxygen | 2 | 6A (or 16) |
| b. Calcium | 4 | 2A (or 2) |
| c. Phosphorus | 3 | 5A (or 15) |
| d. Platinum | 6 | 8B (or 10) |
| e. Iodine | 5 | 7A (or 17) |

**2.18**    Use the definitions from Section 2.4 to identify the element fitting each description.

a. K        c. Ar        e. Zn
b. F        d. Sr        f. Nb

**2.19**

a.  titanium, Ti, group 4B (or 4), period 4, transition metal
b.  phosphorus, P, group 5A (or 15), period 3, main group element
c.  dysprosium, Dy, no group number, period 6, inner transition element

**2.20**    Use Table 2.4 to tell how many electrons are present in each shell, subshell, or orbital.

a. a $2p$ orbital = 2 electrons             c. a $3d$ orbital = 2 electrons
b. the $3d$ subshell = 10 electrons          d. the third shell = 18 electrons

**2.21**    The **electronic configuration** of an individual atom is how the electrons are arranged in an atom's orbitals.

a. $1s^2 2s^2 2p^6 3s^2 3p^2$ = silicon          c. $1s^2 2s^2 2p^6 3s^2 3p^6 4s^2 3d^1$ = scandium
b. $[Ne]3s^2 3p^4$ = sulfur                      d. $[Ar]4s^2 3d^{10}$ = zinc

**2.22**    The **electronic configuration** of an individual atom shows how the electrons are arranged in an atom's orbitals.

a. lithium                                                      c. fluorine
b. beryllium, boron, carbon, nitrogen, oxygen, fluorine, neon    d. oxygen

**2.23**    Use Example 2.5 to help draw the orbital diagram for each element.
[1] Use the atomic number to determine the number of electrons.
[2] Place electrons two at a time into the lowest energy orbitals, using Figure 2.8. When orbitals have the same energy, place electrons one at a time in the orbitals until they are half-filled.

a. magnesium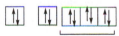

              1s   2s   2p      3s

b. aluminum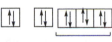

              1s   2s   2p      3s   3p

c. bromine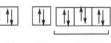

              1s   2s   2p      3s   3p    4s    3d      4p

**2.24** To convert the electronic configuration to noble gas notation, replace the electronic configuration corresponding to the noble gas in the preceding row by the element symbol for the noble gas in brackets.

a. sodium: $1s^2 2s^2 2p^6 3s^1$
  $[Ne]3s^1$
b. silicon: $1s^2 2s^2 2p^6 3s^2 3p^2$
  $[Ne]3s^2 3p^2$

c. iodine: $1s^2 2s^2 2p^6 3s^2 3p^6 4s^2 3d^{10} 4p^6 5s^2 4d^{10} 5p^5$
  $[Kr]5s^2 4d^{10} 5p^5$

**2.25** To obtain the total number of electrons, add up the superscripts. This gives the atomic number and identifies the element. To determine the number of valence electrons, add up the number of electrons in the shell with the highest number.

a. $1s^2 2s^2 2p^6 3s^2$
12 electrons, 2 valence electrons in the 3s orbital, magnesium

b. $1s^2 2s^2 2p^6 3s^2 3p^3$
15 electrons, 5 valence electrons in the 3s and 3p orbitals, phosphorus

c. $1s^2 2s^2 2p^6 3s^2 3p^6 4s^2 3d^{10} 4p^6 5s^2 4d^2$
40 electrons, 2 valence electrons in the 5s orbital, zirconium

d. $[Ar]4s^2 3d^6$
26 electrons, 2 valence electrons in the 4s orbital, iron

**2.26** The group number of a main group element = the number of valence electrons. Use the general electronic configurations in Table 2.6 to write the configuration of the valence electrons.

a. fluorine = 7 valence electrons: $2s^2 2p^5$
b. krypton = 8 valence electrons: $4s^2 4p^6$

c. magnesium = 2 valence electrons: $3s^2$
d. germanium = 4 valence electrons: $4s^2 4p^2$

**2.27**
Se, selenium: $4s^2 4p^4$
Te, tellurium: $5s^2 5p^4$
Po, polonium: $6s^2 6p^4$

**2.28** Write the symbol for each element and use the group number to determine the number of valence electrons for a main group element. Represent each valence electron with a dot.

a. $:\!\overset{\cdot\cdot}{\underset{\cdot\cdot}{Br}}\!\cdot$   c. $\overset{\cdot}{Al}\!\cdot$   e. $:\!\overset{\cdot\cdot}{Ne}\!:$
b. $\overset{\cdot}{Li}$   d. $\cdot\overset{\cdot\cdot}{\underset{\cdot}{S}}\!\cdot$

**2.29** The size of atoms increases down a column of the periodic table, as the valence electrons are farther from the nucleus. The size of atoms decreases across a row of the periodic table as the number of protons in the nucleus increases.

a. neon, carbon, boron
b. beryllium, magnesium, calcium
c. sulfur, silicon, magnesium

d. neon, krypton, xenon
e. oxygen, sulfur, silicon
f. fluorine, sulfur, aluminum

**2.30** Ionization energies decrease down a column of the periodic table as the valence electrons get farther from the positively charged nucleus. Ionization energies generally increase across a row of the periodic table as the number of protons in the nucleus increases.

a. silicon, phosphorus, sulfur  
b. calcium, magnesium, beryllium  
c. beryllium, carbon, fluorine  

d. krypton; argon, neon  
e. tin, silicon, sulfur  
f. calcium, aluminum, nitrogen  

## Solutions to Odd-Numbered End-of-Chapter Problems

**2.31** Use Figure 2.3 to determine which elements are represented in the molecular art.  
a. carbon (black) and oxygen (red)     b. carbon (black), hydrogen (gray), and chlorine (green)

**2.33** Use the periodic table to find the element corresponding to each symbol.

a. Au = gold, At = astatine, Ag = silver  
b. N = nitrogen, Na = sodium, Ni = nickel  
c. S = sulfur, Si = silicon, Sn = tin  

d. Ca = calcium, Cr = chromium, Cl = chlorine  
e. P = phosphorus, Pb = lead, Pt = platinum  
f. Ti = titanium, Ta = tantalum, Tl = thallium  

**2.35** An *element* is a pure substance that cannot be broken down into simpler substances by a chemical reaction. A *compound* is a pure substance formed by combining two or more elements together.

a. $H_2$ = element     c. $S_8$ = element     e. $C_{60}$ = element

b. $H_2O_2$ = compound     d. $Na_2CO_3$ = compound

**2.37**

a. cesium  
b. ruthenium  
c. chlorine  

d. beryllium  
e. fluorine  
f. cerium  

**2.39**

a. sodium: metal, alkali metal, main group element  
b. silver: metal, transition metal  
c. xenon: nonmetal, noble gas, main group element  
d. platinum: metal, transition metal  
e. uranium: metal, inner transition metal  
f. tellurium: metalloid, main group element  

**2.41**

a. 5 protons and 6 neutrons  
b. The atomic number = the number of protons = 5.  
c. The mass number = the number of protons + the number of neutrons = 5 + 6 = 11.  
d. The number of electrons = the number of protons = 5.  
e. element symbol: B

**2.43**

| | Element Symbol | Atomic Number | Mass Number | Number of Protons | Number of Neutrons | Number of Electrons |
|---|---|---|---|---|---|---|
| a. | C | 6 | 12 | 6 | 6 | 6 |
| b. | P | 15 | 31 | 15 | 16 | 15 |
| c. | Zn | 30 | 65 | 30 | 35 | 30 |
| d. | Mg | 12 | 24 | 12 | 12 | 12 |
| e. | I | 53 | 127 | 53 | 74 | 53 |
| f. | Be | 4 | 9 | 4 | 5 | 4 |
| g. | Zr | 40 | 91 | 40 | 51 | 40 |
| h. | S | 16 | 32 | 16 | 16 | 16 |

**2.45**

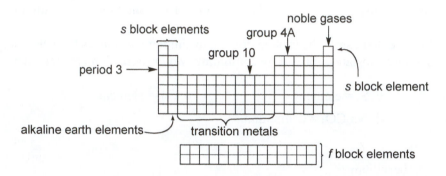

**2.47**    Hydrogen is located in group 1A but is not an alkali metal.

**2.49**    Use Figure 2.1 and the definitions in Answer 2.4 to classify each element in the fourth row of the periodic table as a metal, nonmetal, or metalloid.

K, Ca, Sc, Ti, V, Cr, Mn, Fe, Co, Ni, Cu, Zn, Ga: metals
Ge, As: metalloids
Se, Br, Kr: nonmetals

**2.51**    Group 8A in the periodic table contains only nonmetals.

**2.53**    The atomic number = the number of protons = the number of electrons.
The mass number = the number of protons + the number of neutrons.

| Mass | Protons | Neutrons | Electrons | Group | Symbol |
|---|---|---|---|---|---|
| 16 | 8 | 8 | 8 | 6A | $^{16}_{8}O$ |
| 17 | 8 | 9 | 8 | 6A | $^{17}_{8}O$ |
| 18 | 8 | 10 | 8 | 6A | $^{18}_{8}O$ |

**2.55**    The identity of the element tells us the atomic number.
The number of neutrons = mass number – atomic number.

|   | Symbol | Protons | Neutrons | Electrons |
|---|---|---|---|---|
| a. | $^{27}_{13}\text{Al}$ | 13 | $27 - 13 = 14$ | 13 |
| b. | $^{35}_{17}\text{Cl}$ | 17 | $35 - 17 = 18$ | 17 |
| c. | $^{34}_{16}\text{S}$ | 16 | $34 - 16 = 18$ | 16 |

**2.57**

a. $^{127}_{53}\text{I}$    b. $^{79}_{35}\text{Br}$    c. $^{107}_{47}\text{Ag}$

**2.59**   Multiply the isotopic abundance by the mass of each isotope, and add up the products to give the atomic weight for the element.

Silver

| | | |
|---|---|---|
| Mass due to Ag-107: | $0.5184 \times 106.91$ amu | $= 55.4221$ amu |
| Mass due to Ag-109: | $0.4816 \times 108.90$ amu | $= 52.4462$ amu |
| | Atomic weight | $= 107.8683$ amu rounded to 107.9 amu |

**Answer**

**2.61**   No, the neutral atoms of two different elements cannot have the same number of electrons. Two different elements must have a different number of protons, so in the neutral atoms, they must have a different number of electrons.

**2.63**

a. first shell $(n = 1) = 1$ orbital ($1s$)

b. second shell $(n = 2) = 4$ orbitals ($2s$, three $2p$)

c. third shell $(n = 3) = 9$ orbitals ($3s$, three $3p$, five $3d$)

d. fourth shell $(n = 4) = 16$ orbitals ($4s$, three $4p$, five $4d$, seven $4f$)

**2.65**   Use Example 2.5 to help draw the orbital diagram for each element.

a. B

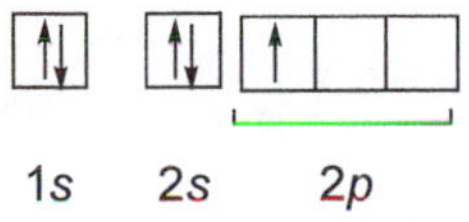

1s   2s   2p

b. K

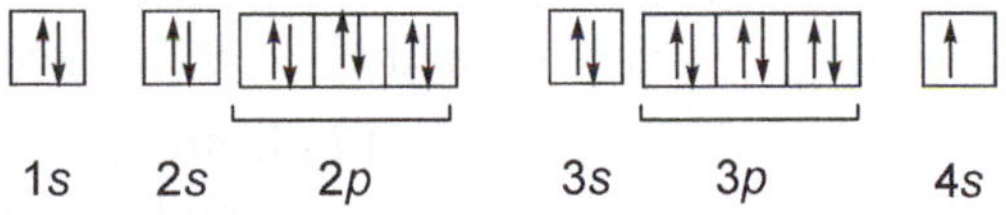

1s   2s   2p   3s   3p   4s

c. Se

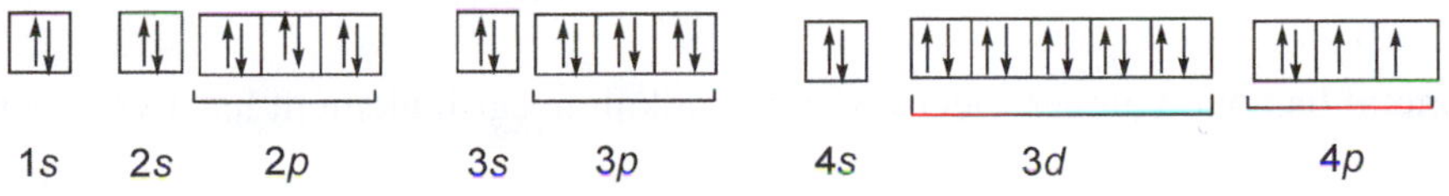

1s   2s   2p   3s   3p   4s   3d   4p

d. Ar

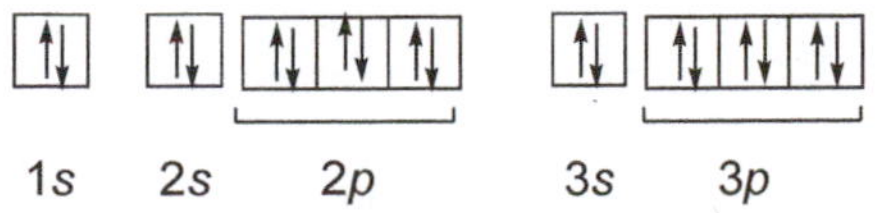

1s   2s   2p   3s   3p

e. Zn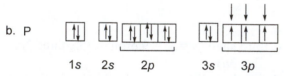

1s  2s  2p  3s  3p  4s  3d

**2.67** To convert the electronic configuration to noble gas notation, replace the electronic configuration corresponding to the noble gas in the preceding row by the symbol for the noble gas in brackets, as in Answer 2.24.

a. B: $1s^2 2s^2 2p^1$ or $[He]2s^2 2p^1$

b. K: $1s^2 2s^2 2p^6 3s^2 3p^6 4s^1$ or $[Ar]4s^1$

c. Se: $1s^2 2s^2 2p^6 3s^2 3p^6 4s^2 3d^{10} 4p^4$ or $[Ar]4s^2 3d^{10} 4p^4$

d. Ar: $1s^2 2s^2 2p^6 3s^2 3p^6$ or $[Ar]$

e. Zn: $1s^2 2s^2 2p^6 3s^2 3p^6 4s^2 3d^{10}$ or $[Ar]4s^2 3d^{10}$

**2.69** To find the number of unpaired electrons, draw the orbital diagram for each element.

one unpaired electron

a. Al

1s  2s  2p  3s  3p

three unpaired electrons

b. P

1s  2s  2p  3s  3p

one unpaired electron

c. Na

1s  2s  2p  3s

**2.71**

a. $1s^2 2s^2 2p^6 3s^2 3p^6 4s^2 3d^{10} 4p^6 5s^2$ = 38 electrons, 2 valence electrons in the 5s orbital, strontium

b. $1s^2 2s^2 2p^6 3s^2 3p^4$ = 16 electrons, 6 valence electrons in the 3s and the 3p orbitals, sulfur

c. $1s^2 2s^2 2p^6 3s^1$ = 11 electrons, 1 valence electron in the 3s orbital, sodium

d. $[Ne]3s^2 3p^5$ = 17 electrons, 7 valence electrons in the 3s and 3p orbitals, chlorine

**2.73** An alkali metal has one valence electron and an alkaline earth element has two valence electrons.

**2.75**

|  | Electrons | Group Number | Valence Electrons | Period | Valence Shell |
|---|---|---|---|---|---|
| a. Carbon | 6 | 4A | 4 | 2 | 2 |
| b. Calcium | 20 | 2A | 2 | 4 | 4 |
| c. Krypton | 36 | 8A | 8 | 4 | 4 |

**2.77**

   a. carbon: $1s^2 2s^2 2p^2$; valence electrons $2s^2 2p^2$

   b. calcium: $1s^2 2s^2 2p^6 3s^2 3p^6 4s^2$; valence electrons $4s^2$

   c. krypton: $1s^2 2s^2 2p^6 3s^2 3p^6 4s^2 3d^{10} 4p^6$; valence electrons $4s^2 4p^6$

**2.79**   The group number of a main group element = the number of valence electrons.

   a. 2A = 2 valence electrons    b. 4A = 4 valence electrons    c. 7A = 7 valence electrons

**2.81**

   a. sulfur: 6, $3s^2 3p^4$    c. barium: 2, $6s^2$    e. tin: 4, $5s^2 5p^2$

   b. chlorine: 7, $3s^2 3p^5$    d. titanium: 2, $4s^2$

**2.83**   Write the element symbol for each element and use the group number to determine the number of valence electrons for a main group element.  Represent each valence electron with a dot.

   a. beryllium    b. silicon    c. iodine    d. magnesium    e. argon

      Be·        ·Ṡi·       ·Ï:       ·Mg·       :Är:

**2.85**   Use the size rules from Answer 2.29.

   a. iodine    b. carbon    c. potassium    d. selenium

**2.87**   Use the rules from Answer 2.30 to decide which has the higher ionization energy.

   a. bromine    b. nitrogen    c. silicon    d. chlorine

**2.89**   Use the size rules from Answer 2.29.

   fluorine, oxygen, sulfur, silicon, magnesium

**2.91**   Use the rules from Answer 2.30 to rank the elements in order of increasing ionization energy.

   sodium, magnesium, phosphorus, nitrogen, fluorine

**2.93**

| | a. Type | b. Block | c,d: Radius | e,f: Ionization Energy | g. Valence Electrons |
|---|---|---|---|---|---|
| Sodium | Metal | $s$ | | | 1 |
| Potassium | Metal | $s$ | Largest | Lowest | 1 |
| Chlorine | Nonmetal | $p$ | Smallest | Highest | 7 |

**2.95**

Carbon-11 has the same number of protons and electrons as carbon-12; that is, six. Carbon-11 has only five neutrons, whereas carbon-12 has six neutrons. The symbol for carbon-11 is $^{11}_{6}C$ .

**2.97** The electron configuration of copper ($1s^2 2s^2 2p^6 3s^2 3p^6 4s^1 3d^{10}$) is unusual because electrons have been added to higher-energy $3d$ orbitals even though there is only one $4s$ electron.

# Chapter 3 Ionic Compounds

## Chapter Review

**[1] What are the basic features of ionic and covalent bonds? (3.1)**

- Both ionic bonding and covalent bonding follow one general rule: Elements gain, lose, or share electrons to attain the electronic configuration of the noble gas closest to them in the periodic table.
- **Ionic bonds** result from the transfer of electrons from one element to another. Ionic bonds form between a metal and a nonmetal. Ionic compounds consist of oppositely charged ions that feel a strong electrostatic attraction for each other.
- **Covalent bonds** result from the sharing of electrons between two atoms. Covalent bonds occur between two nonmetals, or when a metalloid combines with a nonmetal. Covalent bonding forms discrete molecules.

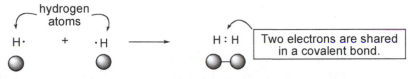

**[2] How can the periodic table be used to determine whether an atom forms a cation or an anion, and its resulting ionic charge? (3.2)**

- Metals form cations and nonmetals form anions.
- By gaining or losing one, two, or three electrons, an atom forms an ion with a completely filled outer shell of electrons.
- The charge on main group ions can be predicted from the position in the periodic table. For metals in groups 1A, 2A, and 3A, the group number = the charge on the cation. For nonmetals in groups 5A, 6A, and 7A, the anion charge = 8 − (the group number).

- **Ionic Charges of the Main Group Elements**

| Group Number | Number of Valence Electrons | Number of Electrons Gained or Lost | General Structure of the Ion |
|---|---|---|---|
| 1A (1) | 1 | 1 e⁻ lost | $M^+$ |
| 2A (2) | 2 | 2 e⁻ lost | $M^{2+}$ |
| 3A (3) | 3 | 3 e⁻ lost | $M^{3+}$ |
| 5A (15) | 5 | 3 e⁻ gained | $X^{3-}$ |
| 6A (16) | 6 | 2 e⁻ gained | $X^{2-}$ |
| 7A (17) | 7 | 1 e⁻ gained | $X^-$ |

**[3] What is the octet rule? (3.2)**

- Main group elements are especially stable when they possess an octet of electrons. Main group elements gain or lose one, two, or three electrons to form ions with eight outer shell electrons.

**[4] What determines the formula of an ionic compound? (3.3)**

- Cations and anions always form ionic compounds that have zero overall charge.
- Ionic compounds are written with the cation first, and then the anion, with subscripts to show how many of each are needed to have zero net charge.

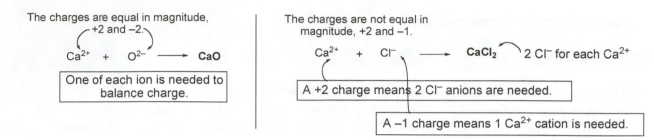

### [5] How are ionic compounds named? (3.4)

- Ionic compounds are always named with the name of the cation first.
- With cations having a fixed charge, the cation has the same name as its neutral element. The name of the anion usually ends in the suffix *-ide* if it is derived from a single atom or *-ate* (or *-ite*) if it is polyatomic.

NaF
sodium fluor*ide*

$Ca_3(PO_4)_2$
calcium phosph*ate*

- When the metal has a variable charge, use the overall anion charge to determine the charge on the cation. Then name the cation using a Roman numeral or the suffix *-ous* (for the ion with the smaller charge) or *-ic* (for the ion with the larger charge).

$CuCl_2$ — Two $Cl^-$ anions mean a $Cu^{2+}$ cation is present.
copper(II) chloride or
cupric chloride

$SnF_2$ — Two $F^-$ anions mean a $Sn^{2+}$ cation is present.
tin(II) fluoride or
stannous fluoride

### [6] Describe the properties of ionic compounds. (3.5)

- Ionic compounds are crystalline solids with the ions arranged to maximize the interactions of the oppositely charged ions.
- Ionic compounds have high melting points and boiling points.
- Most ionic compounds are soluble in water and their aqueous solutions conduct an electric current.

### [7] What are polyatomic ions and how are they named? (3.6)

- Polyatomic ions are charged species that are composed of more than one element.
- The names for polyatomic cations end in the suffix *-onium*.
- Many polyatomic anions have names that end in the suffix *-ate*. The suffix *-ite* is used for an anion that has one fewer oxygen atom than a similar anion named with the *-ate* ending. When two anions differ in the presence of a hydrogen, the word *hydrogen* or prefix *bi-* is added to the name of the anion.

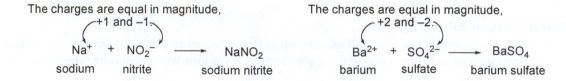

### [8] List useful consumer products and drugs that are composed of ionic compounds.

- Useful ionic compounds that contain alkali metal cations and halogen anions include KI (iodine supplement), NaF (source of fluoride in toothpaste), and KCl (potassium supplement). (3.3)

- Other products contain $SnF_2$ (fluoride source in toothpaste), $Al_2O_3$ (abrasive in toothpaste), and ZnO (sunblock agent). (3.4)
- Useful ionic compounds with polyatomic anions include $CaCO_3$ (antacid and calcium supplement), $Mg(OH)_2$ (antacid), and $FeSO_4$ (iron supplement). (3.6)

## Problem Solving

## [1] Introduction to Bonding (3.1)

**Example 3.1** Predict whether the bonds in the following compounds are ionic or covalent.

    a. $CaCl_2$        b. $CCl_4$

**Analysis**
The position of the elements in the periodic table determines the type of bonds they form. When a metal and nonmetal combine, the bond is ionic. When two nonmetals combine, or a metalloid bonds to a nonmetal, the bond is covalent.

**Solution**
a. Since Ca is a metal on the left side and Cl is a nonmetal on the right side of the periodic table, the bond must be ionic.
b. Since $CCl_4$ contains only the nonmetals carbon and chlorine, the bonds must be covalent.

---

## [2] Ions (3.2)

**Example 3.2** Write the ion symbol for an atom with 11 protons and 10 electrons.

**Analysis**
Since the number of protons equals the atomic number, this quantity identifies the element. The charge is determined by comparing the number of protons and electrons. If the number of electrons is greater than the number of protons, the charge is negative. If the number of protons is greater than the number of electrons, the charge is positive.

**Solution**
An element with 11 protons has an atomic number of 11, identifying it as sodium (Na). Since there is one more proton than electrons (11 vs. 10), the charge is +1. Thus, the answer is $Na^+$.

---

**Example 3.3** How many protons and electrons are present in each ion?

    a. $Mg^{2+}$        b. $Cl^-$

**Analysis**
Use the identity of the element to determine the number of protons. The charge tells how many more or fewer electrons there are compared to the number of protons. A positive charge means more protons than electrons, while a negative charge means more electrons than protons.

**Solution**

a. $Mg^{2+}$: The element magnesium (Mg) has an atomic number of 12, so it has 12 protons. Since the charge is +2, there are two more protons than electrons, giving the ion 10 electrons.

b. $Cl^-$: The element chlorine (Cl) has an atomic number of 17, so it has 17 protons. Since the charge is −1, there is one more electron than protons, giving the ion 18 electrons.

---

**Example 3.4** Use the group number to determine the charge on an ion derived from each element.

a. potassium        b. selenium

**Analysis**

Locate the element in the periodic table. A metal in groups 1A, 2A, or 3A forms a cation equal in charge to the group number. A nonmetal in groups 5A, 6A, and 7A forms an anion whose charge equals 8 – (the group number).

**Solution**

a. Potassium (K) is located in group 1A, so it forms a cation with a +1 charge, $K^+$.

b. Selenium (Se) is located in group 6A, so it forms an anion with a charge of $8 – 6 = 2$; $Se^{2-}$.

---

# [3] Naming Ionic Compounds (3.4)

**Example 3.5** Name each ionic compound.

a. KCl                b. $CaBr_2$

**Analysis**

Name the cation and then the anion.

**Solution**

a. KCl: The cation is potassium and the anion is chloride (derived from chlorine); thus, the name is potassium chloride.

b. $CaBr_2$: The cation is calcium and the anion is bromide (derived from bromine); thus, the name is calcium bromide.

---

# [4] Polyatomic Ions (3.6)

**Example 3.6** Write a formula for an ionic compound formed from calcium and carbonate.

**Analysis**

- Identify the cation and anion and determine the charges.
- When ions of equal charge combine, one of each is needed. When ions of unequal charge combine, use the ionic charges to determine the relative number of each ion.
- Write the formula with the cation first and then the anion, omitting charges. Use parentheses around polyatomic ions when more than one appears in the formula, and use subscripts to indicate the number of each ion.

## Solution

The cation from calcium has a +2 charge since calcium is located in group 2A. The carbonate anion has the formula $CO_3{}^{2-}$. Since the cation and anion have the same charge, there is one of each ion to give an overall charge of zero.

$$Ca^{2+} \ + \ CO_3{}^{2-} \longrightarrow CaCO_3$$

**Answer:** The formula is $CaCO_3$.

---

**Example 3.7** Name each ionic compound.

a. $Fe_2(SO_4)_3$     b. $FeSO_4$

## Analysis

First determine if the cation has fixed or variable charge. To name an ionic compound that contains a cation that always has the same charge, name the cation and then the anion. When the metal has variable charge, use the overall anion charge to determine the charge on the cation. Then name the cation (using a Roman numeral or the suffix *-ous* or *-ic*), followed by the anion.

## Solution

a. $Fe_2(SO_4)_3$: Iron cations have different charges. This cation has a +3 charge = ferric or iron(III). The anion $SO_4{}^{2-}$ is called sulfate.

    **Answer:** iron(III) sulfate or ferric sulfate

b. $FeSO_4$: The iron cation has a charge of +2 = ferrous or iron(II). The anion $SO_4{}^{2-}$ is called sulfate.

    **Answer:** iron(II) sulfate or ferrous sulfate

---

## Self-Test

**[1] Fill in the blank with one of the terms listed below.**

| | | |
|---|---|---|
| Anions (3.2) | Covalent bonds (3.1) | Nomenclature (3.4) |
| Bonding (3.1) | Ionic bonds (3.1) | Octet (3.2) |
| Cations (3.2) | Ionic compounds (3.1) | Polyatomic ion (3.6) |

1. _____ result from the sharing of electrons between two atoms.
2. _____ are positively charged ions. This type of ion has fewer electrons than protons.
3. _____ are composed of cations and anions.
4. _____ is the joining of two atoms in a stable arrangement.
5. A main group element is especially stable when it possesses an _____ of electrons in its outer shell.
6. A _____ is a cation or anion that contains more than one atom.
7. Assigning an unambiguous name to each compound is called chemical _____.
8. _____ result from the transfer of electrons from one element to another.
9. _____ are negatively charged ions. This type of ion has more electrons than protons.

**[2] Fill in the blank with one of the terms listed below.**

| | |
|---|---|
| a. cations | c. ions |
| b. anions | d. molecule |

10. _____ are charged species in which the number of protons and electrons in an atom is *not* equal.

11. A _____ is a discrete group of atoms that share electrons.

12. Metals form _____.

13. Nonmetals form _____.

**[3] Match the name of the anion with its symbol.**

| | | | |
|---|---|---|---|
| a. | $Br^-$ | d. | $I^-$ |
| b. | $Cl^-$ | e. | $O^{2-}$ |
| c. | $F^-$ | f. | $S^{2-}$ |

14. bromide          16. sulfide          18. iodide

15. fluoride          17. chloride          19. oxide

**[4] Match the polyatomic anion symbol to its name.**

| | | | |
|---|---|---|---|
| a. | $HCO_3^-$ | d. | $CH_3CO_2^-$ |
| b. | $SO_3^{2-}$ | e. | $OH^-$ |
| c. | $NO_3^-$ | f. | $HPO_4^{2-}$ |

20. hydroxide

21. sulfite

22. hydrogen carbonate  or bicarbonate

23. hydrogen phosphate

24. nitrate

25. acetate

**[5] Write the molecular formula of the ionic compound derived from the cations on the left and each of the anions across the top.**

| | $^-OH$ | $SO_4^{2-}$ | $PO_4^{3-}$ |
|---|---|---|---|
| $K^+$ | KOH | 27. | 29. |
| $Al^{3+}$ | 26. | 28. | 30. |

## Answers to Self-Test

| | | | | | |
|---|---|---|---|---|---|
| 1. Covalent bonds | 6. polyatomic ion | 11. d | 16. f | 21. b | 26. $Al(OH)_3$ |
| 2. Cations | 7. nomenclature | 12. a | 17. b | 22. a | 27. $K_2SO_4$ |
| 3. Ionic compounds | 8. Ionic bonds | 13. b | 18. d | 23. f | 28. $Al_2(SO_4)_3$ |
| 4. Bonding | 9. Anions | 14. a | 19. e | 24. c | 29. $K_3PO_4$ |
| 5. octet | 10. c | 15. c | 20. e | 25. d | 30. $AlPO_4$ |

## Solutions to In-Chapter Problems

**3.1**    The position of the elements in the periodic table determines the type of bonds they form. When a metal and nonmetal combine, as in (b) and (c), the bond is ionic. When two nonmetals combine, or when a metalloid bonds to a nonmetal, the bond is covalent.

a. CO covalent          c. MgO ionic          e. HF covalent
b. $CaF_2$ ionic          d. $Cl_2$ covalent          f. $C_2H_6$ covalent

**3.2**    An **element** is a pure substance that cannot be broken down into simpler substances by a chemical reaction.
A **compound** is a pure substance formed by combining two or more elements together.
A **molecule** is composed of atoms that are covalently bonded together.

a. $CO_2$ compound, molecule          c. NaF compound          e. $F_2$ element, molecule
b. $H_2O$ compound, molecule          d. $MgBr_2$ compound          f. CaO compound

**3.3**    Vitamin C ($C_6H_8O_6$) is likely to contain covalent bonds because it consists of the nonmetals C, H, and O.

**3.4**    The number of protons equals the atomic number. The charge is determined by comparing the number of protons and electrons. If the number of electrons is greater than the number of protons, the charge is negative. If the number of protons is greater than the number of electrons, the charge is positive.

a. 19 protons and 18 electrons = $K^+$          c. 35 protons and 36 electrons = $Br^-$
b. 7 protons and 10 electrons = $N^{3-}$          d. 23 protons and 21 electrons = $V^{2+}$

**3.5**    Use the identity of the element to determine the number of protons. The charge tells how many more or fewer electrons there are compared to the number of protons. A positive charge means more protons than electrons, while a negative charge means more electrons than protons.

a. $Ni^{2+}$ = 28 protons, 26 electrons          c. $Zn^{2+}$ = 30 protons, 28 electrons
b. $Se^{2-}$ = 34 protons, 36 electrons          d. $Fe^{3+}$ = 26 protons, 23 electrons

**3.6**    Locate the element in the periodic table. A metal in groups 1A, 2A, or 3A forms a cation equal in charge to the group number. A nonmetal in groups 5A, 6A, or 7A forms an anion whose charge equals 8 – (the group number).

a. magnesium          b. iodine          c. selenium          d. rubidium
(group 2A): +2          (group 7A): –1          (group 6A): –2          (group 1A): +1

**3.7**
a. Ne          b. Xe          c. Kr          d. Kr

**3.8**
a. $Au^+$ = 79 protons, 78 electrons          c. $Sn^{2+}$ = 50 protons, 48 electrons
b. $Au^{3+}$ = 79 protons, 76 electrons          d. $Sn^{4+}$ = 50 protons, 46 electrons

**3.9**

    a. Mn = 25 protons, 25 electrons

    b. $Mn^{2+}$ = 25 protons, 23 electrons

    c. $1s^2 2s^2 2p^6 3s^2 3p^6 4s^2 3d^5$         The two $4s^2$ valence electrons would be lost to form $Mn^{2+}$.

**3.10**    Ionic compounds are composed of cations and anions.

    a. lithium (metal) and bromine (nonmetal): yes   c. calcium and magnesium (two metals): no

    b. chlorine and oxygen (two nonmetals): no      d. barium (metal) and chlorine (nonmetal): yes

**3.11**    • Identify the cation and the anion, and use the periodic table to determine the charges.

      • When ions of equal charge combine, one of each ion is needed. When ions of unequal charge combine, use the ionic charges to determine the relative number of each ion.

      • Write the formula with the cation first and then the anion, omitting charges, and using subscripts to indicate the number of each ion.

    a. sodium (+1) and bromine (–1) = NaBr        c. magnesium (+2) and iodine (–1) =

                                            2 $I^-$ anions are needed = $MgI_2$

    b. barium (+2) and oxygen (–2) = BaO       d. lithium (+1) and oxygen (–2) =

                                            2 $Li^+$ cations are needed = $Li_2O$

**3.12**    a. In $Na_2S$, there are twice as many $Na^+$ cations (darker spheres) as there are $S^{2-}$ anions (lighter spheres).

      b. In $MgCl_2$, there are twice as many $Cl^-$ anions (lighter spheres) as there are $Mg^{2+}$ cations (darker spheres).

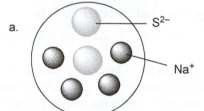
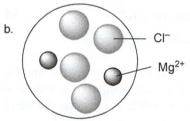

**3.13**    Zinc forms $Zn^{2+}$ and oxygen forms $O^{2-}$; thus, zinc oxide = ZnO.

**3.14**    When a metal forms more than one cation, the cations are named by one of two methods.

        *Method [1]:* Follow the name of the cation by a Roman numeral in parentheses to indicate its charge.

        *Method [2]:* Use the suffix *-ous* for the cation with the lower charge, and the suffix *-ic* for the cation with the higher charge. These suffixes are often added to the Latin names of the elements.

    Anions are named by replacing the ending of the element name by the suffix *-ide*.

    a. $S^{2-}$= sulfide              c. $Cs^+$= cesium             e. $Sn^{4+}$= tin(IV), stannic

    b. $Cu^+$= copper(I), cuprous     d. $Al^{3+}$= aluminum

**3.15**

 a. stannic = $Sn^{4+}$     c. manganese ion = $Mn^{2+}$     e. selenide = $Se^{2-}$

 b. iodide = $I^-$        d. lead(II) = $Pb^{2+}$

**3.16**   Name the cation and then the anion.

 a. NaF = sodium fluoride    c. $SrBr_2$ = strontium bromide    e. $TiO_2$ = titanium oxide

 b. MgO = magnesium oxide   d. $Li_2O$ = lithium oxide      f. $AlCl_3$ = aluminum chloride

**3.17**   First determine if the cation has fixed or variable charge. To name an ionic compound that contains a cation that always has the same charge, name the cation and then the anion (using the suffix *-ide*). When the metal has variable charge, use the overall anion charge to determine the charge on the cation. Then name the cation (using a Roman numeral or the suffix *-ous* or *-ic*), followed by the anion.

 a. $CrCl_3$
Chromium has a variable charge, but here it must have a +3 charge to balance the three chloride ions.
**chromium(III) chloride, chromic chloride**

 d. $PbO_2$
Lead has a variable charge, but here it must have a +4 charge to balance the two oxide ions.
**lead(IV) oxide**

 b. PbS
Lead has a variable charge, but here it must have a +2 charge to balance the sulfide ion.
**lead(II) sulfide**

 e. $FeBr_2$
Iron has a variable charge, but here it must have a +2 charge to balance the two bromide ions.
**iron(II) bromide, ferrous bromide**

 c. $SnF_4$
Tin has a variable charge, but here it must have a +4 charge to balance the four fluoride ions.
**tin(IV) fluoride, stannic fluoride**

 f. $AuCl_3$
Gold has a variable charge, but here it must have a +3 charge to balance the three chloride ions.
**gold(III) chloride**

**3.18**

 a. $Cu_2O$ = copper(I) oxide, cuprous oxide    c. CuCl = copper(I) chloride, cuprous chloride

 b. CuO = copper(II) oxide, cupric oxide     d. $CuCl_2$ = copper(II) chloride, cupric chloride

**3.19**   $Fe_2O_3$ = iron(III) oxide, ferric oxide

**3.20**   Identify the cation and the anion and determine their charges. Balance the charges. Write the formula with the cation first, and use subscripts to show the number of each ion needed to have zero overall charge.

 a. calcium bromide
Calcium is the cation (+2).
Bromide is the anion (−1).
**$CaBr_2$**

 c. ferric bromide
Iron (Fe) is the cation (+3).
Bromide is the anion (−1).
**$FeBr_3$**

 b. copper(I) iodide
Copper(I) is the cation (+1).
Iodide is the anion (−1).
**CuI**

 d. magnesium sulfide
Magnesium is the cation (+2).
Sulfide is the anion (−2).
**MgS**

e. chromium(II) chloride
Chromium is the cation (+2).
Chloride is the anion (–1).
**$CrCl_2$**

f. sodium oxide
Sodium is the cation (+1).
Oxide is the anion (–2).
**$Na_2O$**

**3.21** Ionic compounds have high melting points and high boiling points. They usually dissolve in water. Their solutions conduct electricity and they form crystalline solids.

**3.22** Write the formula formed from polyatomic ions with the cation first and then the anion, omitting charges. Use parentheses around polyatomic ions when more than one appears in the formula, and use subscripts to indicate the number of each ion.

a. magnesium (+2)
   $MgSO_4$

c. nickel (+2)
   $NiSO_4$

e. lithium (+1)
   $Li_2SO_4$

b. sodium (+1)
   $Na_2SO_4$

d. aluminum (+3)
   $Al_2(SO_4)_3$

**3.23** Use Table 3.5 to determine the charge on the polyatomic ions.

a. sodium (+1) and bicarbonate (–1): $NaHCO_3$

c. ammonium (+1) and sulfate (–2): $(NH_4)_2SO_4$

e. calcium (+2) and bisulfate (–1): $Ca(HSO_4)_2$

b. potassium (+1) and nitrate (–1): $KNO_3$

d. magnesium (+2) and phosphate (–3): $Mg_3(PO_4)_2$

f. barium (+2) and hydroxide (–1): $Ba(OH)_2$

**3.24**

a. $OH^- = KOH$
b. $NO_2^- = KNO_2$

c. $SO_4^{2-} = K_2SO_4$
d. $HSO_3^- = KHSO_3$

e. $PO_4^{3-} = K_3PO_4$
f. $CN^- = KCN$

**3.25** First determine if the cation has fixed or variable charge. To name an ionic compound that contains a cation that always has the same charge, name the cation and then the anion. When the metal has variable charge, use the overall anion charge to determine the charge on the cation. Then name the cation (using a Roman numeral or the suffix *-ous* or *-ic*), followed by the anion.

a. $Na_2CO_3$ =
   sodium carbonate

c. $Mg(NO_3)_2$ = magnesium nitrate

e. $Fe(HSO_3)_3$ = iron(III) hydrogen sulfite, ferric bisulfite

b. $Ca(OH)_2$ = calcium hydroxide

d. $Mn(CH_3CO_2)_2$ = manganese acetate

f. $Mg_3(PO_4)_2$ = magnesium phosphate

**3.26** Hydroxyapatite = $Ca_{10}(PO_4)_6(OH)_2$
   Each Ca has a +2 charge; 10 $Ca^{2+}$ = +20
   Each $PO_4$ has a –3 charge; 6 $PO_4^{3-}$ = –18
   Each OH has a –1 charge; 2 $OH^-$ = –2
   Total negative charge of –20 balances a total positive charge of +20.

## Solutions to Odd-Numbered End-of-Chapter Problems

**3.27** Use the criteria in Problem 3.1.
a. $CO_2$ = covalent    b. $H_2SO_4$ = covalent    c. KF = ionic    d. $CH_5N$ = covalent

**3.29** a. potassium (metal) and oxygen (nonmetal) = ionic
b. sulfur and carbon (two nonmetals) = covalent
c. two bromine atoms (two nonmetals) = covalent
d. carbon and oxygen (two nonmetals) = covalent

**3.31** Ionic bonds form between a metal and nonmetal because there is a transfer of electrons from the metal to the nonmetal. A metal gains a noble gas configuration of electrons by giving up electrons. A nonmetal can gain a noble gas configuration by gaining electrons.

**3.33**

a. four protons and two electrons = $Be^{2+}$

b. 22 protons and 20 electrons = $Ti^{2+}$

c. 16 protons and 18 electrons = $S^{2-}$

d. 13 protons and 10 electrons = $Al^{3+}$

e. 17 protons and 18 electrons = $Cl^-$

f. 20 protons and 18 electrons = $Ca^{2+}$

**3.35**

a. a period 2 element that forms a +2 cation = Be
b. an ion from group 7A with 18 electrons = $Cl^-$
c. a cation from group 1A with 36 electrons = $Rb^+$

**3.37** Elements in group 6A gain electrons to form anions because by gaining two electrons they have a filled valence shell.

**3.39**

a. sodium ion = $Na^+$
b. selenide = $Se^{2-}$

c. manganese ion = $Mn^{2+}$
d. gold(III) = $Au^{3+}$

e. stannic = $Sn^{4+}$
f. mercurous = $Hg_2^{2+}$

**3.41** The noble gas with the same number of electrons has the same electronic configuration as each ion.

a. $O^{2-}$ = Ne
b. $Mg^{2+}$ = Ne

c. $Al^{3+}$ = Ne
d. $S^{2-}$ = Ar

e. $F^-$ = Ne
f. $Be^{2+}$ = He

**3.43**

a. lithium = lose one electron (He)
b. iodine = gain one electron (Xe)

c. sulfur = gain two electrons (Ar)
d. strontium = lose two electrons (Kr)

**3.45** Ions that contain an outer shell of eight electrons are likely to form.

a. $S^-$ No, only seven electrons in outer shell
b. $S^{2-}$ Likely to form

c. $S^{3-}$ No, one electron in outer shell
d. $Na^+$ Likely to form

e. $Na^{2+}$ No, only seven electrons in outer shell
f. $Na^-$ No, two electrons in outer shell

**3.47**

| | Number of Valence Electrons | Group Number | Number of Electrons Gained or Lost | Charge | Example |
|---|---|---|---|---|---|
| a. X· | 1 | 1A | Lose 1 | 1+ | Li |
| b. ·Q· | 2 | 2A | Lose 2 | 2+ | Mg |
| c. ·Z̈· | 6 | 6A | Gain 2 | 2− | S |
| d. :Ä· | 7 | 7A | Gain 1 | 1− | Cl |

**3.49**   a. sulfate, $SO_4^{2-}$     b. nitrite, $NO_2^-$     c. sulfide, $S^{2-}$

**3.51**

a. sulfate = $SO_4^{2-}$              c. hydrogen carbonate = $HCO_3^-$
b. ammonium = $NH_4^+$            d. cyanide = $CN^-$

**3.53**

a. $OH^-$ = 9 protons, 10 electrons          c. $PO_4^{3-}$ = 47 protons, 50 electrons
b. $H_3O^+$ = 11 protons, 10 electrons

**3.55**   Transition metals have one or more $d$ electrons. All of these electrons would have to be lost to follow the octet rule, and most transition metals do not lose that many electrons.

**3.57**   Na donates an electron to F; then each atom has eight electrons in its outer shell, which follows the octet rule.

**3.59**

a. calcium ($Ca^{2+}$) and sulfur       c. lithium ($Li^+$) and iodine       e. sodium ($Na^+$) and selenium
   ($S^{2-}$) = CaS                          ($I^-$) = LiI                           ($Se^{2-}$) = $Na_2Se$

b. aluminum ($Al^{3+}$) and          d. nickel ($Ni^{2+}$) and chlorine
   bromine ($Br^-$) = $AlBr_3$            ($Cl^-$) = $NiCl_2$

**3.61**

a. lithium ($Li^+$) and nitrite       c. sodium ($Na^+$) and bisulfite     e. magnesium ($Mg^{2+}$) and
   ($NO_2^-$) = $LiNO_2$                     ($HSO_3^-$) = $NaHSO_3$               hydrogen sulfite ($HSO_3^-$) =
                                                                                    $Mg(HSO_3)_2$

b. calcium ($Ca^{2+}$) and acetate    d. manganese ($Mn^{2+}$) and
   ($CH_3COO^-$) =                         phosphate ($PO_4^{3-}$) =
   $Ca(CH_3COO)_2$                         $Mn_3(PO_4)_2$

**3.63**

| | $Y^-$ | $Y^{2-}$ | $Y^{3-}$ |
|---|---|---|---|
| $X^+$ | XY | $X_2Y$ | $X_3Y$ |
| $X^{2+}$ | $XY_2$ | XY | $X_3Y_2$ |
| $X^{3+}$ | $XY_3$ | $X_2Y_3$ | XY |

**3.65**

|      | $Br^-$ | $OH^-$ | $HCO_3^-$ | $SO_3^{2-}$ | $PO_4^{3-}$ |
|------|--------|--------|-----------|-------------|-------------|
| $Na^+$ | NaBr | NaOH | $NaHCO_3$ | $Na_2SO_3$ | $Na_3PO_4$ |
| $Co^{2+}$ | $CoBr_2$ | $Co(OH)_2$ | $Co(HCO_3)_2$ | $CoSO_3$ | $Co_3(PO_4)_2$ |
| $Al^{3+}$ | $AlBr_3$ | $Al(OH)_3$ | $Al(HCO_3)_3$ | $Al_2(SO_3)_3$ | $AlPO_4$ |

**3.67**

a. $KHSO_4$    b. $Ba(HSO_4)_2$    c. $Al(HSO_4)_3$    d. $Zn(HSO_4)_2$

**3.69**

a. $Ba(CN)_2$    b. $Ba_3(PO_4)_2$    c. $BaHPO_4$    d. $Ba(H_2PO_4)_2$

**3.71**

a. $Na_2O$ = sodium oxide
b. $BaS$ = barium sulfide
c. $PbS_2$ = lead(IV) sulfide
d. $AgCl$ = silver chloride

e. $CoBr_2$ = cobalt(II) bromide
f. $RbBr$ = rubidium bromide
g. $PbBr_2$ = lead(II) bromide

**3.73**

a. $FeCl_2$ = iron(II) chloride, ferrous chloride
b. $FeBr_3$ = iron(III) bromide, ferric bromide

c. $FeS$ = iron(II) sulfide, ferrous sulfide
d. $Fe_2S_3$ = iron(III) sulfide, ferric sulfide

**3.75**    Copper cations can be 1+ or 2+, so the Roman numeral designation is required.  Ca exists only as 2+.

$CuBr_2$: copper(II) bromide or cupric bromide
$CaBr_2$: calcium bromide

**3.77**

a. sodium sulfide and sodium sulfate = $Na_2S$ and $Na_2SO_4$
b. magnesium oxide and magnesium hydroxide = $MgO$ and $Mg(OH)_2$
c. magnesium sulfate and magnesium bisulfate =  $MgSO_4$ and $Mg(HSO_4)_2$

**3.79**

a. $NH_4Cl$ = ammonium chloride

b. $PbSO_4$ = lead(II) sulfate

c. $Cu(NO_3)_2$ = copper(II) nitrate, cupric nitrate

d. $Ca(HCO_3)_2$ = calcium bicarbonate, calcium hydrogen carbonate

e. $Fe(NO_3)_2$ = iron(II) nitrate, ferrous nitrate

**3.81**

a. magnesium carbonate = $MgCO_3$

b. nickel sulfate = $NiSO_4$

c. copper(II) hydroxide = $Cu(OH)_2$

d. potassium hydrogen phosphate = $K_2HPO_4$

e. gold(III) nitrate = $Au(NO_3)_3$

f. lithium phosphate = $Li_3PO_4$

g. aluminum bicarbonate = $Al(HCO_3)_3$

h. chromous cyanide = $Cr(CN)_2$

**3.83**

a. $OH^- = Pb(OH)_4$
   lead(IV) hydroxide

c. $HCO_3^- = Pb(HCO_3)_4$
   lead(IV) bicarbonate

e. $PO_4^{3-} = Pb_3(PO_4)_4$
   lead(IV) phosphate

b. $SO_4^{2-} = Pb(SO_4)_2$
   lead(IV) sulfate

d. $NO_3^- = Pb(NO_3)_4$
   lead(IV) nitrate

f. $CH_3CO_2^- = Pb(CH_3CO_2)_4$
   lead(IV) acetate

**3.85**   The false statements are corrected.

a. True: Ionic compounds have high melting points.
b. False: Ionic compounds are solids at room temperature.
c. False: Most ionic compounds are soluble in water.
d. False: Ionic solids exist as crystalline lattices with the ions arranged to maximize the electrostatic interactions of anions and cations.

**3.87**   NaCl has a higher melting point than $CH_4$ or $H_2SO_4$ because it is an ionic compound, whereas the other two compounds are covalent.  Ionic solids have higher melting points.

**3.89**   a.   A neutral zinc atom has 30 protons and 30 electrons.
b.   The $Zn^{2+}$ cation has 30 protons and 28 electrons.
c.   The electronic configuration of zinc: $1s^2 2s^2 2p^6 3s^2 3p^6 4s^2 3d^{10}$
       The $4s^2$ electrons are lost to form $Zn^{2+}$.

**3.91**

| Cation | a. Number of Protons | b. Number of Electrons | c. Noble Gas | d. Role |
|---|---|---|---|---|
| $Na^+$ | 11 | 10 | Ne | Major cation in extracellular fluids and blood; maintains blood volume and blood pressure |
| $K^+$ | 19 | 18 | Ar | Major intracellular cation |
| $Ca^{2+}$ | 20 | 18 | Ar | Major cation in solid tissues like bone and teeth; required for normal muscle contraction and nerve function |
| $Mg^{2+}$ | 12 | 10 | Ne | Required for normal muscle contraction and nerve function |

**3.93**   silver ($Ag^+$) nitrate ($NO_3^-$) = $AgNO_3$

**3.95**   $CaSO_3$ = calcium sulfite

**3.97**   ammonium ($NH_4^+$) nitrate ($NO_3^-$) = $NH_4NO_3$

**3.99**

a. magnesium oxide (MgO) and potassium iodide (KI)
b. $CaHPO_4$ = calcium hydrogen phosphate
c. $FePO_4$ = iron(III) phosphate, ferric phosphate
d. sodium selenite = $Na_2SeO_3$
e. The name chromium chloride is ambiguous.  Without a designation as chromium(II) or chromium(III) it's impossible to know the ratio of chromium cations to chloride anions.

# Chapter 4 Covalent Compounds

## Chapter Review

**[1] What are the characteristic bonding features of covalent compounds? (4.1)**

- Covalent bonds result from the sharing of electrons between two atoms, forming molecules. Atoms share electrons to attain the electronic configuration of the noble gas nearest them in the periodic table. For many main group elements this results in an octet of electrons.
- Covalent bonds are formed when two nonmetals combine or when a metalloid bonds to a nonmetal. Covalent bonds are preferred with elements that would have to gain or lose too many electrons to form an ion.
- Except for hydrogen, the common elements—C, N, O, and halogens—follow one rule: the number of bonds plus the number of lone pairs equals four.

|  | lone pairs of electrons | | |
|---|---|---|---|
| —H | —C— | —N̈— | —Ö— | —Ẍ: X = F, Cl, Br, I |
| hydrogen | carbon | nitrogen | oxygen | halogen |

| | | | | | |
|---|---|---|---|---|---|
| Number of bonds | 1 | 4 | 3 | 2 | 1 |
| Number of nonbonded electron pairs | 0 | 0 | 1 | 2 | 3 |

**[2] What are Lewis structures and how are they drawn? (4.2)**

- Lewis structures are electron-dot representations of molecules. Two-electron bonds are drawn with a solid line and nonbonded electrons are drawn with dots (:).
- Lewis structures contain only valence electrons. Each H gets two electrons and main group elements generally get eight.

**[3] What are resonance structures? (4.4)**

- Resonance structures are two Lewis structures having the same arrangement of atoms but a different arrangement of electrons.

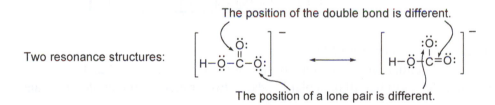

The position of the double bond is different.

Two resonance structures:

The position of a lone pair is different.

- The hybrid is a composite of all resonance structures that spreads out electron pairs in multiple bonds and lone pairs.

**[4] How are covalent compounds with two elements named? (4.5)**

- Name the first nonmetal by its element name and the second using the suffix *-ide*. Add prefixes to show the number of atoms of each element.

silicon tetrafluoride

↓    ↓    ↓      = $SiF_4$

Si    4 F atoms

**[5] How is the molecular shape around an atom determined? (4.6)**

- To determine the shape around an atom, count groups—atoms and lone pairs—and keep the groups as far away from each other as possible.
- Two groups = linear, $180°$ bond angle; three groups = trigonal planar, $120°$ bond angle; four groups = tetrahedral, $109.5°$ bond angle.
- See Table 4.2, Common Molecular Shapes Around Atoms.

**[6] How does electronegativity determine bond polarity? (4.7)**

- Electronegativity is a measure of an atom's attraction for electrons in a bond.
- When two atoms have the same electronegativity value, or the difference is less than 0.5 units, the electrons are equally shared and the bond is nonpolar.
- When two atoms have very different electronegativity values—a difference of 0.5–1.9 units—the electrons are unequally shared and the bond is polar.

**Increasing electronegativity** →

| 1A | 2A | | 3A | 4A | 5A | 6A | 7A | 8A |
|---|---|---|---|---|---|---|---|---|
| H 2.1 | | | | | | | | |
| Li 1.0 | Be 1.5 | | B 2.0 | C 2.5 | N 3.0 | O 3.5 | F 4.0 | |
| Na 0.9 | Mg 1.2 | | Al 1.5 | Si 1.8 | P 2.1 | S 2.5 | Cl 3.0 | |
| K 0.8 | Ca 1.0 | | Ga 1.6 | Ge 1.8 | As 2.0 | Se 2.4 | Br 2.8 | |
| Rb 0.8 | Sr 1.0 | | In 1.7 | Sn 1.8 | Sb 1.9 | Te 2.1 | I 2.5 | |

Increasing electronegativity ↑

**[7] When is a molecule polar or nonpolar? (4.8)**

- A polar molecule has either one polar bond, or two or more bond dipoles that do not cancel.
- A nonpolar molecule has either all nonpolar bonds, or two or more bond dipoles that cancel.

one polar bond      three polar bonds      three polar bonds      two polar bonds

All dipoles cancel.      All dipoles reinforce.      Two dipoles reinforce.

NO net dipole

**polar** molecule      **nonpolar** molecule      **polar** molecule      **polar** molecule

## Problem Solving

## [1] Introduction to Covalent Bonding (4.1)

**Example 4.1** Without referring to Figure 4.1 in the text, how many covalent bonds are predicted for each
atom?

a. C  b. O

**Analysis**
Atoms with one, two, or three valence electrons form one, two, or three bonds, respectively. Atoms with
four or more valence electrons form enough bonds to give an octet; that is, the predicted number
of bonds = 8 – (the number of valence electrons).

**Solution**
a. C has four valence electrons. Thus, it is expected to form 8 – 4 = 4 bonds.
b. O has six valence electrons. Thus, it is expected to form 8 – 6 = 2 bonds.

## [2] Lewis Structures (4.2)

**Example 4.2** Draw a Lewis structure for $CH_2F_2$.

**Analysis and Solution**
**[1] Arrange the atoms.**

H

H  C  F

F

- Place C in the center and 2 H's and 2 F's on the periphery.
- In this arrangement, C is surrounded by four atoms, its usual number.

**[2] Count the electrons.**

| 1 C | x | 4 e⁻ | = | 4 e⁻ |
| 2 H | x | 1 e⁻ | = | 2 e⁻ |
| 2 F | x | 7 e⁻ | = | 14 e⁻ |
| | | | | 20 e⁻ total |

**[3] Add the bonds and lone pairs.**

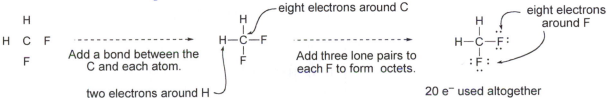

First, add four single bonds—namely, two C–H bonds and two C–F bonds. This uses eight valence
electrons, and gives carbon an octet (four two-electron bonds) and each hydrogen two electrons. Next,
give each F an octet by adding three lone pairs to each F atom. This uses all 20 valence electrons. To
check if a Lewis structure is valid, we must answer YES to three questions.

- Have all the electrons been used?
- Is each H surrounded by two electrons?
- Is every other main group element surrounded by eight electrons?

Since the answer to all three questions is YES, we have drawn a valid Lewis structure for $CH_2F_2$.

---

**Example 4.3** Draw a Lewis structure for molecular formula $C_2H_2Cl_2$ in which each carbon is bonded to one hydrogen and one chlorine.

**Analysis and Solution**

Follow steps [1]–[3] to draw a Lewis structure. Place five bonds between the atoms and add three lone pairs to each Cl. When the two remaining electrons are placed as a lone pair on a carbon, one C still has no octet.

**Step [1]**
Arrange the atoms.

Cl C  C Cl
  H  H

• Each C gets 1 H and 1 Cl.

**Step [2]**
Count the electrons.

$2 C \times 4 e^- = 8 e^-$
$2 H \times 1 e^- = 2 e^-$
$2 Cl \times 7 e^- = 14 e^-$

**24 e⁻ total**

**Step [3]**
Add the bonds and lone pairs.

Add bonds first...   ...then lone pairs.

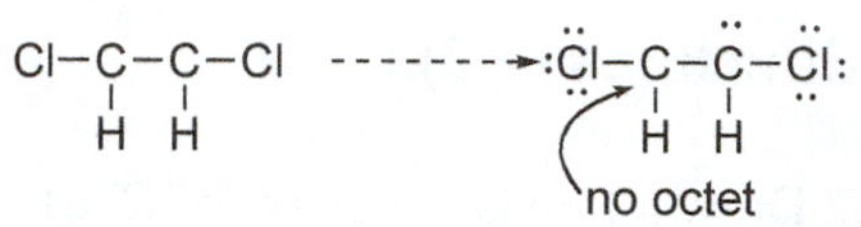

no octet

**Step [4]** To give both C's an octet, change *one* lone pair into *one* bonding pair of electrons between the two C atoms, forming a *double* bond.

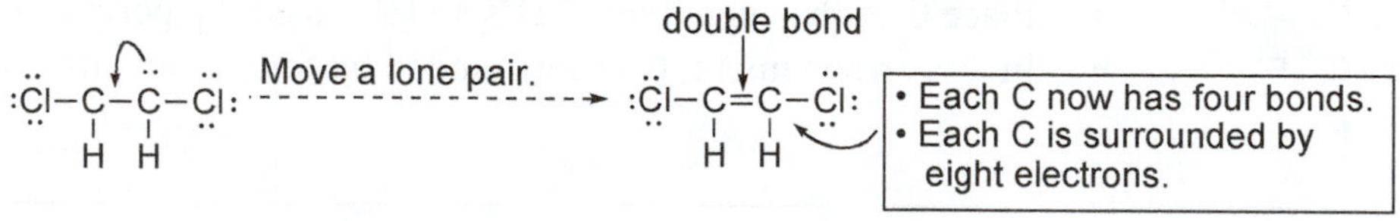

• Each C now has four bonds.
• Each C is surrounded by eight electrons.

This uses all 24 electrons, each C and Cl have an octet, and each H has two electrons. The Lewis structure is valid.

---

# [3] Resonance (4.4)

**Example 4.4** Draw a second resonance structure for the following anion:

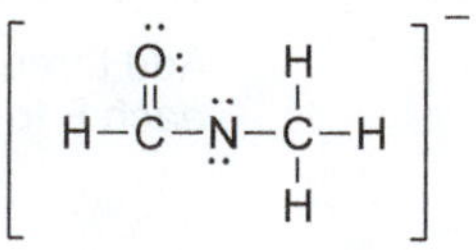

**Analysis**

Concentrate on the multiple bonds and lone pairs. To convert one resonance structure to another, use a lone pair to form a double bond and convert a bonding electron pair in a double bond to a lone pair.

**Solution**

Move an electron pair out on O.

The new resonance structure **B** differs from **A** in the location of a double bond and a lone pair.

---

## [4] Naming Covalent Compounds (4.5)

**Example 4.5** Give the formula for each compound.

a. sulfur dioxide          b. nitrogen trifluoride

**Analysis**
- Write the symbols for the elements in the order given in the name.
- Use the prefixes to write the subscripts.

**Solution**

a. sulfur dioxide

S    2 O atoms

**Answer: SO$_2$**

b. nitrogen trifluoride

N    3 F atoms

**Answer: NF$_3$**

---

## [5] Molecular Shape (4.6)

**Example 4.6** Using the given Lewis structure, determine the shape around the indicated atom in each compound.

a.    H—C—C≡N:

b.    H—C—Cl:

**Analysis**
To predict shape around an atom, we need a valid Lewis structure, which is given in this problem. Then count groups around the atom to determine molecular shape using the information in Table 4.2.

## Solution

a. The indicated C in $CH_3CN$ is surrounded by two atoms and no lone pairs—that is, two groups. An atom surrounded by two groups is linear with a $180°$ bond angle.

b. The C atom in $CH_2Cl_2$ is surrounded by four atoms—that is, four groups. An atom surrounded by four groups is tetrahedral, with $109.5°$ bond angles.

---

## [6] Electronegativity and Bond Polarity (4.7)

**Example 4.7** Use electronegativity values to classify each bond as nonpolar, polar, or ionic.

a. KCl         b. $Br_2$         c. $H_2O$

### Analysis

Calculate the electronegativity difference between the two atoms and use the following rules: less than 0.5 (nonpolar); 0.5–1.9 (polar covalent); greater than 1.9 (ionic).

### Solution

|  | **Electronegativity difference** | **Bond type** |
|---|---|---|
| a. KCl | 3.0 (Cl) – 0.8 (K) = 2.2 | ionic |
| b. $Br_2$ | 2.8 (Br) – 2.8 (Br) = 0 | nonpolar |
| c. $H_2O$ | 3.5 (O) – 2.1 (H) = 1.4 | polar covalent |

---

## [7] Polarity of Molecules (4.8)

**Example 4.8** Using the given Lewis structure, determine whether dimethyl ether is a polar or nonpolar molecule.

dimethyl ether

### Analysis

To determine the overall polarity of a molecule, identify the polar bonds, determine the shape around individual atoms, and then decide if the individual bond dipoles cancel or reinforce.

### Solution

Ignore the C–H bonds since they are all nonpolar. Each C–O bond is polar because the electronegativity difference between O (3.5) and C (2.5) is 1.0. Since the O atom of dimethyl ether has two atoms and two lone pairs around it, dimethyl ether is a bent molecule around the O atom. The two dipoles reinforce (both point *up*), so **dimethyl ether has a net dipole; that is, dimethyl ether is a polar molecule.**

**The two individual dipoles reinforce.**

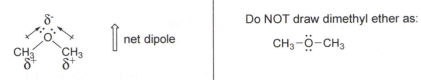

The net dipole bisects the C–O–C bond angle.
The bent representation shows that the dipoles reinforce.

Note: **We must know the geometry to determine if two dipoles cancel or reinforce.** For example, do *not* draw dimethyl ether as a linear molecule around the O atom, because you might think that the two dipoles cancel, when in reality, they reinforce.

## Self-Test

**[1] Fill in the blank with one of the terms listed below.**

| | | |
|---|---|---|
| Covalent bond (4.1) | Hybrid (4.4) | Polar (4.7) |
| Dipole (4.7) | Lewis structures (4.1) | Resonance structures (4.4) |
| Double-headed arrow (4.4) | Molecular formula (4.2) | Valence shell electron pair repulsion theory |
| Electronegativity (4.7) | Nonpolar (4.7) | (4.6) |

1. The electron-dot structures for molecules are called _____.
2. _____ are two or more Lewis structures having the same arrangement of atoms but a different arrangement of electrons.
3. A composite of two or more resonance forms is called a _____.
4. The most stable arrangement keeps groups of electrons as far away from each other as possible; this is the basis for the _____.
5. _____ is a measure of an atom's attraction for electrons in a bond.
6. A _____ results from the sharing of electrons between two atoms.
7. Bonding between atoms of different electronegativity results in the unequal sharing of electrons, which creates a _____ bond.
8. A _____ shows that two Lewis structures are resonance structures.
9. When electrons are equally shared, the bond is said to be _____.
10. A _____ shows the number and identity of all the atoms in a compound.
11. A polar bond is said to have a _____—that is, a separation of charge.

**[2] Fill in the blank with one of the terms listed below.**

a. six electrons   b. four electrons   c. less electronegative   d. more electronegative

12. A double bond contains _____.
13. A triple bond contains _____.
14. The symbol $\delta^+$ is given to the _____ atom in a polar bond.
15. The symbol $\delta^-$ is given to the _____ atom in a polar bond.

**[3] Fill in the blanks with the terms listed below.**

a. trigonal planar                 c. 120°                           e. tetrahedral
b. 109.5°                      d. linear                          f. 180°

16. An atom surrounded by two groups is _____ and has a bond angle of _____.
17. An atom surrounded by three groups is _____ and has bond angles of _____.
18. An atom surrounded by four groups is _____ and has bond angles of _____.

**[4] Match the symbol with its name.**

a. solid line                b. dashed line             c. wedge

19. A _____ is used for bonds in the plane.
20. A _____ is used for a bond in front of the plane.
21. A _____ is used for a bond behind the plane.

**[5] Fill in the table with one of the types of bonds listed below.**

a. ionic                b. nonpolar             c. polar covalent

22. _____          Electrons are equally shared.
23. _____          Electrons are unequally shared, pulled towards the more electronegative element.
24. _____          Electrons are transferred from the less electronegative element to the more electronegative element.

## Answers to Self-Test

| | | | | |
|---|---|---|---|---|
| 1. Lewis structures | 6. covalent bond | 11. dipole | 16. d, f | 21. b |
| 2. Resonance structures | 7. polar | 12. b | 17. a, c | 22. b |
| 3. hybrid | 8. double-headed arrow | 13. a | 18. e, b | 23. c |
| 4. valence shell electron pair repulsion theory | 9. nonpolar | 14. c | 19. a | 24. a |
| 5. Electronegativity | 10. molecular formula | 15. d | 20. c | |

## Solutions to In-Chapter Problems

**4.1**    H is surrounded by two electrons, giving it the noble gas configuration of He. Cl is surrounded by eight electrons, giving it the noble gas configuration of Ar.

$$H\cdot \ + \ \cdot \ddot{\underset{\cdot\cdot}{Cl}}: \ \longrightarrow \ H : \ddot{\underset{\cdot\cdot}{Cl}}:$$

**4.2**    This arrangement gives each Cl an octet and the electronic configuration of Ar.

$$: \ddot{\underset{\cdot\cdot}{Cl}}\cdot \ + \ \cdot \ddot{\underset{\cdot\cdot}{Cl}}: \ \longrightarrow \ : \ddot{\underset{\cdot\cdot}{Cl}} : \ddot{\underset{\cdot\cdot}{Cl}}:$$

**4.3**     Atoms with one, two, or three valence electrons form one, two, or three bonds, respectively. Atoms with four or more valence electrons form enough bonds to give an octet.

a. F forms one bond.
   ($8 - 7$ valence $e^- = 1$ bond)
b. Si forms four bonds.
   ($8 - 4$ valence $e^- = 4$ bonds)
c. Br forms one bond.
   ($8 - 7$ valence $e^- = 1$ bond)
d. O forms two bonds.
   ($8 - 6$ valence $e^- = 2$ bonds)
e. P forms three bonds.
   ($8 - 5$ valence $e^- = 3$ bonds)
f. S forms two bonds.
   ($8 - 6$ valence $e^- = 2$ bonds)

**4.4**

a. 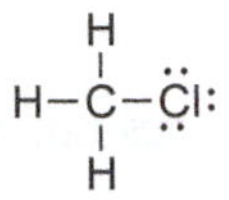   H–C–Cl:

b.   H–N–O–H

c.   H–C–O–H

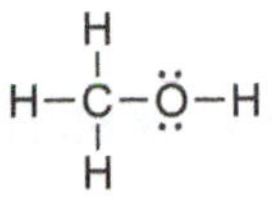

d. 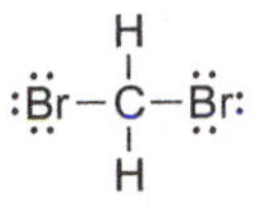   :Br–C–Br:

**4.5**     Ionic bonding is observed in CaO since Ca is a metal and readily transfers electrons to the nonmetal oxygen. Covalent bonding is observed in $CO_2$ since carbon is a nonmetal and does not readily transfer electrons.

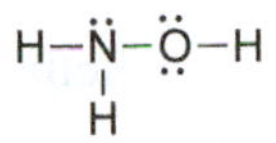

**4.6**

a.

| **Step [1]** | **Step [2]** | **Step [3]** |
| --- | --- | --- |
| Arrange the atoms. | Count the electrons. | Add the bonds and lone pairs. |

H   Br

1 Br x 7 e⁻ = 7 e⁻
1 H x 1 e⁻ = 1 e⁻
**8 e⁻ total**

Add bonds first...          ...then lone pairs.

H–Br          H–Br: 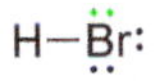
only 2 e⁻ used          8 e⁻ used

b.

| **Step [1]** | **Step [2]** | **Step [3]** |
| --- | --- | --- |
| Arrange the atoms. | Count the electrons. | Add the bonds and lone pairs. |

H
H   C   F
H

• four atoms around C

1 C x 4 e⁻ = 4 e⁻
1 F x 7 e⁻ = 7 e⁻
3 H x 1 e⁻ = 3 e⁻
**14 e⁻ total**

Add bonds first...          ...then lone pairs.

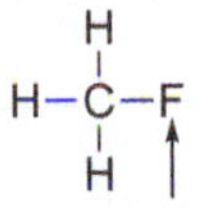

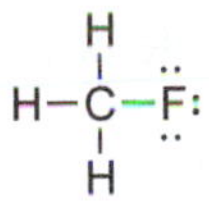

H–C–F          H–C–F:
no octet
only 8 e⁻ used          14 e⁻ used

c. 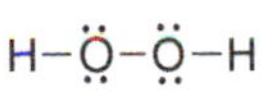   H–O–O–H

d. 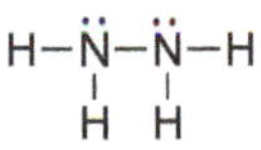   H–N–N–H

e. 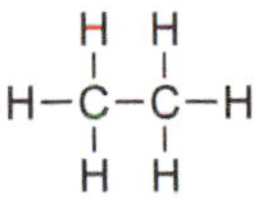   H–C–C–H

f. 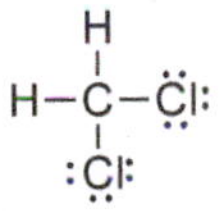   H–C–Cl:
           :Cl:

**4.7**     Use the same steps as in Answer 4.6 to draw the Lewis structure.

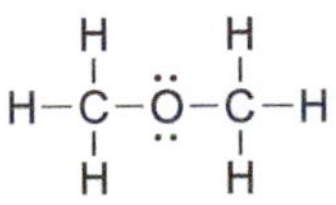

H–C–O–C–H

**4.8**     After placing all electrons in bonds and lone pairs, use a lone pair to form a multiple bond if an atom does not have an octet. Follow the stepwise procedure in Example 4.3.

a. 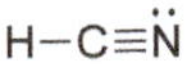   H–C≡N

b.   H–C=O
           H

c.   H–C=C–Cl:
           H  H

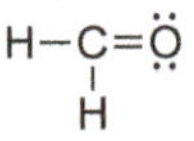

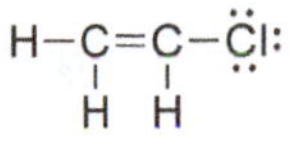

**4.9**    A valid Lewis structure for formic acid contains a carbon–oxygen double bond.  The C atom has four bonds and each O atom has two bonds and two lone pairs.

$$
\begin{array}{c}
:\!\ddot{O}\!: \\
\parallel \\
H\!-\!\overset{}{C}\!-\!\ddot{O}\!-\!H
\end{array}
$$

**4.10**

a. molecular formula: $C_2H_5NO_2$        b. Lewis structure:

$$
\begin{array}{c}
\quad\; H \;\; :\!\ddot{O}\!: \\
\quad\; | \qquad \parallel \\
H\!-\!\ddot{N}\!-\!\overset{|}{C}\!-\!\overset{}{C}\!-\!\ddot{O}\!-\!H \\
\quad\; | \quad | \\
\quad\; H \;\; H
\end{array}
$$

**4.11**    Boron is surrounded by only six electrons, so the Lewis structure for $BBr_3$ does not follow the octet rule.

$$
\begin{array}{c}
\quad\; :\!\ddot{Br}\!: \\
\quad\quad | \\
:\!\ddot{Br}\!-\!B\!-\!\ddot{Br}\!: \\
\end{array}
$$

← six electrons

**4.12**

a.

$$
\begin{array}{c}
:\!\ddot{O}\!: \;\; H \qquad H \;\; :\!\ddot{O}\!: \\
\parallel \quad\; | \qquad\; | \quad\; \parallel \\
H\!-\!\ddot{O}\!-\!P\!-\!\overset{|}{C}\!-\!\ddot{N}\!-\!\overset{|}{C}\!-\!\overset{}{C}\!-\!\ddot{O}\!-\!H \\
\quad\; | \qquad | \quad | \quad | \\
\; H\!-\!\ddot{O}\!: \; H \quad H \;\; H
\end{array}
$$

Each O atom needs two lone pairs and the N atom needs one.

b.  There are 10 electrons around phosphorus.

c.  Phosphorus and hydrogen do not follow the octet rule.

**4.13**    Resonance structures are two or more Lewis structures having the same arrangement of atoms but a different arrangement of electrons.  Draw the second resonance structure as in Sample Problem 4.5 in the text, or Example 4.4.  The two resonance structures in each example differ in the location of a double bond and a lone pair.

a.
$$
\left[\begin{array}{c}
H \;\; :\!\ddot{O}\!: \\
| \qquad \parallel \\
H\!-\!\overset{|}{C}\!-\!\overset{}{C}\!-\!\ddot{O}\!: \\
| \\
H
\end{array}\right]^{-}
\longleftrightarrow
\left[\begin{array}{c}
H \;\; :\!\ddot{O}\!: \\
| \qquad | \\
H\!-\!\overset{|}{C}\!-\!\overset{}{C}\!=\!\ddot{O} \\
| \\
H
\end{array}\right]^{-}
$$

b.
$$
\left[\begin{array}{c}
:\!\ddot{O}\!: \\
\parallel \\
H\!-\!\overset{}{C}\!-\!\ddot{N}\!-\!H
\end{array}\right]^{-}
\longleftrightarrow
\left[\begin{array}{c}
:\!\ddot{O}\!: \\
| \\
H\!-\!\overset{}{C}\!=\!\ddot{N}\!-\!H
\end{array}\right]^{-}
$$

**4.14**

a.
$$
\left[:\!\ddot{O}\!-\!\ddot{N}\!=\!\ddot{O}\right]^{-} \longleftrightarrow \left[\ddot{O}\!=\!\ddot{N}\!-\!\ddot{O}\!:\right]^{-}
$$
        b.
$$
\left[\begin{array}{c}
\ddot{O}\!=\!\overset{}{C}\!-\!H \\
| \\
:\!\ddot{O}\!:
\end{array}\right]^{-} \longleftrightarrow \left[\begin{array}{c}
:\!\ddot{O}\!-\!\overset{}{C}\!-\!H \\
\parallel \\
:\!\ddot{O}\!:
\end{array}\right]^{-}
$$

**4.15**

$$
:\!\ddot{O}\!-\!\ddot{S}\!=\!\ddot{O} \longleftrightarrow \ddot{O}\!=\!\ddot{S}\!-\!\ddot{O}\!:
$$

The location of a lone pair and double bond is different.

**4.16**

Three resonance structures for nitrous oxide:

$$
:\!N\!\equiv\!N\!-\!\ddot{O}\!: \qquad\qquad :\!N\!=\!N\!=\!\ddot{O}\!: \qquad\qquad :\!\ddot{N}\!-\!N\!\equiv\!O\!:
$$

**4.17** Name the compounds using the following two-step method:

Step [1] Name the first nonmetal by its element name and the second using the suffix -*ide*.

Step [2] Add prefixes to show the number of atoms of each element.

a. $CS_2$ = carbon disulfide

b. $SO_2$ = sulfur dioxide

c. $PCl_5$ = phosphorus pentachloride

d. $BF_3$ = boron trifluoride

**4.18**

a. silicon dioxide = $SiO_2$

b. phosphorus trichloride = $PCl_3$

c. sulfur trioxide = $SO_3$

d. dinitrogen trioxide = $N_2O_3$

**4.19** The shape around each carbon atom in ethylene, $H_2C=CH_2$, is trigonal planar, since each C is surrounded by three groups (2 H's and 1 C).

**4.20**

a. $H_2\overset{..}{S}:$

S has two lone pairs.
bent

b. $CH_2Cl_2$

C has four groups.
tetrahedral

c. $\overset{..}{N}Cl_3$

N has one lone pair.
trigonal pyramidal

d. $BBr_3$

B has three groups.
trigonal planar

**4.21**

$\left[ H-\overset{..}{N}-H \right]^{-}$   The ion has a bent geometry (b), because the N is surrounded by four groups.

**4.22** Electronegativity *increases* across a row of the periodic table as the nuclear charge increases (excluding the noble gases).

Electronegativity *decreases* down a column of the periodic table as the atomic radius increases, pushing the valence electrons farther from the nucleus.

a. Na, Li, H    b. Be, C, O    c. I, Cl, F    d. B, N, O

**4.23** Calculate the electronegativity difference between the two atoms and use the following rules: less than 0.5 (nonpolar); 0.5–1.9 (polar covalent); greater than 1.9 (ionic).

a. HF = (4.0 – 2.1) = 1.9 = polar

b. MgO = (3.5 – 1.2) = 2.3 = ionic

c. $F_2$ = (4.0 – 4.0) = 0 = nonpolar

d. ClF = (4.0 – 3.0) = 1.0 = polar

e. $H_2O$ = (3.5 – 2.1) = 1.4 = polar

f. $NH_3$ = (3.0 – 2.1) = 0.9 = polar

**4.24** The head of the bond dipole arrow points towards the more electronegative atom.

a. $\overset{\delta^+}{H}-\overset{\delta^-}{F}$

b. $-\overset{\delta^+}{B}-\overset{\delta^-}{C}-$

c. $-\overset{\delta^-}{C}-\overset{\delta^+}{Li}$

d. $-\overset{\delta^+}{C}-\overset{\delta^-}{Cl}$

**4.25** To determine the overall polarity of a molecule, identify the polar bonds, determine the shape around individual atoms, and then decide if the individual bond dipoles cancel or reinforce.

a. H–Cl $\delta^+ \, \delta^-$

polar

b. H–C–C–H (with H's above and below)

nonpolar
(all nonpolar bonds)

c. (CHF₃ structure) polar

d. H–C≡N $\delta^+ \, \delta^-$

polar

e. (CCl₄ structure)

nonpolar
(The four C–Cl bond dipoles cancel.)

**4.26** All C–H and C–C bonds are nonpolar.

a. H–C–C–O–H structure

polar C–O, O–H
Both C's are tetrahedral
(four atoms around each C).

b. H–C–C–H structure with :O:

trigonal planar
(three groups)

tetrahedral
(four atoms)

## Solutions to Odd-Numbered End-of-Chapter Problems

**4.27** In covalent bonding, atoms share electrons to attain the electronic configuration of the noble gas closest to them in the periodic table. In ionic bonding, one atom donates electrons to the other atom.

a. LiCl: ionic, the metal Li donates electrons to the nonmetal chlorine.
   HCl: covalent, H and Cl share electrons since both atoms are nonmetals and the electronegativity difference is not large enough for electron transfer to occur.
b. KBr: ionic, the metal K donates electrons to the nonmetal bromine.
   HBr: covalent, H and Br share electrons since both atoms are nonmetals and the electronegativity difference is not large enough for electron transfer to occur.

**4.29** Atoms with one, two, or three valence electrons form one, two, or three bonds, respectively. Atoms with four or more valence electrons form enough bonds to give an octet.

a. C = 4 bonds, 0 lone pairs          c. I = 1 bond, 3 lone pairs
b. Se = 2 bonds, 2 lone pairs         d. P = 3 bonds, 1 lone pair

**4.31** Each O and S atom must get two lone pairs and each N atom must get one lone pair.

a. H–C=C–C≡N̈ (with H H below)

b. H–S̈–C–C–C–Ö–H (with :O: above, H :N–H below)

**4.33** Use the common element colors (black = C, red = O, gray = H), and give each O two lone pairs.

H–Ö–C–C–Ö–H (with :O: :O: above)

oxalic acid

**4.35**  Follow the steps in Examples 4.2 and 4.3 to draw the Lewis structures.

a. 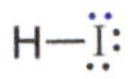  H—Ï:

b. 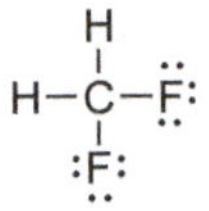  H—C—F:  (with H above C and :F: below C)

c.  H—Së—H

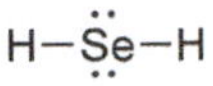

d. 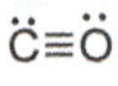  C≡O

e. 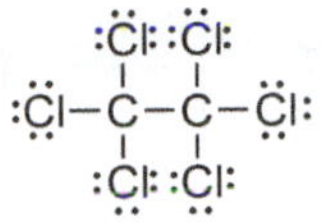  :Cl: :Cl: / :Cl—C—C—Cl: / :Cl: :Cl:

**4.37**  Follow the steps in Examples 4.2 and 4.3.

a. 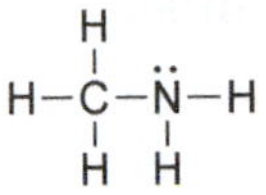  H—C—N—H  (with H above C, H H below C)

b. 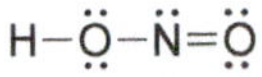  H—Ö—N̈=Ö

c.  H—C—C≡C—H  (with H above and below C)

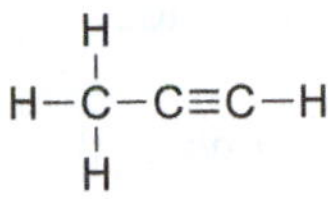

**4.39**  Follow the steps in Example 4.3.

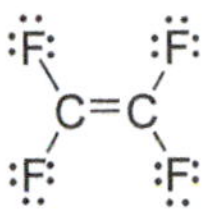

:F:     :F:
 \     /
  C=C
 /     \
:F:     :F:

**4.41**

a. 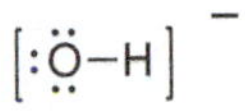  [:Ö—H]⁻

b. 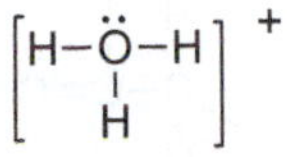  [H—Ö—H]⁺  (with H below O)

**4.43**

a. 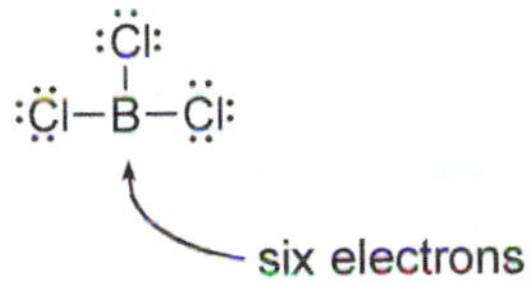  :Cl: / :Cl—B—Cl:  ← six electrons

b.  Ö=S=Ö / :O:  ← 12 electrons

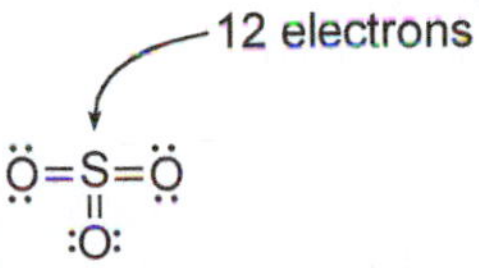

**4.45**  A resonance structure is one representation of electron arrangement with a given placement of atoms.  A resonance hybrid is a composite of two or more resonance structures.

**4.47**  Use Example 4.4.  In this ion, the location of a double bond and a lone pair is different in the two resonance structures.

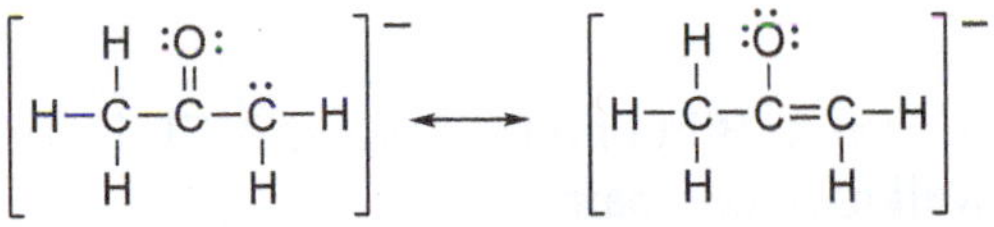

[ H :O: / H—C—C—C—H / H H ]⁻  ←→  [ H :O: / H—C—C=C—H / H H ]⁻

**4.49**  Resonance structures must have the same arrangement of atoms.

a. 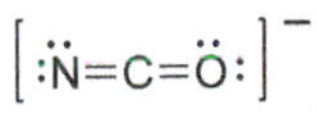  [:N̈=C=Ö:]⁻ and [:N≡C—Ö:]⁻
    resonance structures

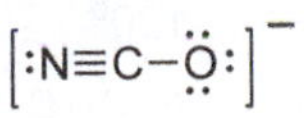

b. 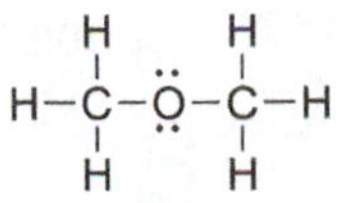  H H / H—C—Ö—C—H / H H   and   H H / H—C—C—Ö—H / H H
    not resonance structures
    The arrangement of atoms is different.

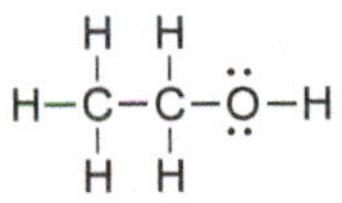

**4.51**    Three resonance structures for the carbonate anion:

$$\left[\ddot{O}=\underset{\underset{\ddot{\underset{..}{O}}:}{|}}{C}-\ddot{O}:\right]^{2-} \longleftrightarrow \left[:\ddot{O}-\underset{\underset{\ddot{\underset{..}{O}}:}{||}}{C}-\ddot{O}:\right]^{2-} \longleftrightarrow \left[:\ddot{O}-\underset{\underset{:\ddot{\underset{..}{O}}:}{|}}{C}=\ddot{O}\right]^{2-}$$

**4.53**    Name each compound as in Answer 4.17.

a. PBr$_3$ = phosphorus tribromide          c. NCl$_3$ = nitrogen trichloride
b. SO$_3$ = sulfur trioxide                          d. P$_2$S$_5$ = diphosphorus pentasulfide

**4.55**    Work backwards to write the formula that corresponds to each name.

a. selenium dioxide = SeO$_2$                    c. dinitrogen pentoxide = N$_2$O$_5$
b. carbon tetrachloride = CCl$_4$

**4.57**    Dihydrogen oxide is water, H$_2$O.

**4.59**    Follow the steps in Example 4.6.

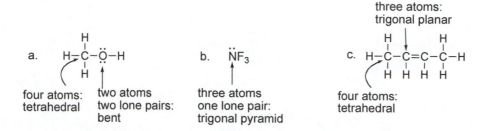

a.  H–C–Ö–H
    (four atoms:   two atoms
    tetrahedral    two lone pairs:
                   bent)

b.  ṄF$_3$
    (three atoms
    one lone pair:
    trigonal pyramid)

three atoms:
trigonal planar

c.  H–C–C=C–C–H
    (four atoms:
    tetrahedral)

**4.61**    Follow the steps in Example 4.6.

a.  H–N–Ö–H
    trigonal pyramid   bent

b.  H–C–C–H
    tetrahedral   trigonal planar

c.  [ H–C–N–H ]$^+$
    tetrahedral

**4.63**

a. [1] A linear geometry means two groups, so the dark red atom has no lone pairs. [2] BeCl$_2$
b. [1] A bent geometry generally means four groups, so the dark red atom has two lone pairs.
   [2] H$_2$O

**4.65**    BCl$_3$ is trigonal planar because it has three Cl's bonded to B but no lone pairs. NCl$_3$ is trigonal pyramidal because it has three Cl's around N as well as a lone pair.

**4.67**

a.  H–C–F:
    tetrahedral:
    both 109.5°

linear:         bent:
180°            109.5°

b.  H–C≡C–C–Ö–H
    tetrahedral:
    109.5°

trigonal planar:
120°            linear:
                180°

c.  H–C=C=O
    trigonal planar:
    120°

**4.69**   CCl₄ is tetrahedral since carbon has four groups around it.  Each C in $C_2Cl_4$ has trigonal planar geometry since each carbon has three groups around it.

**4.71**   Electronegativity *increases* across a row of the periodic table as the nuclear charge increases (excluding the noble gases).
Electronegativity *decreases* down a column of the periodic table as the atomic radius increases, pushing the valence electrons farther from the nucleus.

   a.  Se, S, O        b.  Na, P, Cl            c.  S, Cl, F        d.  P, N, O

**4.73**   In covalent bonding, atoms share electrons to attain the electronic configuration of the noble gas closest to them in the periodic table.  In ionic bonding, one atom donates electrons to the other atom.  When the electronegativity difference between the two atoms in a bond is greater than 1.9, the bond is ionic.

   a. hydrogen and bromine = polar covalent        c. sodium and sulfur = polar covalent
   $(2.8 – 2.1 = 0.7)$                                                      $(2.5 – 0.9 = 1.6)$
   b. nitrogen and carbon = polar covalent          d. lithium and oxygen = ionic
   $(3.0 – 2.5 = 0.5)$                                                      $(3.5 – 1.0 = 2.5)$

**4.75**   Calculate the electronegativity difference between the atoms and carbon and use the following rules: less than 0.5 (nonpolar); 0.5–1.9 (polar covalent); greater than 1.9 (ionic).

   a. C = $(2.5 – 2.5) = 0$ = nonpolar            d. Cl = $(3.0 – 2.5) = 0.5$ = polar
   b. O = $(3.5 – 2.5) = 1.0$ = polar              e. H = $(2.5 – 2.1) = 0.4$ = nonpolar
   c. Li = $(2.5 – 1.0) = 1.5$ = polar

**4.77**   Calculate the electronegativity difference between the atoms to determine which bond is more polar.

   a. C–O = $(3.5 – 2.5) = 1.0$ = more polar      c. Si–C = $(2.5 – 1.8) = 0.7$ = more polar
   C–N = $(3.5 – 3.0) = 0.5$                          P–H = $(2.1 – 2.1) = 0$
   b. C–F = $(4.0 – 2.5) = 1.5$ = more polar
   C–Cl = $(3.0 – 2.5) = 0.5$

**4.79**   The δ⁺ is placed next to the less electronegative atom, and the δ⁻ is placed next to the more electronegative atom.

   δ⁺ δ⁻    δ⁺ δ⁻        δ⁺ δ⁻  δ⁺ δ⁻        δ⁺ δ⁻
   C–O    C–N        C–F    C–Cl        Si–C    P–H
                                                                 no dipole

**4.81** In $CH_3NH_2$ carbon bears a partial positive charge since nitrogen is more electronegative than carbon. In $CH_3MgBr$ carbon bears a partial negative charge since carbon is more electronegative than magnesium.

**4.83** No, a compound with only one polar bond must be polar. The single bond dipole is not cancelled by another bond dipole, so the molecule as a whole remains polar.

**4.85**

a. 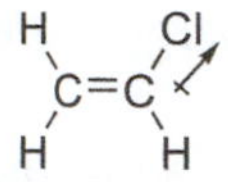
polar bond
polar compound

b. 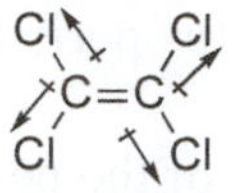
polar bonds
nonpolar compound
Bond dipoles cancel.

c. 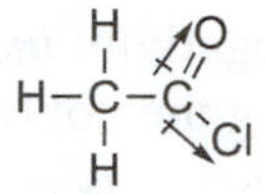
polar bonds
polar compound
Bond dipoles do not cancel.

**4.87** $CHCl_3$ has a net dipole, making it polar. $CCl_4$ has four polar bonds but no net dipole because the four bonds have dipoles that cancel one another.

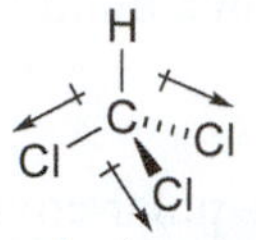

polar bonds
polar compound

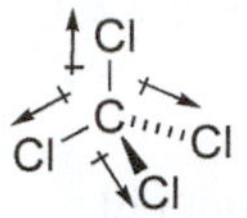

polar bonds
nonpolar compound

**4.89**

a. 20 valence electrons
[1 O atom (6 valence electrons) + 2 Cl atoms (7 valence electrons each)]

b, c:

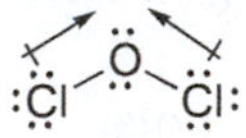

d. $Cl_2O$ has a bent shape since O is surrounded by two atoms and two lone pairs.
e. The compound is polar since the two bond dipoles do not cancel.

**4.91**

a.–c.

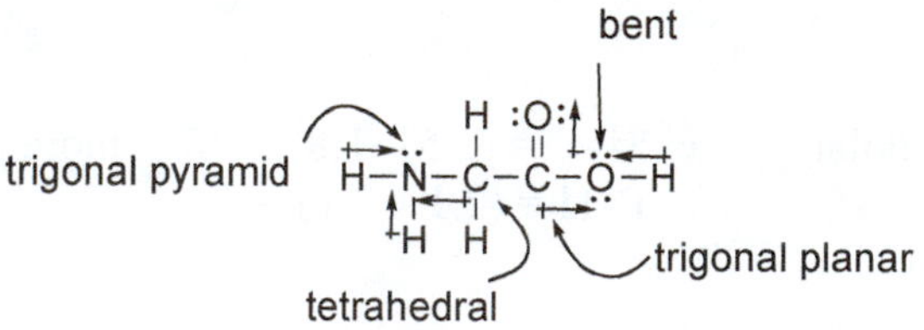

trigonal pyramid
bent
tetrahedral
trigonal planar

| | | |
|---|---|---|
| 2 C x 4 e⁻ | = | 8 e⁻ |
| 5 H x 1 e⁻ | = | 5 e⁻ |
| 1 N x 5 e⁻ | = | 5 e⁻ |
| 2 O x 6 e⁻ | = | 12 e⁻ |
| | | 30 valence e⁻ |

d. Glycine is a polar molecule since the many bond dipoles do not cancel.

**4.93** First, give the O atom two lone pairs and each N atom one lone pair. Then add double bonds to the C's that do not have four bonds.

**4.95**   a.  The predicted shape around each carbon is tetrahedral because each C is surrounded by four atoms.

b.  The interior angle of a triangle with three equal sides is 60°.

c.  The 60° bond angle deviates greatly from the theoretical tetrahedral bond angle (109.5°), making the molecule unstable.

# Chapter 5 Chemical Reactions

## Chapter Review

**[1] What do the terms in a chemical equation mean and how is an equation balanced? (5.1, 5.2)**

- A chemical equation contains the reactants on the left side of an arrow and the products on the right. The coefficients tell how many molecules or moles of a substance react or are formed.
- A chemical equation is balanced by placing coefficients in front of chemical formulas one at a time, beginning with the most complex formula, so that the number of atoms of each element is the same on both sides. When no number precedes a formula, the coefficient is assumed to be "1."

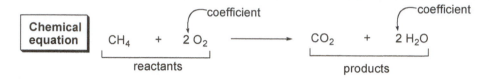

**[2] Define the terms mole and Avogadro's number. (5.3)**

- A **mole** is a quantity that contains $6.02 \times 10^{23}$ items—usually, atoms, molecules, or ions.
- **Avogadro's number** is the number of particles in a mole—$6.02 \times 10^{23}$.
- The number of atoms or molecules in a given number of moles is calculated using Avogadro's number.

Two possible conversion factors:      $\dfrac{1 \text{ mol}}{6.02 \times 10^{23} \text{ atoms}}$      or      $\dfrac{6.02 \times 10^{23} \text{ atoms}}{1 \text{ mol}}$

**[3] How are formula weight and molar mass calculated? (5.4)**

- The **formula weight** is the sum of the atomic weights of all the atoms in a compound, reported in atomic mass units.
- The **molar mass** is the mass of one mole of a substance, reported in g/mol. The molar mass is numerically equal to the formula weight but the units are different (g/mol not amu).

**[4] How are the mass of a substance and its number of moles related? (5.4)**

- The molar mass is used as a conversion factor to determine how many grams are contained in a given number of moles of a substance. Similarly, the molar mass is used to determine how many moles of a substance are contained in a given number of grams.

**[5] How can a balanced equation and molar mass be used to calculate the number of moles and mass of a reaction product? (5.5, 5.6)**

- The coefficients in a balanced chemical equation tell us the number of moles of each reactant that combine and the number of moles of each product formed. Coefficients are used to form mole ratios that serve as conversion factors relating the number of moles of reactants and products.
- When the mass of a substance in a reaction must be calculated, first its number of moles is determined using mole ratios, and then the molar mass is used to convert moles to grams.

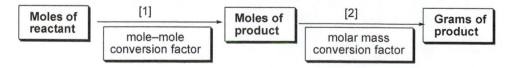

**[6] What is percent yield? (5.7)**
- **Percent yield** is determined by using the following equation:

$$\boxed{\textbf{Percent yield}} \quad = \quad \frac{\text{actual yield (g)}}{\text{theoretical yield (g)}} \quad \times \quad 100\%$$

- The **actual yield** is the amount of product formed in a reaction, determined by weighing a product on a balance. The **theoretical yield** is a quantity calculated from a balanced chemical equation, using mole ratios and molar mass. The theoretical yield is the maximum amount of product that can form in a chemical reaction from the amount of reactants used.

**[7] What is the limiting reactant in a reaction? (5.8)**
- The limiting reactant is the reactant that is completely used up in a reaction. The number of moles of limiting reactant determines the number of moles of product formed using mole ratios in the balanced equation.

**[8] What are oxidation and reduction reactions? (5.9)**
- Oxidation–reduction or redox reactions are electron transfer reactions.
- **Oxidation** results in the loss of electrons. Metals and anions tend to undergo oxidation. In some reactions, oxidation results in the gain of O atoms or the loss of H atoms.
- **Reduction** results in the gain of electrons. Nonmetals and cations tend to undergo reduction. In some reactions, reduction results in the loss of O atoms or the gain of H atoms.

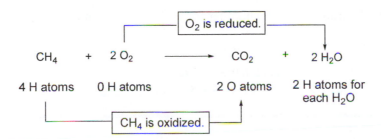

**[9] Give some examples of common or useful redox reactions. (5.9, 5.10)**
- Common examples of redox reactions include the rusting of iron and the combustion of methane. The electric current generated in batteries used for flashlights and pacemakers results from redox reactions.

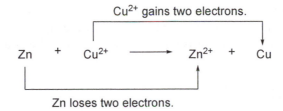

- Zn loses two electrons to form $Zn^{2+}$, so Zn is oxidized.
- $Cu^{2+}$ gains two electrons to form Cu metal, so $Cu^{2+}$ is reduced.

## Problem Solving

## [1] Introduction to Chemical Reactions (5.1)

**Example 5.1** Label the reactants and products, and indicate how many atoms of each type of element are present on each side of the equation.

$$3 \, H_2(g) \quad + \quad N_2(g) \quad \longrightarrow \quad 2 \, NH_3(g)$$

**Analysis**
Reactants are on the left side of the arrow and products are on the right side in a chemical equation. When a formula contains a subscript, multiply its coefficient by the subscript to give the total number of atoms of a given type.

**Solution**
In this equation, the reactants are $H_2$ and $N_2$ while the product is $NH_3$. To determine the number of atoms when a formula has both a coefficient and a subscript, multiply the coefficient by the subscript.

| | | | |
|---|---|---|---|
| $3 \, H_2$ | $=$ | $6 \, H's$ | Multiply the coefficient 3 by the subscript 2. |
| $1 \, N_2$ | $=$ | $2 \, N's$ | There are 2 N's. |
| $2 \, NH_3$ | $=$ | $2 \, N's + 6 \, H's$ | Multiply the coefficient 2 by each subscript: $2 \times 1 \, N = 2 \, N's$; $2 \times 3 \, H's = 6 \, H's$. |

Add up the atoms on each side to determine the total number for each type of element.

$$3 \, H_2(g) \quad + \quad N_2(g) \quad \longrightarrow \quad 2 \, NH_3(g)$$

Atoms in the reactants:     Atoms in the product:
- 6 H's   2 N's       • 6 H's   2 N's

---

## [2] Balancing Chemical Reactions (5.2)

**Example 5.2** Write a balanced equation for the reaction of pentane ($C_5H_{12}$) with oxygen ($O_2$) to form carbon dioxide ($CO_2$) and water ($H_2O$).

**Analysis**
Balance an equation with coefficients, one element at a time, beginning with the most complex formula and starting with an element that appears in only one formula on both sides of the equation. Continue placing coefficients until the number of atoms of each element is equal on both sides of the equation.

**Solution**
**[1] Write the equation with correct formulas.**

$$C_5H_{12} \quad + \quad O_2 \quad \longrightarrow \quad CO_2 \quad + \quad H_2O$$

- None of the elements is balanced in this equation. As an example, there are 5 C's on the left side but only 1 C on the right side.

**[2] Balance the equation with coefficients one element at a time.**

- Begin with $C_5H_{12}$ since its formula is most complex. Balance the 5 C's of pentane by placing the coefficient 5 before $CO_2$. Balance the 12 H's by placing the coefficient 6 before $H_2O$.

Place a 5 to balance C's.

$$C_5H_{12} \quad + \quad O_2 \quad \longrightarrow \quad 5\ CO_2 \quad + \quad 6\ H_2O$$

Place a 6 to balance H's.

- The right side of the equation now has 16 O's. Place a coefficient of 8 in front of $O_2$ to balance the O atoms.

$$C_5H_{12} \quad + \quad 8\ O_2 \quad \longrightarrow \quad 5\ CO_2 \quad + \quad 6\ H_2O$$

**[3] Check.**

- The equation is balanced since the number of atoms of each element is the same on both sides.

**Answer:** $C_5H_{12} \quad + \quad 8\ O_2 \quad \longrightarrow \quad 5\ CO_2 \quad + \quad 6\ H_2O$

Atoms in the reactants:
- 5 C's
- 12 H's
- 16 O's

Atoms in the products:
- 5 C's
- 12 H's (6 x 2)
- 16 O's (5 x 2 O's) + (6 x 1 O)

## [3] The Mole and Avogadro's Number (5.3)

**Example 5.3** How many molecules are contained in 8.00 moles of water ($H_2O$)?

**Analysis and Solution**
**[1] Identify the original quantity and the desired quantity.**

8.00 mol of $H_2O$     ? number of molecules of $H_2O$
original quantity          desired quantity

**[2] Write out the conversion factors.**
- Choose the conversion factor that places the unwanted unit, mol, in the denominator so that the units cancel.

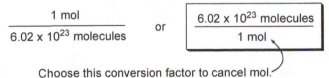

$$\frac{1\ mol}{6.02 \times 10^{23}\ molecules} \quad or \quad \frac{6.02 \times 10^{23}\ molecules}{1\ mol}$$

Choose this conversion factor to cancel mol.

**[3] Set up and solve the problem.**
- Multiply the original quantity by the conversion factor to obtain the desired quantity.
- Multiplication first gives an answer that is not written in scientific notation since the coefficient (48.2) is greater than 10. Moving the decimal one place to the *left* and *increasing* the exponent by one gives the answer written in proper form.

Convert to a number between 1 and 10.

$$8.00 \text{ mol} \quad \times \quad \frac{6.02 \times 10^{23} \text{ molecules}}{1 \text{ mol}} \quad = \quad 48.2 \times 10^{23} \text{ molecules}$$

Moles cancel.

$$= \quad 4.82 \times 10^{24} \text{ molecules of } H_2O$$

**Answer**

## [4] Molar Mass (5.4)

**Example 5.4** What is the molar mass of $C_6H_8O_6$?

**Analysis**
Determine the number of atoms of each element from the subscripts in the chemical formula, multiply the number of atoms of each element by the atomic weight, and add up the results.

**Solution**

| | | | | |
|---|---|---|---|---|
| 6 C atoms | × | 12.01 amu | = | 72.06 amu |
| 8 H atoms | × | 1.008 amu | = | 8.064 amu |
| 6 O atoms | × | 16.00 amu | = | 96.00 amu |
| Formula weight of $C_6H_8O_6$: | | | | 176.124 amu rounded to 176.12 |

**Answer:** Since the formula weight of $C_6H_8O_6$ is 176.12 amu, the molar mass is 176.12 g/mol.

## [5] Mole Calculations in Chemical Equations (5.5)

**Example 5.5** Using the balanced equation, how many moles of $H_2$ are produced from 1.5 mol of HCl?

$$2 \text{ Al} \quad + \quad 6 \text{ HCl} \quad \longrightarrow \quad 2 \text{ AlCl}_3 \quad + \quad 3 \text{ H}_2$$

**Analysis and Solution**
**[1] Identify the original quantity and the desired quantity.**

| 1.5 mol of HCl | ? mol of $H_2$ |
|---|---|
| original quantity | desired quantity |

**[2] Write out the conversion factors.**
- Use the coefficients in the balanced equation to write mole–mole conversion factors for the two compounds, HCl and $H_2$. Choose the conversion factor that places the unwanted unit, moles of HCl, in the denominator so that the units cancel.

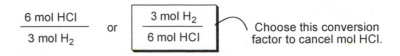

**[3] Set up and solve the problem.**

- Multiply the original quantity by the conversion factor to obtain the desired quantity.

$$1.5 \text{ mol HCl} \quad \times \quad \frac{3 \text{ mol H}_2}{6 \text{ mol HCl}} \quad = \quad 0.75 \text{ mol H}_2 \quad \textbf{Answer}$$

Moles HCl cancel.

## [6] Mass Calculations in Chemical Equations (5.6)

**Example 5.6** How many grams of $Fe_2O_3$ (molar mass 159.7 g/mol) are produced from 3.0 mol of Fe?

$$4 \text{ Fe} + 3 \text{ O}_2 \longrightarrow 2 \text{ Fe}_2\text{O}_3$$

**Analysis and Solution**

**[1] Convert the number of moles of reactant to the number of moles of product using a mole–mole conversion factor.**

- Use the coefficients in the balanced chemical equation to write mole–mole conversion factors for the two compounds—four moles of Fe form two moles of $Fe_2O_3$.
- Multiply the number of moles of reactant (Fe) by the conversion factor to give the number of moles of product ($Fe_2O_3$).

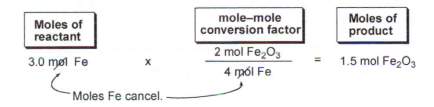

**[2] Convert the number of moles of product to the number of grams of product using the product's molar mass.**

- Use the molar mass of the product ($Fe_2O_3$, molar mass 159.7 g/mol) to write a conversion factor.
- Multiply the number of moles of product (from step [1]) by the conversion factor to give the number of grams of product.

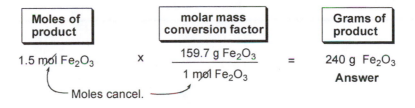

## [7] Percent Yield (5.7)

**Example 5.7** Consider the reaction of hydrogen and nitrogen to form ammonia. If the theoretical yield of ammonia is 10. g in a reaction, what is the percent yield of ammonia if only 4.5 g of ammonia are actually formed?

**Analysis**
Use the formula, percent yield = (actual yield/theoretical yield) × 100% to calculate the percent yield.

**Solution**

$$\text{Percent yield} = \frac{\text{actual yield (g)}}{\text{theoretical yield (g)}} \times 100\%$$

$$= \frac{4.5 \text{ g}}{10. \text{ g}} \times 100\% = 45\% \quad \textbf{Answer}$$

## [8] Oxidation and Reduction (5.9)

**Example 5.8** Identify the species that is oxidized and the species that is reduced in the following reaction. Write out half reactions to show how many electrons are gained or lost by each species.

$$2 \text{ Na} + \text{Cl}_2 \longrightarrow 2 \text{ Na}^+ + 2 \text{ Cl}^-$$

**Analysis**
Metals and anions tend to lose electrons and thus undergo oxidation. Nonmetals and cations tend to gain electrons and thus undergo reduction.

**Solution**
The metal Na is oxidized to $Na^+$, losing one electron for each Na atom. $Cl_2$ gains two electrons, and is thus reduced to 2 $Cl^-$.

$$2 \text{ Na} \longrightarrow 2 \text{ Na}^+ + 2 \text{ e}^- \qquad \text{Cl}_2 + 2 \text{ e}^- \longrightarrow 2 \text{ Cl}^-$$

Na is oxidized.  $\qquad\qquad$  $Cl_2$ is reduced.

We need the same number of electrons lost and gained. Since each $Cl_2$ gains two electrons, 2 Na atoms are needed in order to lose two electrons.

## Self-Test

**[1] Fill in the blank with one of the terms listed below.**

Actual yield (5.7)               Law of conservation of mass (5.1)     Redox reaction (5.9)
Balanced chemical equation (5.2) Molar mass (5.4)                      Reduction (5.9)
Chemical equation (5.1)          Mole (5.3)                            Theoretical yield (5.7)
Formula weight (5.4)

1. The _____ is the mass of one mole of any substance, reported in g/mol.
2. The _____ is the amount of product formed in a reaction.
3. A _____ involves the transfer of electrons from one element to another.
4. The _____ is the sum of the atomic weights of all the atoms in a compound, reported in atomic mass units (amu).
5. _____ results in the loss of oxygen atoms or the gain of hydrogen atoms.
6. A _____ is a quantity that contains $6.02 \times 10^{23}$ items—usually atoms, molecules, or ions.
7. A _____ tells us the number of *moles* of each reactant that combine and the number of *moles* of each product formed.
8. A _____ is the expression used to illustrate what substances constitute the starting materials in a reaction and what products are formed.
9. The _____ states that atoms cannot be created or destroyed in a chemical reaction.
10. The _____ is the amount of product expected from a given amount of reactant based on the coefficients in the balanced chemical equation.

**[2] Label the reactants and products in the equation below.**

$$2\,NH_3 \longrightarrow 2\,NH_2{}^- + 2\,H^+$$

11.             12.

**[3] How many carbons are contained in each representation?**

13. $CH_3CH_2OH$
14. $2\,C_2H_7N$
15. $C_6H_8O_6$

**[4] Match each compound to its molar mass.**

a. 151.9 g/mol          b. 162.3 g/mol          c. 58.4 g/mol

16. $NaCl$
17. $FeSO_4$
18. $C_{10}H_{14}N_2$

**[5] Label each reaction as oxidation or reduction.**

19. $Mg \longrightarrow Mg^{2+} + 2\,e^-$
20. $Cl_2 + 2\,e^- \longrightarrow 2\,Cl^-$
21. $O_2 + 4\,e^- \longrightarrow 2\,O^{2-}$
22. $2\,O^{2-} \longrightarrow O_2 + 4\,e^-$

**[6] Fill in the coefficients to balance the chemical reaction.**

a. 3        b. 4        c. 5

$$C_3H_8 + \underset{23.}{\_\_}\,O_2 \longrightarrow \underset{24.}{\_\_}\,CO_2 + \underset{25.}{\_\_}\,H_2O$$

## Answers to Self-Test

1. molar mass
2. actual yield
3. redox reaction
4. formula weight
5. Reduction

6. mole
7. balanced chemical equation
8. chemical equation
9. law of conservation of mass
10. theoretical yield

11. reactant
12. products
13. two
14. four
15. six

16. c
17. a
18. b
19. oxidation
20. reduction

21. reduction
22. oxidation
23. c
24. a
25. b

## Solutions to In-Chapter Problems

**5.1** The process is a chemical reaction because the reactants contain two gray spheres joined (indicating $H_2$) and two red spheres joined (indicating $O_2$), while the product ($H_2O$) contains a red sphere joined to two gray spheres (indicating O–H bonds).

**5.2** The process is a physical change (freezing) since the particles in the reactants are the same as the particles in the products.

**5.3** Chemical equations are written with the **reactants on the left** and the **products on the right** separated by a **reaction arrow**.

       reactants                          products

a.   $2\,H_2O_2(aq) \longrightarrow 2\,H_2O(l) + O_2(g)$        (4 H, 4 O)

b.   $2\,C_8H_{18} + 25\,O_2 \longrightarrow 16\,CO_2 + 18\,H_2O$        (16 C, 50 O, 36 H)

c.   $2\,Na_3PO_4(aq) + 3\,MgCl_2(aq) \longrightarrow Mg_3(PO_4)_2(s) + 6\,NaCl(aq)$   (3 Mg, 2 P, 8 O, 6 Na, 6 Cl)

**5.4**     To determine the number of each type of atom when a formula has both a coefficient and a subscript, multiply the coefficient by the subscript.

For $3\,Al_2(SO_4)_3$: $Al = 6\ (3 \times 2)$, $S = 9\ (3 \times 3)$, $O = 36\ (3 \times 3 \times 4)$

**5.5** Write the chemical equation for the statement.

$$CH_4(g) + 4\ Cl_2(g) \xrightarrow{\Delta} CCl_4(l) + 4\ HCl(g)$$

**5.6** Balance the equation with coefficients one element at a time to have the same number of atoms on each side of the equation. Follow the steps in Example 5.2.

[1] Place a 2 to balance O's.

a. $2\ H_2 + O_2 \longrightarrow 2\ H_2O$     c. $CH_4 + 2\ Cl_2 \longrightarrow CH_2Cl_2 + 2\ HCl$

[2] Place a 2 to balance H's.

b. $2\ NO + O_2 \longrightarrow 2\ NO_2$

**5.7** Write the balanced chemical equation for carbon monoxide and oxygen reacting to form carbon dioxide. The smallest set of whole numbers must be used.

$$2\ CO + O_2 \longrightarrow 2\ CO_2$$

**5.8** Follow the steps in Example 5.2 to write the balanced chemical equation.

$$2\ C_2H_6 + 7\ O_2 \longrightarrow 4\ CO_2 + 6\ H_2O$$

**5.9** Write a balanced equation for the Haber process.

$$N_2 + 3\ H_2 \longrightarrow 2\ NH_3$$

**5.10** Balance the equations as in Example 5.2.

a. $2\ Al + 3\ H_2SO_4 \longrightarrow Al_2(SO_4)_3 + 3\ H_2$

b. $3\ Na_2SO_3 + 2\ H_3PO_4 \longrightarrow 3\ H_2SO_3 + 2\ Na_3PO_4$

**5.11** One mole, abbreviated as **mol**, always contains an Avogadro's number of particles ($6.02 \times 10^{23}$).

a, b, c, d: $6.02 \times 10^{23}$

**5.12** Multiply the number of moles by Avogadro's number to determine the number of atoms. Avogadro's number is the conversion factor that relates moles to molecules, as in Example 5.3.

a. $2.00\ mol \times 6.02 \times 10^{23}\ atoms/mol = 1.20 \times 10^{24}\ atoms$
b. $6.00\ mol \times 6.02 \times 10^{23}\ atoms/mol = 3.61 \times 10^{24}\ atoms$
c. $0.500\ mol \times 6.02 \times 10^{23}\ atoms/mol = 3.01 \times 10^{23}\ atoms$
d. $25.0\ mol \times 6.02 \times 10^{23}\ atoms/mol = 1.51 \times 10^{25}\ atoms$

**5.13** Multiply the number of moles by Avogadro's number to determine the number of molecules, as in Example 5.3.

a. $2.5 \text{ mol} \times 6.02 \times 10^{23}$ molecules/mol = $1.5 \times 10^{24}$ molecules
b. $0.25 \text{ mol} \times 6.02 \times 10^{23}$ molecules/mol = $1.5 \times 10^{23}$ molecules
c. $0.40 \text{ mol} \times 6.02 \times 10^{23}$ molecules/mol = $2.4 \times 10^{23}$ molecules
d. $55.3 \text{ mol} \times 6.02 \times 10^{23}$ molecules/mol = $3.33 \times 10^{25}$ molecules

**5.14** Use Avogadro's number as a conversion factor to relate molecules to moles.

a. $6.02 \times 10^{25}$ molecules $\times \dfrac{1 \text{ mol}}{6.02 \times 10^{23} \text{ molecules}}$ = 100. mol

b. $3.01 \times 10^{22}$ molecules $\times \dfrac{1 \text{ mol}}{6.02 \times 10^{23} \text{ molecules}}$ = 0.0500 mol

c. $9.0 \times 10^{24}$ molecules $\times \dfrac{1 \text{ mol}}{6.02 \times 10^{23} \text{ molecules}}$ = 15 mol

**5.15** To calculate the formula weight, multiply the number of atoms of each element by the atomic weight and add the results.

a.

| | | | | | |
|---|---|---|---|---|---|
| 1 Ca atom | $\times$ | 40.08 amu | = | 40.08 amu |
| 1 C atom | $\times$ | 12.01 amu | = | 12.01 amu |
| 3 O atoms | $\times$ | 16.00 amu | = | 48.00 amu |

Formula weight of $CaCO_3$      100.09 amu

b.

| | | | | | |
|---|---|---|---|---|---|
| 1 K atom | $\times$ | 39.10 amu | = | 39.10 amu |
| 1 I atom | $\times$ | 126.9 amu | = | 126.9 amu |

Formula weight of KI      166.00 amu rounded to 166.0 amu

**5.16** Calculate the molecular weight in two steps:
[1] Write the correct formula and determine the number of atoms of each element from the subscripts.
[2] Multiply the number of atoms of each element by the atomic weight and add the results.

a.

| | | | | | |
|---|---|---|---|---|---|
| 2 C atoms | $\times$ | 12.01 amu | = | 24.02 amu |
| 6 H atoms | $\times$ | 1.008 amu | = | 6.048 amu |
| 1 O atom | $\times$ | 16.00 amu | = | 16.00 amu |

Molecular weight of ethanol ($C_2H_6O$)      46.068 amu rounded to 46.07 amu

b.

| | | | | | |
|---|---|---|---|---|---|
| 6 H atoms | $\times$ | 1.008 amu | = | 6.048 amu |
| 6 C atoms | $\times$ | 12.01 amu | = | 72.06 amu |
| 1 O atom | $\times$ | 16.00 amu | = | 16.00 amu |

Molecular weight of phenol ($C_6H_6O$)      94.108 amu rounded to 94.11 amu

c.

| | | | | |
|---|---|---|---|---|
| 1 H atom | × | 1.008 amu | = | 1.008 amu |
| 2 C atom | × | 12.01 amu | = | 24.02 amu |
| 1 Br atom | × | 79.90 amu | = | 79.90 amu |
| 1 Cl atom | × | 35.45 amu | = | 35.45 amu |
| 3 F atoms | × | 19.00 amu | = | 57.00 amu |

Molecular weight of halothane      197.378 amu rounded to 197.38 amu
($C_2HBrClF_3$)

**5.17**

$C_{20}H_{24}O_{10}$

| | | | | |
|---|---|---|---|---|
| 20 C atoms | × | 12.01 amu | = | 240.2 amu |
| 24 H atoms | × | 1.008 amu | = | 24.192 amu |
| 10 O atoms | × | 16.00 amu | = | 160.0 amu |

Molecular weight of ginkgolide B:      424.392 amu = 424.4 g/mol

**5.18**   Convert the moles to grams using the molar mass as a conversion factor.

a. 0.500 mol of NaCl × 58.44 g/mol = 29.2 g        c. 3.60 mol of $C_2H_4$ × 28.05 g/mol = 101 g
b. 2.00 mol of KI × 166.0 g/mol = 332 g        d. 0.820 mol of $CH_4O$ × 32.04 g/mol = 26.3 g

**5.19**   Convert grams to moles using the molar mass as a conversion factor.

a.   100. g NaCl   ×   $\dfrac{1\ mol}{58.44\ g}$   =   1.71 mol

b.   25.5 g $CH_4$   ×   $\dfrac{1\ mol}{16.04\ g}$   =   1.59 mol

c.   0.250 g $C_9H_8O_4$   ×   $\dfrac{1\ mol}{180.2\ g}$   =   1.39 x $10^{-3}$ mol

d.   25.0 g $H_2O$   ×   $\dfrac{1\ mol}{18.02\ g}$   =   1.39 mol

**5.20**   Use conversion factors to determine the number of molecules in 1.00 g; two 500.-mg tablets = 1.00 g.

**Answer:**

1.00 g penicillin   ×   $\dfrac{6.02\ x\ 10^{23}\ molecules}{334.4\ g\ penicillin}$   =   1.80 x $10^{21}$ molecules of penicillin

Grams cancel.

**5.21**   Use mole–mole conversion factors as in Example 5.5 and the equation below to solve the problems.

$N_2(g)$        +        $O_2(g)$        $\xrightarrow{\Delta}$        2 NO(g)

a. 3.3 mol $N_2$ × (2 mol NO/1 mol $N_2$) = 6.6 mol NO
b. 0.50 mol $O_2$ × (2 mol NO/1 mol $O_2$) = 1.0 mol NO
c. 1.2 mol $N_2$ × (1 mol $O_2$/1 mol $N_2$) = 1.2 mol $O_2$

**5.22** Use mole–mole conversion factors as in Example 5.5 and the equation below to solve the problems.

$$2\ C_2H_6(g)\ +\ 5\ O_2(g)\ \xrightarrow{\Delta}\ 4\ CO(g)\ +\ 6\ H_2O(g)$$

a. 3.0 mol $C_2H_6$ × (5 mol $O_2$/2 mol $C_2H_6$) = 7.5 mol $O_2$
b. 0.50 mol $C_2H_6$ × (6 mol $H_2O$/2 mol $C_2H_6$) = 1.5 mol $H_2O$
c. 3.0 mol CO × (2 mol $C_2H_6$/4 mol CO) = 1.5 mol $C_2H_6$

**5.23** [1] Convert the number of moles of reactant to the number of moles of product using a mole–mole conversion factor.
[2] Convert the number of moles of product to the number of grams of product using the product's molar mass.

$$C_6H_{12}O_6(aq)\ \longrightarrow\ 2\ C_2H_6O(aq)\ +\ 2\ CO_2(g)$$

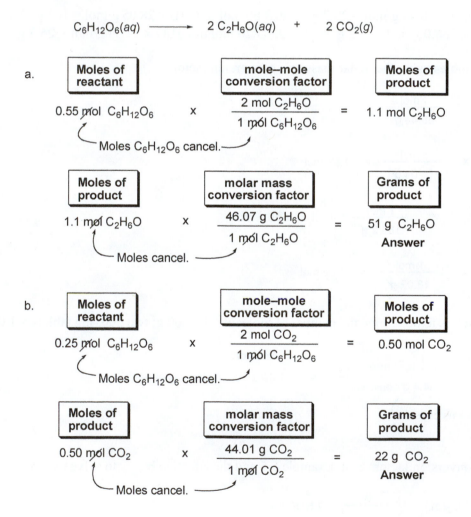

c.

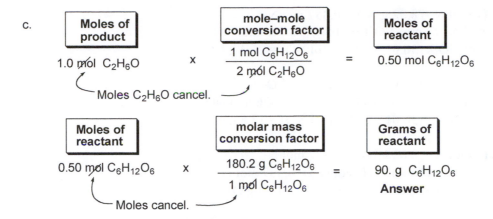

**Moles of product**
1.0 mol $C_2H_6O$ ⟵ Moles $C_2H_6O$ cancel.

×

**mole–mole conversion factor**
$$\frac{1 \text{ mol } C_6H_{12}O_6}{2 \text{ mol } C_2H_6O}$$

=

**Moles of reactant**
0.50 mol $C_6H_{12}O_6$

**Moles of reactant**
0.50 mol $C_6H_{12}O_6$ ⟵ Moles cancel.

×

**molar mass conversion factor**
$$\frac{180.2 \text{ g } C_6H_{12}O_6}{1 \text{ mol } C_6H_{12}O_6}$$

=

**Grams of reactant**
90. g $C_6H_{12}O_6$
**Answer**

**5.24** Use the steps outlined in Answer 5.23 to answer the questions.

$$C_2H_6O(l) \ + \ 3 O_2(g) \longrightarrow 2 CO_2(g) \ + \ 3 H_2O(g)$$

a. 0.50 mol $C_2H_6O \times (2 \text{ mol } CO_2/1 \text{ mol } C_2H_6O) = 1.0$ mol $CO_2$
1.0 mol $CO_2 \times (44.01 \text{ g } CO_2/1 \text{ mol } CO_2) = 44$ g $CO_2$

b. 2.4 mol $C_2H_6O \times (3 \text{ mol } H_2O/1 \text{ mol } C_2H_6O) = 7.2$ mol $H_2O$
7.2 mol $H_2O \times (18.02 \text{ g } H_2O/1 \text{ mol } H_2O) = 130$ g $H_2O$

c. 0.25 mol $C_2H_6O \times (3 \text{ mol } O_2/1 \text{ mol } C_2H_6O) = 0.75$ mol $O_2$
0.75 mol $O_2 \times (32.00 \text{ g } O_2/1 \text{ mol } O_2) = 24$ g $O_2$

**5.25** Use conversion factors to solve the problems. Follow the steps in Sample Problem 5.14.

$$\underset{\text{salicylic acid}}{C_7H_6O_3(s)} \ + \ \underset{\text{acetic acid}}{C_2H_4O_2(l)} \longrightarrow \underset{\text{aspirin}}{C_9H_8O_4(s)} \ + \ H_2O(l)$$

a. 55.5 g $C_7H_6O_3 \times (1 \text{ mol } C_7H_6O_3/138.1 \text{ g } C_7H_6O_3) = 0.402$ mol $C_7H_6O_3$
0.402 mol $C_7H_6O_3 \times (1 \text{ mol } C_9H_8O_4/1 \text{ mol } C_7H_6O_3) = 0.402$ mol $C_9H_8O_4$
0.402 mol $C_9H_8O_4 \times (180.2 \text{ g } C_9H_8O_4/1 \text{ mol } C_9H_8O_4) = 72.4$ g $C_9H_8O_4$

b. 55.5 g $C_7H_6O_3 \times (1 \text{ mol } C_7H_6O_3/138.1 \text{ g } C_7H_6O_3) = 0.402$ mol $C_7H_6O_3$
0.402 mol $C_7H_6O_3 \times (1 \text{ mol } C_2H_4O_2/1 \text{ mol } C_7H_6O_3) = 0.402$ mol $C_2H_4O_2$
0.402 mol $C_2H_4O_2 \times (60.05 \text{ g } C_2H_4O_2/1 \text{ mol } C_2H_4O_2) = 24.1$ g $C_2H_4O_2$

c. 55.5 g $C_7H_6O_3 \times (1 \text{ mol } C_7H_6O_3/138.1 \text{ g } C_7H_6O_3) = 0.402$ mol $C_7H_6O_3$
0.402 mol $C_7H_6O_3 \times (1 \text{ mol } H_2O/1 \text{ mol } C_7H_6O_3) = 0.402$ mol $H_2O$
0.402 mol $H_2O \times (18.02 \text{ g } H_2O/1 \text{ mol } H_2O) = 7.24$ g $H_2O$

**5.26** Use conversion factors to solve the problems. Follow the steps in Sample Problem 5.14.

$$N_2 + O_2 \rightarrow 2 NO$$

a. 10.0 g $N_2 \times$ (1 mol $N_2$/28.02 g $N_2$) = 0.357 mol $N_2$
   0.357 mol $N_2$ (2 mol NO/1 mol $N_2$) = 0.714 mol NO
   0.714 mol NO $\times$ (30.01 g NO/1 mol NO) = 21.4 g NO

b. 10.0 g $O_2 \times$ (1 mol $O_2$/32.00 g $O_2$) = 0.313 mol $O_2$
   0.313 mol $O_2 \times$ (2 mol NO/1 mol $O_2$) = 0.626 mol NO
   0.626 mol NO $\times$ (30.01 g NO/1 mol NO) = 18.8 g NO

c. 10.0 g $N_2 \times$ (1 mol $N_2$/28.02 g $N_2$) = 0.357 mol $N_2$
   0.357 mol $N_2 \times$ (1 mol $O_2$/1 mol $N_2$) = 0.357 mol $O_2$
   0.357 mol $O_2 \times$ (32.00 g $O_2$/1 mol $O_2$) = 11.4 g $O_2$

**5.27**  [1] Convert the number of moles of reactant to the number of moles of product using a mole–mole conversion factor.
[2] Convert the number of moles of product to the number of grams of product—the theoretical yield—using the product's molar mass.

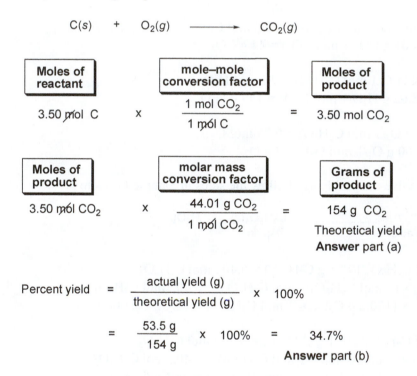

**5.28**    Use the steps in Answer 5.27 to solve the problem.

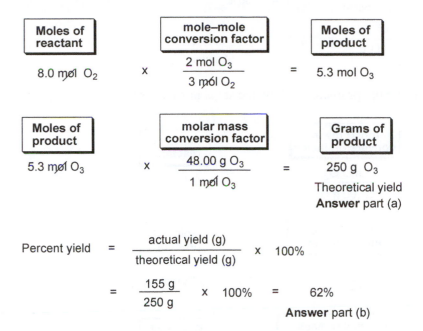

| Moles of reactant | | mole–mole conversion factor | | Moles of product |
|---|---|---|---|---|
| 8.0 mol O$_2$ | X | $\dfrac{2 \text{ mol O}_3}{3 \text{ mol O}_2}$ | = | 5.3 mol O$_3$ |

| Moles of product | | molar mass conversion factor | | Grams of product |
|---|---|---|---|---|
| 5.3 mol O$_3$ | X | $\dfrac{48.00 \text{ g O}_3}{1 \text{ mol O}_3}$ | = | 250 g  O$_3$ |

Theoretical yield
**Answer** part (a)

$$\text{Percent yield} = \frac{\text{actual yield (g)}}{\text{theoretical yield (g)}} \times 100\%$$

$$= \frac{155 \text{ g}}{250 \text{ g}} \times 100\% = 62\%$$

**Answer** part (b)

**5.29**    Use the steps in Sample Problem 5.17 to answer the questions.

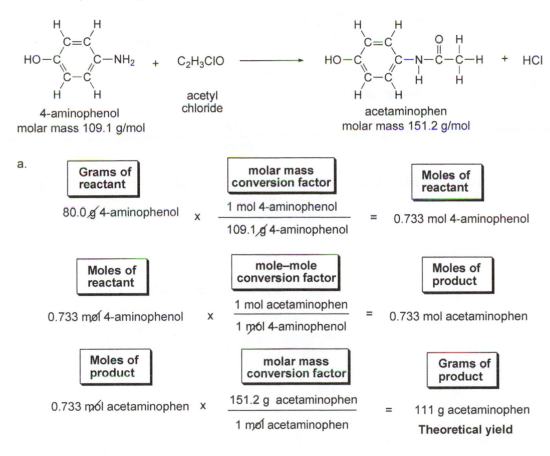

4-aminophenol
molar mass 109.1 g/mol

acetyl chloride

acetaminophen
molar mass 151.2 g/mol

a.

| Grams of reactant | | molar mass conversion factor | | Moles of reactant |
|---|---|---|---|---|
| 80.0 g 4-aminophenol | X | $\dfrac{1 \text{ mol 4-aminophenol}}{109.1 \text{ g 4-aminophenol}}$ | = | 0.733 mol 4-aminophenol |

| Moles of reactant | | mole–mole conversion factor | | Moles of product |
|---|---|---|---|---|
| 0.733 mol 4-aminophenol | X | $\dfrac{1 \text{ mol acetaminophen}}{1 \text{ mol 4-aminophenol}}$ | = | 0.733 mol acetaminophen |

| Moles of product | | molar mass conversion factor | | Grams of product |
|---|---|---|---|---|
| 0.733 mol acetaminophen | X | $\dfrac{151.2 \text{ g  acetaminophen}}{1 \text{ mol acetaminophen}}$ | = | 111 g acetaminophen |

**Theoretical yield**

b.

$$\text{Percent yield} = \frac{\text{actual yield (g)}}{\text{theoretical yield (g)}} \times 100\%$$

$$= \frac{65.5\ g}{111\ g} \times 100\% = 59.0\%$$

**Answer**

**5.30**  Use conversion factors to solve the problem.  Follow the steps in Answer 5.29.

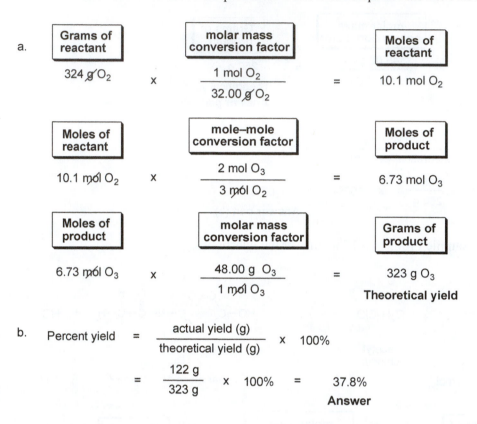

a.

| Grams of reactant | molar mass conversion factor | Moles of reactant |
|---|---|---|

$$324\ \cancel{g}\ O_2 \quad \times \quad \frac{1\ \text{mol}\ O_2}{32.00\ \cancel{g}\ O_2} \quad = \quad 10.1\ \text{mol}\ O_2$$

| Moles of reactant | mole–mole conversion factor | Moles of product |
|---|---|---|

$$10.1\ \cancel{\text{mol}}\ O_2 \quad \times \quad \frac{2\ \text{mol}\ O_3}{3\ \cancel{\text{mol}}\ O_2} \quad = \quad 6.73\ \text{mol}\ O_3$$

| Moles of product | molar mass conversion factor | Grams of product |
|---|---|---|

$$6.73\ \cancel{\text{mol}}\ O_3 \quad \times \quad \frac{48.00\ g\ O_3}{1\ \cancel{\text{mol}}\ O_3} \quad = \quad 323\ g\ O_3$$

**Theoretical yield**

b.    $$\text{Percent yield} = \frac{\text{actual yield (g)}}{\text{theoretical yield (g)}} \times 100\%$$

$$= \frac{122\ g}{323\ g} \times 100\% = 37.8\%$$

**Answer**

**5.31**  To determine the overall percent yield in a synthesis that has more than one step, multiply the percent yield for each step.

a. $(0.90)^{10} \times 100\% = 35\%$

b. $(0.80)^{10} \times 100\% = 11\%$

c. $0.50 \times (0.90)^9 \times 100\% = 19\%$

d. $0.20 \times 0.50 \times 0.50 \times 0.80 \times 0.80 \times 0.80 \times 0.80 \times 0.80 \times 0.80 \times 0.80 \times 100\% = 1.0\%$

**5.32**   Use the steps in Sample Problem 5.18 to answer the questions.

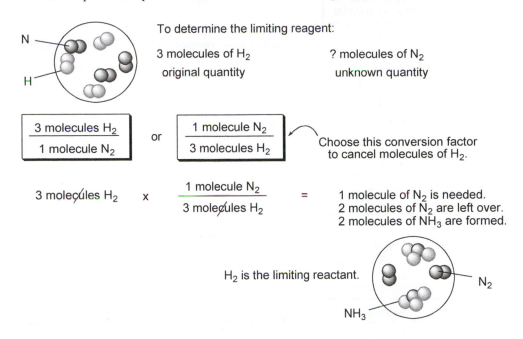

To determine the limiting reagent:

3 molecules of $H_2$          ? molecules of $N_2$

original quantity          unknown quantity

$$\frac{3 \text{ molecules } H_2}{1 \text{ molecule } N_2} \quad \text{or} \quad \frac{1 \text{ molecule } N_2}{3 \text{ molecules } H_2}$$

Choose this conversion factor to cancel molecules of $H_2$.

$$3 \text{ molecules } H_2 \quad \times \quad \frac{1 \text{ molecule } N_2}{3 \text{ molecules } H_2} \quad = \quad$$

1 molecule of $N_2$ is needed.
2 molecules of $N_2$ are left over.
2 molecules of $NH_3$ are formed.

$H_2$ is the limiting reactant.

**5.33**

a.   $5.0 \text{ mol } H_2 \quad \times \quad \dfrac{1 \text{ mol } O_2}{2 \text{ mol } H_2} \quad = \quad$ 2.5 mol of $O_2$ are needed. Since 5.0 mol of $O_2$ are present, $H_2$ is the limiting reactant.

b.   $5.0 \text{ mol } H_2 \quad \times \quad \dfrac{1 \text{ mol } O_2}{2 \text{ mol } H_2} \quad = \quad$ 2.5 mol of $O_2$ are needed. Since 8.0 mol of $O_2$ are present, $H_2$ is the limiting reactant.

c.   $8.0 \text{ mol } H_2 \quad \times \quad \dfrac{1 \text{ mol } O_2}{2 \text{ mol } H_2} \quad = \quad$ 4.0 mol of $O_2$ are needed. Since only 2.0 mol of $O_2$ are present, $O_2$ is the limiting reactant.

d.   $2.0 \text{ mol } H_2 \quad \times \quad \dfrac{1 \text{ mol } O_2}{2 \text{ mol } H_2} \quad = \quad$ 1.0 mol of $O_2$ is needed. Since 5.0 mol of $O_2$ are present, $H_2$ is the limiting reactant.

**5.34**   Calculate the number of moles of product formed as in Sample Problem 5.19.

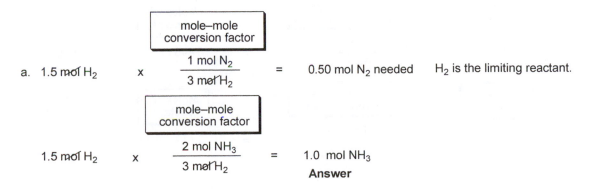

mole–mole conversion factor

a.   $1.5 \text{ mol } H_2 \quad \times \quad \dfrac{1 \text{ mol } N_2}{3 \text{ mol } H_2} \quad = \quad$ 0.50 mol $N_2$ needed     $H_2$ is the limiting reactant.

mole–mole conversion factor

$1.5 \text{ mol } H_2 \quad \times \quad \dfrac{2 \text{ mol } NH_3}{3 \text{ mol } H_2} \quad = \quad$ 1.0 mol $NH_3$

**Answer**

b.  1.0 mol $H_2$   x   $\dfrac{\text{1 mol } N_2}{\text{3 mol } H_2}$   =   0.33 mol $N_2$ needed     $H_2$ is the limiting reactant.

*mole–mole conversion factor*

1.0 mol $H_2$   x   $\dfrac{\text{2 mol } NH_3}{\text{3 mol } H_2}$   =   0.67  mol $NH_3$
**Answer**

*mole–mole conversion factor*

c.  2.0 mol $H_2$   x   $\dfrac{\text{1 mol } N_2}{\text{3 mol } H_2}$   =   0.67 mol $N_2$ needed     $H_2$ is the limiting reactant.

*mole–mole conversion factor*

2.0 mol $H_2$   x   $\dfrac{\text{2 mol } NH_3}{\text{3 mol } H_2}$   =   1.3  mol $NH_3$
**Answer**

*mole–mole conversion factor*

d.  7.5 mol $H_2$   x   $\dfrac{\text{1 mol } N_2}{\text{3 mol } H_2}$   =   2.5 mol $N_2$ needed     $N_2$ is the limiting reactant.

*mole–mole conversion factor*

2.0 mol $N_2$   x   $\dfrac{\text{2 mol } NH_3}{\text{1 mol } N_2}$   =   4.0  mol $NH_3$
**Answer**

**5.35**   Convert the number of grams of each reactant to the number of moles using molar masses. Since the mole ratio of $O_2$ to $N_2$ is 1:1, the limiting reactant has fewer moles.

a.   **Grams of reactant**

12.5 g $N_2$   x   $\dfrac{\text{1 mol } N_2}{\text{28.02 g } N_2}$   =   0.446 mol $N_2$     $N_2$ is the limiting reactant.

**molar mass conversion factor**       **Moles of reactant**

**Grams of reactant**

15.0 g $O_2$   x   $\dfrac{\text{1 mol } O_2}{\text{32.00 g } O_2}$   =   0.469 mol $O_2$

**molar mass conversion factor**       **Moles of reactant**

b.

| Grams of reactant | | molar mass conversion factor | | Moles of reactant |
|---|---|---|---|---|
| 14.0 g $N_2$ | x | $\dfrac{1 \text{ mol } N_2}{28.02 \text{ g } N_2}$ | = | 0.500 mol $N_2$ |

| Grams of reactant | | molar mass conversion factor | | Moles of reactant | |
|---|---|---|---|---|---|
| 13.0 g $O_2$ | x | $\dfrac{1 \text{ mol } O_2}{32.00 \text{ g } O_2}$ | = | 0.406 mol $O_2$ | $O_2$ is the limiting reactant. |

**5.36** Calculate the number of moles of product formed based on the limiting reactant. Then convert moles to grams using molar mass.

a.    0.446 mol $N_2$   x   $\dfrac{2 \text{ mol NO}}{1 \text{ mol } N_2}$   =   0.892 mol NO

       0.892 mol NO   x   $\dfrac{30.01 \text{ g}}{1 \text{ mol NO}}$   =   26.8 g NO

b.    0.406 mol $O_2$   x   $\dfrac{2 \text{ mol NO}}{1 \text{ mol } O_2}$   =   0.812 mol NO

       0.812 mol NO   x   $\dfrac{30.01 \text{ g}}{1 \text{ mol NO}}$   =   24.4 g NO

**5.37**

| Grams of reactant | | molar mass conversion factor | | Moles of reactant | |
|---|---|---|---|---|---|
| 5.00 g $H_2$ | x | $\dfrac{1 \text{ mol } H_2}{2.016 \text{ g } H_2}$ | = | 2.48 mol $H_2$ | |
| 10.0 g $O_2$ | x | $\dfrac{1 \text{ mol } O_2}{32.00 \text{ g } O_2}$ | = | 0.313 mol $O_2$ | $O_2$ is the limiting reactant. |

   0.313 mol $O_2$   x   $\dfrac{2 \text{ mol } H_2O}{1 \text{ mol } O_2}$   =   0.626 mol $H_2O$

   0.626 mol $H_2O$   x   $\dfrac{18.02 \text{ g}}{1 \text{ mol } H_2O}$   =   **11.3 g $H_2O$**

**5.38**  A compound that gains electrons is reduced.
A compound that loses electrons is oxidized.

    (oxidized)    (reduced)                      (oxidized)   (reduced)

a.  $Zn(s) + 2\,H^+(aq) \longrightarrow Zn^{2+}(aq) + H_2(g)$
    c.  $2\,I^- + Br_2 \longrightarrow I_2 + 2\,Br^-$

    $Zn \longrightarrow Zn^{2+} + 2\,e^-$
                         $2\,I^- \longrightarrow I_2 + 2\,e^-$

    $2\,H^+ + 2\,e^- \longrightarrow H_2$
                       $Br_2 + 2\,e^- \longrightarrow 2\,Br^-$

    (reduced)    (oxidized)

b.  $Fe^{3+}(aq) + Al(s) \longrightarrow Al^{3+}(aq) + Fe(s)$
    d.  $2\,AgBr \longrightarrow 2\,Ag + Br_2$

    $Al \longrightarrow Al^{3+} + 3\,e^-$
                $2\,Br^-$ (oxidized)   $2\,Ag^+$ (reduced)

    $Fe^{3+} + 3\,e^- \longrightarrow Fe$
               $2\,Br^- \longrightarrow Br_2 + 2\,e^-$

                                                  $2\,Ag^+ + 2\,e^- \longrightarrow 2\,Ag$

**5.39**  A compound that gains electrons while causing another compound to be oxidized is called an **oxidizing agent**.
A compound that loses electrons while causing another compound to be reduced is called a **reducing agent.**

    a.  Zn reducing agent, $H^+$ oxidizing agent
        c.  $I^-$ reducing agent, $Br_2$ oxidizing agent
    b.  $Fe^{3+}$ oxidizing agent, Al reducing agent
    d.  $Br^-$ reducing agent, $Ag^+$ oxidizing agent

**5.40**

Zn is oxidized, and $Hg^{2+}$ is reduced.

$Zn \longrightarrow Zn^{2+} + 2\,e^-$
         $Hg^{2+} + 2\,e^- \longrightarrow Hg$

Zn loses electrons and is oxidized.
    $Hg^{2+}$ gains electrons and is reduced.

**5.41**  $H_2$ is oxidized since it gains an O atom and $C_2H_4O_2$ is reduced since it gains hydrogen.

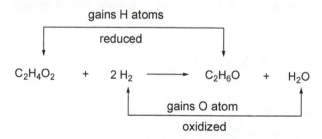

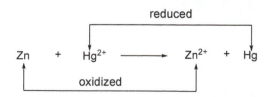

**5.42**  Zn is the reducing agent and $Hg^{2+}$ is the oxidizing agent.

**5.43** The process is a chemical reaction because the spheres in the reactants are joined differently than the spheres in the products.

$$2 CO + 2 O_3 \longrightarrow 2 CO_2 + 2 O_2 \quad \text{(not balanced)}$$

**5.45** The difference between a coefficient and a subscript is that the coefficient indicates the number of molecules or moles undergoing reaction, whereas the subscript indicates the number of atoms of each element in a chemical formula.

**5.47** Add up the number of atoms on each side of the equation and then label the equations as balanced or not balanced.

   a. $2 HCl(aq) + Ca(s) \longrightarrow CaCl_2(aq) + H_2(g)$

         2 H, 2 Cl, 1 Ca: both sides, therefore **balanced**

   b. $TiCl_4 + 2 H_2O \longrightarrow TiO_2 + HCl$

       1 Ti, 4 Cl, 4 H, 2 O            1 Ti, 1 Cl, 1 H, 2 O

                        **NOT balanced**

   c. $Al(OH)_3 + H_3PO_4 \longrightarrow AlPO_4 + 3 H_2O$

       1 Al, 1 P, 7 O, 6 H: both sides, therefore **balanced**

**5.49** Write the balanced equation using the colors of the spheres to identify the atoms (gray = hydrogen and green = chlorine).

$$H_2 + Cl_2 \longrightarrow 2 HCl$$

**5.51** Balance the equation with coefficients one element at a time to have the same number of atoms on each side of the equation. Follow the steps in Example 5.2.

   a. $Ni(s) + 2 HCl(aq) \longrightarrow NiCl_2(aq) + H_2(g)$

   b. $CH_4(g) + 4 Cl_2(g) \longrightarrow CCl_4(g) + 4 HCl(g)$

   c. $2 KClO_3 \longrightarrow 2 KCl + 3 O_2$

   d. $Al_2O_3 + 6 HCl \longrightarrow 2 AlCl_3 + 3 H_2O$

   e. $4 Al(OH)_3 + 6 H_2SO_4 \longrightarrow 2 Al_2(SO_4)_3 + 12 H_2O$

**5.53** Follow the steps in Example 5.2 and balance the equations.

   a. $2 C_6H_6 + 15 O_2 \longrightarrow 12 CO_2 + 6 H_2O$     c. $2 C_8H_{18} + 25 O_2 \longrightarrow 16 CO_2 + 18 H_2O$

   b. $C_7H_8 + 9 O_2 \longrightarrow 7 CO_2 + 4 H_2O$

**5.55** Follow the steps in Example 5.2 and balance the equation.

$$2\ S(s)\ +\ 3\ O_2(g)\ +\ 2\ H_2O(l)\ \longrightarrow\ 2\ H_2SO_4(l)$$

**5.57** Fill in the molecules of the products using the balanced equation and following the law of conservation of mass. Each side must have the same number of O and C atoms.

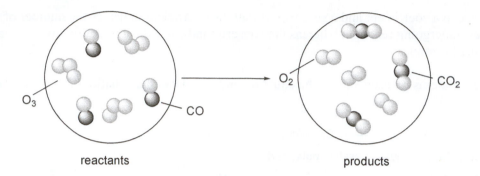

reactants                                products

**5.59** To calculate the formula weight, multiply the number of atoms of each element by the atomic weight and add the results.  The formula weight in amu is equal to the molar mass in g/mol.

a.      1 Na atom      ×      22.99 amu      =      22.99 amu
        1 N atom       ×      14.01 amu      =      14.01 amu
        2 O atoms      ×      16.00 amu      =      32.00 amu
        Formula weight of $NaNO_2$                    69.00 amu = 69.00 g/mol

b.      2 Al atom      ×      26.98 amu      =      53.96 amu
        3 S atoms      ×      32.07 amu      =      96.21 amu
        12 O atoms     ×      16.00 amu      =      192.0   amu
        Formula weight of $Al_2(SO_4)_3$              342.17 amu rounded to 342.2 amu = 342.2 g/mol

c.      6 C atom       ×      12.01 amu      =      72.06   amu
        8 H atoms      ×      1.008 amu      =      8.064 amu
        6 O atoms      ×      16.00 amu      =      96.00   amu
        Formula weight of $C_6H_8O_6$                 176.124 amu rounded to 176.12 amu = 176.12 g/mol

**5.61**    Determine the molecular formula of L-dopa.  Then calculate the formula weight and molar mass as in Answer 5.59.

a. molecular formula = $C_9H_{11}NO_4$
b. formula weight = 197.2 amu
c. molar mass = 197.2 g/mol

L-dopa

**5.63** Convert all of the units to moles, and then compare the atomic mass or formula weight to determine the quantity with the larger mass.

a. 1 mol of Fe atoms (55.85 g/mol) < 1 mol of Sn atoms (118.7 g/mol)
b. 1 mol of C atoms (12.01 g/mol) < $6.02 \times 10^{23}$ N atoms = 1 mol N atoms (14.01 g/mol)
c. 1 mol of N atoms (14.01 g/mol) < 1 mol of $N_2$ molecules = 2 mol N atoms (28.02 g/mol $N_2$)
d. 1 mol of $CO_2$ molecules (44.01 g/mol) > $3.01 \times 10^{23}$ $N_2O$ molecules = 0.500 mol $N_2O$ (44.02 g/mol $N_2O$) = 22.01 g $N_2O$

**5.65** Calculate the molar mass of each compound as in Answer 5.59, and then multiply by 5.00 mol.

a. HCl = 182 g
b. $Na_2SO_4$ = 710. g
c. $C_2H_2$ = 130. g
d. $Al(OH)_3$ = 390. g

**5.67** Convert the grams to moles using the molar mass as a conversion factor.

a.  0.500 g  x  $\dfrac{1\ mol}{342.3\ g}$  =  $1.46 \times 10^{-3}$ mol

c.  25.0 g  x  $\dfrac{1\ mol}{342.3\ g}$  =  0.0730 mol

b.  5.00 g  x  $\dfrac{1\ mol}{342.3\ g}$  =  0.0146 mol

d.  0.0250 g  x  $\dfrac{1\ mol}{342.3\ g}$  =  $7.30 \times 10^{-5}$ mol

**5.69** Multiply the number of moles by Avogadro's number to determine the number of molecules, as in Example 5.3.

a. 2.00 mol × $6.02 \times 10^{23}$ molecules/mol = $1.20 \times 10^{24}$ molecules
b. 0.250 mol × $6.02 \times 10^{23}$ molecules/mol = $1.51 \times 10^{23}$ molecules
c. 26.5 mol × $6.02 \times 10^{23}$ molecules/mol = $1.60 \times 10^{25}$ molecules
d. 222 mol × $6.02 \times 10^{23}$ molecules/mol = $1.34 \times 10^{26}$ molecules
e. $5.00 \times 10^5$ mol × $6.02 \times 10^{23}$ molecules/mol = $3.01 \times 10^{29}$ molecules

**5.71** Use the molar mass as a conversion factor to convert the moles to grams.  Use Avogadro's number to convert the number of molecules to moles.

a.  3.60 mol  x  $\dfrac{90.08\ g}{1\ mol}$  =  324 g

b.  0.580 mol  x  $\dfrac{90.08\ g}{1\ mol}$  =  52.2 g

c.  $7.3 \times 10^{24}$ molecules  x  $\dfrac{1\ mol}{6.02 \times 10^{23}\ molecules}$  x  $\dfrac{90.08\ g}{1\ mol}$  =  $1.1 \times 10^3$ g

d.  $6.56 \times 10^{22}$ molecules  x  $\dfrac{1\ mol}{6.02 \times 10^{23}\ molecules}$  x  $\dfrac{90.08\ g}{1\ mol}$  =  9.82 g

**5.73**

$$2 \text{ H}-\text{C}\equiv\text{C}-\text{H} \quad + \quad 5 \text{ O}_2 \quad \longrightarrow \quad 4 \text{ CO}_2 \quad + \quad 2 \text{ H}_2\text{O}$$
acetylene

a. 12.5 moles of $O_2$ are needed to react completely with 5.00 mol of $C_2H_2$.

5.00 mol $C_2H_2 \times$ (5 mol $O_2$/2 mol $C_2H_2$) = 12.5 mol $O_2$

b. 12 moles of $CO_2$ are formed from 6.0 mol of $C_2H_2$.

6.0 mol $C_2H_2 \times$ (4 mol $CO_2$/2 mol $C_2H_2$) = 12 mol $CO_2$

c. 0.50 moles of $H_2O$ are formed from 0.50 mol of $C_2H_2$.

0.50 mol $C_2H_2 \times$ (2 mol $H_2O$/2 mol $C_2H_2$) = 0.50 mol $H_2O$

d. 0.40 moles of $C_2H_2$ are needed to form 0.80 mol of $CO_2$.

0.80 mol $CO_2 \times$ (2 mol $C_2H_2$/4 mol $CO_2$) = 0.40 mol $C_2H_2$

**5.75** Use conversion factors as in Example 5.6 to solve the problems.

a. 220 g of $CO_2$ are formed from 2.5 mol of $C_2H_2$.
b. 44 g of $CO_2$ are formed from 0.50 mol of $C_2H_2$.
c. 4.5 g of $H_2O$ are formed from 0.25 mol of $C_2H_2$.
d. 240 g of $O_2$ are needed to react with 3.0 mol of $C_2H_2$.

**5.77** Use the equation to determine the percent yield.

$$\text{Percent yield} \quad = \quad \frac{\text{actual yield (g)}}{\text{theoretical yield (g)}} \quad \times \quad 100\%$$

$$= \quad \frac{9.0 \text{ g}}{12.0 \text{ g}} \quad \times \quad 100\% \quad = \quad 75\%$$

**5.79** Use the following equations to determine the percent yield.

a.

| Grams of reactant | | molar mass conversion factor | | Moles of reactant |
|---|---|---|---|---|
| 3.20 g CH$_4$ | $\times$ | $\dfrac{1 \text{ mol CH}_4}{16.04 \text{ g CH}_4}$ | $=$ | 0.200 mol CH$_4$ |

| Moles of reactant | | mole–mole conversion factor | | Moles of product |
|---|---|---|---|---|
| 0.200 mol CH$_4$ | $\times$ | $\dfrac{1 \text{ mol CHCl}_3}{1 \text{ mol CH}_4}$ | $=$ | 0.200 mol CHCl$_3$ |

| Moles of product | | molar mass conversion factor | | Grams of product |
|---|---|---|---|---|
| 0.200 mol CHCl$_3$ | $\times$ | $\dfrac{119.4 \text{ g CHCl}_3}{1 \text{ mol CHCl}_3}$ | $=$ | 23.9 g CHCl$_3$ Theoretical yield |

b. Percent yield $= \dfrac{\text{actual yield (g)}}{\text{theoretical yield (g)}} \times 100\%$

$\qquad = \dfrac{15.0 \text{ g CHCl}_3}{23.9 \text{ g CHCl}_3} \times 100\% = 62.8\%$

**5.81**

a. 4 molecules **A** $\times \dfrac{1 \text{ molecule B}}{1 \text{ molecule A}} =$ 4 molecules of **B** are needed.
Molecule **A** is in excess.
Molecule **B** is the limiting reactant.

b. 4 molecules **A** $\times \dfrac{1 \text{ molecule B}}{2 \text{ molecules A}} =$ 2 molecules of **B** are needed.
Molecule **B** is in excess.
Molecule **A** is the limiting reactant.

c. 4 molecules **A** $\times \dfrac{2 \text{ molecules B}}{1 \text{ molecule A}} =$ 8 molecules of **B** are needed.
Molecule **A** is in excess.
Molecule **B** is the limiting reactant.

**5.83**

a. 1.0 mol NO $\times \dfrac{1 \text{ mol O}_2}{2 \text{ mol NO}} =$ 0.50 mol of $O_2$ is needed.
$O_2$ is in excess.
NO is the limiting reactant.

b. 2.0 mol NO $\times \dfrac{1 \text{ mol O}_2}{2 \text{ mol NO}} =$ 1.0 mol of $O_2$ is needed.
NO is in excess.
$O_2$ is the limiting reactant.

c. 10.0 g NO $\times \dfrac{1 \text{ mol NO}}{30.01 \text{ g NO}} =$ 0.333 mol NO

10.0 g $O_2$ $\times \dfrac{1 \text{ mol O}_2}{32.00 \text{ g O}_2} =$ 0.313 mol $O_2$

0.333 mol NO $\times \dfrac{1 \text{ mol O}_2}{2 \text{ mol NO}} =$ 0.167 mol of $O_2$ is needed.
$O_2$ is in excess.
NO is the limiting reactant.

d. 28.0 g NO $\times \dfrac{1 \text{ mol NO}}{30.01 \text{ g NO}} =$ 0.933 mol NO

16.0 g $O_2$ $\times \dfrac{1 \text{ mol O}_2}{32.00 \text{ g O}_2} =$ 0.500 mol $O_2$

0.933 mol NO $\times \dfrac{1 \text{ mol O}_2}{2 \text{ mol NO}} =$ 0.467 mol of $O_2$ is needed.
$O_2$ is in excess.
NO is the limiting reactant.

**5.85** Use the limiting reactant from Problem 5.83 to determine the amount of product formed. The conversion of moles of limiting reagent to grams of product is combined in a single step.

a.  $1.0 \text{ mol NO} \times \dfrac{2 \text{ mol NO}_2}{2 \text{ mol NO}} \times \dfrac{46.01 \text{ g NO}_2}{1 \text{ mol NO}_2} = 46 \text{ g NO}_2$

b.  $0.50 \text{ mol O}_2 \times \dfrac{2 \text{ mol NO}_2}{1 \text{ mol O}_2} \times \dfrac{46.01 \text{ g NO}_2}{1 \text{ mol NO}_2} = 46 \text{ g NO}_2$

c.  $0.333 \text{ mol NO} \times \dfrac{2 \text{ mol NO}_2}{2 \text{ mol NO}} \times \dfrac{46.01 \text{ g NO}_2}{1 \text{ mol NO}_2} = 15.3 \text{ g NO}_2$

d.  $0.933 \text{ mol NO} \times \dfrac{2 \text{ mol NO}_2}{2 \text{ mol NO}} \times \dfrac{46.01 \text{ g NO}_2}{1 \text{ mol NO}_2} = 42.9 \text{ g NO}_2$

**5.87**

a.  $8.00 \text{ g C}_2\text{H}_4 \times \dfrac{1 \text{ mol C}_2\text{H}_4}{28.05 \text{ g C}_2\text{H}_4} = 0.285 \text{ mol C}_2\text{H}_4$

$12.0 \text{ g HCl} \times \dfrac{1 \text{ mol HCl}}{36.46 \text{ g HCl}} = 0.329 \text{ mol HCl}$

b.  Since the mole ratio in the balanced equation is 1:1, the reactant with the smaller number of moles is the limiting reactant: $C_2H_4$.

c.  $0.285 \text{ mol C}_2\text{H}_4 \times \dfrac{1 \text{ mol C}_2\text{H}_5\text{Cl}}{1 \text{ mol C}_2\text{H}_4} = 0.285 \text{ mol C}_2\text{H}_5\text{Cl}$

d.  $0.285 \text{ mol C}_2\text{H}_5\text{Cl} \times \dfrac{64.51 \text{ g C}_2\text{H}_5\text{Cl}}{1 \text{ mol C}_2\text{H}_5\text{Cl}} = 18.4 \text{ g C}_2\text{H}_5\text{Cl}$

e.  $\dfrac{10.6 \text{ g C}_2\text{H}_5\text{Cl}}{18.4 \text{ g C}_2\text{H}_5\text{Cl}} \times 100\% = 57.6 \% \text{ percent yield}$

**5.89** The difference between a substance that is oxidized and an oxidizing agent is that a substance that is oxidized loses electrons. An oxidizing agent gains electrons.

**5.91** The species that is oxidized loses one or more electrons. The species that is reduced gains one or more electrons.

a.  Fe — Cu$^{2+}$
oxidized — reduced
$Fe \longrightarrow Fe^{2+} + 2 e^-$   $Cu^{2+} + 2 e^- \longrightarrow Cu$

b.  Cl$_2$ — 2 I$^-$
reduced — oxidized
$2 I^- \longrightarrow I_2 + 2 e^-$   $Cl_2 + 2 e^- \longrightarrow 2 Cl^-$

c.  2 Na — Cl$_2$
oxidized — reduced
$2 Na \longrightarrow 2 Na^+ + 2 e^-$   $Cl_2 + 2 e^- \longrightarrow 2 Cl^-$

**5.93** The oxidizing agent gains electrons. The reducing agent loses electrons.

$$Zn \ + \ Ag_2O \ \longrightarrow \ ZnO \ + \ 2\,Ag$$

$$\begin{array}{cc} Zn & Ag^+ \\ \text{oxidized} & \text{reduced} \\ \text{reducing agent} & \text{oxidizing agent} \end{array}$$

**5.95** Acetylene is reduced because it gains hydrogen atoms.

**5.97** Write the balanced equation and the half reactions.

$$2\,Mg \ + \ O_2 \ \longrightarrow \ 2\,MgO$$

$$2\,Mg \longrightarrow 2\,Mg^{2+} + \ 4\,e^- \qquad O_2 \ + \ 4\,e^- \longrightarrow 2\,O^{2-}$$

**5.99** Refer to prior solutions to answer each part.

$$C_{12}H_{22}O_{11}(s) \ + \ H_2O(l) \longrightarrow \ C_2H_6O(l) \ + \ CO_2(g)$$
$$\text{sucrose} \qquad\qquad\qquad \text{ethanol}$$

a. Calculate the molar mass as in Answer 5.59; the molar mass of sucrose = 342.3 g/mol.
b. Follow the steps in Example 5.2.

$$C_{12}H_{22}O_{11}(s) \ + \ H_2O(l) \longrightarrow \ 4\,C_2H_6O(l) \ + \ 4\,CO_2(g)$$

c. 8 mol of ethanol are formed from 2 mol of sucrose.
d. 10 mol of water are needed to react with 10 mol of sucrose.
e. 101 g of ethanol are formed from 0.550 mol of sucrose.
f. 18.4 g of ethanol are formed from 34.2 g of sucrose.
g. 9.21 g ethanol
h. 13.6%

**5.101**

a. $500 \text{ tablets} \times \dfrac{200.\text{ mg ibuprofen}}{1 \text{ tablet}} \times \dfrac{1 \text{ g}}{1000 \text{ mg}} \times \dfrac{1 \text{ mol ibuprofen}}{206.3 \text{ g ibuprofen}} = 0.485 \text{ mol ibuprofen}$

b. $0.485 \text{ mol ibuprofen} \times \dfrac{6.02 \times 10^{23} \text{ molecules}}{1 \text{ mol}} = 2.92 \times 10^{23} \text{ molecules}$

**5.103**

a. $20 \text{ cig} \times \dfrac{1.93 \text{ mg}}{1 \text{ cig}} \times \dfrac{1 \text{ g}}{1000 \text{ mg}} \times \dfrac{1 \text{ mol nicotine}}{162.3 \text{ g nicotine}} = 2.38 \times 10^{-4} \text{ mol nicotine}$

b. $2.38 \times 10^{-4} \text{ mol nicotine} \times \dfrac{6.02 \times 10^{23} \text{ molecules}}{1 \text{ mol}} = 1.43 \times 10^{20} \text{ molecules}$

**5.105**

$$2400 \text{ mg} \times \frac{1 \text{ g}}{1000 \text{ mg}} \times \frac{1 \text{ mol}}{22.99 \text{ g}} \times \frac{6.02 \times 10^{23} \text{ ions}}{1 \text{ mol}} = 6.3 \times 10^{22} \text{ ions}$$

**5.107**

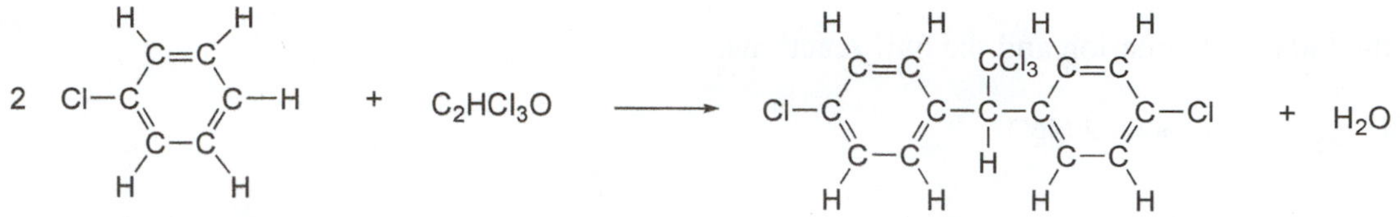

chlorobenzene
112.6 g/mol

DDT
$C_{14}H_9Cl_5$

a. Calculate the molar mass as in Answer 5.59; the molar mass of DDT = 354.5 g/mol.
b. 18 g of DDT would be formed from 0.10 mol of chlorobenzene.
c. 17.8 g is the theoretical yield of DDT in grams from 11.3 g of chlorobenzene.
d. 84.3%

**5.109**  Use conversion factors to answer the questions about dioxin.

a.  $70. \text{ kg} \times \dfrac{3.0 \times 10^{-2} \text{ mg}}{1 \text{ kg}} \times \dfrac{1 \text{ g}}{1000 \text{ mg}} = 2.1 \times 10^{-3} \text{ g dioxin}$

b.  $2.1 \times 10^{-3} \text{ g dioxin} \times \dfrac{1 \text{ mol}}{322.0 \text{ g}} \times \dfrac{6.02 \times 10^{23} \text{ molecules}}{1 \text{ mol}} = 3.9 \times 10^{18} \text{ molecules}$

# Chapter 6 Energy Changes, Reaction Rates, and Equilibrium

## Chapter Review

**[1] What is energy and what units are used to measure energy? (6.1)**
- Energy is the capacity to do work. **Kinetic energy** is the energy of motion, whereas **potential energy** is stored energy.
- Energy is measured in calories (cal) or joules (J), where 1 cal = 4.184 J.
- One nutritional Calorie (Cal) = 1 kcal = 1,000 cal.

$$1 \text{ kcal } = 1,000 \text{ cal}$$
$$1 \text{ kJ } = 1,000 \text{ J}$$
$$1 \text{ kcal } = 4.184 \text{ kJ}$$

**[2] Define bond dissociation energy and describe its relationship to bond strength. (6.2)**
- The **bond dissociation energy** is the energy needed to break a covalent bond by equally dividing the electrons between the two atoms in the bond.
- The higher the bond dissociation energy, the stronger the bond.

**[3] What is the heat of reaction and what is the difference between an endothermic and an exothermic reaction? (6.2)**
- The **heat of reaction,** also called the **enthalpy** and symbolized by $\Delta H$, is the energy absorbed or released in a reaction.
- In an endothermic reaction, energy is absorbed, $\Delta H$ is positive (+), and the products are higher in energy than the reactants. The bonds in the reactants are stronger than the bonds in the products.
- In an exothermic reaction, energy is released, $\Delta H$ is negative (–), and the reactants are higher in energy than the products. The bonds in the products are stronger than the bonds in the reactants.

Bond breaking is **endothermic.**
Energy must be added.

$$\text{H–H} \longrightarrow \text{H}\cdot \ + \ \cdot\text{H} \quad \Delta H = +104 \text{ kcal/mol}$$

$$\text{H}\cdot \ + \ \cdot\text{H} \longrightarrow \text{H–H} \quad \Delta H = -104 \text{ kcal/mol}$$

Bond making is **exothermic.**
Energy is released.

**[4] What are the important features of an energy diagram? (6.3)**
- An energy diagram illustrates the energy changes that occur during the course of a reaction. Energy is plotted on the vertical axis and reaction coordinate is plotted on the horizontal axis. The transition state is located at the top of the energy barrier that separates the reactants and products.
- The **energy of activation** is the energy difference between the reactants and the transition state. The higher the energy of activation, the slower the reaction.
- The difference in energy between the reactants and products is the $\Delta H$.

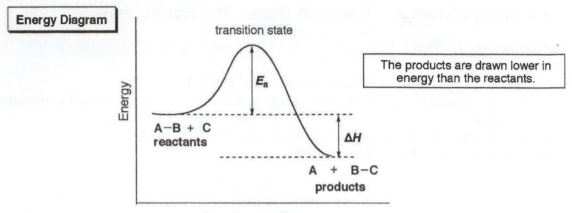

**[5] How do temperature, concentration, and catalysts affect the rate of a reaction? (6.4)**
- Increasing the temperature and concentration increases the reaction rate.
- A catalyst speeds up the rate of a reaction without affecting the energies of the reactants and products. Enzymes are biological catalysts that increase the rate of reactions in living organisms. Catalytic converters use a catalyst to convert automobile engine exhaust to environmentally cleaner products.

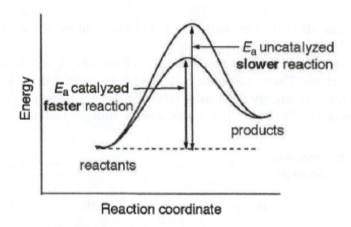

**[6] What are the basic features of equilibrium? (6.5)**
- At equilibrium, the rates of the forward and reverse reactions in a reversible reaction are equal and the net concentrations of all substances do not change.
- The equilibrium constant for a reaction $aA + bB \rightleftharpoons cC + dD$ is written as:

concentration of each product (mol/L)

$$\text{Equilibrium constant} = K = \frac{[\text{products}]}{[\text{reactants}]}$$

concentration of each reactant (mol/L)

$$K = \frac{[C]^c[D]^d}{[A]^a[B]^b}$$

- The magnitude of $K$ tells the relative amount of reactants and products. When $K > 1$, the products are favored; when $K < 1$, the reactants are favored; when $K \approx 1$, both reactants and products are present at equilibrium.

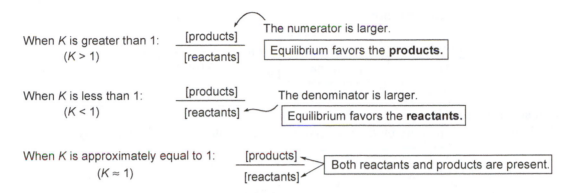

When $K$ is greater than 1:    $\dfrac{[\text{products}]}{[\text{reactants}]}$    The numerator is larger.
($K > 1$)    Equilibrium favors the **products.**

When $K$ is less than 1:    $\dfrac{[\text{products}]}{[\text{reactants}]}$    The denominator is larger.
($K < 1$)    Equilibrium favors the **reactants.**

When $K$ is approximately equal to 1:    $\dfrac{[\text{products}]}{[\text{reactants}]}$    Both reactants and products are present.
($K \approx 1$)

**[7] How does Le Châtelier's principle predict what happens when equilibrium is disturbed? (6.6)**
- Le Châtelier's principle states that a system at equilibrium reacts in such a way as to counteract any disturbance to the equilibrium. How changes in concentration, temperature, and pressure affect equilibrium are summarized in Table 6.5. One example is given below:

Adding more product...

$$2\ CO(g) \ + \ O_2(g) \ \rightleftharpoons \ 2\ CO_2(g)$$

...drives the reaction to the left.

- Catalysts increase the rate at which equilibrium is reached, but do not alter the amount of any substance involved in the reaction.

**[8] How can the principles that describe equilibrium and reaction rates be used to understand the regulation of body temperature? (6.7)**
- Increasing temperature increases the rates of the reactions in the body.
- When temperature is increased, the body dissipates excess heat by dilating blood vessels and sweating. When temperature is decreased, blood vessels constrict to conserve heat, and the body shivers to generate more heat.

## Problem Solving

## [1] Energy (6.1)

**Example 6.1** A reaction releases 573 kJ of energy. How many kilocalories does this correspond to?

**Analysis and Solution**
**[1] Identify the original quantity and the desired quantity.**

573 kJ        ? kcal

original quantity     desired quantity

**[2] Write out the conversion factors.**
- Choose the conversion factor that places the unwanted unit, kilojoules, in the denominator so that the units cancel.

kJ–kcal conversion factors

$$\frac{4.184 \text{ kJ}}{1 \text{ kcal}} \quad \text{or} \quad \boxed{\frac{1 \text{ kcal}}{4.184 \text{ kJ}}} \quad \text{Choose this conversion factor to cancel kJ.}$$

**[3] Set up and solve the problem.**
- Multiply the original quantity by the conversion factor to obtain the desired quantity.

$$573 \text{ kJ} \quad \times \quad \frac{1 \text{ kcal}}{4.184 \text{ kJ}} \quad = \quad 137 \text{ kcal} \quad \textbf{Answer}$$

Kilojoules cancel.

---

**Example 6.2** If a serving of crackers contains 2 g of protein, 5 g of fat, and 22 g of carbohydrates, estimate its number of Calories.

**Analysis**
Use the caloric value (Cal/g) of each class of molecule to form a conversion factor to convert the number of grams to Calories and add up the results.

**Solution**
**[1] Identify the original quantity and the desired quantity.**

2 g protein
5 g fat
22 g carbohydrates

original quantities

? Cal

desired quantity

**[2] Write out the conversion factors.**
- Write out conversion factors that relate the number of grams to the number of Calories for each substance. Each conversion factor must place the unwanted unit, grams, in the denominator so that the units cancel.

Cal–g conversion factor for protein

$$\frac{4 \text{ Cal}}{1 \text{ g protein}}$$

Cal–g conversion factor for carbohydrates

$$\frac{4 \text{ Cal}}{1 \text{ g carbohydrate}}$$

Cal–g conversion factor for fat

$$\frac{9 \text{ Cal}}{1 \text{ g fat}}$$

**[3] Set up and solve the problem.**
- Multiply the original quantity by the conversion factors for protein, fat, and carbohydrates and add up the results to obtain the desired quantity.

| | Calories due to protein | | Calories due to fat | | Calories due to carbohydrate | |
|---|---|---|---|---|---|---|
| Total Calories = | 2 g protein × $\dfrac{4\ Cal}{1\ g\ protein}$ (Grams cancel.) | + | 5 g fat × $\dfrac{9\ Cal}{1\ g\ fat}$ (Grams cancel.) | + | 22 g carbohydrate × $\dfrac{4\ Cal}{1\ g\ carbohydrate}$ (Grams cancel.) | |
| = | 8 Cal | + | 45 Cal | + | 88 Cal | |
| Total Calories = | 141 Cal, rounded to 100 Cal | | | | | |
| | **Answer** | | | | | |

## [2] Energy Changes in Reactions (6.2)

**Example 6.3** Write the equation for the formation of HBr from H and Br atoms. Classify the reaction as endothermic or exothermic, and give the $\Delta H$ using the values in Table 6.2.

**Analysis**

Bond formation is exothermic, so $\Delta H$ is (−). The energy released in forming a bond is (−) the bond dissociation energy.

**Solution**

Energy is released.

$$H\cdot\ +\ \cdot\ddot{\underset{..}{Br}}: \longrightarrow H-\ddot{\underset{..}{Br}}: \qquad \Delta H = -88\ kcal/mol$$

Bond formation is **exothermic.**

**Example 6.4** Considering the indicated carbon–halogen bonds, which bond is predicted to have the higher bond dissociation energy? Which bond is stronger?

**Analysis**

**The higher the bond dissociation energy, the stronger the bond.** In comparing bonds to atoms in the same group of the periodic table, bond dissociation energies and bond strength decrease down a column.

**Solution**

Since I is below Br in the same group of the periodic table, the C–I bond is predicted to have the lower bond dissociation energy, thus making it weaker. The actual values for the bond dissociation energies are given and illustrate that the prediction is indeed true.

lower bond dissociation energy
weaker bond

$\Delta H = +88\ kcal/mol$ $\qquad \Delta H = +71\ kcal/mol$

**Example 6.5** Ammonia ($NH_3$) decomposes to hydrogen and nitrogen and 22.0 kcal/mol of energy is absorbed. How many kilocalories of energy are absorbed when 0.250 mol of ammonia decomposes?

$$2\ NH_3(g) \longrightarrow 3\ H_2(g) + N_2(g) \qquad \Delta H = +22.0\ kcal/mol$$

**Analysis**

Use the given value of $\Delta H$ to set up a conversion factor that relates the kcal of energy absorbed to the number of moles of ammonia. The value for $\Delta H$ means that the given amount of energy is absorbed for the molar quantities shown by the coefficients in the balanced chemical equation.

**Solution**

The given $\Delta H$ value is the amount of energy absorbed when 2 mol of ammonia decompose. Set up a conversion factor that relates kilocalories to moles of ammonia, with moles in the denominator to cancel this unwanted unit.

$$0.250\ \text{mol } NH_3 \quad \times \quad \frac{22.0\ \text{kcal}}{2\ \text{mol } NH_3} \quad = \quad 2.75\ \text{kcal of energy absorbed}$$

kcal–mol conversion factor

Moles cancel.

**Answer**

**Example 6.6** Using the $\Delta H$ and balanced equation for the degradation of ammonia shown in Example 6.5, how many kilocalories of energy are absorbed when 10.0 g of ammonia decomposes?

**Analysis**

To relate the number of grams of ammonia to the number of kilocalories of energy absorbed in the reaction, two operations are needed: [1] Convert the number of grams to the number of moles using the molar mass. [2] Convert the number of moles to the number of kilocalories using $\Delta H$ (kcal/mol) and the coefficients of the balanced chemical equation.

**Solution**

**[1] Convert the number of grams of ammonia to the number of moles of ammonia.**

- Use the molar mass of the reactant ($NH_3$, molar mass 17.03 g/mol) to write a conversion factor. Multiply the number of grams of ammonia by the conversion factor to give the number of moles of ammonia.

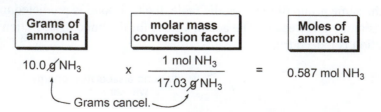

$$10.0\ \text{g } NH_3 \quad \times \quad \frac{1\ \text{mol } NH_3}{17.03\ \text{g } NH_3} \quad = \quad 0.587\ \text{mol } NH_3$$

Grams of ammonia    molar mass conversion factor    Moles of ammonia

Grams cancel.

**[2] Convert the number of moles of ammonia to the number of kilocalories using a kcal–mole conversion factor.**

- Use the $\Delta H$ and the number of moles of ammonia in the balanced chemical equation to write a kcal–mole conversion factor—two moles of ammonia ($NH_3$) absorb 22.0 kcal of energy. Multiply the number of moles of ammonia by the conversion factor to give the number of kilocalories of energy absorbed. This process was illustrated in Example 6.5.

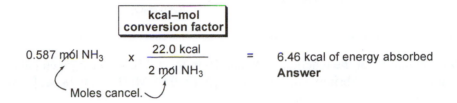

0.587 mol $NH_3$ $\quad$ X $\quad \dfrac{22.0\ kcal}{2\ mol\ NH_3} \quad$ = $\quad$ 6.46 kcal of energy absorbed

**Answer**

Moles cancel.

---

## [3] Energy Diagrams (6.3)

**Example 6.7** Draw an energy diagram for a reaction with a high energy of activation and a $\Delta H$ of +10 kcal/mol. Label the axes, reactants, products, transition state, $E_a$, and $\Delta H$.

**Analysis**

A high energy of activation means a high energy barrier (a large hill) that separates reactants and products. When $\Delta H$ is (+), the products are higher in energy than the reactants.

**Solution**

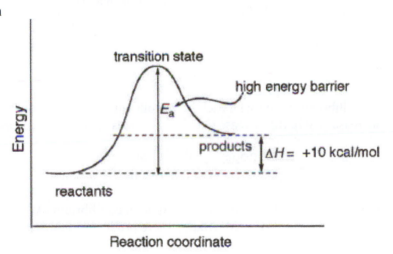

---

## [4] Equilibrium (6.5)

**Example 6.8** Write the expression for the equilibrium constant for the following balanced equation.

$$CO(g) \;+\; H_2O(g) \;\rightleftharpoons\; CO_2(g) \;+\; H_2(g)$$

**Analysis**

To write an expression for the equilibrium constant, multiply the concentration of the products together and divide this number by the product of the concentrations of the reactants. Each concentration term is raised to a power equal to its coefficient in the balanced chemical equation.

## Solution

The concentration terms of the two products are placed in the numerator, multiplied together. The denominator contains concentration terms for the two reactants, multiplied together.

$$\text{Equilibrium constant} \ = \ K \ = \ \frac{[CO_2][H_2]}{[CO][H_2O]}$$

---

**Example 6.9** Calculate $K$ for the reaction of $A_2$ and $B_2$ to form AB, with the given balanced equation and the following equilibrium concentrations: $[A_2] = 0.95$ M; $[B_2] = 0.78$ M; $[AB] = 0.27$ M.

$$A_2 \ + \ B_2 \ \rightleftharpoons \ 2\ AB$$

## Analysis

Write an expression for $K$ using the balanced equation and substitute the equilibrium concentrations of all substances in the expression.

## Solution

$$K \ = \ \frac{[AB]^2}{[A_2][B_2]} \ = \ \frac{(0.27)^2}{(0.95)(0.78)} \ = \ \frac{0.27 \times 0.27}{0.95 \times 0.78}$$

$$= \ \frac{0.0729}{0.741} \ = \ 0.0984 \ \text{rounded to } 0.098$$

**Answer**

---

## [4] Le Châtelier's Principle (6.6)

**Example 6.10** In which direction is the equilibrium shifted with each of the following concentration changes for the given reaction: (a) increase $[H_2O]$; (b) increase $[CO_2]$; (c) decrease $[CO]$; (d) decrease $[H_2]$.

$$CO(g) \ + \ H_2O(g) \ \rightleftharpoons \ CO_2(g) \ + \ H_2(g)$$

## Analysis

Use Le Châtelier's principle to predict the effect of a change in concentration on equilibrium. Adding more reactant or removing product drives the equilibrium to the right. Adding more product or removing reactant drives the equilibrium to the left.

## Solution

a. Increasing $[H_2O]$, a reactant, drives the equilibrium to the right, to form more products.
b. Increasing $[CO_2]$, a product, drives the equilibrium to the left to form more reactants.
c. Decreasing $[CO]$, a reactant, drives the equilibrium to the left to form more reactants.
d. Decreasing $[H_2]$, a product, drives the equilibrium to the right, to form more products.

## Self-Test

**[1] Fill in the blank with one of the terms listed below.**

Bond dissociation energy (6.2)   Energy of activation (6.3)   Le Châtelier's principle (6.6)
Catalyst (6.4)   Exothermic (6.2)   Reaction rate (6.3)
Endothermic (6.2)   Heat of reaction (6.2)   Reversible reaction (6.5)
Energy (6.1)   Kinetic energy (6.1)   Transition state (6.3)

1. The _____ is the energy difference between the reactants and the transition state. The higher the energy difference, the slower the reaction.
2. A _____ increases the rate at which equilibrium is reached, but does not alter the amount of any substance involved in the reaction.
3. _____ is the capacity to do work.
4. The _____, also called the enthalpy and symbolized by $\Delta H$, is the energy absorbed or released in a reaction.
5. The _____ is located at the top of the energy barrier that separates the reactants and products.
6. _____ states that a system at equilibrium reacts in such a way as to counteract any disturbance to the equilibrium.
7. A _____ can occur in either direction, from reactants to products, or from products to reactants.
8. _____ is the energy of motion, whereas potential energy is stored energy.
9. Increasing the temperature and concentration increases _____.
10. When energy is released, a reaction is said to be _____ and $\Delta H$ is negative (–).
11. The _____ is the energy needed to break a covalent bond by equally dividing the electrons between the two atoms in the bond.
12. When energy is absorbed, a reaction is said to be _____ and $\Delta H$ is positive (+).

**[2] Fill in the blank with one of the terms listed below.**

a. $K > 1$      b. $K < 1$      c. $K \approx 1$

13. When _____, the concentration of the reactants is larger than the concentration of the products. We say equilibrium lies to the *left* and favors the *reactants*.
14. When _____, the concentration of the products is larger than the concentration of the reactants. We say equilibrium lies to the *right* and favors the *products*.
15. When _____, anywhere in the range of 0.01–100, both reactants and products are present at equilibrium.

**[3] Pick one of the terms—endothermic or exothermic—for each statement.**

a. endothermic      b. exothermic

16. Heat is released.
17. The products are higher in energy than the reactants.
18. The bonds formed in the products are *stronger* than the bonds broken in the reactants.
19. $\Delta H$ is positive.
20. Heat is absorbed.

**[4] For each term, choose the energy diagram (or diagrams) that best match.**

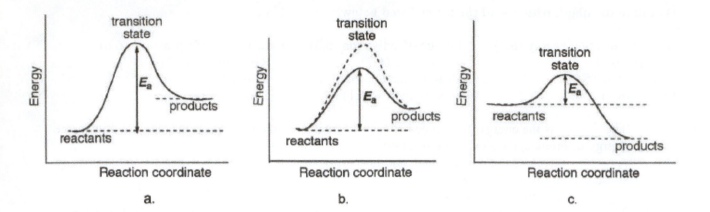

a.        b.        c.

21.     endothermic
22.     exothermic
23.     catalyzed reaction

## Answers to Self-Test

| | | | | |
|---|---|---|---|---|
| 1. energy of activation | 6. Le Châtelier's principle | 11. bond dissociation energy | 16. b | 21. a, b |
| 2. catalyst | 7. reversible reaction | 12. endothermic | 17. a | 22. c |
| 3. Energy | 8. Kinetic energy | 13. b | 18. b | 23. b |
| 4. heat of reaction | 9. reaction rate | 14. a | 19. a | |
| 5. transition state | 10. exothermic | 15. c | 20. a | |

## Solutions to In-Chapter Problems

**6.1** Use conversion factors to solve each problem.

     a. 42 J × (1 cal/4.184 J) = 10. cal          c. 326 kcal × (4.184 kJ/1 kcal) = 1,360 kJ
     b. 55.6 kcal × (1000 cal/1 kcal) = 55,600 cal     d. 25.6 kcal × (4.184 kJ/1 kcal) × (1000 J/1 kJ) =
                                                                        107,000 J

**6.2** Use conversion factors to convert kcal to kJ and J.

     11.5 kcal × (4.184 kJ/1 kcal) = 48.1 kJ
     48.1 kJ × (1000 J/1 kJ) = 48,100 J

**6.3** Use conversion factors to determine the number of Calories in 14 g of olive oil.

     14 g fat × (9 Cal/1 g fat) = 126 Cal, or 100 Cal when rounded to one significant figure

**6.4** Calculate the number of Calories as in Example 6.2.

Total Calories = 6 g fat × (9 Cal/1 g fat) + 20 g carb × (4 Cal/1 g carb) + 2 g protein × (4 Cal/1 g protein)

Total Calories = 54 Cal + 80 Cal + 8 Cal

Total Calories = 142 Cal = rounded to 100 Cal

**6.5** Use Table 6.2 to determine the bond dissociation energy for each reaction. Forming a bond is an **exothermic reaction** and $\Delta H$ is a *negative* number. Breaking a bond is an **endothermic reaction** and $\Delta H$ is a positive number.

a.  H—Br: ⟶ H· + ·Br:
Bond is broken.
+88 kcal/mol
**endothermic**

b.  H· + ·F: ⟶ H—F:
Bond is formed.
−136 kcal/mol
**exothermic**

c.  H—ÖH ⟶ H· + ·ÖH
Bond is broken.
+119 kcal/mol
**endothermic**

**6.6** **The higher the bond dissociation energy, the stronger the bond.** In comparing bonds to atoms in the same group of the periodic table, bond dissociation energies and bond strength decrease down a column.

a.  H—C—I: or H—C—Br:
higher bond
dissociation energy
stronger bond

b.  H—OH or H—SH
higher bond
dissociation energy
stronger bond

**6.7** Use Table 6.3 to answer the questions for a reaction with $\Delta H$ = +22.0 kcal/mol.

a. Heat is absorbed.
b. The bonds in the reactants are stronger.
c. The reactants are lower in energy.
d. The reaction is endothermic.

**6.8** Use a conversion factor to solve the problem.

$C_3H_8(g)$ + 5 $O_2(g)$ ⟶ 3 $CO_2(g)$ + 4 $H_2O(l)$   $\Delta H$ = −531 kcal/mol
propane

kcal–mol
conversion factor

1.00 mol $O_2$ × $\dfrac{531 \text{ kcal}}{5 \text{ mol } O_2}$ = 106 kcal of energy released
**Answer**

Moles cancel.

**6.9** Use conversion factors to solve the problems.

$$C_6H_{12}O_6(s) \longrightarrow 2\ C_2H_6O(l) + 2\ CO_2(g) \qquad \Delta H = -16\ \text{kcal/mol}$$
glucose                    ethanol

a.   $6.0\ \text{mol}\ C_6H_{12}O_6 \times \dfrac{16\ \text{kcal}}{1\ \text{mol}\ C_6H_{12}O_6} = $   96 kcal of energy released
**Answer**

b.   $1.00\ \text{mol}\ C_2H_6O \times \dfrac{16\ \text{kcal}}{2\ \text{mol}\ C_2H_6O} = $   8.0 kcal of energy released
**Answer**

c.   $20.0\ \text{g}\ C_6H_{12}O_6 \times \dfrac{1\ \text{mol}\ C_6H_{12}O_6}{180.2\ \text{g}\ C_6H_{12}O_6} \times \dfrac{16\ \text{kcal}}{1\ \text{mol}\ C_6H_{12}O_6} = $   1.8 kcal of energy released
**Answer**

**6.10**   A high energy of activation means a high energy barrier (a large hill) that separates reactants and products. When $\Delta H$ is positive, the products are higher in energy than the reactants.

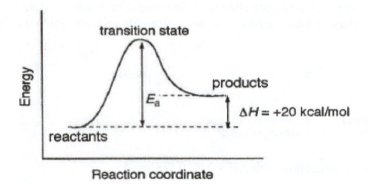

**6.11**   A low energy of activation means a low energy barrier (a small hill) that separates reactants and products. The $\Delta H$ is negative, so the products are lower in energy than the reactants.

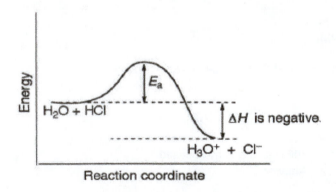

**6.12**   Increasing the concentration of the reactants increases the number of collisions, so the reaction rate increases. Increasing the temperature increases the reaction rate.

a. Increasing the concentration of $O_3$ increases the rate of the reaction.
b. Decreasing the concentration of NO decreases the rate of the reaction.
c. Increasing the temperature increases the rate of the reaction.
d. Decreasing the temperature decreases the rate of the reaction.

**6.13** Yes, $H_2SO_4$ is a catalyst because it speeds up the rate of the reaction. It does not affect the relative energies of reactants and products.

**6.14** The *forward* reaction proceeds from left to right as drawn.
The *reverse* reaction proceeds from right to left as drawn.

a. $2\ SO_2(g)\ +\ O_2(g)\ \longrightarrow\ 2\ SO_3(g)$     forward reaction

    $2\ SO_3(g)\ \longrightarrow\ 2\ SO_2(g)\ +\ O_2(g)$     reverse reaction

b. $N_2(g)\ +\ O_2(g)\ \longrightarrow\ 2\ NO(g)$     forward reaction

    $2\ NO(g)\ \longrightarrow\ N_2(g)\ +\ O_2(g)$     reverse reaction

c. $C_2H_4O_2\ +\ CH_4O\ \longrightarrow\ C_3H_6O_2\ +\ H_2O$     forward reaction

    $C_3H_6O_2\ +\ H_2O\ \longrightarrow\ C_2H_4O_2\ +\ CH_4O$     reverse reaction

**6.15** To write an expression for the equilibrium constant multiply the concentration of the products together and divide this number by the product of the concentrations of the reactants. Each concentration term is raised to a power equal to its coefficient in the balanced chemical equation.

a. $PCl_3(g)\ +\ Cl_2(g)\ \rightleftharpoons\ PCl_5(g)$    $K = \dfrac{[PCl_5]}{[PCl_3][Cl_2]}$

b. $2\ SO_2(g)\ +\ O_2(g)\ \rightleftharpoons\ 2\ SO_3(g)$    $K = \dfrac{[SO_3]^2}{[SO_2]^2[O_2]}$

c. $H_2(g)\ +\ Br_2(g)\ \rightleftharpoons\ 2\ HBr(g)$    $K = \dfrac{[HBr]^2}{[Br_2][H_2]}$

d. $CH_4(g)\ +\ 3\ Cl_2(g)\ \rightleftharpoons\ CHCl_3(g)\ +\ 3\ HCl(g)$    $K = \dfrac{[HCl]^3[CHCl_3]}{[CH_4][Cl_2]^3}$

**6.16** Since $K = 1$, there are equal amounts of **A** and **B** at equilibrium. **A** is represented with grey spheres, and **B** is black.

For $K = 1$:    $K = \dfrac{[B]}{[A]} = \dfrac{1}{1}$   equal amounts of A and B

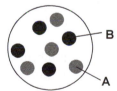

**6.17** When $K > 1$, the equilibrium favors the products.
When $K < 1$, the equilibrium favors the reactants.
When $K \approx 1$, both the reactants and products are present at equilibrium.

a. $5.0 \times 10^{-4}$, $K < 1$, reactants favored
b. $4.4 \times 10^{5}$, $K > 1$, products favored
c. $350$, $K > 1$, products favored
d. $0.35$, $K \approx 1$, reactants and products present

**6.18**  When $K < 1$, the reaction is endothermic. When $K > 1$, the reaction is exothermic.

a. $K < 1$, reactants favored, endothermic
b. $K > 1$, products favored, exothermic
c. $K > 1$, exothermic
d. $K < 1$, (slightly) endothermic

**6.19**  Write the expression for the equilibrium constant as in Answer 6.15.

a.  $K = \dfrac{[H_2S]^2}{[H_2]^2[S_2]}$

b. $K > 1$, products favored
c. $\Delta H$ would be negative since $K > 1$.
d. The products are lower in energy since $\Delta H$ is negative.
e. One cannot predict the rate of reaction without knowing the energy of activation.

**6.20**  First write an expression for $K$ using the balanced equation. Then substitute the equilibrium concentrations of all substances in the expression to calculate $K$.

a.  $K = \dfrac{[CO_2][H_2]}{[CO][H_2O]} = \dfrac{(0.0164)(0.0164)}{(0.0236)(0.00240)} = 4.75$  **Answer**

b.  $K = \dfrac{[N_2O_4]}{[NO_2]^2} = \dfrac{(1.26)}{(0.0760)^2} = \dfrac{1.26}{0.0760 \times 0.0760} = 218$  **Answer**

**6.21**  Use Le Châtelier's principle to predict the effect of a change in concentration on equilibrium. Adding more reactant or removing product drives the equilibrium to the right. Adding more product or removing reactant drives the equilibrium to the left.

a. increase $[H_2]$ (reactant) = right
b. increase [HCl] (product) = left
c. decrease $[Cl_2]$ (reactant) = left
d. decrease [HCl] (product) = right

**6.22**  When temperature is increased, the reaction that removes heat is favored. When temperature is decreased, the reaction that adds heat is favored.

a. The equilibrium shifts to the **right** when the temperature is increased.
b. The equilibrium shifts to the **left** when the temperature is decreased.

**6.23**  When temperature is increased, the reaction that removes heat is favored. When temperature is decreased, the reaction that adds heat is favored.

a. The equilibrium shifts to the **left** when the temperature is increased.
b. The equilibrium shifts to the **right** when the temperature is decreased.

**6.24** When pressure is increased, the equilibrium shifts in the direction that decreases the number of moles. When pressure is decreased, the equilibrium shifts in the direction that increases the number of moles.

a. The equilibrium shifts to the **right** when the pressure is increased.
b. The equilibrium shifts to the **left** when the pressure is decreased.

## Solutions to Odd-Numbered End-of-Chapter Problems

**6.25** Potential energy is stored energy while kinetic energy is the energy of motion. A stationary object on a hill has potential energy, but as it moves down the hill this potential energy is converted to kinetic energy.

**6.27** Use conversion factors to solve the problems.

a. 563 Cal/h × 1000 cal/1 Cal = 563,000 cal/h
b. 563 Cal/h = 563 kcal/h
c. 563 Cal/h × 1000 cal/1 Cal × 4.184 J/1 cal = $2.36 \times 10^6$ J/h
d. 563 Cal/h × 1000 cal/1 Cal × 4.184 J/1 cal × 1 kJ/1000 J = 2,360 kJ/h

**6.29** Use conversion factors to solve the problems.

a. 50 cal × 1 kcal/1000 cal = 0.05 kcal
b. 56 cal × 4.184 J/1 cal × 1 kJ/1000 J = 0.23 kJ
c. 0.96 kJ × 1 cal/4.184 J × 1000 J/1 kJ = 230 cal
d. 4,230 kJ × 1 cal/4.184 J × 1000 J/1 kJ = $1.01 \times 10^6$ cal

**6.31** Calculate the number of Calories as in Example 6.2.

Total Calories = 16 g fat × (9 Cal/1 g fat) + 7 g carb × (4 Cal/1 g carb) + 16 g protein × (4 Cal/1 g protein)

Total Calories = 144 Cal + 28 Cal + 64 Cal

Total Calories = 236 Cal, rounded to 200 Cal

**6.33** Use a conversion factor to solve the problem.

120 Cal × 1 g carbohydrate/4 Cal = 30 g carbohydrates

**6.35** Calculate the number of Calories as in Example 6.2 and then compare.

Total Calories
salmon = 5 g fat × (9 Cal/1 g fat) + 17 g protein × (4 Cal/1 g protein)

Total Calories = 45 Cal + 68 Cal

Total Calories
3 oz **salmon** = **113 Calories**

Total Calories
chicken =       3 g fat × (9 Cal/1 g fat)   +   20 g protein × (4 Cal/1 g protein)

Total Calories =       27 Cal       +       80 Cal

---

Total Calories
3 oz **chicken** =       **107 Calories**

If we take significant figures into account, both answers round to the same value, 100 Calories.

**6.37**   When energy is absorbed, the reaction is endothermic and $\Delta H$ is positive (+).
When energy is released, the reaction is exothermic and $\Delta H$ is negative (–).

**6.39**   In comparing bonds to atoms in the same group of the periodic table, bond dissociation energies and bond strength decrease down a column.

a.  $Cl_2$ has a stronger bond than $Br_2$ because Br is below Cl in the periodic table.
b.  $Cl_2$ has a stronger bond than $I_2$ because I is below Cl in the periodic table.
c.  HF has a stronger bond than HBr because Br is below F in the periodic table.

**6.41**   Use Table 6.3 to answer the questions.

a. $\Delta H$ is a negative value = exothermic.
b. The energy of the reactants is lower than the energy of the products = endothermic.
c. Energy is absorbed in the reaction = endothermic.
d. The bonds in the products are stronger than the bonds in the reactants = exothermic.

**6.43**   Use conversion factors to solve the problems.

$$C(s) \;+\; O_2(g) \longrightarrow CO_2(g) \qquad \Delta H = -94 \text{ kcal/mol}$$

a.   2.5 mol C   ×   $\dfrac{94 \text{ kcal}}{1 \text{ mol C}}$   =   240 kcal of energy released
**Answer**

b.   3.0 mol $O_2$   ×   $\dfrac{94 \text{ kcal}}{1 \text{ mol } O_2}$   =   280 kcal of energy released
**Answer**

c.   25.0 g C   ×   $\dfrac{1 \text{ mol C}}{12.01 \text{ g C}}$   ×   $\dfrac{94 \text{ kcal}}{1 \text{ mol C}}$   =   $2.0 \times 10^2$ kcal of energy released
**Answer**

**6.45**   Use conversion factors to solve the problems.

$$C_6H_{12}O_6(aq) \;+\; 6\,O_2(g) \rightleftharpoons 6\,CO_2(g) \;+\; 6\,H_2O(l) \qquad \Delta H = -678 \text{ kcal/mol}$$
glucose
(molar mass 180.2 g/mol)

a. stronger bonds in products

b. $4.00 \text{ mol } C_6H_{12}O_6 \times \dfrac{678 \text{ kcal}}{1 \text{ mol } C_6H_{12}O_6} = 2{,}710 \text{ kcal}$ **Answer**

c. $3.00 \text{ mol } O_2 \times \dfrac{678 \text{ kcal}}{6 \text{ mol } O_2} = 339 \text{ kcal}$ **Answer**

d. $10.0 \text{ g } C_6H_{12}O_6 \times \dfrac{1 \text{ mol } C_6H_{12}O_6}{180.2 \text{ g } C_6H_{12}O_6} \times \dfrac{678 \text{ kcal}}{1 \text{ mol } C_6H_{12}O_6} = 37.6 \text{ kcal}$ **Answer**

**6.47** The transition state is the unstable intermediate located at the top of the energy hill that separates reactants from products. The difference in energy between the reactants and the transition state is the energy of activation, symbolized by $E_a$.

**6.49** Label the points on the energy diagram.

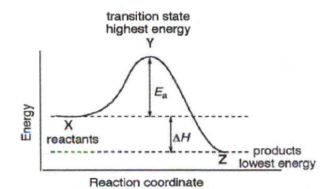

a. X
b. Z
c. Y
d. X, Y
e. X, Z
f. Y
g. Z

**6.51** Draw an energy diagram to fit each description.

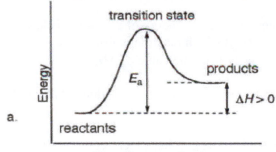

a.

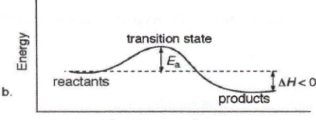

b.

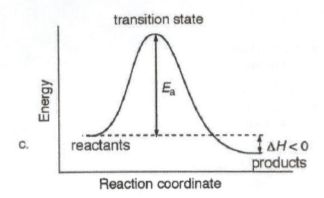

c.

**6.53**    Draw an energy diagram that fits the description.

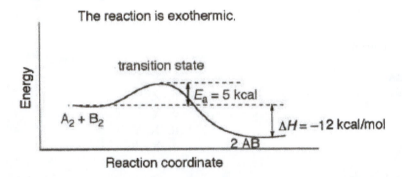

The reaction is exothermic.

**6.55**    Collision orientation affects the rate of reaction because reacting molecules must have the proper orientation for new bonds to form.

**6.57**    Increasing temperature increases the number of collisions, and thereby increases the reaction rate. Since the average kinetic energy of the colliding molecules is larger at higher temperatures, more collisions are effective at causing reaction.

**6.59**    a. The reaction with $E_a = 1$ kcal will proceed faster because the energy of activation is lower.
b. $K$ doesn't affect the reaction rate so we cannot predict which reaction is faster.
c. One cannot predict which reaction will proceed faster from the value of $\Delta H$.

**6.61**    Energy of activation (b) and temperature (c) affect the rate of reaction. $K$ (a) doesn't affect the reaction rate.

**6.63**    A catalyst increases the reaction rate (a) and lowers the $E_a$ (c). It has no effect on $\Delta H$ (b), $K$ (d), or the relative energies of the reactants and products (e).

**6.65**    The forward reaction proceeds from left to right as drawn and the reverse reaction proceeds from right to left as drawn.

**6.67**    When $K > 1$, the equilibrium favors the products.
When $K < 1$, the equilibrium favors the reactants.
When $K \approx 1$, both the reactants and products are present at equilibrium.
When $\Delta H$ is positive, the reactants are favored.
When $\Delta H$ is negative, the products are favored.

a. $K = 5.2 \times 10^3$, $K > 1$, the equilibrium favors the products.
b. $\Delta H = -27$ kcal/mol, the products are favored.
c. $K = 0.002$, $K < 1$, the equilibrium favors the reactants.
d. $\Delta H = +2$ kcal/mol, the reactants are favored.

**6.69**   $K > 1$ is associated with a negative value of $\Delta H$. A $K < 1$ means $\Delta H$ has a positive value.

**6.71**   **A** is represented with black spheres, and **B** is represented with grey spheres.

a.    b. 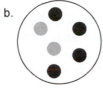   c.

[1] $K = 5$           [3] $K = 0.5$           [2] $K = 1$

**6.73**   Write the expression for the equilibrium constant as in Answer 6.15.

a.  $K = \dfrac{[NO_2]^2}{[NO]^2[O_2]}$     b.  $K = \dfrac{[HBr]^2[CH_2Br_2]}{[CH_4][Br_2]^2}$

**6.75**   Work backwards from the expression for the equilibrium constant to write the chemical equation.

a.  $K = \dfrac{[A_2]}{[A]^2}$   $2\,A \rightleftharpoons A_2$   b.  $K = \dfrac{[AB_3]^2}{[A_2][B_2]^3}$   $A_2 + 3\,B_2 \rightleftharpoons 2\,AB_3$

**6.77**

a.  $K = \dfrac{[Br_2][H_2]}{[HBr]^2}$

b. The reactants are favored at equilibrium since $K < 1$.
c. $\Delta H$ is predicted to be positive since $K < 1$.
d. The reactants are lower in energy since the reactants are favored at equilibrium.
e. You can't predict the reaction rate from the value of $K$.

**6.79**

a.  $K = \dfrac{[CO_2][H_2]}{[CO][H_2O]}$     b.  $\dfrac{(0.15)(0.30)}{(0.090)(0.12)} = K = 4.2$

**6.81**   Use Le Châtelier's principle to predict the effect of a change in concentration on equilibrium. Adding more reactant or removing product drives the equilibrium to the right. Adding more product or removing reactant drives the equilibrium to the left.

a.  When $O_2$ is increased, it drives the equilibrium to the right: increases NO and decreases $N_2$.
b.  When NO is increased, it drives the equilibrium to the left: increases $N_2$ and $O_2$.

**6.83**   Use Le Châtelier's principle to predict the effect of each change.

a. decrease $[O_3]$, shift to right          d. decrease temperature, shift to left
b. decrease $[O_2]$, shift to left           e. add a catalyst, no change
c. increase $[O_3]$, shift to left           f. increase pressure, shift to right

**6.85** Use Le Châtelier's principle to predict the effect of each change.

    a. increase $[C_2H_4]$, shift to right          d. decrease pressure, shift to left
    b. decrease $[Cl_2]$, shift to left            e. increase temperature, shift to left
    c. decrease $[C_2H_4Cl_2]$, shift to right     f. decrease temperature, shift to right

**6.87**

    a.   $K = \dfrac{[C_2H_6]}{[H_2][C_2H_4]}$     b. The reactants are higher in energy when $\Delta H$ is negative.

                                         c. $K > 1$ since $\Delta H$ is negative.

               d. 20.0 g ethylene $\times$ 1 mol ethylene/28.05 g $\times$ 28 kcal/1 mol ethylene = 20. kcal

               e. If ethylene (reactant) concentration is increased, the reaction rate will
                  increase.

               f. 1,2,4: favor shift to right; 3: favors shift to left; 5: no change, but reduces
                  the rate of reaction.

**6.89** Lactase is an enzyme that converts lactose, a naturally occurring sugar in dairy products, into the two simple sugars, glucose and galactose.

**6.91** Use conversion factors to solve the problem.

$$2000 \text{ mL} \times \frac{5 \text{ g glucose}}{100 \text{ mL}} \times \frac{4 \text{ Calories}}{1 \text{ g glucose}} = 400 \text{ Calories}$$

**6.93**

Total Calories =     29 g fat $\times$ (9 Cal/1 g fat)    +    34 g carb $\times$ (4 Cal/1 g carb)    + 32 g protein $\times$ (4 Cal/1 g protein)

| Total Calories = | 261 Cal | + | 136 Cal | + | 128 Cal |
|---|---|---|---|---|---|

Total Calories =    525 Calories

     525 Cal $\times$ 1 h/280 Cal = 1.9 h rounded to 2 h

**6.95** Use conversion factors to solve the problem.

$$\frac{531 \text{ kcal}}{1 \text{ mol } C_3H_8} \times \frac{1 \text{ mol } C_3H_8}{44.09 \text{ g } C_3H_8} = 12.0 \text{ kcal/g propane} \qquad \frac{688 \text{ kcal}}{1 \text{ mol } C_4H_{10}} \times \frac{1 \text{ mol } C_4H_{10}}{58.12 \text{ g } C_4H_{10}} = 11.8 \text{ kcal/g butane}$$

**6.97** Use conversion factors to solve the problem.

$$1 \text{ gal gas} \times \frac{4 \text{ qt}}{1 \text{ gal}} \times \frac{946 \text{ mL}}{1 \text{ qt}} \times \frac{0.700 \text{ g}}{1 \text{ mL}} \times \frac{1 \text{ mol } C_8H_{18}}{114.3 \text{ g } C_8H_{18}} \times \frac{1303 \text{ kcal}}{1 \text{ mol}} = 30,200 \text{ kcal}$$

# Chapter 7 Gases, Liquids, and Solids

## Chapter Review

**[1] What is pressure and what units are used to measure it? (7.2)**
- Pressure is the force per unit area. The pressure of a gas is the force exerted when gas particles strike a surface. Pressure is measured by a barometer and recorded in atmospheres (atm), millimeters of mercury (mm Hg), or pounds per square inch (psi).
- 1 atm = 760 mm Hg = 14.7 psi.

$$\text{Pressure} \quad = \quad \frac{\text{Force}}{\text{Area}} \quad = \quad \frac{F}{A}$$

**[2] What are gas laws and how are they used to describe the relationship between the pressure, volume, and temperature of a gas? (7.3)**
- Because gas particles are far apart and behave independently, a set of gas laws describes the behavior of all gases regardless of their identity. Three gas laws—**Boyle's law, Charles's law,** and **Gay–Lussac's law**—describe the relationship between the pressure, volume, and temperature of a gas. These gas laws are summarized in "Key Equations—The Gas Laws" at the end of the Chapter Review.
- For a constant amount of gas, the following relationships exist.
  - [1] The pressure and volume of a gas are inversely related, so increasing the pressure decreases the volume at constant temperature.
  - [2] The volume of a gas is proportional to its Kelvin temperature, so increasing the temperature increases the volume at constant pressure.
  - [3] The pressure of a gas is proportional to its Kelvin temperature, so increasing the temperature increases the pressure at constant volume.

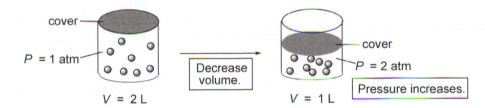

**[3] Describe the relationship between the volume and number of moles of a gas. (7.4)**
- **Avogadro's law** states that when temperature and pressure are held constant, the volume of a gas is proportional to its number of moles.
- One mole of any gas has the same volume, the standard molar volume of 22.4 L, at 1 atm and 273 K (STP).

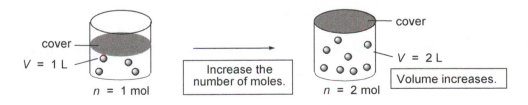

**[4] What is the ideal gas law? (7.5)**

- The **ideal gas law** is an equation that relates the pressure ($P$), volume ($V$), temperature ($T$), and number of moles ($n$) of a gas; $PV = nRT$, where $R$ is the universal gas constant. The ideal gas law can be used to calculate any one of the four variables, as long as the other three variables are known.

$$\boxed{PV = nRT}$$

Ideal gas law

For atm: $R = 0.0821 \dfrac{L \cdot atm}{mol \cdot K}$

For mm Hg: $R = 62.4 \dfrac{L \cdot mm\ Hg}{mol \cdot K}$

**[5] What is Dalton's law and how is it used to relate partial pressures and the total pressure of a gas mixture? (7.6)**

- **Dalton's law** states that the total pressure of a gas mixture is the sum of the partial pressures of its component gases. The partial pressure is the pressure exerted by each component of a mixture.

$$P_{total} = P_A + P_B + P_C$$

total pressure        partial pressures of **A**, **B**, and **C**

**[6] What types of intermolecular forces exist and how do they determine a compound's boiling point and melting point? (7.7)**

- Intermolecular forces are the forces of attraction between molecules. Three types of intermolecular forces exist in covalent compounds.
  - **London dispersion forces** are due to momentary changes in electron density in a molecule.
  - **Dipole–dipole interactions** are due to permanent dipoles.

- **Hydrogen bonding**, the strongest intermolecular force, results when a H atom bonded to an O, N, or F is attracted to an O, N, or F atom in another molecule.

- The stronger the intermolecular forces, the higher the boiling point and melting point of a compound.

**[7] Describe three features of the liquid state—vapor pressure, viscosity, and surface tension. (7.8)**

- **Vapor pressure** is the pressure exerted by gas molecules in equilibrium with the liquid phase. Vapor pressure increases with increasing temperature. The higher the vapor pressure at a given temperature, the lower the boiling point of a compound.

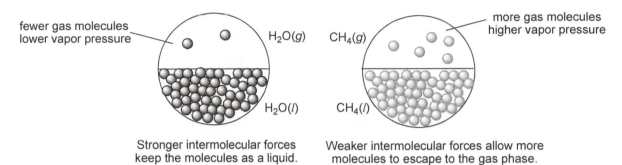

fewer gas molecules
lower vapor pressure

$H_2O(g)$    $CH_4(g)$

more gas molecules
higher vapor pressure

$H_2O(l)$    $CH_4(l)$

Stronger intermolecular forces
keep the molecules as a liquid.

Weaker intermolecular forces allow more
molecules to escape to the gas phase.

- **Viscosity** measures a liquid's resistance to flow. More viscous compounds tend to have stronger intermolecular forces or they have high molecular weights.
- **Surface tension** measures a liquid's resistance to spreading out. The stronger the intermolecular forces, the higher the surface tension.

**[8] Describe the features of different types of solids. (7.9)**

- Solids can be amorphous or crystalline. An **amorphous solid** has no regular arrangement of particles. A **crystalline solid** has a regular arrangement of particles in a repeating pattern. There are four types of crystalline solids: **Ionic solids** are composed of ions. **Molecular solids** are composed of individual molecules. **Network solids** are composed of vast arrays of covalently bonded atoms in a regular three-dimensional arrangement. **Metallic solids** are composed of metal cations with a cloud of delocalized electrons.

**[9] Describe the energy changes that accompany changes of state. (7.10)**

- A phase change converts one state to another. Energy is absorbed when a more organized state is converted to a less organized state. Thus, energy is absorbed when a solid melts to form a liquid, or when a liquid vaporizes to form a gas.
- Energy is released when a less organized state is converted to a more organized state. Thus, energy is released when a gas condenses to form a liquid, or a liquid freezes to form a solid.
- The **heat of fusion** is the energy needed to melt 1 g of a substance, while the **heat of vaporization** is the energy needed to vaporize 1 g of a substance.

**[10] What changes are depicted on heating and cooling curves? (7.11)**

- A heating curve shows how the temperature of a substance changes as heat is added. Diagonal lines show the temperature increase of a single phase. Horizontal lines correspond to phase changes—solid to liquid or liquid to gas.
- A cooling curve shows how the temperature of a substance changes as heat is removed. Diagonal lines show the temperature decrease of a single phase. Horizontal lines correspond to phase changes—gas to liquid or liquid to solid.

## Key Equations—The Gas Laws

| Name | Equation | Variables Related | Constant Terms |
|------|----------|-------------------|----------------|
| Boyle's law | $P_1V_1 = P_2V_2$ | $P, V$ | $T, n$ |
| Charles's law | $\dfrac{V_1}{T_1} = \dfrac{V_2}{T_2}$ | $V, T$ | $P, n$ |
| Gay–Lussac's law | $\dfrac{P_1}{T_1} = \dfrac{P_2}{T_2}$ | $P, T$ | $V, n$ |
| Combined gas law | $\dfrac{P_1V_1}{T_1} = \dfrac{P_2V_2}{T_2}$ | $P, V, T$ | $n$ |
| Avogadro's law | $\dfrac{V_1}{n_1} = \dfrac{V_2}{n_2}$ | $V, n$ | $P, T$ |
| Ideal gas law | $PV = nRT$ | $P, V, T, n$ | $R$ |

## Problem Solving

## [1] Gases and Pressure (7.2)

**Example 7.1** Convert 200. psi to:
       a. atmospheres          b. mm Hg

**Analysis**
To solve each part, set up conversion factors that relate the two units under consideration. Use conversion factors that place the unwanted unit, psi, in the denominator to cancel.

In part (a), the conversion factor must relate psi and atm:

psi–atm conversion factor

$$\frac{1 \text{ atm}}{14.7 \text{ psi}} \leftarrow \text{unwanted unit}$$

In part (b), the conversion factor must relate psi and mm Hg:

psi–mm Hg conversion factor

$$\frac{760. \text{ mm Hg}}{14.7 \text{ psi}} \leftarrow \text{unwanted unit}$$

**Solution**
a. Convert the original unit (200. psi) to the desired unit (atm) using the conversion factor:

$$200. \text{ psi} \times \frac{1 \text{ atm}}{14.7 \text{ psi}} = 13.6 \text{ atm}$$
**Answer**

Psi cancels.

b. Convert the original unit (200. psi) to the desired unit (mm Hg) using the conversion factor:

$$200. \text{ psi} \times \frac{760. \text{ mm Hg}}{14.7 \text{ psi}} = 10,300 \text{ mm Hg}$$
**Answer**

Psi cancels.

# [2] Gas Laws (7.3)

**Example 7.2** A tank of compressed air contains 20. L of gas at 204 atm pressure. What volume does this gas occupy at 1.0 atm?

**Analysis**

Boyle's law can be used to solve this problem since an initial pressure and volume ($P_1$ and $V_1$) and a final pressure ($P_2$) are known, and a final volume ($V_2$) must be determined.

**Solution**

**[1] Identify the known quantities and the desired quantity.**

$$P_1 = 204 \text{ atm} \qquad P_2 = 1.0 \text{ atm}$$
$$V_1 = 20. \text{ L} \qquad\qquad\qquad V_2 = ?$$

known quantities          desired quantity

---

**[2] Write the equation and rearrange it to isolate the desired quantity, $V_2$, on one side.**

$$P_1V_1 = P_2V_2 \qquad \text{Solve for } V_2 \text{ by dividing both sides by } P_2.$$

$$\frac{P_1V_1}{P_2} = V_2$$

---

**[3] Solve the problem.**

- Substitute the three known quantities in the equation and solve for $V_2$.

$$V_2 = \frac{P_1V_1}{P_2} = \frac{(204 \text{ atm})(20. \text{ L})}{1.0 \text{ atm}} = 4{,}080 \text{ rounded to } 4{,}100 \text{ L}$$

Atm cancels.          **Answer**

- Thus, the volume increased as the pressure decreased.

---

**Example 7.3** A balloon that contains 0.75 L of air at 28 °C is cooled to –98 °C. What volume does the balloon now occupy?

**Analysis**

Since this question deals with volume and temperature, Charles's law is used to determine a final volume because three quantities are known—the initial volume and temperature ($V_1$ and $T_1$), and the final temperature ($T_2$).

**Solution**

**[1] Identify the known quantities and the desired quantity.**

$$V_1 = 0.75 \text{ L}$$
$$T_1 = 28 \text{ °C} \qquad T_2 = -98 \text{ °C} \qquad V_2 = ?$$

known quantities          desired quantity

- Both temperatures must be converted to Kelvin temperatures using the equation
  **$K = °C + 273$.**
- $T_1 = 28 °C + 273 = 301 K$
- $T_2 = -98 °C + 273 = 175 K$

---

**[2] Write the equation and rearrange it to isolate the desired quantity, $V_2$, on one side.**
- Use Charles's law.

$$\frac{V_1}{T_1} = \frac{V_2}{T_2} \quad \text{Solve for } V_2 \text{ by multiplying both sides by } T_2.$$

$$\frac{V_1 T_2}{T_1} = V_2$$

---

**[3] Solve the problem.**
- Substitute the three known quantities in the equation and solve for $V_2$.

$$V_2 = \frac{V_1 T_2}{T_1} = \frac{(0.75 \text{ L})(175 \text{ K})}{301 \text{ K}} = 0.44 \text{ L}$$

Answer

Kelvins cancel.

- Since temperature has decreased, the volume of gas must decrease as well.

---

**Example 7.4** A balloon contains 2.0 L of helium at 25 °C and 760 mm Hg. What is the volume of the balloon when it ascends to an altitude where the temperature is –40. °C and the pressure is 540 mm Hg?

**Analysis**
Since this question deals with pressure, volume, and temperature, the combined gas law is used to determine a final volume ($V_2$) because five quantities are known—the initial pressure, volume, and temperature ($P_1$, $V_1$, and $T_1$), and the final pressure and temperature ($P_2$ and $T_2$).

**Solution**
**[1] Identify the known quantities and the desired quantity.**

$P_1 = 760 \text{ mm Hg}$     $P_2 = 540 \text{ mm Hg}$
$T_1 = 25 °C$     $T_2 = -40. °C$
$V_1 = 2.0 \text{ L}$                                  $V_2 = ?$

known quantities                  desired quantity

- Both temperatures must be converted to Kelvin temperatures.
- $T_1 = K = °C + 273 = 25 °C + 273 = 298 K$
- $T_2 = K = °C + 273 = -40. °C + 273 = 233 K$

---

**[2] Write the equation and rearrange it to isolate the desired quantity, $V_2$, on one side.**
- Use the combined gas law.

$$\frac{P_1V_1}{T_1} = \frac{P_2V_2}{T_2} \qquad \text{Solve for } V_2 \text{ by multiplying both sides by } \frac{T_2}{P_2}.$$

$$\frac{P_1V_1T_2}{T_1P_2} = V_2$$

**[3] Solve the problem.**

- Substitute the five known quantities in the equation and solve for $V_2$.

$$V_2 = \frac{P_1V_1T_2}{T_1P_2} = \frac{(760 \text{ mm Hg})(2.0 \text{ L})(233 \text{ K})}{(298 \text{ K})(540 \text{ mm Hg})} = \begin{array}{c} 2.2 \text{ L} \\ \textbf{Answer} \end{array}$$

Kelvins and mm Hg cancel.

## [3] Avogadro's Law (7.4)

**Example 7.5** The lungs of an average dog hold 0.15 mol of air in a volume of 3.6 L. How many moles of air do the lungs of an average cat hold if the volume is 1.2 L? Assume the temperature and pressure are constant.

**Analysis**

This question deals with volume and number of moles, so Avogadro's law is used to determine a final number of moles when three quantities are known—the initial volume and number of moles ($V_1$ and $n_1$), and the final volume ($V_2$).

**Solution**

**[1] Identify the known quantities and the desired quantity.**

$$V_1 = 3.6 \text{ L} \qquad V_2 = 1.2 \text{ L}$$
$$n_1 = 0.15 \text{ mol} \qquad\qquad\qquad n_2 = ?$$

known quantities          desired quantity

**[2] Write the equation and rearrange it to isolate the desired quantity, $n_2$, on one side.**

- Use Avogadro's law. To solve for $n_2$, we must invert the numerator and denominator on *both* sides of the equation, and then multiply by $V_2$.

$$\boxed{\frac{V_1}{n_1} = \frac{V_2}{n_2}} \xrightarrow{\text{Switch } V \text{ and } n \text{ on both sides.}} \frac{n_1}{V_1} = \frac{n_2}{V_2} \qquad \text{Solve for } n_2 \text{ by multiplying both sides by } V_2.$$

$$\frac{n_1V_2}{V_1} = n_2$$

**[3] Solve the problem.**

- Substitute the three known quantities in the equation and solve for $n_2$.

$$n_2 \quad = \quad \frac{n_1 V_2}{V_1} \quad = \quad \frac{(0.15 \text{ mol})(1.2 \cancel{L})}{(3.6 \cancel{L})} \quad = \quad \begin{array}{c} 0.050 \text{ mol} \\ \textbf{Answer} \end{array}$$

Liters cancel.

## [4] The Ideal Gas Law (7.5)

**Example 7.6** What volume does 15.0 g of $O_2$ occupy at 1.00 atm and 32 °C?

**Analysis**
Use the ideal gas law to calculate $V$, since $P$ and $T$ are known and $n$ can be determined by using the molar mass of $O_2$ (32.00 g/mol).

**Solution**
**[1] Identify the known quantities and the desired quantity.**

$P$ = 1.00 atm
$T$ = 32 °C      15.0 g $O_2$          $V$ = ? L
       known quantities          desired quantity

**[2] Convert all values to proper units and choose the value of $R$ that contains these units.**
- Convert °C to K. K = °C + 273 = 32 °C + 273 = 305 K.
- Use the value of $R$ with atm since the pressure is given in atm; that is, $R = 0.0821$ L•atm/(mol•K).
- Convert the number of grams of $O_2$ to the number of moles of $O_2$ using the molar mass (32.00 g/mol).

$$15.0 \, \cancel{g} \, O_2 \quad \times \quad \frac{1 \text{ mol } O_2}{32.00 \, \cancel{g} \, O_2} \quad = \quad 0.469 \text{ mol } O_2$$

molar mass conversion factor

Grams cancel.

**[3] Write the equation and rearrange it to isolate the desired quantity, $V$, on one side.**
- Use the ideal gas law and solve for $V$ by dividing both sides by $P$.

$$PV \quad = \quad nRT \qquad \text{Solve for } V \text{ by dividing both sides by } P.$$

$$V \quad = \quad \frac{nRT}{P}$$

**[4] Solve the problem.**
- Substitute the three known quantities in the equation and solve for $V$.

$$V \;=\; \frac{nRT}{P} \;=\; \frac{(0.469 \text{ mol})\left(0.0821 \;\dfrac{L \cdot atm}{mol \cdot K}\right)(305 \text{ K})}{1.00 \text{ atm}} \;=\; \begin{array}{c} 11.7 \text{ L} \\ \textbf{Answer} \end{array}$$

---

## [5] Dalton's Law and Partial Pressures (7.6)

**Example 7.7** A sample of air contains four gases with the following partial pressures: $N_2$ (530. mm Hg), $O_2$ (129 mm Hg), $CO_2$ (48 mm Hg), and $H_2O$ (52 mm Hg). What is the total pressure of the sample?

**Analysis**
Using Dalton's law, the total pressure is the sum of the partial pressures.

**Solution**
Adding up the four partial pressures gives the total:

$$530. + 129 + 48 + 52 = 759 \text{ mm Hg (total pressure)}$$

---

**Example 7.8** A mixture of gas contains 31% $O_2$, 68% $N_2$, and 1% $CO_2$ by volume. What is the partial pressure of each gas at sea level, where the total pressure is 760 mm Hg?

**Analysis**
Convert each percent to a decimal by moving the decimal point two places to the left. Multiply each decimal by the total pressure to obtain the partial pressure for each component.

**Solution**

|  |  | Partial pressure |
|---|---|---|
| Fraction $O_2$: 31% = 0.31 | 0.31 × 760 mm Hg = | 240 mm Hg ($O_2$) |
| Fraction $N_2$: 68% = 0.68 | 0.68 × 760 mm Hg = | 520 mm Hg ($N_2$) |
| Fraction $CO_2$: 1% = 0.01 | 0.01 × 760 mm Hg = | 8 mm Hg ($CO_2$) |
|  |  | 768 rounded to 770 mm Hg |

---

## [6] Intermolecular Forces, Boiling Point, and Melting Point (7.7)

**Example 7.9** What types of intermolecular forces are present in each compound?
         a. HBr          b. $CH_4$          c. $H_2O$

**Analysis**
- London dispersion forces are present in all covalent compounds.
- Dipole–dipole interactions are present only in polar compounds with a permanent dipole.
- Hydrogen bonding occurs only in compounds that contain an O–H, N–H, or H–F bond.

## Solution

a.

- HBr has London forces like all covalent compounds.
- HBr has a polar bond so it exhibits dipole–dipole interactions.
- HBr has no H atom on an O, N, or F, so it has no intermolecular hydrogen bonding.

b.

nonpolar molecule

- $CH_4$ is a nonpolar molecule since it has only nonpolar C–H bonds. Thus, it exhibits only London forces.

c.

net dipole

- $H_2O$ has London forces like all covalent compounds.
- $H_2O$ has a net dipole from its two polar bonds (Section 4.8), so it exhibits dipole–dipole interactions.
- $H_2O$ has a H atom bonded to O, so it exhibits intermolecular hydrogen bonding.

---

## [7] The Liquid State (7.8)

**Example 7.10** Using the given boiling points, predict which compound has the higher vapor pressure at a given temperature: ethanol ($C_2H_6O$, bp 78 °C) or 1-propanol ($C_3H_8O$, bp 97 °C).

## Analysis
A higher boiling point means a compound has a lower vapor pressure and stronger forces.

## Solution
Since ethanol has a lower boiling point than 1-propanol, ethanol has a higher vapor pressure, and 1-propanol has a lower vapor pressure at a given temperature.

---

## Self-Test

**[1] Fill in the blank with one of the terms listed below.**

| | | |
|---|---|---|
| Amorphous solid (7.9) | Freezing (7.10) | Surface tension (7.8) |
| Boiling point (7.7) | London dispersion forces (7.7) | Universal gas constant (7.5) |
| Condensation (7.8, 7.10) | Melting (7.10) | Vapor pressure (7.8) |
| Crystalline solid (7.9) | Pressure (7.2) | Vaporization (7.10) |
| Dipole–dipole interactions (7.7) | Standard molar volume (7.4) | Viscosity (7.8) |

1. Converting a liquid to a gas is called _____.
2. _____ is the pressure exerted by gas molecules in equilibrium with the liquid phase.
3. _____ are very weak interactions due to the momentary changes in electron density in a molecule.
4. The _____ of a compound is the temperature at which a liquid is converted to the gas phase.

5.  Converting a solid to a liquid is called _____.
6.  The product of pressure and volume divided by the product of moles and Kelvin temperature is a constant, called the _____ and symbolized by $R$.
7.  _____ is a measure of a fluid's resistance to flow freely.
8.  At STP, one mole of any gas has the same volume, 22.4 L, called the _____.
9.  _____ converts a gas to a liquid.
10. A _____ has a regular arrangement of particles—atoms, molecules, or ions—with a repeating structure.
11. An _____ has no regular arrangement of its closely packed particles.
12. _____ are the attractive forces between the permanent dipoles of two polar molecules.
13. _____ is the force exerted per unit area.
14. _____ converts a liquid to a solid.
15. _____ is a measure of the resistance of a liquid to spread out.

**[2] Fill in each blank with one of the terms listed below.**

  a. pressure  b. volume  c. temperature

16. Boyle's law relates _____ and _____.
17. Charles's law relates _____ and _____.
18. Gay–Lussac's law relates _____ and _____.

**[3] Pick the appropriate type of solid for each compound.**

  a. ionic solid  c. network solid
  b. molecular solid  d. metallic solid

19. graphite
20. silver
21. NaBr
22. $H_2O$

**[4] Fill in each blank with one of the terms listed below.**

  a. vaporization  b. sublimation  c. deposition

23. Condensation is the opposite of _____.
24. When a gas is converted directly to a solid, this is called _____.
25. When a solid forms a gas without passing through a liquid phase, this is called _____.

## Answers to Self-Test

| | | | | |
|---|---|---|---|---|
| 1. vaporization | 6. universal gas constant | 11. amorphous solid | 16. a and b | 21. a |
| 2. Vapor pressure | 7. Viscosity | 12. Dipole–dipole | 17. b and c | 22. b |
| 3. London | 8. standard molar volume |    interactions | 18. a and c | 23. a |
|    dispersion forces | 9. Condensation | 13. Pressure | 19. c | 24. c |
| 4. boiling point | 10. crystalline solid | 14. Freezing | 20. d | 25. b |
| 5. melting | | 15. Surface tension | | |

## Solutions to In-Chapter Problems

**7.1** Use Table 7.1 to compare the features of different states of methanol.

|         | a. Density | b. Intermolecular Spacing | c. Intermolecular Attraction |
|---------|------------|---------------------------|------------------------------|
| Gas     | Lowest     | Greatest                  | Lowest                       |
| Liquid  | Higher     | Smaller                   | Higher                       |
| Solid   | Highest    | Smallest                  | Highest                      |

**7.2** Use conversion factors to convert mm Hg to atm and psi.

a. $630 \text{ mm Hg} \times \dfrac{1 \text{ atm}}{760. \text{ mm Hg}} = 0.83 \text{ atm}$   **Answer**

(mm Hg cancels.)

b. $630 \text{ mm Hg} \times \dfrac{14.7 \text{ psi}}{760. \text{ mm Hg}} = 12 \text{ psi}$   **Answer**

(mm Hg cancels.)

**7.3** Use conversion factors to solve the problems.

a. $3.0 \text{ atm} \times \dfrac{760. \text{ mm Hg}}{1 \text{ atm}} = 2{,}300 \text{ mm Hg}$   **Answer**

(Atm cancels.)

c. $424 \text{ mm Hg} \times \dfrac{1 \text{ atm}}{760. \text{ mm Hg}} = 0.558 \text{ atm}$   **Answer**

(mm Hg cancels.)

b. $720 \text{ mm Hg} \times \dfrac{14.7 \text{ psi}}{760. \text{ mm Hg}} = 14 \text{ psi}$   **Answer**

(mm Hg cancels.)

**7.4** Use conversion factors to solve the problems.

$120 \text{ mm Hg} \times (1 \text{ atm}/760 \text{ mm Hg}) = 0.16 \text{ atm}$      $80 \text{ mm Hg} \times (1 \text{ atm}/760 \text{ mm Hg}) = 0.1 \text{ atm}$

**7.5** Use Boyle's law to solve the problems as in Example 7.2.

a. $V_2 = \dfrac{P_1 V_1}{P_2} = \dfrac{(4.0 \text{ atm})(2.0 \text{ L})}{5.0 \text{ atm}} = 1.6 \text{ L}$   **Answer**

(Atm cancels.)

b. $V_2 = \dfrac{P_1 V_1}{P_2} = \dfrac{(4.0 \text{ atm})(2.0 \text{ L})}{2.5 \text{ atm}} = 3.2 \text{ L}$   **Answer**

(Atm cancels.)

c. $V_2 = \dfrac{P_1 V_1}{P_2} = \dfrac{(4.0 \text{ atm})(2.0 \text{ L})}{10.0 \text{ atm}} = 0.80 \text{ L}$   **Answer**

(Atm cancels.)

d. $4.0 \text{ atm} \times \dfrac{760. \text{ mm Hg}}{1 \text{ atm}} = 3.0 \times 10^3 \text{ mm Hg}$   $V_2 = \dfrac{P_1 V_1}{P_2} = \dfrac{(3.0 \times 10^3 \text{ mm Hg})(2.0 \text{ L})}{380 \text{ mm Hg}} = 16 \text{ L}$   **Answer**

(mm Hg cancels.)

**7.6** Use Boyle's law to solve the problems as in Example 7.2.

a. $P_2 = \dfrac{P_1V_1}{V_2} = \dfrac{(0.50 \text{ atm})(15.0 \text{ mL})}{30.0 \text{ mL}} = $    0.25 atm   **Answer**

mL cancels.

b. $P_2 = \dfrac{P_1V_1}{V_2} = \dfrac{(0.50 \text{ atm})(15.0 \text{ mL})}{5.0 \text{ mL}} = $    1.5 atm   **Answer**

mL cancels.

c. $P_2 = \dfrac{P_1V_1}{V_2} = \dfrac{(0.50 \text{ atm})(15.0 \text{ mL})}{100. \text{ mL}} = $    0.075 atm   **Answer**

mL cancels.

d. $1.0 \text{ L} = 1.0 \times 10^3 \text{ mL}$    $P_2 = \dfrac{P_1V_1}{V_2} = \dfrac{(0.50 \text{ atm})(15.0 \text{ mL})}{1.0 \times 10^3 \text{ mL}} = $    0.0075 atm   **Answer**

mL cancels.

**7.7** Use Charles's law to solve the problem as in Example 7.3. Remember to convert temperature to K.

$K = °C + 273$
$T_1 = 37 °C + 273 = 310. \text{ K}$
$T_2 = 0.0 °C + 273 = 273 \text{ K}$

$V_2 = \dfrac{V_1T_2}{T_1} = \dfrac{(0.50 \text{ L})(273 \text{ K})}{310. \text{ K}} = $    0.44 L   **Answer**

Kelvins cancel.

**7.8** Use Charles's law to solve the problems as in Example 7.3.

a. $V_2 = \dfrac{V_1T_2}{T_1} = \dfrac{(25.0 \text{ L})(450 \text{ K})}{45 \text{ K}} = $    250 L   **Answer**

Kelvins cancel.

b. $K = °C + 273$
$T_1 = 400. °C + 273 = 673 \text{ K}$
$T_2 = 50. °C + 273 = 323 \text{ K}$

$V_2 = \dfrac{V_1T_2}{T_1} = \dfrac{(50.0 \text{ mL})(323 \text{ K})}{673 \text{ K}} = $    24 mL   **Answer**

Kelvins cancel.

**7.9** Use Gay–Lussac's law to solve the problem.

$$\dfrac{P_1}{T_1} = \dfrac{P_2}{T_2}$$

$$\dfrac{P_2T_1}{P_1} = T_2$$

$K = {}^\circ C + 273$
$T_1 = 100.\ {}^\circ C + 273 = 373\ K$

$T_2 = \dfrac{P_2 T_1}{P_1} = \dfrac{(1.05\ atm)(373\ K)}{1.00\ atm} = 392\ K\ or\ 119\ {}^\circ C$

**Answer**

Atm cancels.

**7.10** Use Gay–Lussac's law to solve the problem as in Sample Problem 7.4. $T_1 = 25\ {}^\circ C + 273 = 298\ K$.

a. $P_2 = \dfrac{P_1 T_2}{T_1} = \dfrac{(1.00\ atm)(310.\ K)}{298\ K} = 1.04\ atm$

**Answer**

Kelvins cancel.

b. $P_2 = \dfrac{P_1 T_2}{T_1} = \dfrac{(1.00\ atm)(150.\ K)}{298\ K} = 0.503\ atm$

**Answer**

Kelvins cancel.

c. $P_2 = \dfrac{P_1 T_2}{T_1} = \dfrac{(1.00\ atm)(323\ K)}{298\ K} = 1.08\ atm$

**Answer**

Kelvins cancel.

d. $P_2 = \dfrac{P_1 T_2}{T_1} = \dfrac{(1.00\ atm)(473\ K)}{298\ K} = 1.59\ atm$

**Answer**

Kelvins cancel.

**7.11** As a sealed container is heated in the microwave, the gases inside expand and increase the container's internal pressure, eventually popping the lid off.

**7.12** Use the combined gas law to solve the problem as in Example 7.4.

$$\dfrac{P_1 V_1}{T_1} = \dfrac{P_2 V_2}{T_2}$$

$$\dfrac{P_1 V_1 T_2}{T_1 V_2} = P_2$$

$$P_2 = \dfrac{P_1 V_1 T_2}{T_1 V_2} = \dfrac{(750\ mm\ Hg)(1.0\ L)(233\ K)}{(298\ K)(2.0\ L)} = 290\ mm\ Hg$$

**Answer**

**7.13** Use Avogadro's law to solve the problems as in Example 7.5.

$$\dfrac{V_1}{n_1} = \dfrac{V_2}{n_2} \qquad \dfrac{n_2 V_1}{n_1} = V_2 \quad \text{Solve for } V_2 \text{ by multiplying both sides by } n_2.$$

a. $V_2 = \dfrac{n_2 V_1}{n_1} = \dfrac{(2.5\ mol)(3.5\ L)}{(5.0\ mol)} = 1.8\ L$

**Answer**

Mol cancel.

b. $V_2 = \dfrac{n_2 V_1}{n_1} = \dfrac{(3.65\ \text{mol})(3.5\ \text{L})}{(5.0\ \text{mol})} = 2.6\ \text{L}$ **Answer**

Mol cancel.

c. $V_2 = \dfrac{n_2 V_1}{n_1} = \dfrac{(21.5\ \text{mol})(3.5\ \text{L})}{(5.0\ \text{mol})} = 15\ \text{L}$ **Answer**

Mol cancel.

**7.14** Convert moles to volume at STP as in Sample Problem 7.7.

a. $4.5\ \text{mol}\ O_2 \quad \times \quad \dfrac{22.4\ \text{L}}{1\ \text{mol}} = 1.0 \times 10^2\ \text{L}$ **Answer**

b. $0.35\ \text{mol}\ O_2 \quad \times \quad \dfrac{22.4\ \text{L}}{1\ \text{mol}} = 7.8\ \text{L}$ **Answer**

c. $18.0\ \text{g}\ O_2 \quad \times \quad \dfrac{1\ \text{mol}}{32.00\ \text{g}} \quad \times \quad \dfrac{22.4\ \text{L}}{1\ \text{mol}} = 12.6\ \text{L}$ **Answer**

**7.15** Use conversion factors to solve the problems.

a. $1.5\ \text{L} \quad \times \quad \dfrac{1\ \text{mol}}{22.4\ \text{L}} = 0.067\ \text{mol}$ **Answer**

b. $8.5\ \text{L} \quad \times \quad \dfrac{1\ \text{mol}}{22.4\ \text{L}} = 0.38\ \text{mol}$ **Answer**

c. $25\ \text{mL} \quad \times \quad \dfrac{1\ \text{L}}{1000\ \text{mL}} \quad \times \quad \dfrac{1\ \text{mol}}{22.4\ \text{L}} = 1.1 \times 10^{-3}\ \text{mol}$ **Answer**

**7.16** Use the ideal gas law to solve the problem as in Example 7.6.

$P = 175\ \text{atm}$
$T = 20.\ °C \qquad V = 5.0\ \text{L} \qquad\qquad n = ?\ \text{mol}$

known quantities          desired quantity

Convert °C to K. $K = °C + 273 = 20.\ °C + 273 = 293\ K$

Use the value of $R$ with atm since the pressure is given in atm; that is, $R = 0.0821\ \text{L·atm/(mol·K)}$.

$PV = nRT$   Solve for $n$ by dividing both sides by $RT$.

$n = \dfrac{PV}{RT} = \dfrac{(5.0\ \text{L})(175\ \text{atm})}{\left(0.0821\ \dfrac{\text{L·atm}}{\text{mol·K}}\right)(293\ \text{K})} = 36\ \text{mol}$ **Answer**

**7.17** Use the ideal gas law to solve the problems as in Example 7.6 and Answer 7.16.

a.   $PV = nRT$   Solve for $P$ by dividing both sides by $V$.

$R = 0.0821$ L·atm/(mol·K)

K = °C + 273 = 25 °C + 273 = 298 K

$$P = \frac{nRT}{V} = \frac{0.45 \text{ mol} \left(0.0821 \ \frac{\text{L} \cdot \text{atm}}{\text{mol} \cdot \text{K}}\right) (298 \text{ K})}{(10.0 \text{ L})} = 1.1 \text{ atm} \quad \textbf{Answer}$$

b.   $PV = nRT$   Solve for $P$ by dividing both sides by $V$.

$R = 0.0821$ L·atm/(mol·K)

K = °C + 273 = 20. °C + 273 = 293 K

$$P = \frac{nRT}{V}$$

molar mass conversion factor

$$10.0 \text{ g N}_2 \ \times \ \frac{1 \text{ mol N}_2}{28.02 \text{ g N}_2} = 0.357 \text{ mol N}_2$$

Grams cancel.

$$P = \frac{nRT}{V} = \frac{0.357 \text{ mol} \left(0.0821 \ \frac{\text{L} \cdot \text{atm}}{\text{mol} \cdot \text{K}}\right) (293 \text{ K})}{(5.0 \text{ L})} = 1.7 \text{ atm} \quad \textbf{Answer}$$

**7.18** Use Dalton's law to solve the problem as in Example 7.7.

$P_{\text{total}} = P_{O_2} + P_{CO_2}$
4.0 atm = 2.5 atm + $P_{CO_2}$
$CO_2$: 1.5 atm    $O_2$: 2.5 atm

**7.19** Use Dalton's law to solve the problem as in Example 7.8.

Methane: $0.85 \times 750$ mm Hg = 637.5 rounded to 640 mm Hg
Ethane: $0.10 \times 750$ mm Hg = 75 mm Hg
Propane: $0.050 \times 750$ mm Hg = 37.5 rounded to 38 mm Hg

**7.20** Use Dalton's law to solve the problem.

Chamber **A:** pure $O_2$ = 2.5 atm = higher pressure of $O_2$.
Chamber **B:** 40.% $O_2$ = $0.40 \times 5.5$ atm = 2.2 atm for $O_2$

**7.21** London dispersion forces are very weak interactions due to the momentary changes in electron density in a molecule. All molecules exhibit these forces (a–d).

**7.22** With dipole–dipole interactions, molecules align so that partial positive and partial negative charges are close together.

$$\delta^+ \ \delta^- \quad \delta^+ \ \delta^-$$
$$H\text{–}Cl \quad H\text{–}Cl$$
$$\longmapsto \quad \longmapsto$$

**7.23** Use Table 7.3 to decide which types of forces are present in each molecule. "+" means the force is present.

|  | London Dispersion | Dipole–Dipole | Hydrogen Bonding |
|---|---|---|---|
| a. $Cl_2$ | + | no net dipole | – |
| b. HCN | + | + | – |
| c. HF | + | + | + (the only molecule listed with an O–H, N–H, or H–F bond) |
| d. $CH_3Cl$ | + | + | – |
| e. $H_2$ | + | no net dipole | – |

**7.24** London dispersion forces are present in all covalent compounds.
Dipole–dipole interactions are present only in polar compounds with a permanent dipole.
Hydrogen bonding occurs only in compounds that contain an O–H, N–H, or H–F bond.

    a. $H_2O$ has hydrogen bonding, which is stronger than the London forces in $CO_2$.
    b. HBr has dipole–dipole interactions, which are stronger than the London forces in $CO_2$.
    c. $H_2O$ has hydrogen bonding, which is stronger than the dipole–dipole interactions in HBr.
    d. $C_2H_6$ has stronger London dispersion forces because it is larger than $CH_4$.

**7.25** The *stronger* the intermolecular forces, the *higher* the boiling point and melting point.

    a. $C_2H_6$ has stronger London dispersion forces, so it has a higher boiling point and melting point than $CH_4$.
    b. $CH_3OH$ has hydrogen bonding, so it has a higher boiling point and melting point than $C_2H_6$.
    c. Br is larger than Cl, so HBr has a higher boiling point and melting point than HCl.
    d. $CH_3Br$ has dipole–dipole interactions, so it has a higher boiling point and melting point than $C_2H_6$, which has only London dispersion forces.

**7.26** Water has stronger intermolecular forces since it can hydrogen bond, whereas $CO_2$ has only London dispersion forces. This explains why water is a liquid at room temperature, whereas $CO_2$ is a gas.

**7.27** The stronger the intermolecular forces, the lower the vapor pressure at a given temperature.

    a. $CH_4$ has a higher vapor pressure because it has only London dispersion forces, whereas $NH_3$ has hydrogen bonding.
    b. $CH_4$ has a higher vapor pressure because it has only London dispersion forces and is a smaller molecule than $C_2H_6$.
    c. $C_2H_6$ has a higher vapor pressure because it has only London dispersion forces, whereas $CH_3OH$ has hydrogen bonding.

**7.28**    When you get out of a pool, the water on your body evaporates and this cools your skin. When you re-enter the water, the water feels warmer because your skin is cooler.

**7.29**    Benzene cannot hydrogen bond, whereas water can, so benzene has much *weaker* intermolecular forces than water and is thus *less* viscous. Ethylene glycol, on the other hand, has two OH groups capable of hydrogen bonding, so it has *stronger* intermolecular forces than water and is thus *more* viscous.

**7.30**    The *stronger* the intermolecular forces, the stronger surface molecules are pulled down toward the interior of a liquid and the *higher* the **surface tension**.

The surface tension of gasoline should be lower than that of water because gasoline cannot hydrogen bond, whereas water can, giving it higher surface tension.

**7.31**    An **ionic solid** is composed of oppositely charged ions.
A **molecular solid** is composed of individual molecules arranged regularly.
A **network solid** is composed of a vast number of atoms covalently bonded together forming sheets or three-dimensional arrays.
A **metallic solid,** such as copper or silver, can be treated as a lattice of metal cations surrounded by a cloud of electrons that move freely.

a. $CaCl_2$ = ionic          c. sugar ($C_{12}H_{22}O_{11}$) = molecular
b. Fe = metallic           d. $NH_3(s)$ = molecular

**7.32**    Use the heat of fusion as a conversion factor to determine the amount of energy, as in Sample Problem 7.14.

a. $50.0 \text{ g} \times \dfrac{79.7 \text{ cal}}{1 \text{ g}} = $ 3,985 cal rounded to 3,990 cal **Answer**

b. $35.0 \text{ g} \times \dfrac{79.7 \text{ cal}}{1 \text{ g}} = $ 2,790 cal **Answer**

c. $35.0 \text{ g} \times \dfrac{79.7 \text{ cal}}{1 \text{ g}} \times \dfrac{1 \text{ kcal}}{1000 \text{ cal}} = $ 2.79 kcal **Answer**

d. $1.00 \text{ mol} \times \dfrac{18.02 \text{ g}}{1 \text{ mol}} \times \dfrac{79.7 \text{ cal}}{1 \text{ g}} = $ 1,430 cal **Answer**

**7.33**    Use the heat of vaporization to convert grams to an energy unit, calories, as in Sample Problem 7.15.

a. $42 \text{ g} \times \dfrac{540 \text{ cal}}{1 \text{ g}} = $ 23,000 cal **Answer**

b. $42 \text{ g} \times \dfrac{540 \text{ cal}}{1 \text{ g}} = $ 23,000 cal **Answer**

c. $1.00 \text{ mol} \times \dfrac{18.02 \text{ g}}{1 \text{ mol}} \times \dfrac{540 \text{ cal}}{1 \text{ g}} \times \dfrac{1 \text{ kcal}}{1000 \text{ cal}} = 9.7 \text{ kcal}$ **Answer**

d. $3.5 \text{ mol} \times \dfrac{18.02 \text{ g}}{1 \text{ mol}} \times \dfrac{540 \text{ cal}}{1 \text{ g}} \times \dfrac{1 \text{ kcal}}{1000 \text{ cal}} = 34 \text{ kcal}$ **Answer**

**7.34** The first graphic represents a gas (randomly placed spheres that are far apart) and the second graphic represents a liquid (closely packed but randomly arranged spheres). The molecular art therefore represents condensation. This is exothermic.

**7.35** a. Since heat is added, the graph is a heating curve.
b. The melting point of the substance is at the plateau **B→C,** which is 40 °C.
c. The boiling point of the substance is at the plateau **D→E,** which is 77 °C.
d. Solid and liquid are present at the plateau **B→C.**
e. Only liquid is present along the **C→D** diagonal.

**7.36** a. gas      b. solid and liquid      c. liquid      d. solid      e. liquid and gas

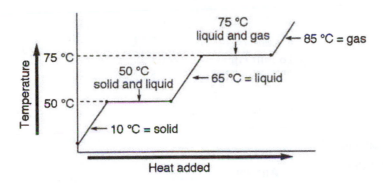

## Solutions to Odd-Numbered End-of-Chapter Problems

**7.37** Use a conversion factor to convert from mm Hg to atm.

$814.3 \text{ mm Hg} \times \dfrac{1 \text{ atm}}{760. \text{ mm Hg}} = 1.071 \text{ atm}$ **Answer**

**7.39** Use conversion factors to solve the problems.

a. $2.8 \text{ atm} \times \dfrac{14.7 \text{ psi}}{1 \text{ atm}} = 41 \text{ psi}$ **Answer**      c. $20.0 \text{ atm} \times \dfrac{760. \text{ torr}}{1 \text{ atm}} = 15,200 \text{ torr}$ **Answer**

b. $520 \text{ mm Hg} \times \dfrac{1 \text{ atm}}{760. \text{ mm Hg}} = 0.68 \text{ atm}$ **Answer**      d. $100. \text{ mm Hg} \times \dfrac{101,325 \text{ Pa}}{760. \text{ mm Hg}} = 13,300 \text{ Pa}$ **Answer**

**7.41** a. The volume decreases from **X** to (a) with the same number of gas particles; the pressure increases.

b. The volume remains the same, but the number of particles decreases; the pressure decreases.

c. The volume remains the same, but the number of particles increases; the pressure increases.

**7.43**    Use the gas laws to draw the balloons.

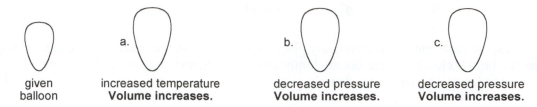

| given balloon | a. increased temperature **Volume increases.** | b. decreased pressure **Volume increases.** | c. decreased pressure **Volume increases.** |

**7.45**    a. Volume increases as outside atmospheric pressure decreases as the balloon floats to a higher altitude.

b. Volume decreases at the lower temperature when the balloon is placed in liquid nitrogen.

c. Volume decreases as external pressure increases in the hyperbaric chamber.

d. Volume increases as temperature increases in the microwave.

**7.47**    Use Boyle's law to fill in the table as in Example 7.2.

|    | $P_1$ | $V_1$ | $P_2$ | $V_2$ |
|----|-------|-------|-------|-------|
| a. | 2.0 atm | 3.0 L | 8.0 atm | **0.75 L** |
| b. | 55 mm Hg | 0.35 L | 18 mm Hg | **1.1 L** |
| c. | 705 mm Hg | 215 mL | **99.7 mm Hg** | 1.52 L |

**7.49**    Use Boyle's law to solve the problem as in Example 7.2.

$$V_2 \;=\; \frac{P_1 V_1}{P_2} \;=\; \frac{(3.5\ \text{atm})(10.\ \text{mL})}{1.0\ \text{atm}} \;=\; \underset{\textbf{Answer}}{35\ \text{mL}}$$

**7.51**    Use Charles's law to fill in the table as in Example 7.3.

|    | $V_1$ | $T_1$ | $V_2$ | $T_2$ |
|----|-------|-------|-------|-------|
| a. | 5.0 L | 310 K | **4.0 L** | 250 K |
| b. | 150 mL | 45 K | **1.1 L** | 45 °C |
| c. | 60.0 L | 0.0 °C | 180 L | **820 K** |

**7.53**    Use Charles's law to solve the problem as in Example 7.3.

K = °C + 273
$T_1$ = 25 °C + 273 = 298 K
$T_2$ = −78 °C + 273 = 195 K

$$V_2 \;=\; \frac{V_1 T_2}{T_1} \;=\; \frac{(2.2\ \text{L})(195\ \text{K})}{298\ \text{K}} \;=\; \underset{\textbf{Answer}}{1.4\ \text{L}}$$

**7.55** Use Gay–Lussac's law to fill in the table as in Sample Problem 7.4.

|   | $P_1$ | $T_1$ | $P_2$ | $T_2$ |
|---|---|---|---|---|
| a. | 3.25 atm | 298 K | **4.34 atm** | 398 K |
| b. | 550 mm Hg | 273 K | **350 mm Hg** | −100. °C |
| c. | 0.50 atm | 250 °C | 955 mm Hg | **1,300 K** |

**7.57** Use Gay–Lussac's law to solve the problem as in Sample Problem 7.4.

$K = °C + 273$
$T_1 = 100. °C + 273 = 373 \text{ K}$
$T_2 = 150. °C + 273 = 423 \text{ K}$

$$P_2 = \frac{P_1 T_2}{T_1} = \frac{(1.0 \text{ atm})(423 \text{ K})}{373 \text{ K}} = 1.1 \text{ atm} \quad \textbf{Answer}$$

**7.59** Use the combined gas law to solve the problems as in Example 7.4.

|   | $P_1$ | $V_1$ | $T_1$ | $P_2$ | $V_2$ | $T_2$ |
|---|---|---|---|---|---|---|
| a. | 0.90 atm | 4.0 L | 265 K | **1.4 atm** | 3.0 L | 310 K |
| b. | 1.2 atm | 75 L | 5.0 °C | 700. mm Hg | **110 L** | 50 °C |
| c. | 200. mm Hg | 125 mL | 298 K | 100. mm Hg | 0.62 L | **740 K** |

**7.61** Use the combined gas law to solve the problem as in Example 7.4.

$K = °C + 273$
$T_1 = 18 °C + 273 = 291 \text{ K}$
$T_2 = 25 °C + 273 = 298 \text{ K}$

$$\frac{P_1 V_1}{T_1} = \frac{P_2 V_2}{T_2} \quad \text{Solve for } V_2 \text{ by multiplying both sides by } \frac{T_2}{P_2}.$$

$$V_2 = \frac{P_1 V_1 T_2}{T_1 P_2} = \frac{(200. \text{ atm})(6.0 \text{ L})(298 \text{ K})}{(291 \text{ K})(1.0 \text{ atm})} = 1,200 \text{ L} \quad \textbf{Answer}$$

**7.63** STP is "standard temperature and pressure," or 0 °C at 760 mm Hg. The standard molar volume is the volume that one mole of a gas occupies at STP, or 22.4 liters.

**7.65** Convert volume to moles at STP as in Sample Problem 7.7.

a. $5.0 \text{ L He} \times \dfrac{1 \text{ mol}}{22.4 \text{ L}} = 0.22 \text{ mol}$ **Answer**

b. $11.2 \text{ L He} \times \dfrac{1 \text{ mol}}{22.4 \text{ L}} = 0.500 \text{ mol}$ **Answer**

c. $50.0 \text{ mL He} \times \dfrac{1 \text{ L}}{1000 \text{ mL}} \times \dfrac{1 \text{ mol}}{22.4 \text{ L}} = 0.002\,23 \text{ mol}$ **Answer**

**7.67** Convert moles and grams to volume at STP as in Sample Problem 7.7.

a. $4.2 \text{ mol Ar} \times \dfrac{22.4 \text{ L}}{1 \text{ mol}} = 94 \text{ L}$ **Answer**

b. $3.5 \text{ g } CO_2 \quad \times \quad \dfrac{1 \text{ mol}}{44.01 \text{ g}} \quad \times \quad \dfrac{22.4 \text{ L}}{1 \text{ mol}} \quad = \quad 1.8 \text{ L}$ **Answer**

c. $2.1 \text{ g } N_2 \quad \times \quad \dfrac{1 \text{ mol}}{28.02 \text{ g}} \quad \times \quad \dfrac{22.4 \text{ L}}{1 \text{ mol}} \quad = \quad 1.7 \text{ L}$ **Answer**

**7.69** Use the ideal gas law to solve the problem as in Example 7.6.

Convert °C to K. K = °C + 273 = 37 °C + 273 = 310. K

Use the value of $R$ with mm Hg since the pressure is given in mm Hg; that is, $R$ = 62.4 L•mm Hg/(mol•K).

$PV = nRT$   Solve for $n$ by dividing both sides by $RT$.

$$n = \dfrac{PV}{RT} = \dfrac{(747 \text{ mm Hg})(0.45 \text{ L})}{\left(62.4 \dfrac{\text{L} \cdot \text{mm Hg}}{\text{mol} \cdot \text{K}}\right)(310. \text{ K})} = 0.017 \text{ mol}$$ **Answer**

**7.71** Use the ideal gas law to solve the problem as in Example 7.6.

Convert °C to K. K = °C + 273 = 37 °C + 273 = 310. K

Use the value of $R$ with mm Hg since the pressure is given in mm Hg; that is, $R$ = 62.4 L•mm Hg/(mol•K).

$PV = nRT$   Solve for $n$ by dividing both sides by $RT$.

$$n = \dfrac{PV}{RT} = \dfrac{(740 \text{ mm Hg})(5.0 \text{ L})}{\left(62.4 \dfrac{\text{L} \cdot \text{mm Hg}}{\text{mol} \cdot \text{K}}\right)(310. \text{ K})} = 0.19 \text{ mol}$$ **Answer**

$$0.19 \text{ mol} \times \dfrac{6.02 \times 10^{23} \text{ molecules}}{1 \text{ mol}} = 1.1 \times 10^{23} \text{ molecules}$$ **Answer**

**7.73** Use the ideal gas law to solve the problem as in Example 7.6.

Use the value of $R$ with mm Hg since the pressure is given in mm Hg; that is, $R$ = 62.4 L•mm Hg/(mol•K).

$PV = nRT$   Solve for $n$ by dividing both sides by $RT$.

$$n = \dfrac{PV}{RT} = \dfrac{(500. \text{ mm Hg})(2.0 \text{ L})}{\left(62.4 \dfrac{\text{L} \cdot \text{mm Hg}}{\text{mol} \cdot \text{K}}\right)(273 \text{ K})} = 0.059 \text{ mol} \times \dfrac{32.00 \text{ g}}{1 \text{ mol}} = 1.9 \text{ g } O_2$$

$O_2$ **has more moles and more mass.**

$$n = \dfrac{PV}{RT} = \dfrac{(650 \text{ mm Hg})(1.5 \text{ L})}{\left(62.4 \dfrac{\text{L} \cdot \text{mm Hg}}{\text{mol} \cdot \text{K}}\right)(298 \text{ K})} = 0.052 \text{ mol} \times \dfrac{28.02 \text{ g}}{1 \text{ mol}} = 1.5 \text{ g } N_2$$

**7.75** Use Dalton's law.

A: $\dfrac{3 \text{ red spheres}}{9 \text{ spheres total}}$ x 630 mm Hg = 210 mm Hg

B: $\dfrac{6 \text{ blue spheres}}{9 \text{ spheres total}}$ x 630 mm Hg = 420 mm Hg

**7.77** Use Dalton's law to solve the problem as in Example 7.8.

Oxygen: $0.21 \times 460$ mm Hg = 97 mm Hg for $O_2$
Nitrogen: $0.78 \times 460$ mm Hg = 360 mm Hg for $N_2$

**7.79** If the overall pressure is three times as great, the partial pressure is three times higher.

593 mm Hg $\times$ 3 = 1,780 mm Hg

**7.81** Water is a liquid at room temperature because it is capable of hydrogen bonding and these strong intermolecular attractive forces give it a higher boiling point than $H_2S$, which can't hydrogen bond.

**7.83** Use Table 7.3 to decide which types of forces are present in each molecule.
a. chloroethane: London forces, dipole–dipole
b. cyclopropane: London forces only

**7.85** Hydrogen bonding occurs only in compounds that contain an O–H, N–H, or H–F bond. Only compound (d) can hydrogen bond.

**7.87** No, $H_2C=O$ has no H on the O atom, so it cannot form hydrogen bonds with another molecule of formaldehyde.

**7.89** a. Ethylene has London forces only, whereas methanol has London forces, dipole–dipole forces, and hydrogen bonding.
b. Methanol has a higher boiling point because it can hydrogen bond.
c. Ethylene has a higher vapor pressure at any given temperature because it has weaker intermolecular forces.

**7.91** Vapor pressure and boiling point are inversely related: lower vapor pressure corresponds to a higher boiling point.

Increasing boiling point: butane, acetaldehyde, Freon-113

**7.93** Glycerol is more viscous than water since it has three OH groups and many opportunities for hydrogen bonding. Acetone cannot hydrogen bond, so its intermolecular forces are weaker and it has low viscosity.

**7.95** Use the definitions from Answer 7.31 to classify each solid.

      a. KI: ionic                                   d. diamond: network

      b. $CO_2$: molecular                          e. the plastic polyethylene: amorphous

      c. bronze, an alloy of Cu and Sn:  metallic

**7.97**     **Evaporation** is an endothermic process by which a liquid enters the gas phase.  **Condensation** is an exothermic process that occurs when a gas enters the liquid phase.

**7.99**     The first graphic represents a liquid (closely packed but randomly arranged spheres) and the second graphic represents a gas (randomly placed spheres that are far apart).  The molecular art therefore represents vaporization.  Energy is absorbed.

**7.101**    a. Melting 100 g of ice is endothermic: energy is absorbed.
            b. Freezing 25 g of water is exothermic: energy is released.
            c. Condensing 20 g of steam is exothermic: energy is released.
            d. Vaporizing 30 g of water is endothermic: energy is absorbed.

**7.103**    Use the heat of fusion and heat of the vaporization as conversion factors to determine the amount of energy required, as in Sample Problem 7.14.

$$250 \text{ g} \times \frac{79.7 \text{ cal}}{1 \text{ g}} = 2.0 \times 10^4 \text{ cal} \quad\quad 50.0 \text{ g} \times \frac{540 \text{ cal}}{1 \text{ g}} = 2.7 \times 10^4 \text{ cal}$$

                                            melting                                       vaporizing

                                                                                      **Answer**

**7.105**

     a.

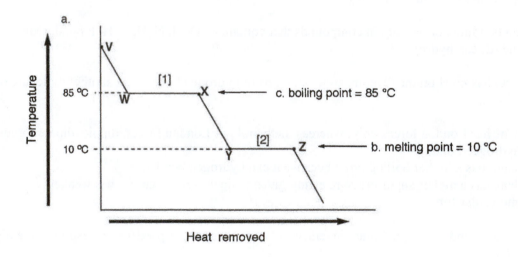

**7.107**

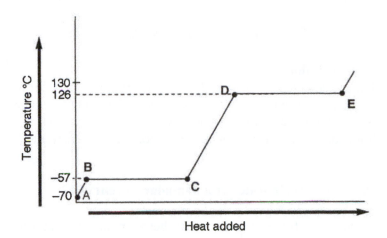

**7.109** The gases inside the bag had a volume of 250 mL at 760 mm Hg and take up a greater volume at the reduced pressure of 650 mm Hg.

$$P_1V_1 = P_2V_2 \qquad V_2 = \frac{P_1V_1}{P_2} = \frac{(760\ \text{mm Hg})(250\ \text{mL})}{650\ \text{mm Hg}} = 290\ \text{mL}$$

**7.111** Water is one of the few substances that expands as it enters the solid phase. This causes the bottle to crack because the water occupies a larger volume when it freezes.

**7.113** a. As a person breathes faster, he eliminates more $CO_2$ from the lungs; therefore, the measured value of $CO_2$ is lower than the normal value of 40 mm Hg.

b. Many people with advanced lung disease have lost lung tissue over time and therefore cannot exchange adequate amounts of oxygen through the lungs and into the blood, leading to a lower-than-normal partial pressure of oxygen. In addition, they often breathe more slowly and with lower volumes than normal, so they cannot eliminate enough $CO_2$ and therefore the partial pressure of $CO_2$ climbs.

**7.115** Use the ideal gas law to solve the problem.

Use the value of $R$ with atm since the pressure is given in atm; that is, $R = 0.0821$ L•atm/(mol•K).

$PV = nRT$ — Solve for $n$ by dividing both sides by $RT$.

$$n = \frac{PV}{RT} = \frac{(2\ \text{atm})(11.2\ \text{L})}{\left(0.0821\ \dfrac{\text{L} \cdot \text{atm}}{\text{mol} \cdot \text{K}}\right)(273\ \text{K})} = 1\ \text{mol} = \text{molar mass } 4.0\ \text{g/mol} = \text{Helium}$$

# Chapter 8 Solutions

## Chapter Review

**[1] What are the fundamental features of a solution? (8.1)**

- A solution is a homogeneous mixture that contains small dissolved particles. Any phase of matter can form solutions. When two substances form a solution, the substance present in the lesser amount is the solute and the substance present in the larger amount is the solvent.
- A solution conducts electricity if it contains dissolved ions, but does not conduct electricity if it contains atoms or neutral molecules.

**[2] What determines whether a substance is soluble in water or a nonpolar solvent? (8.2)**

- One rule summarizes solubility: "Like dissolves like."
- Most ionic compounds are soluble in water. If the attractive forces between the ions and water are stronger than the attraction between the ions in the crystal, an ionic compound dissolves in water.
- Small polar compounds that can hydrogen bond are soluble in water.
- Nonpolar compounds are soluble in nonpolar solvents. Compounds with many nonpolar C–C and C–H bonds are soluble in nonpolar solvents.

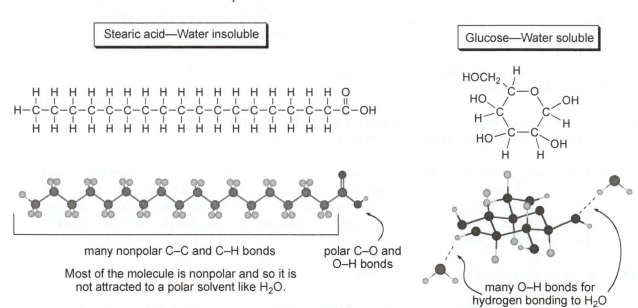

**[3] What effect do temperature and pressure have on solubility? (8.3)**

- The solubility of a solid in a liquid solvent generally increases with increasing temperature. The solubility of gases decreases with increasing temperature.
- Increasing pressure increases the solubility of a gas in a solvent. Pressure changes do not affect the solubility of liquids and solids.

**[4] How is the concentration of a solution expressed? (8.4, 8.5)**

- Concentration is a measure of how much solute is dissolved in a given amount of solution, and can be measured using mass, volume, or moles.
- Weight/volume percent (w/v)% concentration is the number of grams of solute dissolved in 100 mL of solution:

$$\boxed{\begin{array}{c} \textbf{Weight/volume} \\ \textbf{percent concentration} \end{array}} \quad \text{(w/v)\%} \quad = \quad \frac{\text{mass of solute (g)}}{\text{volume of solution (mL)}} \quad \text{x} \quad 100\%$$

- Volume/volume percent (v/v)% concentration is the number of milliliters of solute dissolved in 100 mL of solution:

$$\boxed{\begin{array}{c} \textbf{Volume/volume} \\ \textbf{percent concentration} \end{array}} \quad \text{(v/v)\%} \quad = \quad \frac{\text{volume of solute (mL)}}{\text{volume of solution (mL)}} \quad \text{x} \quad 100\%$$

- Parts per million (ppm) is the number of parts of solute in 1,000,000 parts of solution, where the units for both the solute and the solution are the same.

$$\boxed{\textbf{Parts per million}} \quad \text{ppm} \quad = \quad \frac{\text{mass of solute (g)}}{\text{mass of solution (g)}} \quad \text{x} \quad 10^6$$

or

$$\text{ppm} \quad = \quad \frac{\text{volume of solute (mL)}}{\text{volume of solution (mL)}} \quad \text{x} \quad 10^6$$

- Molarity (M) is the number of moles per liter of solution.

$$\boxed{\textbf{Molarity} \quad = \quad \textbf{M} \quad = \quad \frac{\textbf{moles of solute (mol)}}{\textbf{liter of solution (L)}}}$$

## [5] How are dilutions performed? (8.6)

- Dilution is the addition of solvent to decrease the concentration of a solute. Since the number of moles of solute is constant in carrying out a dilution, a new molarity ($M_2$) or volume ($V_2$) can be calculated from a given molarity ($M_1$) and volume ($V_1$) using the equation $M_1V_1 = M_2V_2$, as long as three of the four quantities are known.

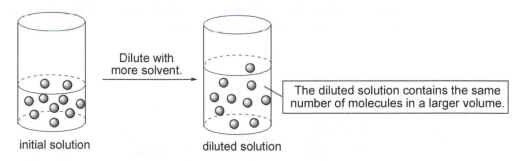

initial solution        diluted solution

## [6] How do dissolved particles affect the boiling point and melting point of a solution? (8.7)

- A nonvolatile solute lowers the vapor pressure above a solution, thus increasing its boiling point.
- A nonvolatile solute makes it harder for solvent molecules to form a crystalline solid, thus decreasing its melting point.

**[7] What is osmosis? (8.8)**
- Osmosis is the selective diffusion of solvent, usually water, across a semipermeable membrane. Solvent always moves from the less concentrated solution to the more concentrated solution, until the osmotic pressure prevents additional flow of solvent.
- Since living cells contain and are surrounded by biological solutions separated by a semipermeable membrane, the osmotic pressure must be the same on both sides of the membrane. Dialysis is similar to osmosis in that it involves the selective passage of several substances—water, small molecules, and ions—across a dialyzing membrane.

## Problem Solving

## [1] Solubility–General Features (8.2)

**Example 8.1** Predict the water solubility of each compound:

  a. NaCl  b. ethanol ($CH_3CH_2OH$)  c. $C_8H_{18}$

**Analysis**
Use the general solubility rule: "Like dissolves like." Generally, ionic and polar compounds are soluble in water. Nonpolar compounds are soluble in nonpolar solvents, but insoluble in water.

**Solution**
a. NaCl is an ionic compound, so it dissolves in water, a polar solvent.
b. $CH_3CH_2OH$ is a small polar molecule that contains an OH group. As a result, it can hydrogen bond to water, making it soluble.
c. $C_8H_{18}$ has only nonpolar C–C and C–H bonds, making it a nonpolar molecule that is therefore water insoluble.

## [2] Concentration Units–Percent Concentration (8.4)

**Example 8.2** A solution contains 0.80 g of NaCl dissolved in a total volume of 25 mL. What is the weight/volume percent concentration of NaCl?

**Analysis**
Use the formula (w/v)% = (grams of solute)/(mL of solution) × 100%.

**Solution**

$$(w/v)\% = \frac{0.80 \text{ g NaCl}}{25 \text{ mL solution}} \times 100\% = 3.2\% \text{ (w/v) NaCl}$$

**Answer**

**Example 8.3** A 1-L bottle of mouthwash contains 80 mL of ethanol. What is the volume/volume percent concentration of ethanol?

**Analysis**
Use the formula (v/v)% = (mL of solute)/(mL of solution) × 100%. Recall that 1 L = 1,000 mL.

**Solution**

$$(v/v)\% = \frac{80 \text{ mL ethanol}}{1000 \text{ mL mouthwash}} \times 100\% = 8\% \text{ (v/v) ethanol}$$
**Answer**

---

**Example 8.4** A solution contains 0.47% (w/v) of the antibiotic Abx in water. How many grams of antibiotic (Abx) are contained in 500. mL of this solution?

**Analysis and Solution**
**[1] Identify the known quantities and the desired quantity.**

0.47% (w/v) Abx solution
500. mL                    ? g Abx

known quantities           desired quantity

---

**[2] Write out the conversion factors.**
- Set up conversion factors that relate grams of antibiotic to the volume of the solution using the weight/volume percent concentration. Choose the conversion factor so that the unwanted unit, mL solution, cancels.

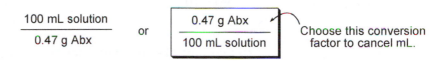

$$\frac{100 \text{ mL solution}}{0.47 \text{ g Abx}} \quad \text{or} \quad \frac{0.47 \text{ g Abx}}{100 \text{ mL solution}}$$

Choose this conversion factor to cancel mL.

---

**[3] Solve the problem.**
- Multiply the original quantity by the conversion factor to obtain the desired quantity.

$$500. \text{ mL} \times \frac{0.47 \text{ g Abx}}{100 \text{ mL solution}} = 2.4 \text{ g antibiotic}$$
**Answer**

---

**Example 8.5** What volume of a 1.0% (w/v) solution of lidocaine (anesthetic) contains 20. mg?

**Analysis and Solution**
**[1] Identify the known quantities and the desired quantity.**

1.0% (w/v) lidocaine solution
20. mg                         ? mL lidocaine

known quantities               desired quantity

---

**[2] Write out the conversion factors.**
- Use the weight/volume percent concentration to set up conversion factors that relate grams of lidocaine to mL of solution. Since percent concentration is expressed in grams, a mg–g conversion factor is needed as well. Choose the conversion factors that place the unwanted units, mg and g, in the denominator to cancel.

mg–g conversion factors

$$\frac{1000 \text{ mg}}{1 \text{ g}} \quad \text{or} \quad \boxed{\frac{1 \text{ g}}{1000 \text{ mg}}}$$

g–mL solution conversion factors

$$\frac{1.0 \text{ g lidocaine}}{100 \text{ mL solution}} \quad \text{or} \quad \boxed{\frac{100 \text{ mL solution}}{1.0 \text{ g lidocaine}}}$$

Choose the conversion factors with the unwanted units—mg and g—in the denominator.

---

**[3] Solve the problem.**

- Multiply the original quantity by the conversion factors to obtain the desired quantity.

$$20. \text{ m\!g lidocaine} \quad \times \quad \frac{1 \text{ g}}{1000 \text{ m\!g}} \quad \times \quad \frac{100 \text{ mL solution}}{1.0 \text{ g lidocaine}} \quad = \quad 2.0 \text{ mL solution} \quad \textbf{Answer}$$

---

**Example 8.6** What is the concentration of **X** in parts per million in a solution that contains 45 mg of **X** in 5,000 g of solution?

**Analysis**

Use the formula, ppm = (g of solute)/(g of solution) $\times 10^6$.

**Solution**

**[1] Convert milligrams of X to grams of X so that both the solute and solution have the same unit.**

$$45 \text{ m\!g X} \quad \times \quad \frac{1 \text{ g}}{1000 \text{ m\!g}} \quad = \quad 0.045 \text{ g X}$$

---

**[2] Use the formula to calculate parts per million.**

$$\frac{0.045 \text{ g X}}{5{,}000 \text{ g solution}} \quad \times \quad 10^6 \quad = \quad 9 \text{ ppm X} \quad \textbf{Answer}$$

---

## [3] Concentration Units–Molarity (8.5)

**Example 8.7** What is the molarity of one ampoule of sodium bicarbonate solution prepared from 4.2 g of sodium bicarbonate ($NaHCO_3$) in 50. mL of solution?

**Analysis and Solution**

**[1] Identify the known quantities and the desired quantity.**

$$\begin{array}{cc} 4.2 \text{ g } NaHCO_3 & \\ 50. \text{ mL solution} & ? \text{ M (mol/L)} \\ \text{known quantities} & \text{desired quantity} \end{array}$$

---

**[2] Convert the number of grams of sodium bicarbonate to the number of moles using the molar mass (84.01 g/mol).**

$$4.2\text{ g NaHCO}_3 \quad \times \quad \frac{1 \text{ mol}}{84.01 \text{ g}} \quad = \quad 0.050 \text{ mol NaHCO}_3$$

Grams cancel.

- Since the volume of the solution is given in mL, it must be converted to L.

$$50.\text{ mL solution} \quad \times \quad \frac{1 \text{ L}}{1000 \text{ mL}} \quad = \quad 0.050 \text{ L solution}$$

Milliliters cancel.

**[3] Divide the number of moles of solute by the number of liters of solution to obtain the molarity.**

$$M = \frac{\text{moles of solute (mol)}}{V \text{ (L)}} = \frac{0.050 \text{ mol NaHCO}_3}{0.050 \text{ L solution}} = 1.0 \text{ M}$$

molarity

**Answer**

**Example 8.8** How many grams of epinephrine are contained in 5.0 mL of a 0.050 M solution?

**Analysis**
Use the molarity to convert the volume of the solution to moles of solute. Then use the molar mass to convert moles to grams.

**Solution**
**[1] Identify the known quantities and the desired quantity.**

0.050 M
5.0 mL solution                    ? g epinephrine

known quantities                    desired quantity

**[2] Determine the number of moles of epinephrine using the molarity.**

| volume | | molarity | | mL–L conversion factor |

$$5.0 \text{ mL solution} \quad \times \quad \frac{0.050 \text{ mol epinephrine}}{1 \text{ L}} \quad \times \quad \frac{1 \text{ L}}{1000 \text{ mL}} \quad = \quad 0.000\,25 \text{ mol epinephrine}$$

**[3] Convert the number of moles of epinephrine to grams using the molar mass (183.2 g/mol).**

molar mass
conversion factor

$$0.000\,25 \text{ mol epinephrine} \quad \times \quad \frac{183.2 \text{ g epinephrine}}{1 \text{ mol epinephrine}} \quad = \quad 0.046 \text{ g epinephrine}$$

**Answer**

## [4] Dilution (8.6)

**Example 8.9** What is the concentration of a solution formed by diluting 4.0 mL of a 1.2 M glucose solution to 60.0 mL?

### Analysis
Since we know an initial molarity and volume ($M_1$ and $V_1$) and a final volume ($V_2$), we can calculate a final molarity ($M_2$) using the equation, $M_1V_1 = M_2V_2$.

### Solution
**[1] Identify the known quantities and the desired quantity.**

$M_1 = 1.2$ M

$V_1 = 4.0$ mL          $V_2 = 60.0$ mL          $M_2 = ?$

known quantities                          desired quantity

**[2] Write the equation and rearrange it to isolate the desired quantity, $M_2$, on one side.**

$$M_1V_1 \quad = \quad M_2V_2 \quad \text{Solve for } M_2 \text{ by dividing both sides by } V_2.$$

$$\frac{M_1V_1}{V_2} \quad = \quad M_2$$

**[3] Solve the problem.**
- Substitute the three known quantities in the equation and solve for $M_2$.

$$M_2 \quad = \quad \frac{M_1V_1}{V_2} \quad = \quad \frac{(1.2 \text{ M})(4.0 \text{ mL})}{(60.0 \text{ mL})} \quad = \quad \text{0.080 M glucose solution}$$
**Answer**

## [5] Colligative Properties (8.7)

**Example 8.10** What is the boiling point of a solution that contains 0.25 mol of NaCl in 1.00 kg of water?

### Analysis
Determine the number of "particles" contained in the solute. Use 0.51 °C/mol as a conversion factor to relate temperature change to the number of moles of solute particles.

### Solution
Each NaCl provides two "particles," $Na^+$ and $Cl^-$.

$$\text{temperature increase} \quad = \quad \frac{0.51 \text{ °C}}{\text{mol particles}} \quad \times \quad 0.25 \text{ mol NaCl} \quad \times \quad \frac{2 \text{ mol particles}}{\text{mol NaCl}} \quad = \quad 0.26 \text{ °C}$$

The boiling point of the solution is $100.0$ °C $+ 0.26$ °C $= 100.26$ °C, rounded to $100.3$ °C.

## Self-Test

**[1] Fill in the blank with one of the terms listed below.**

Colligative properties (8.7)     Heterogeneous mixture (8.1)     Nonvolatile (8.7)
Colloid (8.1)                    Homogeneous mixture (8.1)       Solubility (8.2)
Dilution (8.6)                   Molarity (8.5)                  Solution (8.1)
Electrolyte (8.1)                Nonelectrolyte (8.1)            Volatile (8.7)

1.  A _____ is a homogeneous mixture that contains small particles.
2.  _____ is the amount of solute that dissolves in a given amount of solvent.
3.  _____ is the number of moles of solute per liter of solution.
4.  A _____ does not have a uniform composition throughout the sample.
5.  A _____ is a homogeneous mixture with larger particles, often having an opaque appearance.
6.  _____ is the addition of solvent to decrease the concentration of solute.
7.  A _____ solute does not readily escape into the vapor phase, and thus it has a negligible vapor pressure at a given temperature.
8.  A substance that conducts an electric current in water is called an _____.
9.  _____ are properties of a solution that depend on the concentration of the solute but not its identity.
10. A substance that does not conduct an electric current in water is called a _____.
11. A _____ has a uniform composition throughout the sample.
12. A _____ solute readily escapes into the vapor phase.

**[2] Fill in the blank with one of the terms listed below.**

   a. exothermic          b. endothermic          c. raises          d. lowers

13. When solvation releases more energy than that required to separate particles, the overall process is _____.
14. When the separation of particles requires more energy than is released during solvation, the process is _____.
15. One mole of any nonvolatile solute _____ the freezing point of one kilogram of water the same amount, 1.86 $^{\circ}$C.
16. One mole of any nonvolatile solute _____ the boiling point of one kilogram of water the same amount, 0.51 $^{\circ}$C.

**[3] Match the abbreviation with the definition below.**

   a. (v/v)%          b. (w/v)%

17. The number of grams of solute dissolved in 100 mL of solution
18. The number of milliliters of solute dissolved in 100 mL of solution

**[4] Decide if the compounds are water soluble or water insoluble.**

a. water soluble          b. water insoluble

19. $CH_3CH_2CH_2CH_2CH_3$
20. NaBr
21. $NH_3$
22. $C_6H_{12}$

**[5] Fill in the blank with one of the terms below.**

a. lower       b. higher

23. A hypertonic solution has a _____ osmotic pressure than body fluids.
24. A hypotonic solution has a _____ osmotic pressure than body fluids.
25. A fewer number of dissolved particles means a _____ osmotic pressure.

## Answers to Self-Test

| | | | | |
|---|---|---|---|---|
| 1. solution | 6. Dilution | 11. homogeneous mixture | 16. c | 21. a |
| 2. Solubility | 7. nonvolatile | 12. volatile | 17. b | 22. b |
| 3. Molarity | 8. electrolyte | 13. a | 18. a | 23. b |
| 4. heterogeneous mixture | 9. Colligative properties | 14. b | 19. b | 24. a |
| 5. colloid | 10. nonelectrolyte | 15. d | 20. a | 25. a |

## Solutions to In-Chapter Problems

8.1     A **heterogeneous mixture** does not have a uniform composition throughout a sample.
       A **solution** is a homogeneous mixture that contains small particles. Liquid solutions are transparent.
       A **colloid** is a homogeneous mixture with larger particles, often having an opaque appearance.

       a. Cherry Garcia ice cream: heterogeneous mixture
       b. mayonnaise: colloid
       c, d, e. seltzer water, nail polish remover, and brass: solutions

8.2     A substance that conducts an electric current in water is called an **electrolyte.** If a solution contains
       *ions*, it will conduct electricity.
       A substance that does not conduct an electric current in water is called a **nonelectrolyte.** If a solution
       contains *neutral molecules*, it will not conduct electricity.

       a. KCl in $H_2O$: electrolyte                         c. KI in $H_2O$: electrolyte
       b. sucrose ($C_{12}H_{22}O_{11}$) in $H_2O$: nonelectrolyte

**8.3**    Use the general solubility rule: "Like dissolves like." Generally, ionic and polar compounds are soluble in water. Nonpolar compounds are soluble in nonpolar solvents.

a.

$NaNO_3$

ionic compound
water soluble

c.

$$HO-\overset{\overset{\displaystyle H}{|}}{\underset{\underset{\displaystyle H}{|}}{C}}-\overset{\overset{\displaystyle H}{|}}{\underset{\underset{\displaystyle H}{|}}{C}}-OH$$

polar compound
water soluble

e.

$NH_2OH$

polar compound
water soluble

b.    $CH_4$

nonpolar compound
NOT water soluble

d.    KBr

ionic compound
water soluble

**8.4**    a. Benzene ($C_6H_6$) and hexane ($C_6H_{14}$): both nonpolar compounds—would form a solution.
b. $Na_2SO_4$ and $H_2O$: one ionic compound and one polar compound—would form a solution.
c. NaCl and hexane ($C_6H_{14}$): one ionic and one nonpolar compound—cannot form a solution.
d. $H_2O$ and $CCl_4$: one polar and one nonpolar compound—cannot form a solution.

**8.5**    Identify the cation and anion and use the solubility rules to predict if the ionic compound is water soluble.

a. $Li_2CO_3$: cation $Li^+$ with anion $CO_3{}^{2-}$—water soluble.
b. $MgCO_3$: cation $Mg^{2+}$ with anion $CO_3{}^{2-}$—water insoluble.
c. KBr: cation $K^+$ with anion $Br^-$—water soluble.
d. $PbSO_4$: cation $Pb^{2+}$ with anion $SO_4{}^{2-}$—water insoluble.
e. $CaCl_2$: cation $Ca^{2+}$ with anion $Cl^-$—water soluble.
f. $MgCl_2$: cation $Mg^{2+}$ with anion $Cl^-$—water soluble.

**8.6**    Magnesium salts are not water soluble unless the anion is a halide, nitrate, acetate, or sulfate. The interionic forces between $Mg^{2+}$ and $OH^-$ must be stronger than the forces of attraction between the ions and water. Therefore, milk of magnesia is a heterogeneous mixture, rather than a solution.

**8.7**    A soft drink becomes "flat" when $CO_2$ escapes from the solution. $CO_2$ comes out of solution faster at room temperature than at the cooler temperature of the refrigerator because the solubility of gases in liquids decreases as the temperature increases.

**8.8**    For most ionic and molecular **solids**, solubility generally *increases* as temperature increases.
The solubility of **gases** *decreases* with increasing temperature.
Pressure changes do not affect the solubility of liquids and solids.
The *higher* the pressure, the *higher* the solubility of a **gas** in a solvent.

|  | $Na_2CO_3(s)$ | $N_2(g)$ |
|---|---|---|
| a.  increasing temperature | increased solubility | decreased solubility |
| b.  decreasing temperature | decreased solubility | increased solubility |
| c.  increasing pressure | no change | increased solubility |
| d.  decreasing pressure | no change | decreased solubility |

**8.9**     Use the formula in Example 8.2 to solve the problem; 525 mg = 0.525 g bismuth subsalicylate.

$$(w/v)\% \ = \ \frac{0.525 \text{ g bismuth}}{15 \text{ mL solution}} \ \times \ 100\% \ = \ 3.5\% \text{ (w/v) bismuth subsalicylate}$$
**Answer**

**8.10**    Use the formula in Example 8.2 to solve the problems.

$$(w/v)\% \ = \ \frac{4.3 \text{ g ethanol}}{30. \text{ mL solution}} \ \times \ 100\% \ = \ 14\% \text{ (w/v) ethanol}$$
**Answer**

$$(w/v)\% \ = \ \frac{0.021 \text{ g antiseptic}}{30. \text{ mL solution}} \ \times \ 100\% \ = \ 0.070\% \text{ (w/v) antiseptic}$$
**Answer**

**8.11**    Use the formula in Example 8.3 to solve the problem.

$$(v/v)\% \ = \ \frac{21 \text{ mL ethanol}}{250 \text{ mL mouthwash}} \ \times \ 100\% \ = \ 8.4\% \text{ (v/v) ethanol}$$
**Answer**

**8.12**    Use the stepwise analysis in Example 8.4 to solve the problem.

| 8.0% (v/v) mouthwash 30. mL | ? mL ethanol |
|---|---|
| known quantities | desired quantity |

$$\frac{8.0 \text{ mL ethanol}}{100 \text{ mL mouthwash}} \quad \text{or} \quad \frac{100 \text{ mL mouthwash}}{8.0 \text{ mL ethanol}}$$

$$30. \text{ mL mouthwash} \ \times \ \frac{8.0 \text{ mL ethanol}}{100 \text{ mL mouthwash}} \ = \ 2.4 \text{ mL ethanol}$$
**Answer**

**8.13**    Use the stepwise analysis in Example 8.5 to solve the problem.

| 0.50% (w/v) vitamin C 1,000. mg vitamin C known quantities | ? mL solution desired quantity |
|---|---|

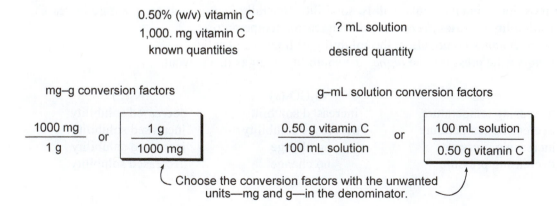

mg–g conversion factors

$$\frac{1000 \text{ mg}}{1 \text{ g}} \quad \text{or} \quad \frac{1 \text{ g}}{1000 \text{ mg}}$$

g–mL solution conversion factors

$$\frac{0.50 \text{ g vitamin C}}{100 \text{ mL solution}} \quad \text{or} \quad \frac{100 \text{ mL solution}}{0.50 \text{ g vitamin C}}$$

Choose the conversion factors with the unwanted units—mg and g—in the denominator.

$$1000. \text{ mg vitamin C} \quad \times \quad \frac{1 \text{ g}}{1000 \text{ mg}} \quad \times \quad \frac{100 \text{ mL solution}}{0.50 \text{ g vitamin C}} \quad = \quad 2.0 \times 10^2 \text{ mL solution}$$

**Answer**

**8.14** Use the stepwise analysis in Example 8.4 to solve the problem.

0.20% (w/v) dextromethorphan
2.0% (w/v) guaifenisin
3.0 tsp = 15 mL cough syrup
known quantities

? mg of each drug
desired quantity

$$15 \text{ mL cough syrup} \quad \times \quad \frac{0.20 \text{ g dextromethorphan}}{100 \text{ mL cough syrup}} \quad \times \quad \frac{1000 \text{ mg}}{1 \text{ g}} \quad = \quad 30. \text{ mg dextromethorphan}$$

**Answer**

$$15 \text{ mL cough syrup} \quad \times \quad \frac{2.0 \text{ g guaifenisin}}{100 \text{ mL cough syrup}} \quad \times \quad \frac{1000 \text{ mg}}{1 \text{ g}} \quad = \quad 3.0 \times 10^2 \text{ mg guaifenisin}$$

**Answer**

**8.15** Use the formula, ppm = (g of solute)/(g of solution) $\times 10^6$ to calculate parts per million as in Example 8.6.

a. $0.042 \text{ mg DDT} \quad \times \quad \frac{1 \text{ g}}{1000 \text{ mg}} \quad = \quad 0.000\,042 \text{ g DDT}$   $\frac{0.000\,042 \text{ g DDT}}{1,400 \text{ g plankton}} \quad \times \quad 10^6 \quad = \quad 0.030 \text{ ppm DDT}$

**Answer**

b. $1.0 \text{ kg tissue} \quad \times \quad \frac{1000 \text{ g}}{1.0 \text{ kg}} \quad = \quad 1,000 \text{ g tissue}$   $\frac{5 \times 10^{-4} \text{ g DDT}}{1,000 \text{ g tissue}} \quad \times \quad 10^6 \quad = \quad 0.5 \text{ ppm DDT}$

**Answer**

c. $2.0 \text{ mg DDT} \quad \times \quad \frac{1 \text{ g}}{1000 \text{ mg}} \quad = \quad 0.0020 \text{ g DDT}$   $\frac{0.0020 \text{ g DDT}}{1,000 \text{ g tissue}} \quad \times \quad 10^6 \quad = \quad 2.0 \text{ ppm DDT}$

**Answer**

d. $225 \text{ µg DDT} \quad \times \quad \frac{1 \text{ g}}{1,000,000 \text{ µg}} \quad = \quad 0.000\,225 \text{ g DDT}$   $\frac{0.000\,225 \text{ g DDT}}{1.0 \times 10^3 \text{ g breast milk}} \quad \times \quad 10^6 \quad = \quad 0.23 \text{ ppm DDT}$

**Answer**

**8.16** Calculate the molarity of each aqueous solution as in Example 8.7.

a. $M = \frac{\text{moles of solute (mol)}}{V \text{ (L)}} = \frac{1.0 \text{ mol NaCl}}{0.50 \text{ L solution}} = 2.0 \text{ M}$

molarity

**Answer**

b. $250 \text{ mL} \times \frac{1 \text{ L}}{1000 \text{ mL}} = 0.25 \text{ L}$   $M = \frac{2.0 \text{ mol NaCl}}{0.25 \text{ L solution}} = 8.0 \text{ M}$

molarity

**Answer**

c. $5.0 \text{ mL} \times \frac{1 \text{ L}}{1000 \text{ mL}} = 0.0050 \text{ L}$   $M = \frac{0.050 \text{ mol NaCl}}{0.0050 \text{ L solution}} = 10. \text{ M}$

molarity

**Answer**

d.  12.0 g NaCl  x  $\dfrac{1 \text{ mol NaCl}}{58.44 \text{ g NaCl}}$ = 0.205 mol NaCl     M = $\dfrac{0.205 \text{ mol NaCl}}{2.0 \text{ L solution}}$ = 0.10 M

molarity **Answer**

e.  24.4 g NaCl  x  $\dfrac{1 \text{ mol NaCl}}{58.44 \text{ g NaCl}}$ = 0.418 mol NaCl

M = $\dfrac{0.418 \text{ mol NaCl}}{0.35 \text{ L solution}}$ = 1.2 M

350 mL  x  $\dfrac{1 \text{ L}}{1000 \text{ mL}}$ = 0.35 L      molarity **Answer**

f.  60.0 g NaCl  x  $\dfrac{1 \text{ mol NaCl}}{58.44 \text{ g NaCl}}$ = 1.03 mol NaCl

M = $\dfrac{1.03 \text{ mol NaCl}}{0.75 \text{ L solution}}$ = 1.4 M

750 mL  x  $\dfrac{1 \text{ L}}{1000 \text{ mL}}$ = 0.75 L      molarity **Answer**

**8.17**  Calculate the molarity of each solution to determine which has the higher concentration.

10.0 g NaOH  x  $\dfrac{1 \text{ mol NaOH}}{40.00 \text{ g NaOH}}$ = 0.250 mol NaOH

M = $\dfrac{0.250 \text{ mol NaOH}}{0.15 \text{ L solution}}$ = 1.7 M

150 mL  x  $\dfrac{1 \text{ L}}{1000 \text{ mL}}$ = 0.15 L      molarity

**Answer**
**more concentrated**

15.0 g NaOH  x  $\dfrac{1 \text{ mol NaOH}}{40.00 \text{ g NaOH}}$ = 0.375 mol NaOH

M = $\dfrac{0.375 \text{ mol NaOH}}{0.25 \text{ L solution}}$ = 1.5 M

250 mL  x  $\dfrac{1 \text{ L}}{1000 \text{ mL}}$ = 0.25 L      molarity **Answer**

**8.18**  Use the molarity as a conversion factor to determine the number of milliliters as in Sample Problem 8.9.

a.    1.5 M solution        ? V (L) solution      V (L) = $\dfrac{\text{moles of solute (mol)}}{\text{M}}$
      0.15 mol glucose

    known quantities      desired quantity      = $\dfrac{0.15 \text{ mol glucose}}{1.5 \text{ mol/L}}$  = 0.10 L solution

0.10 L solution      x  $\dfrac{1000 \text{ mL}}{1 \text{ L}}$  =  $1.0 \times 10^2$ mL glucose solution
**Answer**

b.   V (L)  =  $\dfrac{\text{moles of solute (mol)}}{\text{M}}$

=  $\dfrac{0.020 \text{ mol glucose}}{1.5 \text{ mol/L}}$  =  0.013 L solution  x  $\dfrac{1000 \text{ mL}}{1 \text{ L}}$  =  13 mL glucose solution
**Answer**

c.     $V(L) = \dfrac{\text{moles of solute (mol)}}{M}$

   $= \dfrac{0.0030 \text{ mol glucose}}{1.5 \text{ mol/L}} = 0.0020 \text{ L solution} \times \dfrac{1000 \text{ mL}}{1 \text{ L}} = $ **2.0 mL glucose solution**
   **Answer**

d.     $V(L) = \dfrac{\text{moles of solute (mol)}}{M}$

   $= \dfrac{3.0 \text{ mol glucose}}{1.5 \text{ mol/L}} = 2.0 \text{ L solution} \times \dfrac{1000 \text{ mL}}{1 \text{ L}} = $ **2.0 x 10³ mL glucose solution**
   **Answer**

**8.19**   Use the molarity as a conversion factor to determine the number of moles.

a.   mol  =  $V(L) \cdot M$  =  (2.0 L)(2.0 mol/L)  =  4.0 mol NaCl

b.   mol  =  $V(L) \cdot M$  =  (2.5 L)(0.25 mol/L)  =  0.63 mol NaCl

c.   mol  =  $V(L) \cdot M$  =  (0.025 L)(2.0 mol/L)  =  0.050 mol NaCl

d.   mol  =  $V(L) \cdot M$  =  (0.25 L)(0.25 mol/L)  =  0.063 mol NaCl

**8.20**   Use the molarity to convert the volume of the solution to moles of solute. Then use the molar mass to convert moles to grams as in Example 8.8.

| volume | | molarity | | | molar mass conversion factor | | |
|---|---|---|---|---|---|---|---|

a.  0.10 L solution  ×  $\dfrac{1.25 \text{ mol NaCl}}{1 \text{ L}}$  =  0.125 mol NaCl  ×  $\dfrac{58.44 \text{ g NaCl}}{1 \text{ mol NaCl}}$  =  **7.3 g NaCl**
   **Answer**

b.  2.0 L solution  ×  $\dfrac{1.25 \text{ mol NaCl}}{1 \text{ L}}$  =  2.5 mol NaCl  ×  $\dfrac{58.44 \text{ g NaCl}}{1 \text{ mol NaCl}}$  =  **150 g NaCl**
   **Answer**

c.  0.55 L solution  ×  $\dfrac{1.25 \text{ mol NaCl}}{1 \text{ L}}$  =  0.69 mol NaCl  ×  $\dfrac{58.44 \text{ g NaCl}}{1 \text{ mol NaCl}}$  =  **40. g NaCl**
   **Answer**

d.  50. mL solution  ×  $\dfrac{1 \text{ L}}{1000 \text{ mL}}$  ×  $\dfrac{1.25 \text{ mol NaCl}}{1 \text{ L}}$  ×  $\dfrac{58.44 \text{ g NaCl}}{1 \text{ mol NaCl}}$  =  **3.7 g NaCl**
   **Answer**

**8.21**   Use the molar mass to convert grams to moles. Then use the molarity to convert moles of solute to volume of solution. Sample Problem 8.10 is similar, but the order of steps is different.

|  | molar mass<br>conversion factor |  | molarity<br>conversion factor |  |  |  |
|---|---|---|---|---|---|---|

a.   0.500 g sucrose  x   $\dfrac{1 \text{ mol sucrose}}{342.3 \text{ g sucrose}}$   x   $\dfrac{1 \text{ L}}{0.25 \text{ mol sucrose}}$   x   $\dfrac{1000 \text{ mL}}{1 \text{ L}}$   =   5.8 mL solution

**Answer**

b.   2.0 g sucrose   x   $\dfrac{1 \text{ mol sucrose}}{342.3 \text{ g sucrose}}$   x   $\dfrac{1 \text{ L}}{0.25 \text{ mol sucrose}}$   x   $\dfrac{1000 \text{ mL}}{1 \text{ L}}$   =   23 mL solution

**Answer**

c.   1.25 g sucrose   x   $\dfrac{1 \text{ mol sucrose}}{342.3 \text{ g sucrose}}$   x   $\dfrac{1 \text{ L}}{0.25 \text{ mol sucrose}}$   x   $\dfrac{1000 \text{ mL}}{1 \text{ L}}$   =   15 mL solution

**Answer**

d. 50.0 mg sucrose  x  $\dfrac{1 \text{ g}}{1000 \text{ mg}}$  x  $\dfrac{1 \text{ mol sucrose}}{342.3 \text{ g sucrose}}$  x  $\dfrac{1 \text{ L}}{0.25 \text{ mol sucrose}}$  x  $\dfrac{1000 \text{ mL}}{1 \text{ L}}$  =  0.58 mL solution

**Answer**

**8.22**   Calculate a new molarity ($M_2$) using the equation, $M_1V_1 = M_2V_2$.

$$M_2 \ = \ \frac{M_1V_1}{V_2} \ = \ \frac{(3.8 \text{ M})(25.0 \text{ mL})}{(275 \text{ mL})} \ = \ 0.35 \text{ M glucose solution}$$

**Answer**

**8.23**   Calculate the volume needed ($V_1$) using the equation, $M_1V_1 = M_2V_2$.

a.   $V_1 \ = \ \dfrac{M_2V_2}{M_1} \ = \ \dfrac{(2.5 \text{ M})(525 \text{ mL})}{6.0 \text{ M}} \ = \ 220 \text{ mL}$

**Answer**

b.   $V_1 \ = \ \dfrac{M_2V_2}{M_1} \ = \ \dfrac{(4.0 \text{ M})(750 \text{ mL})}{6.0 \text{ M}} \ = \ 5.0 \times 10^2 \text{ mL}$

**Answer**

c.   $V_1 \ = \ \dfrac{M_2V_2}{M_1} \ = \ \dfrac{(0.10 \text{ M})(450 \text{ mL})}{6.0 \text{ M}} \ = \ 7.5 \text{ mL}$

**Answer**

d.   $V_1 \ = \ \dfrac{M_2V_2}{M_1} \ = \ \dfrac{(3.5 \text{ M})(25 \text{ mL})}{6.0 \text{ M}} \ = \ 15 \text{ mL}$

**Answer**

**8.24**   Calculate the new concentration of ketamine, and determine the volume needed to supply 75 mg.

$$C_2 \ = \ \frac{C_1V_1}{V_2} \ = \ \frac{(100. \text{ mg/mL})(2.0 \text{ mL})}{10.0 \text{ mL}} \ = \ 20. \text{ mg/mL}$$

75 mg  x  $\dfrac{1 \text{ mL}}{20. \text{ mg}}$  =  3.8 mL

**Answer**

**8.25** Determine the number of "particles" contained in the solute. Use 0.51 $^{\circ}$C/mol as a conversion factor to relate the temperature change to the number of moles of solute particles, as in Example 8.10.

    a.  2.0 mol of sucrose molecules

$$\frac{0.51\ ^{\circ}C}{mol\ particles} \times 2.0\ mol = 1.0\ ^{\circ}C \qquad \text{Boiling point} = 100.0\ ^{\circ}C + 1.0\ ^{\circ}C = 101.0\ ^{\circ}C$$

    b.  2.0 mol of $KNO_3$

$$\frac{0.51\ ^{\circ}C}{mol\ particles} \times 2.0\ mol\ KNO_3 \times \frac{2\ mol\ particles}{mol\ KNO_3} = 2.0\ ^{\circ}C$$

$$\text{Boiling point} = 100.0\ ^{\circ}C + 2.0\ ^{\circ}C = 102.0\ ^{\circ}C$$

    c.  2.0 mol of $CaCl_2$

$$\frac{0.51\ ^{\circ}C}{mol\ particles} \times 2.0\ mol\ CaCl_2 \times \frac{3\ mol\ particles}{mol\ CaCl_2} = 3.1\ ^{\circ}C$$

$$\text{Boiling point} = 100.0\ ^{\circ}C + 3.1\ ^{\circ}C = 103.1\ ^{\circ}C$$

    d.  20.0 g of NaCl

$$20.0\ g\ NaCl \times \frac{1\ mol\ NaCl}{58.44\ g\ NaCl} = 0.342\ mol\ NaCl$$

$$\frac{0.51\ ^{\circ}C}{mol\ particles} \times 0.342\ mol\ NaCl \times \frac{2\ mol\ particles}{mol\ NaCl} = 0.35\ ^{\circ}C$$

$$\text{Boiling point} = 100.0\ ^{\circ}C + 0.35\ ^{\circ}C = 100.35\ ^{\circ}C \text{ rounded to } 100.4\ ^{\circ}C$$

**8.26** Determine the number of "particles" contained in the solute. Use 1.86 $^{\circ}$C/mol as a conversion factor to relate temperature change to the number of moles of solute particles, as in Sample Problem 8.14.

    a.  2.0 mol of sucrose molecules

$$\frac{1.86\ ^{\circ}C}{mol\ particles} \times 2.0\ mol = 3.7\ ^{\circ}C \qquad \text{Melting point} = -3.7\ ^{\circ}C$$

    b.  2.0 mol of $KNO_3$

$$\frac{1.86\ ^{\circ}C}{mol\ particles} \times 2.0\ mol\ KNO_3 \times \frac{2\ mol\ particles}{mol\ KNO_3} = 7.4\ ^{\circ}C \qquad \text{Melting point} = -7.4\ ^{\circ}C$$

    c.  2.0 mol of $CaCl_2$

$$\frac{1.86\ ^{\circ}C}{mol\ particles} \times 2.0\ mol\ CaCl_2 \times \frac{3\ mol\ particles}{mol\ CaCl_2} = 11\ ^{\circ}C \qquad \text{Melting point} = -11\ ^{\circ}C$$

d. 20.0 g of NaCl

$$20.0 \, \text{g NaCl} \quad \times \quad \frac{1 \, \text{mol NaCl}}{58.44 \, \text{g NaCl}} \quad = \quad 0.342 \, \text{mol NaCl}$$

$$\frac{1.86 \, ^\circ\text{C}}{\text{mol particles}} \quad \times \quad 0.342 \, \text{mol NaCl} \quad \times \quad \frac{2 \, \text{mol particles}}{\text{mol NaCl}} \quad = \quad 1.27 \, ^\circ\text{C} \qquad \text{Melting point} = -1.27 \, ^\circ\text{C}$$

**8.27** Determine the number of moles of ethylene glycol. Use 1.86 $^\circ$C/mol as a conversion factor to relate temperature change to the number of moles of solute particles, as in Sample Problem 8.14.

$$250 \, \text{g ethylene glycol} \quad \times \quad \frac{1 \, \text{mol}}{62.07 \, \text{g}} \quad = \quad 4.0 \, \text{mol ethylene glycol}$$

$$\text{temperature decrease} \quad = \quad \frac{1.86 \, ^\circ\text{C}}{\text{mol particles}} \quad \times \quad 4.0 \, \text{mol } C_2H_6O_2 \quad = \quad -7.4 \, ^\circ\text{C}$$

**8.28** The solvent (water) flows from the less concentrated solution to the more concentrated solution.

a. A 5.0% sugar solution has greater osmotic pressure than a 1.0% sugar solution.
b. 4.0 M NaCl has greater osmotic pressure than 3.0 M NaCl.
c. 0.75 M NaCl has greater osmotic pressure than 1.0 M glucose solution because NaCl has two particles for each NaCl, whereas glucose has only one.

**8.29** a. Water flows across the membrane.
b. More water initially flows from the 1.0 M side to the more concentrated 1.5 M side. When equilibrium is reached there is equal flow of water in both directions.
c. The height of the 1.5 M side will be higher and the height of the 1.0 M NaCl will be lower.

**8.30** A *hypotonic* solution has a lower osmotic pressure than body fluids.
A *hypertonic* solution has a higher osmotic pressure than body fluids.

a. A 3% (w/v) glucose solution is hypotonic, so it will cause hemolysis.
b. A 0.15 M KCl solution is isotonic, so no change results.
c. A 0.15 M $Na_2CO_3$ solution is hypertonic, so it will cause crenation.

## Solutions to Odd-Numbered End-of-Chapter Problems

**8.31** **A** shows KI dissolved in water, since the $K^+$ and $I^-$ ions are separated when KI is dissolved. The solution contains ions, so it conducts an electric current. **B** shows KI as a molecule, without being separated into ions, a situation that does not occur.

**8.33** Use the definitions from Answer 8.1 to classify each substance as a heterogeneous mixture, a solution, or a colloid.

a. bronze (alloy of Sn and Cu): solution
b. diet soda: solution
c. orange juice with pulp: heterogeneous mixture

d. household ammonia: solution
e. gasoline : solution
f. fog: colloid

**8.35**    A solution that has less than the maximum number of grams of solute is said to be **unsaturated.**
A solution that has the maximum number of grams of solute that can dissolve is said to be **saturated.**
A solution that has more than the maximum number of grams of solute is said to be **supersaturated.**

a. Adding 30 g to 100 mL of $H_2O$ at 20 °C: unsaturated
b. Adding 65 g to 100 mL of $H_2O$ at 50 °C: saturated
c. Adding 20 g to 50 mL of $H_2O$ at 20 °C: saturated
d. Adding 42 g to 100 mL of $H_2O$ at 50 °C and slowly cooling to 20 °C to give a clear solution with
     no precipitate: supersaturated

**8.37**    Use the general solubility rule: "Like dissolves like." Generally, ionic and polar compounds are
soluble in water. Nonpolar compounds are soluble in nonpolar solvents.

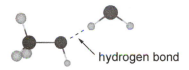

a.   LiCl

ionic
water soluble

b.

nonpolar
NOT water soluble

c.

polar
water soluble

d.   $Na_3PO_4$

ionic
water soluble

**8.39**    Methanol will hydrogen bond to water.

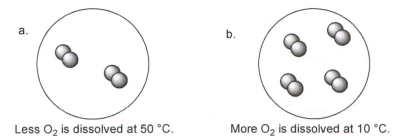

hydrogen bond

**8.41**    Water-soluble compounds are ionic or are small polar molecules that can hydrogen bond with the
water solvent, but nonpolar compounds, such as oil, are soluble in nonpolar solvents.

**8.43**    Iodine would not be soluble in water but is soluble in $CCl_4$ since $I_2$ is nonpolar and $CCl_4$ is a nonpolar
solvent.

**8.45**    Cholesterol is not water soluble because it is a large nonpolar molecule with a single OH group.

**8.47**    The solubility of gases decreases with increasing temperature. The $H_2O$ molecules are omitted in
each representation.

a.

Less $O_2$ is dissolved at 50 °C.

b.

More $O_2$ is dissolved at 10 °C.

**8.49**    For most ionic and molecular **solids,** solubility generally *increases* as temperature increases.
The solubility of **gases** *decreases* as temperature increases.
Pressure changes do not affect the solubility of liquids and solids.
The *higher* the pressure, the *higher* the solubility of a **gas** in a solvent.

|  | NaCl(*s*) |
|---|---|
| a. increasing temperature | increased solubility |
| b. decreasing temperature | decreased solubility |
| c. increasing pressure | no change |
| d. decreasing pressure | no change |

**8.51**    A decrease in temperature (a) increases the solubility of a gas and (b) decreases the solubility of a solid.

**8.53**    Many ionic compounds are soluble in water because the ion–dipole interactions between ions and water provide the energy needed to break apart the ions from the crystal lattice. The water molecules form a loose shell of solvent around each ion.

**8.55**    Identify the cation and anion and use the solubility rules to predict if the ionic compound is water soluble.

a. $K_2SO_4$: cation $K^+$ with anion $SO_4{}^{2-}$—water soluble.
b. $MgSO_4$: cation $Mg^{2+}$ with anion $SO_4{}^{2-}$—water soluble.
c. $ZnCO_3$: cation $Zn^{2+}$ with anion $CO_3{}^{2-}$—not water soluble.
d. KI: cation $K^+$ with anion $I^-$—water soluble.
e. $Fe(NO_3)_3$: cation $Fe^{3+}$ with anion $NO_3{}^-$—water soluble.
f. $PbCl_2$: cation $Pb^{2+}$ with anion $Cl^-$—not water soluble.
g. CsCl : cation $Cs^+$ with anion $Cl^-$—water soluble.
h. $Ni(HCO_3)_2$: cation $Ni^{2+}$ with anion $HCO_3{}^-$—not water soluble.

**8.57**    Write two conversion factors for each concentration.

a.  5% (w/v)    $\dfrac{5\ g}{100\ mL}$    $\dfrac{100\ mL}{5\ g}$        c.  10 ppm    $\dfrac{10\ g}{10^6\ g}$    $\dfrac{10^6\ g}{10\ g}$

b.  6.0 M    $\dfrac{6.0\ mol}{1.0\ L}$    $\dfrac{1.0\ L}{6.0\ mol}$

**8.59**    Use the formula in Example 8.2 to solve the problems.

a.    (w/v)%  =  $\dfrac{10.0\ g\ LiCl}{750\ mL\ solution}$  x  100%  =  1.3% (w/v) LiCl
                                                                    **Answer**

b.    (w/v)%  =  $\dfrac{25\ g\ NaNO_3}{150\ mL\ solution}$  x  100%  =  17% (w/v) $NaNO_3$
                                                                    **Answer**

c.    (w/v)%  =  $\dfrac{40.0\ g\ NaOH}{500.\ mL\ solution}$  x  100%  =  8.00% (w/v) NaOH
                                                                    **Answer**

**8.61**    Use the formula in Example 8.3 to solve the problem.

$$(v/v)\% = \frac{25 \text{ mL ethyl acetate}}{150 \text{ mL solution}} \times 100\% = 17\% \text{ (v/v) ethyl acetate}$$
**Answer**

**8.63**    Calculate the molarity of each aqueous solution as in Example 8.7.

a.    $$M = \frac{\text{moles of solute (mol)}}{V \text{ (L)}} = \frac{3.5 \text{ mol KCl}}{1.50 \text{ L solution}} = 2.3 \text{ M}$$
molarity
**Answer**

b.    $$855 \text{ mL solution} \times \frac{1 \text{ L}}{1000 \text{ mL}} = 0.855 \text{ L solution}$$

$$M = \frac{\text{moles of solute (mol)}}{V \text{ (L)}} = \frac{0.44 \text{ mol NaNO}_3}{0.855 \text{ L solution}} = 0.51 \text{ M}$$
molarity
**Answer**

c.    $$25.0 \text{ g NaCl} \times \frac{1 \text{ mol NaCl}}{58.44 \text{ g}} = 0.428 \text{ mol NaCl}$$

$$650 \text{ mL solution} \times \frac{1 \text{ L}}{1000 \text{ mL}} = 0.65 \text{ L solution}$$

$$M = \frac{\text{moles of solute (mol)}}{V \text{ (L)}} = \frac{0.428 \text{ mol NaCl}}{0.65 \text{ L solution}} = 0.66 \text{ M}$$
molarity
**Answer**

d.    $$10.0 \text{ g NaHCO}_3 \times \frac{1 \text{ mol NaHCO}_3}{84.01 \text{ g}} = 0.119 \text{ mol NaHCO}_3$$

$$M = \frac{\text{moles of solute (mol)}}{V \text{ (L)}} = \frac{0.119 \text{ mol NaHCO}_3}{3.3 \text{ L solution}} = 0.036 \text{ M}$$
molarity
**Answer**

**8.65**

a.  Add 12 g of acetic acid to the flask and then water to bring the volume to 250 mL.

$$(w/v)\% = \frac{x \text{ g acetic acid}}{250 \text{ mL solution}} \times 100\% = 4.8\% \text{ (w/v) acetic acid}$$
$x = 12$ g acetic acid
**Answer**

b.  Add 55 mL of ethyl acetate to the flask and then water to bring the volume to 250 mL.

$$(v/v)\% = \frac{x \text{ mL ethyl acetate}}{250 \text{ mL solution}} \times 100\% = 22\% \text{ (v/v) ethyl acetate}$$
$x = 55$ mL ethyl acetate
**Answer**

c.  Add 37 g of NaCl to the flask and then water to bring the volume to 250 mL.

$$M = \frac{\text{moles of solute (mol)}}{V\ (L)}$$
molarity

moles of solute (mol) = (M)(V) = (2.5 M) (0.25 L solution) = 0.63 mol NaCl

0.63 mol NaCl $\times \dfrac{58.44\ g}{1\ mol\ NaCl}$ = 37 g NaCl

**Answer**

**8.67** Calculate the moles of solute in each solution.

a. moles of solute (mol) = (M)(V) = (0.25 M) (0.15 L solution) = 0.038 mol NaNO$_3$

**Answer**

b. moles of solute (mol) = (M)(V) = (2.0 M) (0.045 L solution) = 0.090 mol HNO$_3$

**Answer**

c. moles of solute (mol) = (M)(V) = (1.5 M) (2.5 L solution) = 3.8 mol HCl

**Answer**

**8.69** Calculate the number of grams of each solute.

a. 0.038 mol NaNO$_3$ $\times \dfrac{85.00\ g\ NaNO_3}{1\ mol\ NaNO_3}$ = 3.2 g NaNO$_3$

**Answer**

b. 0.090 mol HNO$_3$ $\times \dfrac{63.02\ g\ HNO_3}{1\ mol\ HNO_3}$ = 5.7 g HNO$_3$

**Answer**

c. 3.8 mol HCl $\times \dfrac{36.46\ g\ HCl}{1\ mol\ HCl}$ = 140 g HCl

**Answer**

**8.71** Calculate the number of milliliters of ethanol in the bottle of wine.

(v/v)% = $\dfrac{x\ mL\ ethanol}{750\ mL\ solution}$ $\times$ 100% = 11.0% (v/v) ethanol       x = 83 mL ethanol

**Answer**

**8.73** Calculate the amount of acetic acid (number of grams and number of moles) in the solution and convert to molarity.

a. (w/v)% = $\dfrac{x\ g\ acetic\ acid}{1890\ mL\ solution}$ $\times$ 100% = 5.0% (w/v) acetic acid       x = 95 g acetic acid

**Answer**

b. 95 g acetic acid $\times \dfrac{1\ mol\ acetic\ acid}{60.05\ g\ acetic\ acid}$ = 1.6 mol acetic acid

**Answer**

c. $M = \dfrac{\text{moles of solute (mol)}}{V\ (L)}$ = $\dfrac{1.6\ mol\ acetic\ acid}{1.89\ L\ solution}$ = 0.85 M
molarity

**Answer**

**8.75**   Use the formula, ppm = (g of solute)/(g of solution) × 10⁶, to calculate parts per million as in Example 8.6.

a.   $80 \ \mu g \ CHCl_3 \ \times \ \dfrac{1 \ g}{1{,}000{,}000 \ \mu g} \ = \ 0.000 \ 08 \ g \ CHCl_3$   $\dfrac{0.000 \ 08 \ g \ CHCl_3}{1{,}000 \ g} \ \times \ 10^6 \ = \ 0.08 \ ppm \ CHCl_3$
   **Answer**

b.   $700 \ \mu g \ glyphosate \ \times \ \dfrac{1 \ g}{1{,}000{,}000 \ \mu g} \ = \ 0.0007 \ g \ glyphosate$

   $\dfrac{0.0007 \ g \ glyphosate}{1{,}000 \ g} \ \times \ 10^6 \ = \ 0.7 \ ppm \ glyphosate$
   **Answer**

**8.77**   When solution **X** is diluted, the volume will increase, but the amount of solute will stay the same (**B**).

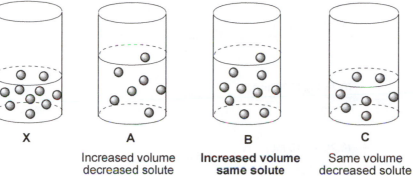

|   X   |   A   |   B   |   C   |
|-------|-------|-------|-------|
|       | Increased volume decreased solute | **Increased volume same solute** | Same volume decreased solute |

**8.79**

a.

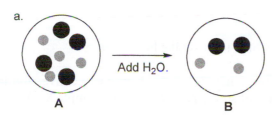

Add H₂O.

B has half as much NaCl, and is therefore half the molarity: 0.05 M.

b.   $M_1 V_1 \ = \ M_2 V_2 \ = \ (0.1 \ M)(50.0 \ mL) \ = \ (0.05 \ M)(x \ mL)$

   $x \ = \ 100 \ mL$          Since the original volume was 50 mL, **50 mL** of water is added.

**8.81**   Concentration is the amount of solute per unit volume in a solution. Dilution is the addition of solvent to decrease the concentration of solute.

**8.83**   Use the equation, $(C_1)(V_1) = (C_2)(V_2)$, to calculate the new concentration after dilution.

a.   $C_2 \ = \ \dfrac{C_1 V_1}{V_2} \ = \ \dfrac{(30.0\%)(100. \ mL)}{200. \ mL} \ = \ 15.0\% \ (w/v)$

b.   $C_2 \ = \ \dfrac{C_1 V_1}{V_2} \ = \ \dfrac{(30.0\%)(100. \ mL)}{500. \ mL} \ = \ 6.00\% \ (w/v)$

c. $C_2 = \dfrac{C_1 V_1}{V_2} = \dfrac{(30.0\%)(250 \text{ mL})}{1500 \text{ mL}} = 5.0\% \text{ (w/v)}$

d. $C_2 = \dfrac{C_1 V_1}{V_2} = \dfrac{(30.0\%)(350 \text{ mL})}{750 \text{ mL}} = 14\% \text{ (w/v)}$

**8.85**  Use the equation, $(M_1)(V_1) = (M_2)(V_2)$, to calculate the new molarity after dilution.

$M_2 = \dfrac{M_1 V_1}{V_2} = \dfrac{(12.0 \text{ M})(125 \text{ mL})}{850 \text{ mL}} = 1.8 \text{ M}$

**8.87**  Use the equation, $(M_1)(V_1) = (M_2)(V_2)$, to calculate the volume needed to prepare each solution.

a. $V_1 = \dfrac{M_2 V_2}{M_1} = \dfrac{(1.0 \text{ M})(25 \text{ mL})}{2.5 \text{ M}} = 10. \text{ mL}$

b. $V_1 = \dfrac{M_2 V_2}{M_1} = \dfrac{(0.75 \text{ M})(1500 \text{ mL})}{2.5 \text{ M}} = 450 \text{ mL}$

c. $V_1 = \dfrac{M_2 V_2}{M_1} = \dfrac{(0.25 \text{ M})(15 \text{ mL})}{2.5 \text{ M}} = 1.5 \text{ mL}$

d. $V_1 = \dfrac{M_2 V_2}{M_1} = \dfrac{(0.025 \text{ M})(250 \text{ mL})}{2.5 \text{ M}} = 2.5 \text{ mL}$

**8.89**  Ocean water contains nonvolatile dissolved salts, which increase the boiling point.

**8.91**  Determine the number of "particles" contained in the solute. Use 0.51 °C/mol as a conversion factor to relate temperature change to the number of moles of solute particles, as in Example 8.10.

a. 3.0 mol of fructose molecules

$\dfrac{0.51 \text{ °C}}{\text{mol particles}} \times 3.0 \text{ mol} = 1.5 \text{ °C}$   100.0 °C + 1.5 °C = 101.5 °C

b. 1.2 mol of KI

$\dfrac{0.51 \text{ °C}}{\text{mol particles}} \times 1.2 \text{ mol KI} \times \dfrac{2 \text{ mol particles}}{\text{mol KI}} = 1.2 \text{ °C}$   100.0 °C + 1.2 °C = 101.2 °C

c. 1.5 mol $Na_3PO_4$

$\dfrac{0.51 \text{ °C}}{\text{mol particles}} \times 1.5 \text{ mol Na}_3\text{PO}_4 \times \dfrac{4 \text{ mol particles}}{\text{mol Na}_3\text{PO}_4} = 3.1 \text{ °C}$   100.0 °C + 3.1 °C = 103.1 °C

**8.93**   Determine the number of "particles" contained in the solute. Use 1.86 $^{\circ}$C/mol as a conversion factor to relate temperature change to the number of moles of solute particles, as in Sample Problem 8.14.

$$\frac{150\ g\ \text{ethylene glycol}}{(C_2H_6O_2)} \times \frac{1\ \text{mol ethylene glycol}}{62.07\ g\ \text{ethylene glycol}} = 2.4\ \text{mol ethylene glycol}$$

$$\frac{1.86\ ^{\circ}C}{\text{mol particles}} \times 2.4\ \text{mol} = 4.5\ ^{\circ}C \qquad \text{Melting point} = -4.5\ ^{\circ}C$$

**8.95**   a. The NaCl solution has a higher boiling point because it contains two particles per mole, whereas the glucose solution contains one per mole.
b. The glucose solution has the higher melting point because it contains one particle per mole, whereas the NaCl solution contains two particles per mole.
c. The NaCl solution has the higher osmotic pressure because it contains two particles per mole, whereas the glucose solution contains one per mole.
d. The glucose solution has a higher vapor pressure at a given temperature because it contains one particle per mole, whereas the NaCl solution contains two particles per mole.

**8.97**   Determine the number of "particles" contained in the solute. If necessary, use 1.86 $^{\circ}$C/mol as a conversion factor to relate the temperature change to the number of moles of solute particles.

a. A 0.10 M glucose solution has a higher melting point than 0.10 M NaOH even though the solutions have the same molar concentration, because the NaOH solution contains twice as many particles. Therefore the NaOH solution will have its melting point reduced by twice as much as the glucose solution.

b. A 0.20 M NaCl solution has a higher melting point than a 0.15 M $CaCl_2$ solution.

$$\frac{1.86\ ^{\circ}C}{\text{mol particles}} \times 0.20\ \text{mol NaCl} \times \frac{2\ \text{mol particles}}{\text{mol NaCl}} = 0.74\ ^{\circ}C \qquad \text{Melting point} = -0.74\ ^{\circ}C$$

**higher melting point**

$$\frac{1.86\ ^{\circ}C}{\text{mol particles}} \times 0.15\ \text{mol } CaCl_2 \times \frac{3\ \text{mol particles}}{\text{mol } CaCl_2} = 0.84\ ^{\circ}C \qquad \text{Melting point} = -0.84\ ^{\circ}C$$

c. A 0.10 M $Na_2SO_4$ solution has a higher melting point than 0.10 M $Na_3PO_4$ even though the solutions have the same molar concentration, because the $Na_2SO_4$ solution contains fewer particles (three vs. four).

d. A 0.10 M glucose solution has a higher melting point than a 0.20 M glucose solution since it has a lower molar concentration, which causes less melting point depression.

**8.99**   a. **A > B.** Since solution A contains a dissolved solute, water will flow from compartment **B** to compartment **A**.
b. **B > A.** Since solution B is more concentrated, water will flow from compartment **A** to compartment **B**.

    c. **No change** will occur since the solutions have an equal number of dissolved particles.

    d. **A > B.** Solution **A** contains more dissolved particles than solution **B.** Therefore, water will flow from compartment **B** to compartment **A.**

    e. **No change** will occur since the solutions have an equal number of dissolved particles (NaCl has two particles per mole, making it equivalent to the glucose solution).

**8.101**  At warmer temperatures, $CO_2$ is less soluble in water and more is in the gas phase and escapes as the can is opened and pressure is reduced.

**8.103**  Calculate the weight/volume percent concentration of glucose, and the molarity.

$$(w/v)\% \; = \; \frac{0.09 \text{ g glucose}}{100 \text{ mL blood}} \; \times \; 100\% \; = \; \begin{array}{c} 0.09\% \text{ (w/v) glucose} \\ \textbf{Answer} \end{array}$$

$$0.09 \text{ g glucose} \times \frac{1 \text{ mol glucose}}{180.2 \text{ g glucose}} = 0.0005 \text{ mol glucose} \qquad \frac{0.0005 \text{ mol glucose}}{0.1 \text{ L blood}} = \begin{array}{c} 0.005 \text{ M} \\ \textbf{Answer} \end{array}$$

**8.105**  a.  280 mL of mannitol solution would have to be given.

$$(w/v)\% \; = \; \frac{70. \text{ g mannitol}}{x \text{ mL solution}} \; \times \; 100\% \; = \; 25\% \text{ (w/v) mannitol} \qquad \begin{array}{c} x = 280 \text{ mL} \\ \textbf{Answer} \end{array}$$

    b.  The hypertonic mannitol solution draws water out of swollen brain cells and thus reduces the pressure on the brain.

**8.107**  When a cucumber is placed in a concentrated salt solution, water moves out of the cells of the cucumber to the hypertonic salt solution, so the cucumber shrinks and loses its crispness.

**8.109**  NaCl, KCl, and glucose are found in the bloodstream. If they weren't in the dialyzer fluid, they would move out of the bloodstream into the dialyzer, and their concentrations in the bloodstream would fall.

**8.111**  Convert ounces to milliliters, and then calculate the weight/volume percent concentration.

$$8.0 \text{ oz} \times \frac{29.6 \text{ mL}}{1 \text{ oz}} = 240 \text{ mL}$$

$$(w/v)\% \; = \; \frac{15 \text{ g complex carbohydrates}}{240 \text{ mL solution}} \; \times \; 100\% \; = \; \begin{array}{c} 6.3\% \text{ (w/v) complex carbohydrates} \\ \textbf{Answer} \end{array}$$

**8.113**  Use the molarity to determine the number of moles of HCl, and then convert this to grams.

$$\text{mol} = M \times V = 0.10 \text{ M HCl} \times 2.0 \text{ L} = 0.20 \text{ mol HCl}$$

$$0.20 \text{ mol HCl} \times \frac{36.46 \text{ g HCl}}{1 \text{ mol HCl}} = \begin{array}{c} 7.3 \text{ g HCl} \\ \textbf{Answer} \end{array}$$

**8.115**

$5.0 \text{ L} = 5.0 \times 10^3 \text{ mL blood}$

$(w/v)\% = \dfrac{x \text{ g ethanol}}{5.0 \times 10^3 \text{ mL blood}} \times 100\% = 0.08\% \text{ (w/v) ethanol}$    $x = 4 \text{ g ethanol}$

$4 \text{ g ethanol} \times \dfrac{1000 \text{ mg}}{1 \text{ g}} = 4{,}000 \text{ mg ethanol}$
**Answer**

**8.117**

a.    $\dfrac{x \text{ g acetaminophen}}{1 \text{ mL}} \times 10^6 = 15 \text{ ppm acetaminophen}$    $x = 0.000\ 015 \text{ g acetaminophen}$

$0.000\ 015 \text{ g acetaminophen} \times \dfrac{10^6 \text{ µg}}{1 \text{ g}} = 15 \text{ µg acetaminophen, which is in the therapeutic range}$

b.    $\dfrac{0.000\ 015 \text{ g acetaminophen}}{1 \text{ mL blood}} \times 5000 \text{ mL blood} \times \dfrac{1 \text{ mol}}{151.2 \text{ g}} = 0.000\ 50 \text{ mol acetaminophen}$
**Answer**

# Chapter 9 Acids and Bases

## Chapter Review

**[1] Describe the principal features of acids and bases. (9.1)**

- A Brønsted–Lowry acid is a proton donor, often symbolized by HA. A Brønsted–Lowry acid must contain one or more hydrogen atoms.

This proton is donated to $H_2O$.

$$HCl(g) \quad + \quad H_2O(l) \quad \longrightarrow \quad H_3O^+(aq) \quad + \quad Cl^-(aq)$$

**Brønsted–Lowry acid**

- A Brønsted–Lowry base is a proton acceptor, often symbolized by B:. To form a bond to a proton, a Brønsted–Lowry base must contain a lone pair of electrons.

This electron pair forms a new bond to a H from $H_2O$.

$$H{-}\overset{\displaystyle H}{\underset{\displaystyle H}{N}}{-}H \quad + \quad H_2O(l) \quad \longrightarrow \quad \left[ H{-}\overset{\displaystyle H}{\underset{\displaystyle H}{N}}{-}H \right]^+ \quad + \quad OH^-(aq)$$

**Brønsted–Lowry base**

**[2] What are the principal features of an acid–base reaction? (9.2)**

- In a Brønsted–Lowry acid–base reaction, a proton is transferred from the acid (HA) to the base (B:). In this reaction, the acid loses a proton to form its conjugate base and the base gains a proton to form its conjugate acid.

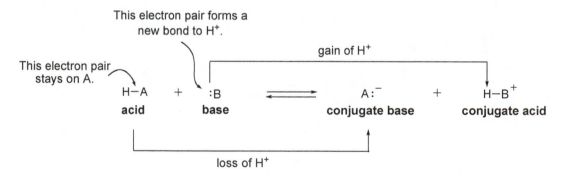

**[3] How is acid strength related to the direction of equilibrium in an acid–base reaction? (9.3)**

- A strong acid readily donates a proton, and when dissolved in water, 100% of the acid dissociates into ions.

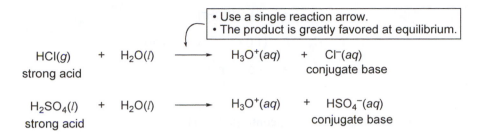

- A strong base readily accepts a proton. When a strong base like NaOH is dissolved in water, 100% of the base dissociates into ions.

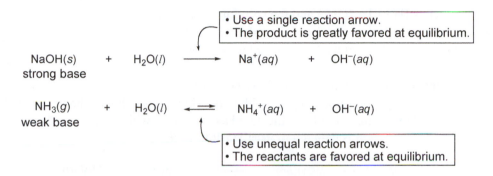

- An inverse relationship exists between acid and base strength. **A strong acid forms a weak conjugate base, whereas a weak acid forms a strong conjugate base.**
- **In an acid–base reaction, the stronger acid reacts with the stronger base to form the weaker acid and weaker base.**

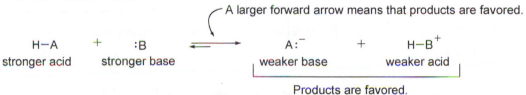

**[4] What is the acid dissociation constant and how is it related to acid strength? (9.4)**
- For a general acid HA, the acid dissociation constant $K_a$ is defined by the following equation:

$$K_a = \frac{[H_3O^+][A:^-]}{[HA]}$$

- **The _stronger_ the acid, the _larger_ the $K_a$.** Equilibrium in an acid–base reaction favors formation of the acid with the smaller $K_a$ value.

**[5] What is the ion–product of water and how is it used to calculate hydronium or hydroxide ion concentration? (9.5)**
- The ion–product of water, $K_w$, is a constant for all aqueous solutions; $K_w = [H_3O^+][OH^-] = 1.0 \times 10^{-14}$ at 25 °C. If either $[H_3O^+]$ or $[OH^-]$ is known, the other value can be calculated from $K_w$.

$$K_w = [H_3O^+][OH^-]$$

$$K_w = (1.0 \times 10^{-7}) \times (1.0 \times 10^{-7})$$

$$K_w = 1.0 \times 10^{-14}$$

## [6] What is pH? (9.6)

- The pH of a solution measures the concentration of $H_3O^+$.

$$pH = -\log[H_3O^+]$$

- A pH = 7 means $[H_3O^+] = [OH^-]$ and the solution is neutral.
- A pH < 7 means $[H_3O^+] > [OH^-]$ and the solution is acidic.
- A pH > 7 means $[H_3O^+] < [OH^-]$ and the solution is basic.

## [7] Draw the products of some common acid–base reactions. (9.7)

- In a Brønsted–Lowry acid–base reaction with hydroxide bases (MOH), the acid HA donates a proton to $OH^-$ to form $H_2O$. The anion from the acid HA combines with the cation $M^+$ of the base to form the salt MA. This reaction is called a **neutralization** reaction.

$$HA(aq) + MOH(aq) \longrightarrow H-OH(l) + MA(aq)$$

$$\text{acid} \qquad \text{base} \qquad\qquad \text{water} \qquad \text{salt}$$

- In acid–base reactions with bicarbonate ($HCO_3^-$) or carbonate ($CO_3^{2-}$) bases, carbonic acid ($H_2CO_3$) is formed, which decomposes to form $H_2O$ and $CO_2$.

$$H^+(aq) + HCO_3^-(aq) \longrightarrow [H_2CO_3(aq)] \longrightarrow H_2O(l) + CO_2(g)$$

bicarbonate $\qquad\qquad\qquad\qquad\qquad\qquad\qquad\qquad$ bubbles of $CO_2$

## [8] What happens to the pH of an aqueous solution when a salt is dissolved? (9.8)

- A salt can form an acidic, basic, or neutral solution depending on whether its cation and anion are derived from strong or weak acids and bases.

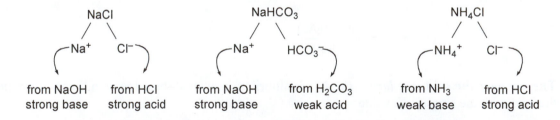

- A salt derived from a strong acid and strong base forms a neutral solution with pH = 7. When one ion of a salt is derived from a weak acid or base, the ion derived from the stronger acid or base determines whether the solution is acidic or basic.

For $NaHCO_3$: $HCO_3^-(aq) + H_2O(l) \rightleftharpoons H_2CO_3(aq) + \boxed{OH^-(aq)}$

Hydroxide makes the solution basic,
so the pH > 7.

For NH₄Cl:   $NH_4^+(aq)$   +   $H_2O(l)$   ⇌   $NH_3(aq)$   +   $\boxed{H_3O^+(aq)}$

$H_3O^+$ makes the solution acidic,
so the pH < 7.

**[9] How is a titration used to determine the concentration of an acid or base? (9.9)**
- A titration is a procedure that uses a base of known volume and molarity to react with a known volume of acid of unknown molarity. The volume and molarity of the base are used to calculate the number of moles of base that react, and from this value, the molarity of the acid can be determined. A titration can also be used to determine the concentration of a base by using an acid of known molarity.

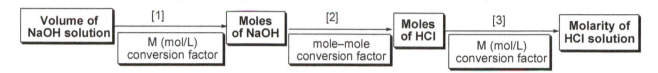

**[10] What is a buffer? (9.10)**
- A buffer is a solution whose pH changes very little when acid or base is added. Most buffers are composed of approximately equal amounts of a weak acid and the salt of its conjugate base.

**[11] What is the principal buffer present in the blood? (9.11)**
- The principal buffer in the blood is carbonic acid/bicarbonate. Since carbonic acid ($H_2CO_3$) is in equilibrium with dissolved $CO_2$, the amount of $CO_2$ in the blood affects its pH, which is normally maintained in the range of 7.35–7.45. When the $CO_2$ concentration in the blood is higher than normal, the acid–base equilibrium shifts to form more $H_3O^+$ and the pH decreases. When the $CO_2$ concentration in the blood is lower than normal, the acid–base equilibrium shifts to consume $[H_3O^+]$, so $[H_3O^+]$ decreases, and the pH increases.

$$CO_2(g) \; + \; H_2O(l) \; \rightleftharpoons \; \underset{\text{carbonic acid}}{H_2CO_3(aq)} \; \overset{H_2O}{\rightleftharpoons} \; H_3O^+(aq) \; + \; \underset{\text{bicarbonate}}{HCO_3^-(aq)}$$

principal buffer in the blood

## Problem Solving

## [1] Introduction to Acids and Bases (9.1)

**Example 9.1** Which of the following species can be Brønsted–Lowry acids?

a. HBr                b. $H_2SO_4$                c. $I_2$

**Analysis**
A Brønsted–Lowry acid must contain a hydrogen atom, but it may be neutral or contain a net positive or negative charge.

**Solution**
a. HBr is a Brønsted–Lowry acid since it contains a H.
b. $H_2SO_4$ is a Brønsted–Lowry acid since it contains a H.
c. $I_2$ is not a Brønsted–Lowry acid because it does not contain a H.

---

**Example 9.2** Which of the following species can be Brønsted–Lowry bases?

a. KOH          b. $Br^-$          c. $C_2H_6$

**Analysis**
A Brønsted–Lowry base must contain a lone pair of electrons, but it may be neutral or have a net negative charge.

**Solution**
a. KOH is a base since it contains hydroxide, $OH^-$, which has three lone pairs on its O atom.
b. $Br^-$ is a base since it has four lone pairs.
c. $C_2H_6$ is not a base since it has no lone pairs.

---

**Example 9.3** Classify each reactant as a Brønsted–Lowry acid or base.

a. $Cl^-(aq)$   +   $HSO_4^-(aq)$  ⇌  $HCl(aq)$   +   $SO_4^{2-}(aq)$

b. $HPO_4^{2-}(aq)$   +   $OH^-(aq)$  ⇌  $PO_4^{3-}(aq)$   +   $H_2O(l)$

**Analysis**
In each equation, the Brønsted–Lowry acid is the species that loses a proton and the Brønsted–Lowry base is the species that gains a proton.

**Solution**
a. $HSO_4^-$ is the acid since it loses a proton ($H^+$) to form $SO_4^{2-}$, and $Cl^-$ is the base since it gains a proton to form HCl.

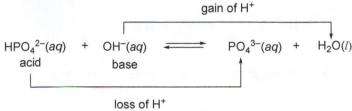

b. $HPO_4^{2-}$ is the acid since it loses a proton ($H^+$) to form $PO_4^{3-}$, and $OH^-$ is the base since it gains a proton to form $H_2O$.

## [2] The Reaction of a Brønsted–Lowry Acid with a Brønsted–Lowry Base (9.2)

**Example 9.4** Draw the conjugate acid of each base:

a. $Br^-$    b. $HPO_4^{2-}$

### Analysis
Conjugate acid–base pairs differ by the presence of a proton. To draw a conjugate acid from a base, *add* a proton, $H^+$. Then add +1 to the charge of the base to give the charge on the conjugate acid.

### Solution
a. $Br^- + H^+$ gives HBr as the conjugate acid. HBr has no charge since a proton with a +1 charge is added to an anion with a –1 charge.
b. $HPO_4^{2-} + H^+$ gives $H_2PO_4^-$ as the conjugate acid. $H_2PO_4^-$ has a –1 charge since a proton with a +1 charge is added to an anion with a –2 charge.

**Example 9.5** Draw the conjugate base of each acid:

a. $H_3O^+$    b. $H_2Se$

### Analysis
Conjugate acid–base pairs differ by the presence of a proton. To draw a conjugate base from an acid, *remove* a proton, $H^+$. Then, add –1 to the charge of the acid to give the charge on the conjugate base.

### Solution
a. Remove $H^+$ from $H_3O^+$ to form $H_2O$, the conjugate base. $H_2O$ has no charge since –1 is added to a cation that has a +1 charge to begin with.
b. Remove $H^+$ from $H_2Se$ to form $HSe^-$, the conjugate base. $HSe^-$ has a –1 charge since –1 is added to a molecule that had no charge to begin with.

## [3] Acid and Base Strength (9.3)

**Example 9.6** Using Table 9.1:

    a.  Is $H_2SO_4$ or $NH_4^+$ the stronger acid?
    b.  Draw the conjugate base of each acid and predict which base is stronger.

### Analysis
Use Table 9.1 to determine the stronger acid. The stronger the acid, the weaker the conjugate base.

### Solution
a. $H_2SO_4$ is located above $NH_4^+$ in Table 9.1, making it the stronger acid.
b. To draw each conjugate base, remove a proton ($H^+$). Since $NH_4^+$ is the weaker acid, $NH_3$ is the stronger conjugate base.

$$H_2SO_4 \xrightarrow{\text{lose } H^+} HSO_4^- \quad \boxed{\text{weaker base}}$$

stronger acid

$$NH_4^+ \xrightarrow{\text{lose } H^+} NH_3 \quad \boxed{\text{stronger base}}$$

weaker acid

## [4] Equilibrium and Acid Dissociation Constants (9.4)

**Example 9.7** Which acid is stronger: HF or $H_3PO_4$?

**Analysis**
Use Table 9.2 to find the $K_a$ for each acid. The acid with the larger $K_a$ is the stronger acid.

**Solution**

| HF | $H_3PO_4$ |
|---|---|
| $K_a = 7.2 \times 10^{-4}$ | $K_a = 7.5 \times 10^{-3}$ |
| | larger $K_a$ |
| | **stronger acid** |

## [5] The Dissociation of Water (9.5)

**Example 9.8** Calculate the value of $[H_3O^+]$ and $[OH^-]$ in a 0.001 M KOH solution.

**Analysis**
Since KOH is a strong base that completely dissociates to form $K^+$ and $OH^-$, the concentration of KOH gives the concentration of $OH^-$ ions. The $[OH^-]$ can then be used to calculate $[H_3O^+]$ from the expression for $K_w$.

**Solution**
The value of $[OH^-]$ in a 0.001 M KOH solution is 0.001 M $= 1 \times 10^{-3}$ M.

$$[H_3O^+] = \frac{K_w}{[OH^-]} = \frac{1 \times 10^{-14}}{1 \times 10^{-3}} = 1 \times 10^{-11} \text{ M}$$

concentration of $H_3O^+$

concentration of $OH^-$

## [6] Common Acid–Base Reactions (9.7)

**Example 9.9** Write a balanced equation for the reaction of $HNO_3$ with LiOH.

**Analysis**
The acid and base react to form a salt and water.

**Solution**
$HNO_3$ is the acid and LiOH is the base. $H^+$ from the acid reacts with $OH^-$ from the base to give $H_2O$. A salt ($LiNO_3$) is also formed from the cation of the base ($Li^+$) and the anion of the acid ($NO_3^-$).

$$HNO_3(aq) \quad + \quad LiOH(aq) \quad \longrightarrow \quad LiNO_3(aq) \quad + \quad H_2O(l)$$

acid        base            salt

## [7] The Acidity and Basicity of Salt Solutions (9.8)

**Example 9.10** Determine whether each salt forms an acidic, basic, or neutral solution when dissolved in water.

       a. NaBr                  b. $KCH_3COO$

**Analysis**
Determine what type of acid and base (strong or weak) are used to form the salt. When the ions in the salt come from a strong acid and strong base, the solution is neutral. When the ions come from acids and bases of different strength, the ion derived from the stronger reactant determines the acidity.

**Solution**

a.

NaBr

$Na^+$    $Br^-$

from NaOH      from HBr
**strong base**     **strong acid**

**neutral** solution

b.

$KCH_3COO$

$K^+$    $CH_3COO^-$

from KOH      from $CH_3COOH$
**strong base**     weak acid

**basic** solution

## [8] Titration (9.9)

**Example 9.11** What is the molarity of an HCl solution if 17.2 mL of 0.15 M NaOH are needed to neutralize 5.00 mL of the sample?

$$HCl(aq) \quad + \quad NaOH(aq) \quad \longrightarrow \quad NaCl(aq) \quad + \quad H_2O(l)$$

**Analysis and Solution**
**[1] Determine the number of moles of base used to neutralize the acid.**
- Use the molarity (M) and volume ($V$) of the base to calculate the number of moles (mol = $MV$).

$$17.2 \text{ mL NaOH} \quad \times \quad \frac{1 \text{ L}}{1000 \text{ mL}} \quad \times \quad \frac{0.15 \text{ mol NaOH}}{1 \text{ L}} \quad = \quad 0.0026 \text{ mol NaOH}$$

**[2] Determine the number of moles of acid that react from the balanced chemical equation.**
- Since each HCl molecule contains one proton, *one* mole of the acid HCl reacts with *one* mole of the base NaOH in the neutralization reaction. The coefficients in the balanced equation form a mole ratio to calculate the number of moles of acid that react.

$$0.0026 \text{ mol NaOH} \quad \times \quad \frac{1 \text{ mol HCl}}{1 \text{ mol NaOH}} \quad = \quad 0.0026 \text{ mol HCl}$$

**[3] Determine the molarity of the acid from the number of moles and known volume.**

$$\begin{array}{ccc} M \\ \text{molarity} \end{array} = \frac{\text{mol}}{L} = \frac{0.0026 \text{ mol HCl}}{5.00 \text{ mL solution}} \times \frac{1000 \text{ mL}}{1 L} = \begin{array}{c} 0.52 \text{ M HCl} \\ \textbf{Answer} \end{array}$$

## [9] Buffers (9.10)

**Example 9.12** What is the pH of a buffer that contains 0.15 M $NaH_2PO_4$ and 0.15 M $Na_2HPO_4$?

**Analysis**
Use the $K_a$ of the weak acid of the buffer from Table 9.5 and the expression $[H_3O^+] = K_a([HA]/[A:^-])$ to calculate $[H_3O^+]$. Calculate the pH using the expression $pH = -\log [H_3O^+]$.

**Solution**
**[1] Substitute the given concentrations of $NaH_2PO_4$ and $Na_2HPO_4$ for [HA] and [A:$^-$], respectively. $K_a$ for $NaH_2PO_4$ is $6.2 \times 10^{-8}$.**

$$[H_3O^+] = K_a \times \frac{[NaH_2PO_4]}{[Na_2HPO_4]} = (6.2 \times 10^{-8}) \times \frac{[0.15 \text{ M}]}{[0.15 \text{ M}]}$$

$$[H_3O^+] = 6.2 \times 10^{-8} \text{ M}$$

**[2] Use an electronic calculator to convert the $H_3O^+$ concentration to pH.**

$$pH = -\log [H_3O^+] = -\log(6.2 \times 10^{-8})$$

$$pH = 7.21$$

## Self-Test

**[1] Fill in the blank with one of the terms listed below.**

| | | |
|---|---|---|
| Acid (9.1) | Brønsted–Lowry base (9.1) | Monoprotic acid (9.1) |
| Amphoteric (9.2) | Buffer (9.10) | Net ionic equation (9.7) |
| Base (9.1) | Conjugate acid (9.2) | Proton transfer reaction (9.2) |
| Brønsted–Lowry acid (9.1) | Conjugate base (9.2) | Triprotic acid (9.1) |

1. A _____ contains only the species involved in a reaction.
2. A _____ is a proton acceptor.
3. According to the Arrhenius definition, an _____ contains a hydrogen atom and dissolves in water to form a hydrogen ion, $H^+$.

4.  A _____ is a solution whose pH changes very little when acid or base is added.
5.  A compound that contains both a hydrogen atom and a lone pair of electrons can be either an acid or a base, depending on the particular reaction. Such a compound is said to be _____.
6.  The product formed by loss of a proton from an acid is called its _____.
7.  A Brønsted–Lowry acid–base reaction is a _____ since it always results in the transfer of a proton from an acid to a base.
8.  A _____ contains *three* acidic protons.
9.  The product formed by gain of a proton by a base is called its _____.
10. According to the Arrhenius definition, a _____ contains hydroxide and dissolves in water to form OH⁻.
11. A _____ is a proton donor.
12. A _____ contains *one* acidic proton.

**[2] Fill in the blank with one of the terms listed below.**

> a. strong acid          c. strong base
> b. weak acid            d. weak base

13. A _____ donates a proton, but when dissolved in water, only a small fraction of it dissociates into ions.
14. A _____ accepts a proton, but when dissolved in water, only a small fraction of it forms ions.
15. A _____ readily accepts a proton, forming a *weak* conjugate acid.
16. A _____ readily donates a proton, forming a *weak* conjugate base.

**[3] Fill in the blank with one of the terms listed below.**

> a. an acidic solution          b. a basic solution          c. a neutral solution

17. A salt derived from a *strong base* and a weak acid forms _____.
18. A salt derived from a weak base and a *strong acid* forms _____.
19. A solution with a pH = 7 is _____.
20. A solution with a pH > 7 is _____.
21. A solution with a pH < 7 is _____.

**[4] Decide if each compound is a Brønsted–Lowry acid, a Brønsted–Lowry base, or both.**

> a. Brønsted–Lowry acid          b. Brønsted–Lowry base

22. $Ca(OH)_2$          23. $H_2SO_4$          24. $H_2\ddot{O}:$          25.

## Answers to Self-Test

| | | | | |
|---|---|---|---|---|
| 1. net ionic equation | 6.  conjugate base | 11. Brønsted–Lowry acid | 16. a | 21. a |
| 2. Brønsted–Lowry base | 7.  proton transfer reaction | 12. monoprotic acid | 17. b | 22. b |
| 3. acid | 8.  triprotic acid | 13. b | 18. a | 23. a |
| 4. buffer | 9.  conjugate acid | 14. d | 19. c | 24. a, b |
| 5. amphoteric | 10. base | 15. c | 20. b | 25. a, b |

## Solutions to In-Chapter Problems

**9.1** A Brønsted–Lowry acid must contain a hydrogen atom, but it may be neutral or contain a net positive or negative charge. Use Example 9.1 to help determine which of the compounds are Brønsted–Lowry acids.

    a. HI: contains a H atom,              c. $H_2PO_4^-$: contains a H atom,
       a Brønsted–Lowry acid                 a Brønsted–Lowry acid
    b. $SO_4^{2-}$: no H atom                  d. $Cl^-$: no H atom

**9.2** A Brønsted–Lowry base must contain a lone pair of electrons, but it may be neutral or have a net negative charge. Use Example 9.2 to help determine which of the compounds are Brønsted–Lowry bases.

    a. $Al(OH)_3$: lone pairs on OH,         c. $NH_4^+$: no lone pair of electrons
       a Brønsted–Lowry base
    b. $Br^-$: lone pairs on Br,             d. $CN^-$: lone pairs on C and N,
       a Brønsted–Lowry base                 a Brønsted–Lowry base

**9.3** In each equation, the Brønsted–Lowry acid is the species that loses a proton and the Brønsted–Lowry base is the species that gains a proton. Use Example 9.3 to help determine which reactant is an acid and which is a base.

    a.     $HCl(g)$   +   $NH_3(g)$  ⟶  $Cl^-(aq)$   +   $NH_4^+(aq)$
             acid             base

    b.    $CH_3COOH(l)$  +  $H_2O(l)$  ⟶  $CH_3COO^-(aq)$  +  $H_3O^+(aq)$
             acid             base

    c.     $OH^-(aq)$  +  $HSO_4^-(aq)$  ⟶  $H_2O(l)$   +   $SO_4^{2-}(aq)$
             base            acid

**9.4** Use Example 9.4 to draw the conjugate acid of each species. Conjugate acid–base pairs differ by the presence of a proton. To draw a conjugate acid from a base, *add* a proton, $H^+$. Then add +1 to the charge of the base to give the charge on the conjugate acid.

    a. $H_2O$: add one $H^+$ to make $H_3O^+$.
    b. $I^-$: add one $H^+$ to make HI.
    c. $HCO_3^-$: add one $H^+$ to make $H_2CO_3$.

**9.5** Use Example 9.5 to draw the conjugate base of each species. Conjugate acid–base pairs differ by the presence of a proton. To draw a conjugate base from an acid, *remove* a proton, $H^+$. Then, add –1 to the charge of the acid to give the charge on the conjugate base.

    a. $H_2S$: remove one $H^+$ to make $HS^-$.
    b. HCN: remove one $H^+$ to make $CN^-$.
    c. $HSO_4^-$: remove one $H^+$ to make $SO_4^{2-}$.

**9.6** Use Sample Problem 9.6 to label the acid and base in each reaction. The Brønsted–Lowry acid loses a proton to form its conjugate base. The Brønsted–Lowry base gains a proton to form its conjugate acid.

        base          acid                 conjugate base    conjugate acid

  a.  $H_2O(l)$  +  $HI(g)$  $\rightleftharpoons$  $I^-(aq)$  +  $H_3O^+(aq)$

          acid                 base            conjugate base    conjugate acid

  b.  $CH_3COOH(l)$  +  $NH_3(g)$  $\rightleftharpoons$  $CH_3COO^-(aq)$  +  $NH_4^+(aq)$

        base          acid               conjugate acid    conjugate base

  c.  $Br^-(aq)$  +  $HNO_3(aq)$  $\rightleftharpoons$  $HBr(aq)$  +  $NO_3^-(aq)$

**9.7** Draw the conjugate acid and base of ammonia as in Examples 9.4 and 9.5.

    a. $NH_3$: add one $H^+$ to make $NH_4^+$.               b. $NH_3$: remove one $H^+$ to make $NH_2^-$.

**9.8** The stronger the acid, the more readily it dissociates to form its conjugate base. In molecular art, the strongest acid has the most $A^-$ and $H_3O^+$ ions, and the fewest molecules of undissociated HA.

    **D** (smallest amount of $A^-$ and $H_3O^+$) < **F** < **E** (largest amount of $A^-$ and $H_3O^+$)

**9.9** Use Table 9.1 to determine which acid is stronger. The stronger the acid, the weaker the conjugate base.

    a. $H_2SO_4$ is the stronger acid; $H_3PO_4$ has the stronger conjugate base.
    b. HCl is the stronger acid; HF has the stronger conjugate base.
    c. $H_2CO_3$ is the stronger acid; $NH_4^+$ has the stronger conjugate base.
    d. HF is the stronger acid; HCN has the stronger conjugate base.

**9.10** Draw the conjugate acid of each species as in Example 9.4. Then compare the acids.

    a. $NO_2^-$: The conjugate acid is $HNO_2$.
       $NO_3^-$: The conjugate acid is $HNO_3$.
    b. Since $NO_2^-$ is the stronger base, it has a weaker conjugate acid. Therefore, $HNO_3$ is the stronger acid.

**9.11** To determine if the reactants or products are favored at equilibrium:
- Identify the acid in the reactants and the conjugate acid in the products.
- Determine the relative strength of the acid and the conjugate acid.
- Equilibrium favors the formation of the weaker acid.

  a.  $HF(g)$  +  $OH^-(aq)$  $\rightleftharpoons$  $F^-(aq)$  +  $H_2O(l)$

       acid                                      conjugate acid
                                            weaker acid
                                    **Products are favored.**

  b.  $NH_4^+(aq)$  +  $Cl^-(aq)$  $\rightleftharpoons$  $NH_3(g)$  +  $HCl(aq)$

        acid                                     conjugate acid
       weaker acid
  **Reactants are favored.**

c. $HCO_3^-(aq)$ + $H_3O^+(aq)$ ⇌ $H_2CO_3(aq)$ + $H_2O(l)$

                  acid                               conjugate acid
                                              weaker acid
**Products are favored.**

**9.12**    Compare the acid and conjugate acid to determine if the reactants or products are favored.

a. $C_3H_6O_3(aq)$ + $H_2O(l)$ ⇌ $C_3H_5O_3^-(aq)$ + $H_3O^+(aq)$

          acid                                      conjugate acid
      weaker acid
**Reactants are favored.**

b. $C_3H_6O_3(aq)$ + $HCO_3^-(aq)$ ⇌ $C_3H_5O_3^-(aq)$ + $H_2CO_3(aq)$

         acid                                    conjugate acid
                                             weaker acid
**Products are favored.**

**9.13**    Use Table 9.2 to find the $K_a$ for each acid as in Example 9.7. The acid with the larger $K_a$ is the stronger acid.

a. increasing acid strength: $HPO_4^{2-}$, $H_2PO_4^-$, $H_3PO_4$
b. increasing acid strength: HCN, $CH_3COOH$, HF

**9.14**    The stronger acid has the larger $K_a$. The weaker acid has the stronger conjugate base.

a. $H_3PO_4$ is a stronger acid than $CH_3COOH$.
b. Remove one $H^+$ to draw the conjugate base: $H_2PO_4^- < CH_3COO^-$ (the stronger base is formed from the weaker acid).

**9.15**    To determine the direction of equilibrium, identify the acid in the reactants and the conjugate acid in the products as in Sample Problem 9.10. Then compare their $K_a$ values. **Equilibrium favors the formation of the acid with the smaller $K_a$ value.**

                    reactants favored

$HCO_3^-(aq)$ + $NH_3(aq)$ ⇌ $CO_3^{2-}(aq)$ + $NH_4^+(aq)$

        acid                                  conjugate acid
$K_a = 5.6 \times 10^{-11}$                          $K_a = 5.6 \times 10^{-10}$
     smaller $K_a$                                 larger $K_a$
    **weaker acid**                             **stronger acid**

**9.16**    Compare the acids by comparing their $K_a$ values from Table 9.2.
a. $H_2CO_3$ has the larger $K_a$.
b. $H_2CO_3$ is stronger.
c. HCN has the stronger conjugate base.
d. $H_2CO_3$ has the weaker conjugate base.
e. When $H_2CO_3$ is dissolved in water, the equilibrium lies further to the right.

**9.17** Use the equation $[OH^-] = K_w/[H_3O^+]$ to calculate the hydroxide ion concentration as in Sample Problem 9.11. When $[OH^-] > [H_3O^+]$, the solution is basic. When $[OH^-] < [H_3O^+]$, the solution is acidic.

a. $[OH^-] = \dfrac{K_w}{[H_3O^+]} = \dfrac{1.0 \times 10^{-14}}{10^{-3}} = 10^{-11}$ M

        acidic

b. $[OH^-] = \dfrac{K_w}{[H_3O^+]} = \dfrac{1.0 \times 10^{-14}}{10^{-11}} = 10^{-3}$ M

        basic

c. $[OH^-] = \dfrac{K_w}{[H_3O^+]} = \dfrac{1.0 \times 10^{-14}}{2.8 \times 10^{-10}} = 3.6 \times 10^{-5}$ M

        basic

d. $[OH^-] = \dfrac{K_w}{[H_3O^+]} = \dfrac{1.0 \times 10^{-14}}{5.6 \times 10^{-4}} = 1.8 \times 10^{-11}$ M

        acidic

**9.18** Use the equation $[H_3O^+] = K_w/[OH^-]$ to calculate the hydronium ion concentration as in Sample Problem 9.11.

a. $[H_3O^+] = \dfrac{K_w}{[OH^-]} = \dfrac{1.0 \times 10^{-14}}{10^{-6}} = 10^{-8}$ M

        basic

b. $[H_3O^+] = \dfrac{K_w}{[OH^-]} = \dfrac{1.0 \times 10^{-14}}{10^{-9}} = 10^{-5}$ M

        acidic

c. $[H_3O^+] = \dfrac{K_w}{[OH^-]} = \dfrac{1.0 \times 10^{-14}}{5.2 \times 10^{-11}} = 1.9 \times 10^{-4}$ M

        acidic

d. $[H_3O^+] = \dfrac{K_w}{[OH^-]} = \dfrac{1.0 \times 10^{-14}}{7.3 \times 10^{-4}} = 1.4 \times 10^{-11}$ M

        basic

**9.19** Since NaOH is a strong base that completely dissociates to form $Na^+$ and $OH^-$, the concentration of NaOH gives the concentration of $OH^-$ ions. The $[OH^-]$ can then be used to calculate $[H_3O^+]$ from the expression for $K_w$. Similarly, since HCl is a strong acid and completely dissociates, the $[H_3O^+]$ can then be used to calculate $[OH^-]$ from the expression for $K_w$. See Example 9.8.

a. $[H_3O^+] = \dfrac{K_w}{[OH^-]} = \dfrac{1 \times 10^{-14}}{10^{-3}} = 10^{-11}$ M

        concentration of $H_3O^+$

    concentration of $OH^-$

b. $[OH^-] = \dfrac{K_w}{[H_3O^+]} = \dfrac{1 \times 10^{-14}}{10^{-3}} = 10^{-11}$ M

concentration of $H_3O^+$

concentration of $OH^-$

c. $[OH^-] = \dfrac{K_w}{[H_3O^+]} = \dfrac{1 \times 10^{-14}}{1.5} = 6.7 \times 10^{-15}$ M

concentration of $H_3O^+$

concentration of $OH^-$

d. $[H_3O^+] = \dfrac{K_w}{[OH^-]} = \dfrac{1 \times 10^{-14}}{3.0 \times 10^{-1}} = 3.3 \times 10^{-14}$ M

concentration of $OH^-$

concentration of $H_3O^+$

**9.20** When the coefficient of a number written in scientific notation is one, the pH equals the value of $x$ in $10^{-x}$.

a. $1 \times 10^{-6}$ M: pH = 6

b. $1 \times 10^{-12}$ M: pH = 12

c. 0.000 01 M = $10^{-5}$: pH = 5

d. 0.000 000 000 01 M = $10^{-11}$: pH = 11

**9.21** The pH equals the value of $x$ in $10^{-x}$, and this gives the $H_3O^+$ concentration. An *acidic* solution has a pH < 7. A *basic* solution has a pH > 7. A *neutral* solution has a pH = 7.

a. pH = 13, $[H_3O^+] = 1 \times 10^{-13}$ M: basic

b. pH = 7, $[H_3O^+] = 1 \times 10^{-7}$ M: neutral

c. pH = 3, $[H_3O^+] = 1 \times 10^{-3}$ M: acidic

**9.22** Use a calculator to determine the antilogarithm of (– pH); $[H_3O^+] =$ antilog(–pH).

a. $[H_3O^+]$ = antilog(–pH) = antilog(–10.2)

$[H_3O^+]$ = $6 \times 10^{-11}$ M

b. $[H_3O^+]$ = antilog(–pH) = antilog(–7.8)

$[H_3O^+]$ = $2 \times 10^{-8}$ M

c. $[H_3O^+]$ = antilog(–pH) = antilog(–4.3)

$[H_3O^+]$ = $5 \times 10^{-5}$ M

**9.23** Use a calculator to determine the logarithm of a number that contains a coefficient other than one in scientific notation; pH = –log $[H_3O^+]$.

a. pH = –log $[H_3O^+]$ = –log($1.8 \times 10^{-6}$)

= –(–5.74) = 5.74

b. pH = –log $[H_3O^+]$ = –log($9.21 \times 10^{-12}$)

= –(–11.036) = 11.036

c. pH = –log $[H_3O^+]$ = –log($8.8 \times 10^{-5}$)

= –(–4.06) = 4.06

d. pH = –log $[H_3O^+]$ = –log($7.62 \times 10^{-11}$)

= –(–10.118) = 10.118

**9.24**   Label the organs based on the definitions listed in Answer 9.21.

saliva—acidic to slightly basic                   blood—basic
pancreas—basic                                    stomach—acidic
small intestines—basic                            large intestines—acidic to neutral
urine—acidic to basic

**9.25**   Write the balanced equations as in Section 9.7A.

a.   $HNO_3(aq)$  +   $NaOH(aq)$  $\longrightarrow$   $H_2O(l)$  +  $NaNO_3(aq)$
                                                        water          salt

b.   $H_2SO_4(aq)$  +   $KOH(aq)$  $\longrightarrow$   $H_2O(l)$  +  $K_2SO_4(aq)$
                                                        water          salt

     $H_2SO_4(aq)$  +   2 $KOH(aq)$  $\longrightarrow$   2 $H_2O(l)$  +  $K_2SO_4(aq)$

           Place a 2 to balance K.                      Place a 2 to balance H and O.

**9.26**   A net ionic equation contains only the species involved in a reaction.

$H^+(aq)$  +   $OH^-(aq)$  $\longrightarrow$   $H_2O(l)$ for Problem 9.25a, b

**9.27**   The acid and base react to form a salt and carbonic acid ($H_2CO_3$), which decomposes to $CO_2$ and $H_2O$ as in Sample Problem 9.17.

$H_2SO_4(aq)$  +   $CaCO_3(s)$  $\longrightarrow$   $CaSO_4(aq)$   +   $H_2O(l)$ + $CO_2(g)$
                                                                   from $H_2CO_3$

**9.28**   The acid and base react to form a salt and carbonic acid ($H_2CO_3$), which decomposes to $CO_2$ and $H_2O$ as in Sample Problem 9.17.

a.   $HNO_3(aq)$  +   $NaHCO_3(aq)$  $\longrightarrow$   $NaNO_3(aq)$   +   $H_2O(l)$ + $CO_2(g)$
                                                                         from $H_2CO_3$

b.   2 $HNO_3(aq)$  +  $MgCO_3(s)$  $\longrightarrow$   $Mg(NO_3)_2(aq)$   +   $H_2O(l)$ + $CO_2(g)$
                                                                           from $H_2CO_3$

**9.29**   Follow Example 9.10 to determine the acidity when a salt is dissolved in water. Determine what types of acid and base (strong or weak) are used to form the salt. When the ions in the salt come from a strong acid and strong base, the solution is neutral. When the ions come from acids and bases of different strength, the ion derived from the stronger reactant determines the acidity.

a.  KI

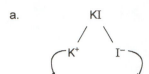

from KOH          from HI
**strong base   strong acid**

**neutral** solution

c.  Ca(NO₃)₂

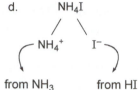

from Ca(OH)₂    from HNO₃
**strong base    strong acid**

**neutral** solution

e.  BaCl₂

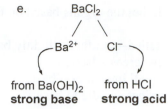

from Ba(OH)₂    from HCl
**strong base    strong acid**

**neutral** solution

b.  K₂CO₃

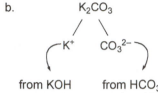

from KOH          from HCO₃⁻
**strong base**    weak acid

**basic** solution

d.  NH₄I

from NH₃          from HI
weak base        **strong acid**

**acidic** solution

f.  Na₃PO₄

from NaOH        from HPO₄²⁻
**strong base**    weak acid

**basic** solution

**9.30**   A basic solution has a pH > 7.

a.  LiCl

from LiOH         from HCl
**strong base   strong acid**

**neutral** solution

c.  NH₄Br

from NH₃          from HBr
weak base        **strong acid**

**acidic** solution

b.  K₂CO₃

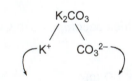

from KOH          from HCO₃⁻
**strong base**    weak acid

**basic** solution
**pH > 7**

d.  MgCO₃

from Mg(OH)₂    from HCO₃⁻
**strong base**    weak acid

**basic** solution
**pH > 7**

**9.31**   Calculate the molarity of the solution using a titration as in Example 9.11.

$$25.5 \text{ mL NaOH} \quad \times \quad \frac{1 \text{ L}}{1000 \text{ mL}} \quad \times \quad \frac{0.24 \text{ mol NaOH}}{1 \text{ L}} \quad = \quad 0.0061 \text{ mol NaOH}$$

$$0.0061 \text{ mol NaOH} \quad \times \quad \frac{1 \text{ mol HCl}}{1 \text{ mol NaOH}} \quad = \quad 0.0061 \text{ mol HCl}$$

$$M = \frac{\text{mol}}{\text{L}} = \frac{0.0061 \text{ mol HCl}}{15 \text{ mL solution}} \quad \times \quad \frac{1000 \text{ mL}}{1 \text{ L}} \quad = \quad 0.41 \text{ M HCl}$$

molarity                                                              **Answer**

**9.32**  Calculate the number of milliliters of NaOH solution needed.

$$5.0 \text{ mL } H_2SO_4 \quad \times \quad \frac{1 \text{ L}}{1000 \text{ mL}} \quad \times \quad \frac{6.0 \text{ mol } H_2SO_4}{1 \text{ L}} \quad = \quad 0.030 \text{ mol } H_2SO_4$$

$$0.030 \text{ mol } H_2SO_4 \quad \times \quad \frac{2 \text{ mol NaOH}}{1 \text{ mol } H_2SO_4} \quad = \quad 0.060 \text{ mol NaOH}$$

$$0.060 \text{ mol NaOH} \quad \times \quad \frac{1 \text{ L}}{2.0 \text{ mol NaOH}} \quad \times \quad \frac{1000 \text{ mL}}{1 \text{ L}} \quad = \quad 30. \text{ mL NaOH solution}$$

**9.33**  **A *buffer* is a solution whose pH changes very little when acid or base is added.** Most buffers are solutions composed of approximately equal amounts of a weak acid and the salt of its conjugate base.

    a.  A solution containing HBr and NaBr is not a buffer, because it contains a strong acid, HBr.
    b.  A solution containing HF and KF is a buffer since HF is a weak acid and $F^-$ is its conjugate base.
    c.  A solution containing $CH_3COOH$ alone is not a buffer since it contains a weak acid only.

**9.34**

    a.  $HCO_3^-$ and $CO_3^{2-}$ are both needed because the $H_3O^+$ concentration depends on two terms—$K_a$, which is a constant, and the ratio of the concentrations of the weak acid and its conjugate base. If these concentrations do not change much, the concentration of $H_3O^+$ and therefore the pH do not change much.
    b.  When a small amount of acid is added, $[HCO_3^-]$ increases and $[CO_3^{2-}]$ decreases.
    c.  When a small amount of base is added, $[HCO_3^-]$ decreases and $[CO_3^{2-}]$ increases.

**9.35**  Calculate the pH of the buffer as in Example 9.12.  The pH is the same if equal concentrations of the weak acid and conjugate base are present.

    a.  $[H_3O^+] \;=\; K_a \times \dfrac{[H_2PO_4^-]}{[HPO_4^{2-}]} \;=\; (6.2 \times 10^{-8}) \times \dfrac{[0.10 \text{ M}]}{[0.10 \text{ M}]}$

$$[H_3O^+] \;=\; 6.2 \times 10^{-8} \text{ M}$$

$$pH \;=\; -\log [H_3O^+] \;=\; -\log(6.2 \times 10^{-8})$$

$$pH \;=\; 7.21$$

    b.  $[H_3O^+] \;=\; K_a \times \dfrac{[H_2PO_4^-]}{[HPO_4^{2-}]} \;=\; (6.2 \times 10^{-8}) \times \dfrac{[1.0 \text{ M}]}{[1.0 \text{ M}]}$

$$[H_3O^+] \;=\; 6.2 \times 10^{-8} \text{ M}$$

$$pH \;=\; -\log [H_3O^+] \;=\; -\log(6.2 \times 10^{-8})$$

$$pH \;=\; 7.21$$

c. $[H_3O^+] = K_a \times \dfrac{[H_2PO_4^-]}{[HPO_4^{2-}]} = (6.2 \times 10^{-8}) \times \dfrac{[0.50\ M]}{[0.50\ M]}$

$[H_3O^+] = 6.2 \times 10^{-8}\ M$

$pH = -\log[H_3O^+] = -\log(6.2 \times 10^{-8})$

$pH = 7.21$

**9.36**   Calculate the pH of the buffer as in Example 9.12.

$[H_3O^+] = K_a \times \dfrac{[CH_3COOH]}{[CH_3COO^-]} = (1.8 \times 10^{-5}) \times \dfrac{[0.20\ M]}{[0.15\ M]}$

$[H_3O^+] = 2.4 \times 10^{-5}\ M$

$pH = -\log[H_3O^+] = -\log(2.4 \times 10^{-5})$

$pH = 4.62$

## Solutions to Odd-Numbered End-of-Chapter Problems

**9.37**  A Brønsted–Lowry acid must contain a hydrogen atom, but it may be neutral or contain a net positive or negative charge.  Use Example 9.1 to help determine which of the compounds are Brønsted–Lowry acids.

a. HBr: contains a H atom,
   a Brønsted–Lowry acid
b. $Br_2$: no H atom
c. $AlCl_3$: no H atom

d. HCOOH: contains a H atom,
   a Brønsted–Lowry acid
e. $NO_2^-$: no H atom
f. $HNO_2$: contains a H atom,
   a Brønsted–Lowry acid

**9.39** A Brønsted–Lowry base must contain a lone pair of electrons, but it may be neutral or have a net negative charge. Use Example 9.2 to help determine which of the compounds are Brønsted–Lowry bases.

a. $OH^-$: lone pairs on OH,
   a Brønsted–Lowry base
b. $Ca^{2+}$: no lone pair of electrons

c. $C_2H_6$: no lone pair of electrons

d. $PO_4^{3-}$: lone pairs on O,
   a Brønsted–Lowry base
e. $OCl^-$: lone pairs on O and Cl,
   a Brønsted–Lowry base
f. $MgCO_3$: lone pairs on O,
   a Brønsted–Lowry base

**9.41** Draw the conjugate acid of each species as in Example 9.4.

a. $HS^-$: add one $H^+$ to make $H_2S$.

b. $CO_3^{2-}$: add one $H^+$ to make $HCO_3^-$.

c. $NO_2^-$: add one $H^+$ to make $HNO_2$.

d.   H—C—N—H     add one $H^+$     $\left[ \text{H—C—N—H} \right]^+$
        to make

9.43 Draw the conjugate base of each species as in Example 9.5.

a. $HNO_2$: remove one $H^+$ to make $NO_2^-$.     c. $H_2O_2$: remove one $H^+$ to make $HO_2^-$.
b. $NH_4^+$: remove one $H^+$ to make $NH_3$.

9.45 The acid loses a proton (gray sphere) to form its conjugate base, while the base gains a proton to form its conjugate acid.

a.     acid     +     base     ⟶     conjugate base     +     conjugate acid

b.     base     +     acid     ⟶     conjugate acid     +     conjugate base

9.47 Label the conjugate acid–base pairs in each reaction as in Answer 9.6.

a.     $HI(g)$ acid                     $I^-(aq)$        conjugate base
       $NH_3(g)$ base                   $NH_4^+(aq)$     conjugate acid

b.     $HCOOH(l)$ acid                  $HCOO^-(aq)$     conjugate base
       $H_2O(l)$ base                   $H_3O^+(aq)$     conjugate acid

c.     $HSO_4^-(aq)$ base               $H_2SO_4(aq)$    conjugate acid
       $H_2O(l)$     acid               $OH^-(aq)$       conjugate base

9.49 Draw the conjugate acid and base of $HCO_3^-$.

a. conjugate acid: $H_2CO_3$        b. conjugate base: $CO_3^{2-}$

9.51 Draw the acid–base reaction.

$HNO_3(aq)$ + $H_2O(l)$ ⟶ $H_3O^+(aq)$ + $NO_3^-(aq)$

9.53 **A** represents HCl because it shows a fully dissociated acid.  **B** represents HF, only partially dissociated.

9.55 a.  **B** represents a strong acid because HZ is completely dissociated, forming $H_3O^+$ and $Z^-$.
     b.  **A** represents a weak acid because most of the acid HZ remains, and few ions ($H_3O^+$ and $Z^-$) are formed.

**9.57** Use Tables 9.1 and 9.2 to determine the stronger acid as in Example 9.7. The acid with the larger $K_a$ is the stronger acid.

    a. $CH_3COOH$ is stronger than $H_2O$.         c. $H_2SO_4$ is stronger than $HSO_4^-$.
    b. $H_3PO_4$ is stronger than $HCO_3^-$.

**9.59** The weaker acid has the stronger conjugate base.

    a. $H_2O$                 b. $HCO_3^-$              c. $HSO_4^-$

**9.61** a. An acid that dissociates to a greater extent in water is a stronger acid: **A is the stronger acid.**
    b. An acid with a smaller $K_a$ is weaker (**A** is weaker): **B is the stronger acid.**
    c. An acid with a stronger conjugate base is a weaker acid (**A** is weaker): **B is the stronger acid.**

**9.63** The acid with the larger $K_a$ is the stronger acid. The stronger acid has a weaker conjugate base.

a.     $HSO_4^-$        $SO_4^{2-}$                    $H_2PO_4^-$       $HPO_4^{2-}$

      stronger acid    conjugate base         $K_a = 6.2 \times 10^{-8}$    conjugate base
     $K_a = 1.2 \times 10^{-2}$                                     **stronger base**
       larger $K_a$
    **stronger acid**

b.     $CH_3COOH$      $CH_3COO^-$         $CH_3CH_2COOH$     $CH_3CH_2COO^-$

      stronger acid    conjugate base         $K_a = 1.3 \times 10^{-5}$    conjugate base
     $K_a = 1.8 \times 10^{-5}$                                      **stronger base**
       larger $K_a$
    **stronger acid**

**9.65** The stronger acid has more dissociated ions. The stronger acid has the larger $K_a$.

    a. **B** has more dissociated ions ($A^-$ and $H_3O^+$) and is therefore the stronger acid.
    b. **B** has the larger $K_a$ since it is the stronger acid.

**9.67** To determine if the reactants or products are favored at equilibrium:
    •   Identify the acid in the reactants and the conjugate acid in the products.
    •   Determine the relative strength of the acid and the conjugate acid.
    •   Equilibrium favors the formation of the weaker acid.

a.     $H_3PO_4(aq)$    +    $CN^-(aq)$    $\rightleftharpoons$    $H_2PO_4^-(aq)$    +    $HCN(aq)$
         acid                                              conjugate acid
                                                    weaker acid
                                      **Products are favored.**

b.     $Br^-(aq)$    +    $HSO_4^-(aq)$   $\rightleftharpoons$   $SO_4^{2-}(aq)$    +    $HBr(aq)$
                          acid                                  conjugate acid
                          weaker acid
                **Reactants are favored.**

c. $CH_3COO^-(aq)$  +  $H_2CO_3(aq)$  $\rightleftharpoons$  $CH_3COOH(aq)$  +  $HCO_3^-(aq)$

                   acid                        conjugate acid

             weaker acid

  **Reactants are favored.**

**9.69** Use the equation $[OH^-] = K_w/[H_3O^+]$ to calculate the hydroxide ion concentration as in Answer 9.17. When $[OH^-] > [H_3O^+]$, the solution is basic. When $[OH^-] < [H_3O^+]$, the solution is acidic.

a. $[OH^-] = \dfrac{K_w}{[H_3O^+]} = \dfrac{1.0 \times 10^{-14}}{10^{-8}} = 10^{-6}$ M

                                                          basic

b. $[OH^-] = \dfrac{K_w}{[H_3O^+]} = \dfrac{1.0 \times 10^{-14}}{10^{-10}} = 10^{-4}$ M

                                                          basic

c. $[OH^-] = \dfrac{K_w}{[H_3O^+]} = \dfrac{1.0 \times 10^{-14}}{3.0 \times 10^{-4}} = 3.3 \times 10^{-11}$ M

                                                              acidic

d. $[OH^-] = \dfrac{K_w}{[H_3O^+]} = \dfrac{1.0 \times 10^{-14}}{2.5 \times 10^{-11}} = 4.0 \times 10^{-4}$ M

                                                              basic

**9.71** Use the equation $[H_3O^+] = K_w/[OH^-]$ to calculate the hydronium ion concentration as in Sample Problem 9.11.

a. $[H_3O^+] = \dfrac{K_w}{[OH^-]} = \dfrac{1.0 \times 10^{-14}}{10^{-2}} = 10^{-12}$ M

                                                          basic

b. $[H_3O^+] = \dfrac{K_w}{[OH^-]} = \dfrac{1.0 \times 10^{-14}}{4.0 \times 10^{-8}} = 2.5 \times 10^{-7}$ M

                                                              acidic

c. $[H_3O^+] = \dfrac{K_w}{[OH^-]} = \dfrac{1.0 \times 10^{-14}}{6.2 \times 10^{-7}} = 1.6 \times 10^{-8}$ M

                                                          basic

d. $[H_3O^+] = \dfrac{K_w}{[OH^-]} = \dfrac{1.0 \times 10^{-14}}{8.5 \times 10^{-13}} = 1.2 \times 10^{-2}$ M

                                                              acidic

**9.73** Use a calculator to determine the logarithm of a number that contains a coefficient other than one in scientific notation; $pH = -\log [H_3O^+]$ as in Answer 9.23.

a. $pH = -\log[H_3O^+] = -\log(10^{-12})$

$\qquad\qquad\qquad\qquad = -(-12) = 12$

c. $pH = -\log[H_3O^+] = -\log(1.6 \times 10^{-8})$

$\qquad\qquad\qquad\qquad\qquad = -(-7.80) = 7.80$

b. $pH = -\log[H_3O^+] = -\log(2.5 \times 10^{-7})$

$\qquad\qquad\qquad\qquad = -(-6.60) = 6.60$

d. $pH = -\log[H_3O^+] = -\log(1.2 \times 10^{-2})$

$\qquad\qquad\qquad\qquad\qquad = -(-1.92) = 1.92$

**9.75**

| $[H_3O^+]$ | $[OH^-]$ | pH | Classification |
|---|---|---|---|
| $5.3 \times 10^{-3}$ | $1.9 \times 10^{-12}$ | 2.28 | acidic |
| $5.0 \times 10^{-7}$ | $2.0 \times 10^{-8}$ | 6.30 | acidic |
| $4 \times 10^{-5}$ | $2.5 \times 10^{-10}$ | 4.4 | acidic |
| $1.5 \times 10^{-5}$ | $6.8 \times 10^{-10}$ | 4.82 | acidic |

**9.77** Use a calculator to determine the antilogarithm of $(-pH)$; $[H_3O^+] = $ antilog$(-pH)$.

a. $[H_3O^+] = $ antilog$(-pH) = $ antilog$(-12)$

$\qquad [H_3O^+] = 1 \times 10^{-12}$ M

c. $[H_3O^+] = $ antilog$(-pH) = $ antilog$(-1.80)$

$\qquad [H_3O^+] = 1.6 \times 10^{-2}$ M

b. $[H_3O^+] = $ antilog$(-pH) = $ antilog$(-1)$

$\qquad [H_3O^+] = 1 \times 10^{-1}$ M

d. $[H_3O^+] = $ antilog$(-pH) = $ antilog$(-8.90)$

$\qquad [H_3O^+] = 1.3 \times 10^{-9}$ M

**9.79** Use the equations in Answers 9.77 and 9.69 to calculate the concentrations of $H_3O^+$ and $OH^-$ in the sample.

$[H_3O^+] = $ antilog$(-pH) = $ antilog$(-5.90)$

$\qquad\qquad [H_3O^+] = 1.3 \times 10^{-6}$ M

$[OH^-] = \dfrac{K_w}{[H_3O^+]} = \dfrac{1.0 \times 10^{-14}}{1.3 \times 10^{-6}} = 7.7 \times 10^{-9}$ M

**9.81** Use the equations in Answers 9.77 and 9.69 to calculate the concentrations of $H_3O^+$ and $OH^-$ in the sample.

$[H_3O^+] = $ antilog$(-pH) = $ antilog$(-4.10)$

$\qquad\qquad [H_3O^+] = 7.9 \times 10^{-5}$ M

$[OH^-] = \dfrac{K_w}{[H_3O^+]} = \dfrac{1.0 \times 10^{-14}}{7.9 \times 10^{-5}} = 1.3 \times 10^{-10}$ M

**9.83** Use a calculator to determine the logarithm of a number that contains a coefficient other than one in scientific notation; $pH = -\log[H_3O^+]$.

a. $pH = -\log[H_3O^+] = -\log(2.5 \times 10^{-3})$

$\qquad\qquad\qquad\qquad = -(-2.60) = 2.60$

b. $[H_3O^+] = \dfrac{K_w}{[OH^-]} = \dfrac{1.0 \times 10^{-14}}{1.5 \times 10^{-2}} = 6.7 \times 10^{-13}$ M

$pH = -\log[H_3O^+] = -\log(6.7 \times 10^{-13})$

$\qquad\qquad\qquad\qquad = -(-12.17) = 12.17$

**9.85** The pH of 0.10 M HCl is lower than the pH of 0.1 M CH₃COOH solution (1.0 vs. 2.9) because HCl is fully dissociated to $H_3O^+$ and $Cl^-$, whereas CH₃COOH is not fully dissociated.

**9.87** Write the balanced equations as in Section 9.7.

a. $HBr(aq)$ + $KOH(aq)$ ⟶ $KBr(aq)$ + $H_2O(l)$

b. $2\ HNO_3(aq)$ + $Ca(OH)_2(aq)$ ⟶ $2\ H_2O(l)$ + $Ca(NO_3)_2(aq)$

c. $HCl(aq)$ + $NaHCO_3(aq)$ ⟶ $NaCl(aq)$ + $H_2O(l)$ + $CO_2(g)$

d. $H_2SO_4(aq)$ + $Mg(OH)_2(aq)$ ⟶ $2\ H_2O(l)$ + $MgSO_4(aq)$

**9.89** Write the balanced equation.

$2\ HNO_3(aq)$ + $CaCO_3(s)$ ⟶ $Ca(NO_3)_2(aq)$ + $CO_2(g)$ + $H_2O(l)$

**9.91** Follow Example 9.10 to determine the acidity when a salt is dissolved in water. Determine what types of acid and base (strong or weak) are used to form the salt. When the ions in the salt come from a strong acid and strong base, the solution is neutral. When the ions come from acids and bases of different strength, the ion derived from the stronger reactant determines the acidity.

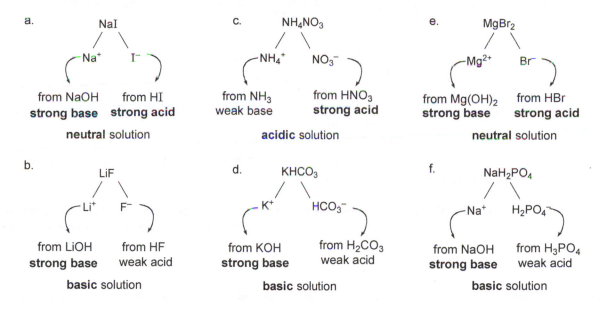

**9.93** Calculate the molarity of the solution using a titration as in Example 9.11.

35.5 m̶L̶ NaOH   ×   $\dfrac{1\ \cancel{L}}{1000\ \cancel{mL}}$   ×   $\dfrac{0.10\ \text{mol NaOH}}{1\ \cancel{L}}$   =   0.0036 mol NaOH

$$0.0036 \text{ mol NaOH} \quad \times \quad \frac{1 \text{ mol HCl}}{1 \text{ mol NaOH}} \quad = \quad 0.0036 \text{ mol HCl}$$

$$M = \frac{\text{mol}}{\text{L}} = \frac{0.0036 \text{ mol HCl}}{25 \text{ mL solution}} \times \frac{1000 \text{ mL}}{1 \text{ L}} = 0.14 \text{ M HCl}$$
molarity                                                                 **Answer**

**9.95** Calculate the molarity of the solution using a titration as in Example 9.11.

$$15.5 \text{ mL NaOH} \quad \times \quad \frac{1 \text{ L}}{1000 \text{ mL}} \quad \times \quad \frac{0.20 \text{ mol NaOH}}{1 \text{ L}} \quad = \quad 0.0031 \text{ mol NaOH}$$

$$0.0031 \text{ mol NaOH} \quad \times \quad \frac{1 \text{ mol CH}_3\text{COOH}}{1 \text{ mol NaOH}} \quad = \quad 0.0031 \text{ mol CH}_3\text{COOH}$$

$$M = \frac{\text{mol}}{\text{L}} = \frac{0.0031 \text{ mol CH}_3\text{COOH}}{25 \text{ mL solution}} \times \frac{1000 \text{ mL}}{1 \text{ L}} = 0.12 \text{ M CH}_3\text{COOH}$$
molarity                                                                      **Answer**

**9.97** Calculate the number of milliliters of solution needed.

$$10.0 \text{ mL CH}_3\text{COOH} \quad \times \quad \frac{1 \text{ L}}{1000 \text{ mL}} \quad \times \quad \frac{2.5 \text{ mol CH}_3\text{COOH}}{1 \text{ L}} \quad = \quad 0.025 \text{ mol CH}_3\text{COOH}$$

$$0.025 \text{ mol CH}_3\text{COOH} \quad \times \quad \frac{1 \text{ mol NaOH}}{1 \text{ mol CH}_3\text{COOH}} \quad = \quad 0.025 \text{ mol NaOH}$$

$$0.025 \text{ mol NaOH} \quad \times \quad \frac{1 \text{ L}}{1.0 \text{ mol NaOH}} \quad \times \quad \frac{1000 \text{ mL}}{1 \text{ L}} \quad = \quad 25 \text{ mL NaOH solution}$$

**9.99**  **B** is a buffer because it contains equal amounts of the acid and its conjugate base.  **A** is not a buffer since it contains only acid.

**9.101**  A buffer is most effective at minimizing pH changes when the concentrations of the weak acid and its conjugate base are equal because when acid or base is added to a buffer, a proton donor and a proton acceptor are available to react with either of them.  Since the ratio of the concentration of the acid and conjugate base is one to begin with, the ratio stays close to one after acid or base is added.

**9.103**  Yes, a buffer can be prepared from equal amounts of NaCN and HCN.  HCN is a weak acid and $CN^-$ is its conjugate base, so in equal amounts they form a buffer.

**9.105**  a.  Both $HNO_2$ and $NO_2^-$ are needed to prepare the buffer since $HNO_2$ is a proton donor that will react with added base and $NO_2^-$ is a proton acceptor that will react with added acid.
b.  When a small amount of acid is added to the buffer, the concentration of $HNO_2$ increases and the concentration of $NO_2^-$ decreases.
c.  The concentration of $HNO_2$ decreases and the concentration of $NO_2^-$ increases when a small amount of base is added to the buffer.

**9.107** Calculate the pH of the buffer as in Example 9.12.

a. $[H_3O^+]$ = $K_a$ x $\dfrac{[Na_2HPO_4]}{[Na_3PO_4]}$ = $2.2 \times 10^{-13}$ x $\dfrac{[0.10\ M]}{[0.10\ M]}$

$[H_3O^+]$ = $2.2 \times 10^{-13}\ M$

pH = $-\log[H_3O^+]$ = $-\log(2.2 \times 10^{-13})$

pH = 12.66

b. $[H_3O^+]$ = $K_a$ x $\dfrac{[NaHCO_3]}{[Na_2CO_3]}$ = $5.6 \times 10^{-11}$ x $\dfrac{[0.22\ M]}{[0.22\ M]}$

$[H_3O^+]$ = $5.6 \times 10^{-11}\ M$

pH = $-\log[H_3O^+]$ = $-\log(5.6 \times 10^{-11})$

pH = 10.25

**9.109** Calculate the pH of the buffer as in Example 9.12.

a. $[H_3O^+]$ = $K_a$ x $\dfrac{[CH_3COOH]}{[NaCH_3COO]}$ = $1.8 \times 10^{-5}$ x $\dfrac{[0.20\ M]}{[0.20\ M]}$

$[H_3O^+]$ = $1.8 \times 10^{-5}\ M$

pH = $-\log[H_3O^+]$ = $-\log(1.8 \times 10^{-5})$

pH = 4.74

b. $[H_3O^+]$ = $K_a$ x $\dfrac{[CH_3COOH]}{[NaCH_3COO]}$ = $1.8 \times 10^{-5}$ x $\dfrac{[0.20\ M]}{[0.40\ M]}$

$[H_3O^+]$ = $9.0 \times 10^{-6}\ M$

pH = $-\log[H_3O^+]$ = $-\log(9.0 \times 10^{-6})$

pH = 5.05

c. $[H_3O^+]$ = $K_a$ x $\dfrac{[CH_3COOH]}{[NaCH_3COO]}$ = 1.8 x 10$^{-5}$ x $\dfrac{[0.20\ M]}{[0.10\ M]}$

$[H_3O^+]$ = 3.6 x 10$^{-5}$ M

pH = $-\log [H_3O^+]$ = $-\log(3.6 \times 10^{-5})$

pH = 4.44

**9.111** The pH of unpolluted rainwater is lower than that of pure water because $CO_2$ combines with rainwater to form $H_2CO_3$, which is acidic, lowering the pH.

**9.113** Calculate the concentrations.

$[H_3O^+]$ = antilog($-$pH) = antilog($-7.50$)

$[H_3O^+]$ = 3.2 x 10$^{-8}$ M

$[OH^-]$ = $\dfrac{K_w}{[H_3O^+]}$ = $\dfrac{1.0 \times 10^{-14}}{3.2 \times 10^{-8}}$ = 3.1 x 10$^{-7}$ M

**9.115** By breathing into a bag, the individual breathes in air with a higher $CO_2$ concentration. Thus, the $CO_2$ concentration in the lungs and the blood increases, thereby lowering the pH.

**9.117** A rise in $CO_2$ concentration leads to an increase in $H^+(aq)$ concentration by the following equilibria:

$CO_2(g) + H_2O(l) \rightleftharpoons H_2CO_3(aq) \rightleftharpoons HCO_3^-(aq) + H^+(aq)$

**9.119** Write the acid–base reaction that occurs when $OCl^-$ dissolves in water.

$OCl^-(aq) + H_2O(l) \longrightarrow HOCl(aq) + OH^-(aq)$

This reaction forms $OH^-$, which makes the pool water basic.

# Chapter 10 Nuclear Chemistry

## Chapter Review

**[1] Describe the different types of radiation emitted by a radioactive nucleus. (10.1)**
- A radioactive nucleus can emit alpha ($\alpha$) particles, beta ($\beta$) particles, positrons, or gamma ($\gamma$) rays.
- An $\alpha$ particle is a high-energy nucleus that contains two protons and two neutrons.

      alpha particle:     $\alpha$   or   $_2^4\text{He}$

- A $\beta$ particle is a high-energy electron that has a –1 charge and a negligible mass.

      beta particle:     $\beta$   or   $_{-1}^{\phantom{-}0}e$

- A positron is an antiparticle of a $\beta$ particle. A positron has a +1 charge and negligible mass.

      Symbol:    $_{+1}^{\phantom{+}0}e$ or $\beta^+$     Formation:   $_1^1p \longrightarrow \;_0^1n \;+\; _{+1}^{\phantom{+}0}e$

             positron                       proton      neutron     positron

- A gamma ray is high-energy radiation with no mass or charge.

      gamma ray:     $\gamma$

**[2] How are equations for nuclear reactions written? (10.2)**
- In an equation for a nuclear reaction, the sum of the mass numbers ($A$) must be equal on both sides of the equation. The sum of the atomic numbers ($Z$) must be equal on both sides of the equation as well.
- For example, in the given equation, the sum of the mass numbers on each side of the equation is 18, and the sum of the atomic numbers on each side of the equation is 9.

$$_9^{18}\text{F} \longrightarrow \;_{+1}^{\phantom{+}0}e \;+\; _8^{18}\text{O}$$

**[3] What is the half-life of a radioactive isotope? (10.3)**
- The half-life ($t_{1/2}$) is the time it takes for one-half of a radioactive sample to decay. Knowing the half-life and the amount of a radioactive substance, one can calculate how much of a sample remains after a period of time. For example, P-32 decays to S-32 with a half-life of 14 days.

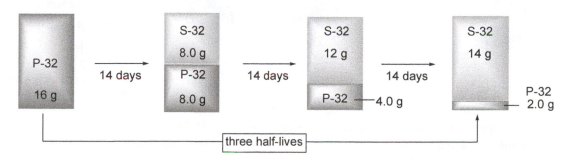

- The half-life of radioactive C-14 can be used to date archaeological artifacts.

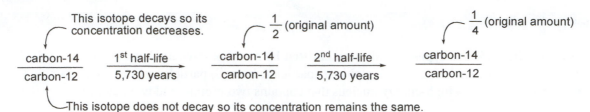

## [4] What units are used to measure radioactivity? (10.4)

- Radiation in a sample is measured by the number of disintegrations per second, most often using the curie; $1\ \text{Ci} = 3.7 \times 10^{10}$ disintegrations/s. The becquerel is also used; $1\ \text{Bq} = 1$ disintegration/s; $1\ \text{Ci} = 3.7 \times 10^{10}\ \text{Bq}$.
- The exposure of a substance to radioactivity is measured with the rad (radiation absorbed dose) or the rem (radiation equivalent for man).

## [5] Give examples of common radioisotopes used in medicine. (10.5)

- Iodine-131 is used to diagnose and treat thyroid disease.
- Technetium-99m is used to evaluate the functioning of the gall bladder and bile ducts, and in bone scans to look for the spread of cancer to other sites in the body.
- Red blood cells tagged with technetium-99m are used to find the site of a gastrointestinal bleed.
- Thallium-201 is used to diagnose coronary artery disease.
- Cobalt-60 is used as an external source of radiation for cancer treatment.
- Iodine-125 and iridium-192 are used in the internal radiation treatment of prostate cancer and breast cancer, respectively.
- Carbon-11, oxygen-15, nitrogen-13, and fluorine-18 are used in positron emission tomography.

## [6] What are nuclear fission and nuclear fusion? (10.6)

- Nuclear fission is the splitting apart of a heavy nucleus into lighter nuclei and neutrons.
- Nuclear fusion is the joining together of two light nuclei to form a larger nucleus.
- Both nuclear fission and nuclear fusion release a great deal of energy. Nuclear fission is used in nuclear power plants to generate electricity. Nuclear fusion occurs in stars.
- One common nuclear reaction is the fission of uranium-235 into krypton-91 and barium-142.

$$^{235}_{92}\text{U} + {}^{1}_{0}\text{n} \longrightarrow {}^{91}_{36}\text{Kr} + {}^{142}_{56}\text{Ba} + 3\,{}^{1}_{0}\text{n}$$

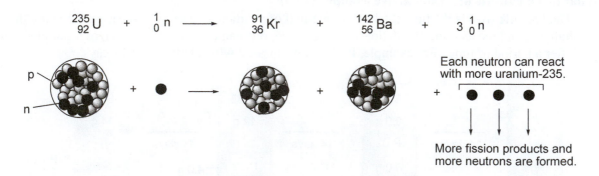

## [7] What medical imaging techniques do not use radioactivity? (10.7)

- X-rays and CT scans both use X-rays, a high-energy form of electromagnetic radiation.
- MRIs use low-energy radio waves to image soft tissue.

## Problem Solving

## [1] Introduction (10.1)

**Example 10.1** Thallium-201 and iridium-192 are radioactive isotopes that can be produced using a nuclear reactor. Complete the following table for both isotopes.

|  | Atomic Number | Mass Number | Number of Protons | Number of Neutrons | Isotope Symbol |
|---|---|---|---|---|---|
| Thallium-201 |  |  |  |  |  |
| Iridium-192 |  |  |  |  |  |

### Analysis
- The atomic number ($Z$) = the number of protons.
- The mass number ($A$) = the number of protons + the number of neutrons.
- Isotopes are written with the mass number to the upper left of the element symbol and the atomic number to the lower left of the element symbol.

### Solution

|  | Atomic Number | Mass Number | Number of Protons | Number of Neutrons | Isotope Symbol |
|---|---|---|---|---|---|
| Thallium-201 | 81 | 201 | 81 | $201 - 81 = 120$ | $^{201}_{81}\text{Tl}$ |
| Iridium-192 | 77 | 192 | 77 | $192 - 77 = 115$ | $^{192}_{77}\text{Ir}$ |

## [2] Nuclear Reactions (10.2)

**Example 10.2** Write a balanced nuclear equation for the $\beta$ emission of strontium-90, a radioisotope used to produce $\beta$ particles.

### Analysis
Balance the atomic numbers and mass numbers on both sides of a nuclear equation. With $\beta$ emission, treat the $\beta$ particle as an electron with zero mass in balancing mass numbers, and a $-1$ charge when balancing the atomic numbers.

### Solution
**[1] Write an incomplete equation with the original nucleus on the left and the particle emitted on the right.**
- Use the identity of the element to determine the atomic number; strontium has an atomic number of 38.

$$^{90}_{38}\text{Sr} \longrightarrow \ ^{0}_{-1}\text{e} \ + \ ?$$

**[2] Calculate the mass number and the atomic number of the newly formed nucleus on the right.**
- Mass number: Since a $\beta$ particle has no mass, the masses of the new particle and the original particle are the same, 90.

- Atomic number: Since β emission converts a neutron into a proton, the new nucleus has one more proton than the original nucleus; $38 = -1 + ?$. Thus, the new nucleus has an atomic number of 39.

---

**[3] Use the atomic number to identify the new nucleus and complete the equation.**
- From the periodic table, the element with an atomic number of 39 is yttrium, Y.
- Write the mass number and the atomic number with the element symbol to complete the equation.

$$^{90}_{38}\text{Sr} \longrightarrow \,^{0}_{-1}\text{e} \ + \ ^{90}_{39}\text{Y}$$

---

**Example 10.3** Write a balanced nuclear equation for the positron emission of rubidium-77.

**Analysis**
Balance the atomic numbers and mass numbers on both sides of a nuclear equation. With positron emission, treat the positron as a particle with zero mass when balancing mass numbers, and a +1 charge when balancing the atomic numbers.

**Solution**
**[1] Write an incomplete equation with the original nucleus on the left and the particle emitted on the right.**
- Use the identity of the element to determine the atomic number; rubidium has an atomic number of 37.

$$^{77}_{37}\text{Rb} \longrightarrow \,^{0}_{+1}\text{e} \ + \ ?$$

---

**[2] Calculate the mass number and the atomic number of the newly formed nucleus on the right.**
- Mass number: Since a $\beta^+$ particle has no mass, the masses of the new particle and the original particle are the same, 77.
- Atomic number: Since $\beta^+$ emission converts a proton into a neutron, the new nucleus has one fewer proton than the original nucleus; $37 - 1 = 36$. Thus, the new nucleus has an atomic number of 36.

---

**[3] Use the atomic number to identify the new nucleus and complete the equation.**
- From the periodic table, the element with an atomic number of 36 is krypton, Kr.
- Write the mass number and the atomic number with the element symbol to complete the equation.

$$^{77}_{37}\text{Rb} \longrightarrow \,^{0}_{+1}\text{e} \ + \ ^{77}_{36}\text{Kr}$$

---

## [3] Detecting and Measuring Radiation (10.4)

**Example 10.4** A patient must be given a 7.5-mCi dose of iodine-131, which is available as a solution that contains 2.5 mCi/mL. What volume of solution must be administered?

**Analysis**
Use the amount of radioactivity (mCi/mL) as a conversion factor to convert the dose of radioactivity from millicuries to a volume in milliliters.

**Solution**

The dose of radioactivity is known in millicuries, and the amount of radioactivity per unit volume (2.5 mCi/mL) is also known. Use 2.5 mCi/mL as a millicurie–milliliter conversion factor.

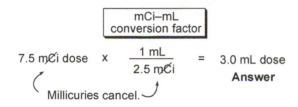

---

## Self-Test

**[1] Fill in the blank with one of the terms listed below.**

Alpha (α) particle (10.1)     Nuclear fission (10.6)     Radioactivity (10.1)
Beta (β) particle (10.1)      Nuclear fusion (10.6)      Radiocarbon dating (10.3)
Gamma (γ) ray (10.1)          Positron (10.1)            Rem (10.4)
Geiger counter (10.4)         Rad (10.4)                 X-rays (10.7)
Half-life (10.3)              Radioactive isotope (10.1)

1. The _____ of a radioactive isotope is the time it takes for one-half of the sample to decay.
2. _____ is the joining together of two light nuclei to form a larger nucleus.
3. A _____ is a small portable device used for measuring radioactivity.
4. An _____ is a high-energy particle that contains two protons and two neutrons.
5. A _____ is unstable and spontaneously emits energy to form a more stable nucleus.
6. _____ is based on the fact that the ratio of radioactive carbon-14 to stable carbon-12 is a constant value in a living organism.
7. A _____ is a high-energy electron.
8. _____ are a high-energy type of electromagnetic radiation.
9. A _____ is called an antiparticle of a β particle, since their charges are different but their masses are the same.
10. A _____ is the amount of radiation absorbed by one gram of a substance. The amount of energy absorbed varies with both the nature of the substance and the type of radiation.
11. _____ is the splitting apart of a heavy nucleus into lighter nuclei and neutrons.
12. A _____ is high-energy radiation released from a radioactive nucleus.
13. _____ is any form of nuclear radiation emitted by a radioactive isotope.
14. A _____ is the amount of radiation that also factors in its energy and potential to damage tissue.

**[2] Fill in the blanks in the table below.**

|  | Atomic Number | Mass Number | Number of Protons | Number of Neutrons |
|---|---|---|---|---|
| $^{60}_{27}\text{Co}$ | 15. _____ | 60 | 17. _____ | 18. _____ |
| $^{99m}_{43}\text{Tc}$ | 43 | 16. _____ | 43 | 19. _____ |

**[3] Pick the type of radioactivity that matches each phrase.  A question may have more than one answer.**

a. alpha particle                         c. positron
b. beta particle                          d. gamma ray

20. Has a positive charge
21. Has a negative charge
22. Has no charge
23. Has a mass equivalent to that of helium
24. Has no mass
25. Is formed by the decay of a radioactive nucleus

## Answers to Self-Test

1. half-life            6. Radiocarbon dating   11. Nuclear fission   16. 99      21. b
2. Nuclear fusion       7. beta particle        12. gamma ray         17. 27      22. d
3. Geiger counter       8. X-rays               13. Radioactivity     18. 33      23. a
4. alpha particle       9. positron             14. rem               19. 56      24. b, c, d
5. radioactive isotope  10. rad                 15. 27                20. a, c    25. a, b, c, d

## Solutions to In-Chapter Problems

10.1    Refer to Example 10.1 to answer the question.
- The atomic number ($Z$) = the number of protons.
- The mass number ($A$) = the number of protons + the number of neutrons.
- Isotopes are written with the mass number to the upper left of the element symbol and the atomic number to the lower left of the element symbol.

|  | Atomic Number | Mass Number | Number of Protons | Number of Neutrons | Isotope Symbol |
|---|---|---|---|---|---|
| Cobalt-59 | 27 | 59 | 27 | 32 | $^{59}_{27}\text{Co}$ |
| Cobalt-60 | 27 | 60 | 27 | 33 | $^{60}_{27}\text{Co}$ |

10.2    Fill in the table as in Example 10.1, using the rules in Answer 10.1.

|  |  | Atomic Number | Mass Number | Number of Protons | Number of Neutrons |
|---|---|---|---|---|---|
| a. | $^{85}_{38}\text{Sr}$ | 38 | 85 | 38 | 47 |
| b. | $^{67}_{31}\text{Ga}$ | 31 | 67 | 31 | 36 |
| c. | Selenium-75 | 34 | 75 | 34 | 41 |

10.3    An $\alpha$ particle has no electrons around the nucleus and a +2 charge, whereas a helium atom has two electrons around the nucleus and is neutral.

**10.4**  Use Table 10.1 to identify Q in each of the following symbols.

a.  $_{-1}^{0}Q$  
β particle

b.  $_{2}^{4}Q$  
α particle

c.  $_{+1}^{0}Q$  
positron

**10.5**  Write a balanced nuclear equation as in Example 10.2.

**[1]  Write an incomplete equation with the original nucleus on the left and the particle emitted on the right.**  Radon-222 has an atomic number of 86.

$$_{86}^{222}Rn \longrightarrow \ _{2}^{4}He \ + \ ?$$

**[2] Calculate the mass number and the atomic number of the newly formed nucleus on the right.**
- Radon-222 emits an α particle.
- Mass number: Subtract the mass of an α particle (4) to obtain the mass of the new nucleus; $222 - 4 = 218$.
- Atomic number: Subtract the two protons of an α particle to obtain the atomic number of the new nucleus; $86 - 2 = 84$.

**[3] Use the atomic number to identify the new nucleus and complete the equation.**
- From the periodic table, the element with an atomic number of 84 is polonium, Po.
- Write the mass number and the atomic number with the element symbol to complete the equation.

$$_{86}^{222}Rn \longrightarrow \ _{2}^{4}He \ + \ _{84}^{218}Po$$

**10.6**  Work backwards to write the equation that produces radon-222.

$$? \longrightarrow \ _{2}^{4}He \ + _{86}^{222}Rn$$

Mass number = 222 + 4 = 226

$$_{88}^{226}Ra \longrightarrow \ _{2}^{4}He \ + _{86}^{222}Rn$$

Atomic number = 86 + 2 = 88

**10.7**  Write the balanced nuclear equation for each isotope as in Example 10.2 or Answer 10.5.

a.  $_{84}^{218}Po \longrightarrow \ _{2}^{4}He \ + \ _{82}^{214}Pb$

c.  $_{99}^{252}Es \longrightarrow \ _{2}^{4}He \ + \ _{97}^{248}Bk$

b.  $_{90}^{230}Th \longrightarrow \ _{2}^{4}He \ + \ _{88}^{226}Ra$

**10.8**  Write a balanced nuclear equation for the β emission of iodine-131 as in Example 10.2.

**[1] Write an incomplete equation with the original nucleus on the left and the particle emitted on the right.**
- Use the identity of the element to determine the atomic number; iodine has an atomic number of 53.

$$_{53}^{131}I \longrightarrow \ _{-1}^{0}e \ + \ ?$$

**[2] Calculate the mass number and the atomic number of the newly formed nucleus on the right.**

- Mass number: Since a β particle has no mass, the masses of the new particle and the original particle are the same, 131.
- Atomic number: Since β emission converts a neutron into a proton, the new nucleus has one more proton than the original nucleus; $53 = -1 + ?$. Thus, the new nucleus has an atomic number of 54.

**[3] Use the atomic number to identify the new nucleus and complete the equation.**

- From the periodic table, the element with an atomic number of 54 is xenon, Xe.
- Write the mass number and the atomic number with the element symbol to complete the equation.

$$^{131}_{53}\text{I} \longrightarrow \; ^{\;0}_{-1}\text{e} \; + \; ^{131}_{\;54}\text{Xe}$$

**10.9** Write a balanced nuclear equation for the β emission of each isotope as in Example 10.2 and Answer 10.8.

a. $^{20}_{\;9}\text{F} \longrightarrow \; ^{\;0}_{-1}\text{e} \; + \; ^{20}_{10}\text{Ne}$     c. $^{55}_{24}\text{Cr} \longrightarrow \; ^{\;0}_{-1}\text{e} \; + \; ^{55}_{25}\text{Mn}$

b. $^{92}_{38}\text{Sr} \longrightarrow \; ^{\;0}_{-1}\text{e} \; + \; ^{92}_{39}\text{Y}$

**10.10** Write a balanced nuclear equation for positron emission as in Example 10.3.

a. **[1] Write an incomplete equation with the original nucleus on the left and the particle emitted on the right.**

- Use the identity of the element to determine the atomic number; arsenic has an atomic number of 33.

$$^{74}_{33}\text{As} \longrightarrow \; ^{\;0}_{+1}\text{e} \; + \; ?$$

**[2] Calculate the mass number and the atomic number of the newly formed nucleus on the right.**

- Mass number: Since a $\beta^+$ particle has no mass, the masses of the new particle and the original particle are the same, 74.
- Atomic number: Since $\beta^+$ emission converts a proton into a neutron, the new nucleus has one fewer proton than the original nucleus; $33 - 1 = 32$. Thus, the new nucleus has an atomic number of 32.

**[3] Use the atomic number to identify the new nucleus and complete the equation.**

- From the periodic table, the element with an atomic number of 32 is germanium, Ge.
- Write the mass number and the atomic number with the element symbol to complete the equation.

$$^{74}_{33}\text{As} \longrightarrow \; ^{\;0}_{+1}\text{e} \; + \; ^{74}_{32}\text{Ge}$$

b. Use the same steps as in part (a).

$$^{15}_{8}O \longrightarrow {}^{0}_{+1}e + {}^{15}_{7}N$$

**10.11** When β and γ emission occur together, the atomic number increases by one (due to the β particle), and a γ ray is released.

$$^{192}_{77}Ir \longrightarrow {}^{192}_{78}Pt + {}^{0}_{-1}e + \gamma$$

Both β particles and γ rays are emitted.

**10.12** When γ emission occurs alone, there is no change in the atomic number or the mass number. When β and γ emission occur together, the atomic number increases by one (due to the β particle), and a γ ray is released.

a. $$^{11}_{5}B \longrightarrow {}^{11}_{5}B + \gamma$$

γ emission alone
no change in atomic number or mass number

b. $$^{40}_{19}K \longrightarrow {}^{40}_{20}Ca + {}^{0}_{-1}e + \gamma$$

Atomic number increases, but no change in the mass number.    Both β particles and γ rays are emitted.

**10.13** To calculate the amount of radioisotope present after the given number of half-lives, multiply the initial mass by ½ for each half-life.

a. 1.00 g      ×   $\dfrac{1}{2}$  ×  $\dfrac{1}{2}$  =   0.250 g of phosphorus-32 remains.
initial mass

The mass is halved two times.

b. 1.00 g      ×   $\dfrac{1}{2}$  ×  $\dfrac{1}{2}$  ×  $\dfrac{1}{2}$  ×  $\dfrac{1}{2}$   =   0.0625 g of phosphorus-32 remains.
initial mass

The mass is halved four times.

c. 1.00 g      ×   $\dfrac{1}{2}$  ×  $\dfrac{1}{2}$  ×  $\dfrac{1}{2}$  ×  $\dfrac{1}{2}$  ×  $\dfrac{1}{2}$  ×  $\dfrac{1}{2}$  ×  $\dfrac{1}{2}$  ×  $\dfrac{1}{2}$
initial mass

The mass is halved eight times.

=   0.00391 g of phosphorus-32 remains.

d. 1.00 g  ×  $\dfrac{1}{2}$×$\dfrac{1}{2}$×$\dfrac{1}{2}$×$\dfrac{1}{2}$×$\dfrac{1}{2}$×$\dfrac{1}{2}$×$\dfrac{1}{2}$×$\dfrac{1}{2}$×$\dfrac{1}{2}$×$\dfrac{1}{2}$×$\dfrac{1}{2}$×$\dfrac{1}{2}$×$\dfrac{1}{2}$×$\dfrac{1}{2}$×$\dfrac{1}{2}$×$\dfrac{1}{2}$×$\dfrac{1}{2}$×$\dfrac{1}{2}$×$\dfrac{1}{2}$×$\dfrac{1}{2}$
initial mass

The mass is halved twenty times.    =   $9.54 \times 10^{-7}$ g of phosphorus-32 remains.

**10.14** To calculate the amount of radioisotope present, first determine the number of half-lives that occur in the given amount of time. Then multiply the initial mass by ½ for each half-life to determine the amount present.

a. $6.0 \text{ hours} \times \dfrac{1 \text{ half-life}}{6.0 \text{ hours}} = 1.0 \text{ half-life}$    $\underset{\text{initial mass}}{160. \text{ mg}} \times \dfrac{1}{2} = 80. \text{ mg of Tc-99m}$

b. $18.0 \text{ hours} \times \dfrac{1 \text{ half-life}}{6.0 \text{ hours}} = 3.0 \text{ half-lives}$    $\underset{\text{initial mass}}{160. \text{ mg}} \times \dfrac{1}{2} \times \dfrac{1}{2} \times \dfrac{1}{2} = 20.0 \text{ mg of Tc-99m}$

c. $24.0 \text{ hours} \times \dfrac{1 \text{ half-life}}{6.0 \text{ hours}} = 4.0 \text{ half-lives}$    $\underset{\text{initial mass}}{160. \text{ mg}} \times \dfrac{1}{2} \times \dfrac{1}{2} \times \dfrac{1}{2} \times \dfrac{1}{2} = 10.0 \text{ mg of Tc-99m}$

d. $2 \text{ days} \times \dfrac{24 \text{ hours}}{1 \text{ day}} \times \dfrac{1 \text{ half-life}}{6.0 \text{ hours}} = 8 \text{ half-lives}$

$\underset{\text{initial mass}}{160. \text{ mg}} \times \dfrac{1}{2} \times \dfrac{1}{2} \times \dfrac{1}{2} \times \dfrac{1}{2} \times \dfrac{1}{2} \times \dfrac{1}{2} \times \dfrac{1}{2} \times \dfrac{1}{2} = 0.6 \text{ mg of Tc-99m}$

**10.15**  If an artifact has 1/8 of the amount of C-14 compared to living organisms, it has decayed by three half-lives ($\frac{1}{2} \times \frac{1}{2} \times \frac{1}{2}$).

$3 \text{ half-lives} \times \dfrac{5{,}730 \text{ years}}{1 \text{ half-life}} = 17{,}200 \text{ years}$

**10.16**  Use the amount of radioactivity (mCi/mL) as a conversion factor to convert the dose of radioactivity from millicuries to a volume in milliliters.

$\boxed{\begin{array}{c} \text{mCi–mL} \\ \text{conversion factor} \end{array}}$

$110 \text{ mCi dose} \times \dfrac{1 \text{ mL}}{25 \text{ mCi}} = 4.4 \text{ mL}$  **Answer**

Millicuries cancel.

**10.17**  Use the conversion factor given to convert mrem to rem.

a. $200 \text{ mrem} \times \dfrac{1 \text{ rem}}{1000 \text{ mrem}} = 0.2 \text{ rem}$    b. $0.014 \text{ rem} \times \dfrac{1000 \text{ mrem}}{1 \text{ rem}} = 14 \text{ mrem}$

larger dose

**10.18**  Calculate the number of half-lives that pass in nine days.

$9 \text{ days} \times \dfrac{1 \text{ half-life}}{3 \text{ days}} = 3 \text{ half-lives}$    $\dfrac{1}{2} \times \dfrac{1}{2} \times \dfrac{1}{2} = \dfrac{1}{8}$

**10.19**

a. $^{153}_{62}$Sm — 153 – 62 = 91 neutrons
62 protons, 62 electrons

b. $^{153}_{62}$Sm $\longrightarrow$ $^{153}_{63}$Eu + $^{0}_{-1}$e

Atomic number increases.    One β particle is emitted.

c. $\dfrac{150 \text{ mCi}}{\text{initial activity}}$ x $\dfrac{1}{2}$ x $\dfrac{1}{2}$ x $\dfrac{1}{2}$ x $\dfrac{1}{2}$ = 9.4 mCi

**10.20**    The emission of a positron decreases the atomic number by one, but the mass number stays the same. Use Example 10.3.

$$^{13}_{7}\text{N} \longrightarrow \ ^{0}_{+1}\text{e} + \ ^{13}_{6}\text{C}$$

**10.21**    Nuclear fission splits an atom into two lighter nuclei. Write the equation with the information given, and then balance the equation.

$$^{235}_{92}\text{U} + \ ^{1}_{0}\text{n} \longrightarrow \ ^{133}_{51}\text{Sb} + \ ? + 3\,^{1}_{0}\text{n}$$

The atomic number must be 41 = niobium:
92 = 51 + X
The mass number of niobium must be 100:
235 + 1 = 133 + X + 3, X = 100

$$^{235}_{92}\text{U} + \ ^{1}_{0}\text{n} \longrightarrow \ ^{133}_{51}\text{Sb} + \ ^{100}_{41}\text{Nb} + 3\,^{1}_{0}\text{n}$$

**10.22**    Balance each reaction.

a.    $^{1}_{1}$H + $^{1}_{1}$H $\longrightarrow$ $^{2}_{1}$H + $^{0}_{+1}$e

b.    $^{1}_{1}$H + $^{2}_{1}$H $\longrightarrow$ $^{3}_{2}$He

c.    $^{1}_{1}$H + $^{3}_{2}$He $\longrightarrow$ $^{4}_{2}$He + $^{0}_{+1}$e

## Solutions to Odd-Numbered End-of-Chapter Problems

**10.23**    Refer to Example 10.1 to answer the question.
- The atomic number ($Z$) = the number of protons.
- The mass number ($A$) = the number of protons + the number of neutrons.
- Isotopes are written with the mass number to the upper left of the element symbol and the atomic number to the lower left of the element symbol.

| | a. Atomic Number | b. Number of Protons | c. Number of Neutrons | d. Mass Number | Isotope Symbol |
|---|---|---|---|---|---|
| Fluorine-18 | 9 | 9 | 9 | 18 | $^{18}_{9}F$ |
| Fluorine-19 | 9 | 9 | 10 | 19 | $^{19}_{9}F$ |

**10.25** Fill in the table using Example 10.1 and the definitions in Answer 10.23.

| | Atomic Number | Mass Number | Number of Protons | Number of Neutrons | Isotope Symbol |
|---|---|---|---|---|---|
| a. Chromium-51 | 24 | 51 | 24 | 27 | $^{51}_{24}Cr$ |
| b. Palladium-103 | 46 | 103 | 46 | 57 | $^{103}_{46}Pd$ |
| c. Potassium-42 | 19 | 42 | 19 | 23 | $^{42}_{19}K$ |
| d. Xenon-133 | 54 | 133 | 54 | 79 | $^{133}_{54}Xe$ |

**10.27** Use the definitions in Section 10.1B and Table 10.1 to fill in the table.

| | Change in Mass | Change in Charge |
|---|---|---|
| a. α particle | −4 | −2 |
| b. β particle | 0 | +1 |
| c. γ ray | 0 | 0 |
| d. positron | 0 | −1 |

**10.29** Use the definitions in Section 10.1B and Table 10.1 to fill in the table.

| | Mass | Charge |
|---|---|---|
| a. α | 4 | +2 |
| b. n | 1 | 0 |
| c. γ | 0 | 0 |
| d. β | 0 | −1 |

**10.31** Complete the equation. The white circles represent neutrons and the black circles represent protons.

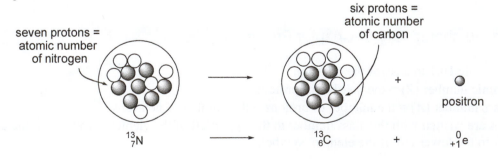

**10.33** Complete the nuclear equation as in Example 10.2.

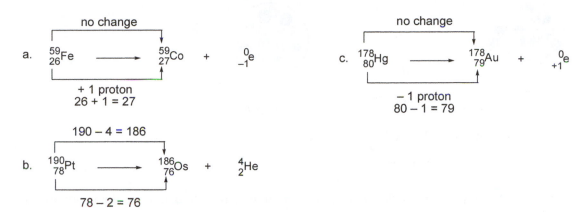

a.   no change

$$^{59}_{26}Fe \longrightarrow ^{59}_{27}Co + ^{0}_{-1}e$$

+ 1 proton
26 + 1 = 27

c.   no change

$$^{178}_{80}Hg \longrightarrow ^{178}_{79}Au + ^{0}_{+1}e$$

− 1 proton
80 − 1 = 79

b.   190 − 4 = 186

$$^{190}_{78}Pt \longrightarrow ^{186}_{76}Os + ^{4}_{2}He$$

78 − 2 = 76

**10.35** Complete each nuclear equation.

a.

$$^{90}_{39}Y \longrightarrow ^{90}_{40}Zr + ^{0}_{-1}e$$

+ 1 proton        β particle

b.

$$^{135}_{60}Nd \longrightarrow ^{135}_{59}Pr + ^{0}_{+1}e$$

Work backwards:
59 + 1 = 60        positron

c.   210 − 4 = 206

$$^{210}_{83}Bi \longrightarrow ^{206}_{81}Tl + ^{4}_{2}He$$

83 − 2 = 81        α particle

**10.37** Write the two nuclear equations for the decay of bismuth-214.

$$^{214}_{83}Bi \longrightarrow ^{214}_{84}Po + ^{0}_{-1}e$$

β particle

$$^{214}_{83}Bi \longrightarrow ^{210}_{81}Tl + ^{4}_{2}He$$

α particle

**10.39** Write the chemical equation for each nuclear reaction.

a.

$$^{232}_{90}Th \longrightarrow ^{4}_{2}He + ^{228}_{88}Ra$$

α particle

b.

$$^{25}_{11}Na \longrightarrow ^{25}_{12}Mg + ^{0}_{-1}e$$

β particle

c.

$$^{118}_{54}Xe \longrightarrow ^{118}_{53}I + ^{0}_{+1}e$$

positron

d.

$$^{243}_{96}Cm \longrightarrow ^{4}_{2}He + ^{239}_{94}Pu$$

α particle

**10.41**

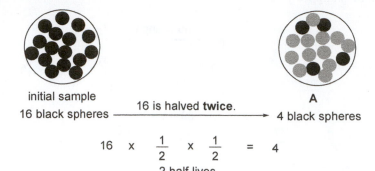

initial sample
16 black spheres ——— 16 is halved **twice**. ——→ 4 black spheres

$$16 \quad \times \quad \frac{1}{2} \quad \times \quad \frac{1}{2} \quad = \quad 4$$

2 half-lives

**10.43**  Determine the number of half-lives that have passed in 12 days to determine the length of each half-life.

$$\underset{\text{initial mass}}{2.4 \text{ g}} \quad \times \quad \frac{1}{2} \quad \times \quad \frac{1}{2} \quad \times \quad \frac{1}{2} \quad = \quad 0.30 \text{ g} \qquad \frac{12 \text{ days}}{3 \text{ half-lives}} = 4.0 \text{ days in one half-life}$$

The mass is halved three times.

**10.45**  To calculate the amount of iodine present after the given number of half-lives, multiply the initial mass by ½ for each half-life. The total mass is always 64 mg.

a.  8.0 days = 1 half-life

$$\underset{\text{initial mass}}{64 \text{ mg}} \quad \times \quad \frac{1}{2} \quad = \quad 32 \text{ mg of iodine-131} \qquad 64 - 32 = 32 \text{ mg of xenon-131}$$

b.  16 days = 2 half-lives

$$\underset{\text{initial mass}}{64 \text{ mg}} \quad \times \quad \frac{1}{2} \quad \times \quad \frac{1}{2} \quad = \quad 16 \text{ mg of iodine-131} \qquad 64 - 16 = 48 \text{ mg of xenon-131}$$

c.  24 days = 3 half-lives

$$\underset{\text{initial mass}}{64 \text{ mg}} \quad \times \quad \frac{1}{2} \quad \times \quad \frac{1}{2} \quad \times \quad \frac{1}{2} \quad = \quad 8.0 \text{ mg of iodine-131} \qquad 64 - 8.0 = 56 \text{ mg of xenon-131}$$

d.  32 days = 4 half-lives

$$\underset{\text{initial mass}}{64 \text{ mg}} \quad \times \quad \frac{1}{2} \quad \times \quad \frac{1}{2} \quad \times \quad \frac{1}{2} \quad \times \quad \frac{1}{2} \quad = \quad 4.0 \text{ mg of iodine-131} \qquad 64 - 4.0 = 60. \text{ mg of xenon-131}$$

**10.47**  If the half-life of an isotope is 24 hours, then two half-lives have passed in 48 hours.  Therefore, 25% of the initial amount of isotope still remains.

$$\underset{\text{initial mass}}{100\%} \quad \times \quad \frac{1}{2} \quad \times \quad \frac{1}{2} \quad = \quad 25\%$$

**10.49**  In artifacts over 50,000 years old, the percentage of carbon-14 is too small to accurately measure.

**10.51**   Calculate the amount of radioactivity present after each amount of time has elapsed.

a.   20 mCi          x   $\dfrac{1}{2}$   =   10 mCi
    initial activity

             6 h = one half-life

b.   20 mCi          x   $\dfrac{1}{2}$   x   $\dfrac{1}{2}$   =   5 mCi
    initial activity
           12 h = two half-lives

c.   20 mCi          x   $\dfrac{1}{2}$   x   $\dfrac{1}{2}$   x   $\dfrac{1}{2}$   x   $\dfrac{1}{2}$   =   1 mCi
    initial activity
           24 h = four half-lives

**10.53**   Use the conversion factors in Table 10.3 to convert mCi to disintegrations/second.

5.0 mCi   x   $\dfrac{3.7 \times 10^{10}\ \text{disintegrations/s}}{1\ \text{Ci}}$   x   $\dfrac{1\ \text{Ci}}{1000\ \text{mCi}}$   =   $1.9 \times 10^{8}$ disintegrations/second

**10.55**   Use the concentration as a conversion factor as in Example 10.4.

28 mCi   x   $\dfrac{1\ \text{mL}}{12\ \text{mCi}}$   =   2.3 mL solution

**10.57**   Use conversion factors to solve the problem; 1 mCi = 1,000 μCi.

68 kg   x   $\dfrac{190\ \mu\text{Ci}}{1\ \text{kg}}$   x   $\dfrac{1\ \text{mCi}}{1000\ \mu\text{Ci}}$   x   $\dfrac{1\ \text{mL}}{5\ \text{mCi}}$   =   2.6 mL solution

**10.59**   Curie refers to the number of disintegrations per second and the number of rads indicates the radiation absorbed by one gram of a substance.

**10.61**   This represents a fatal dose because 600 rem is uniformly fatal.

20 Sv   x   $\dfrac{100\ \text{rem}}{1\ \text{Sv}}$   =   2,000 rem

**10.63**   Nuclear fission is the splitting of nuclei and nuclear fusion is the joining of small nuclei to form larger ones.

**10.65**   a. Nuclear fusion is a reaction that occurs in the sun.
    b. Nuclear fission occurs when a neutron is used to bombard a nucleus.
    c. In both nuclear fission and nuclear fusion a large amount of energy is released.
    d. Nuclear fusion reactions require very high temperatures.

**10.67** Nuclear fission splits an atom into two lighter nuclei. Balance each equation.

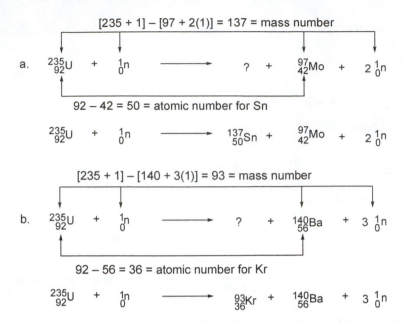

a.

$[235 + 1] - [97 + 2(1)] = 137 =$ mass number

$$^{235}_{92}U + ^{1}_{0}n \longrightarrow ? + ^{97}_{42}Mo + 2 ^{1}_{0}n$$

$92 - 42 = 50 =$ atomic number for Sn

$$^{235}_{92}U + ^{1}_{0}n \longrightarrow ^{137}_{50}Sn + ^{97}_{42}Mo + 2 ^{1}_{0}n$$

b.

$[235 + 1] - [140 + 3(1)] = 93 =$ mass number

$$^{235}_{92}U + ^{1}_{0}n \longrightarrow ? + ^{140}_{56}Ba + 3 ^{1}_{0}n$$

$92 - 56 = 36 =$ atomic number for Kr

$$^{235}_{92}U + ^{1}_{0}n \longrightarrow ^{93}_{36}Kr + ^{140}_{56}Ba + 3 ^{1}_{0}n$$

**10.69**

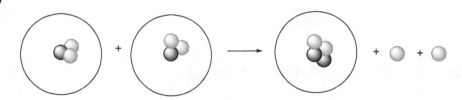

**10.71** Write out the equation from the information given. Then balance the equation.

$$^{2}_{1}H + ^{2}_{1}H \longrightarrow ^{3}_{1}H + ^{1}_{1}H$$
tritium

**10.73** Two problems with generating electricity from nuclear power plants include the containment of radiation leaks and the disposal of radioactive waste.

**10.75** Balance the nuclear equation.

$209 + 58 - 1 = 266 =$ mass number

$$^{209}_{83}Bi + ^{58}_{26}Fe \longrightarrow ? + ^{1}_{0}n$$

$83 + 26 = 109 =$ atomic number

$$^{209}_{83}Bi + ^{58}_{26}Fe \longrightarrow ^{266}_{109}Mt + ^{1}_{0}n$$

meitnerium-266

**10.77**

a. $^{74}_{33}\text{As} \longrightarrow \ ^{0}_{+1}\text{e} \ + \ ^{74}_{32}\text{Ge}$

b. $90 \text{ days} \times \dfrac{1 \text{ half-life}}{18 \text{ days}} = 5 \text{ half-lives}$   $\underset{\text{initial mass}}{120 \text{ mg}} \times \dfrac{1}{2} \times \dfrac{1}{2} \times \dfrac{1}{2} \times \dfrac{1}{2} \times \dfrac{1}{2} = 4 \text{ mg of As-74}$

c. $7.5 \text{ mCi} \times \dfrac{2.0 \text{ mL}}{10.0 \text{ mCi}} = 1.5 \text{ mL}$

**10.79**

a. $^{89}_{38}\text{Sr}$  $\begin{array}{l} 89 - 38 = 51 \text{ neutrons} \\ 38 \text{ protons, } 38 \text{ electrons} \end{array}$

c. $\underset{\text{initial acitivity}}{10.0 \ \mu\text{g}} \times \dfrac{1}{2} \times \dfrac{1}{2} \times \dfrac{1}{2} \times \dfrac{1}{2} = 0.625 \ \mu\text{g}$

b. $^{89}_{38}\text{Sr} \longrightarrow \ ^{89}_{39}\text{Y} \ + \ ^{0}_{-1}\text{e}$

**10.81**   a. Iodine-131 is used for the treatment and diagnosis of thyroid diseases.
b. Iridium-192 is used for the treatment of breast cancer.
c. Thallium-201 is used for the diagnosis of heart disease.

**10.83**   a. Iodine-125, with its longer half-life, is used for the treatment of prostate cancers with implanted radioactive seeds.
b. Iodine-131, with its shorter half-life, is used for the diagnostic and therapeutic treatment of thyroid diseases and tumors. A patient is administered radioactive iodine-131, which is then incorporated into the thyroid hormone, thyroxine. Since its half-life is short, the radioactive iodine isotope decays so that little remains after a month or so.

**10.85**   Radiology technicians must be shielded to avoid exposure to excessive and dangerous doses of radiation.

**10.87**   High doses of stable iodine will prevent the absorption and uptake of the radioactive iodine-131.

**10.89**   Use conversion factors to solve the problem.

$1.0 \text{ g coal} \times \dfrac{1 \text{ kg}}{1000 \text{ g}} \times \dfrac{2.2 \text{ lb}}{1 \text{ kg}} \times \dfrac{3.4 \times 10^8 \text{ kcal}}{2000 \text{ lb}} = 370 \text{ kcal}$

# Chapter 11 Introduction to Organic Molecules and Functional Groups

## Chapter Review

**[1] What are the characteristic features of organic compounds? (11.2)**

- Organic compounds contain carbon atoms and most contain hydrogen atoms. Carbon forms four bonds.

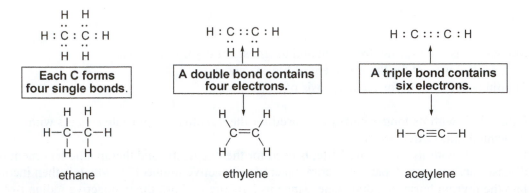

- Carbon forms single, double, and triple bonds to itself and other atoms.

| | | |
|---|---|---|
| Each C forms four single bonds. | A double bond contains four electrons. | A triple bond contains six electrons. |
| ethane | ethylene | acetylene |

- Carbon atoms can bond to form chains or rings.

propane          cyclopropane

- Organic compounds often contain heteroatoms, commonly N, O, and the halogens.

**[2] How can we predict the shape around an atom in an organic molecule? (11.3)**

- The shape around an atom is determined by counting groups, and then arranging the groups to keep them as far away from each other as possible.

| Number of Groups | Number of Atoms | Number of Lone Pairs | Shape | Bond Angle | Example |
|---|---|---|---|---|---|
| 2 | 2 | 0 | linear | 180° | HC≡CH |
| 3 | 3 | 0 | trigonal planar | 120° | $CH_2=CH_2$ |
| 4 | 4 | 0 | tetrahedral | 109.5° | $CH_3CH_3$ |
| 4 | 3 | 1 | trigonal pyramidal | ~109.5° | $CH_3NH_2$ (N atom) |
| 4 | 2 | 2 | bent | ~109.5° | $CH_3OH$ (O atom) |

**[3] What shorthand methods are used to draw organic molecules? (11.4)**

- In condensed structures, atoms are drawn in but the two-electron bonds are generally omitted. Lone pairs are omitted as well. Parentheses are used around like groups bonded together or to the same atom.

| | | |
|---|---|---|
| | | Keep the vertical bond line to show where the $CH_3$ group is bonded. |

- Three assumptions are used in drawing skeletal structures: [1] There is a carbon at the intersection of two lines or at the end of any line. [2] Each carbon has enough hydrogens to give it four bonds. [3] Heteroatoms and the hydrogens bonded to them are drawn in.

**[4] What is a functional group and why are functional groups important? (11.5)**

- A functional group is an atom or a group of atoms with characteristic chemical and physical properties.
- A functional group determines all of the properties of a molecule—its shape, physical properties, and the type of reactions it undergoes.

**[5] How do organic compounds differ from ionic inorganic compounds? (11.6)**

- The main difference between organic compounds and ionic inorganic compounds is their bonding. Organic compounds are composed of discrete molecules with covalent bonds. Ionic inorganic compounds are composed of ions held together by the strong attraction of oppositely charged ions. Other properties that are consequences of these bonding differences are summarized in Table 11.5.

**[6] How can we determine if an organic molecule is polar or nonpolar? (11.6A)**

- A nonpolar molecule has either no polar bonds or two or more bond dipoles that cancel.
- A polar molecule has either one polar bond, or two or more bond dipoles that do not cancel.

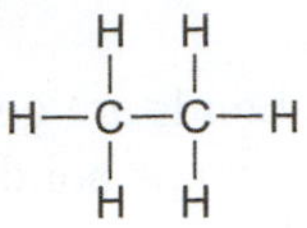

ethane

no polar bonds
nonpolar molecule

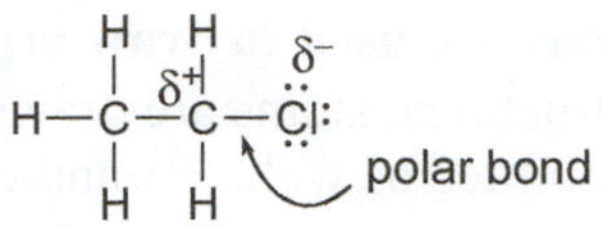

chloroethane

one polar bond
polar molecule

Two bond dipoles cancel.

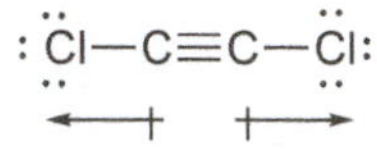

Two bond dipoles do *not* cancel.

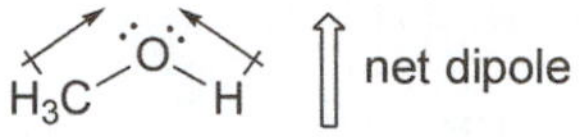

dichloroacetylene

nonpolar molecule

methanol

polar molecule

**[7] What are the solubility properties of organic compounds? (11.6B)**
- Most organic molecules are soluble in organic solvents.
- Hydrocarbons and other nonpolar organic molecules are insoluble in water.
- Polar organic molecules are water soluble only if they are small and contain a nitrogen or oxygen atom that can hydrogen bond with water.

$CH_3CH_2$—OH

ethanol

$H_2O$ soluble

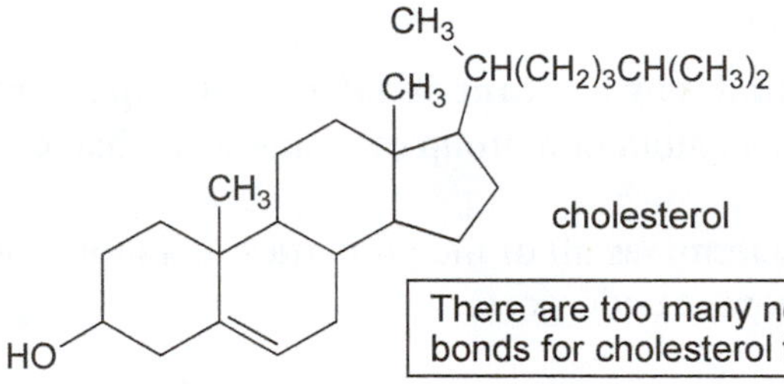

cholesterol

There are too many nonpolar C–C and C–H bonds for cholesterol to be soluble in water.

$H_2O$ insoluble

**[8] What are vitamins and when is a vitamin fat soluble or water soluble? (11.7)**
- A vitamin is an organic compound that cannot be synthesized by the body but is needed in small amounts for cell function.
- A fat-soluble vitamin has many nonpolar C–C and C–H bonds and few polar bonds, making it insoluble in water. Vitamin A is fat soluble.
- A water-soluble vitamin has many polar bonds so it dissolves in water. Vitamin C is water soluble.

## Problem Solving

## [1] Characteristic Features of Organic Compounds (11.2)

**Example 11.1** Draw in all H's and lone pairs in each compound.

a. Br—C—C

b. C—C≡N

**Analysis**

**Each C and heteroatom must be surrounded by eight electrons.** Use the common bonding patterns in Table 11.1 to fill in the needed H's and lone pairs. C needs four bonds; Br needs one bond and three lone pairs; N needs three bonds and one lone pair.

**Solution**

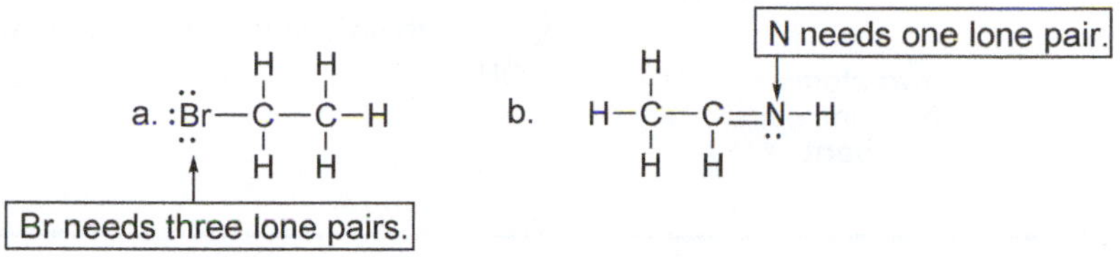

---

## [2] Shapes of Organic Molecules (11.3)

**Example 11.2** Determine the shape around each atom in acetonitrile.

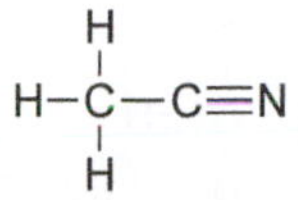

acetonitrile

**Analysis**

First, draw in all lone pairs on the heteroatom. A heteroatom needs eight electrons around it, so add one lone pair to N. Then count groups to determine the molecular shape.

**Solution**

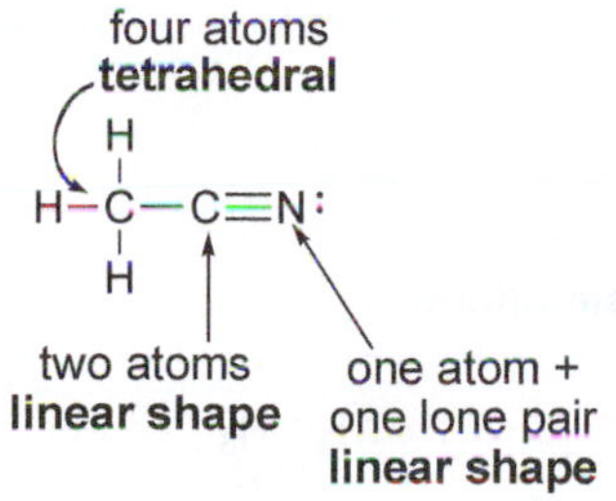

- One C is surrounded by four atoms (3 H's and 1 C), making it tetrahedral.
- One C is surrounded by two atoms (1 C and 1 N), making it linear.
- The N atom is surrounded by 1 C and one lone pair (two groups), giving it a linear shape.

---

**Example 11.3** Determine the shape around each indicated atom in Vitamin C.

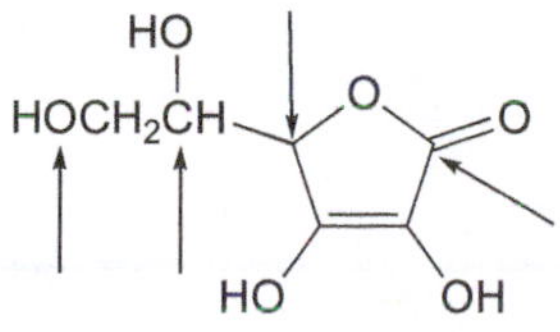

vitamin C
(ascorbic acid)

**Analysis**

First, add lone pairs of electrons to each heteroatom so that it is surrounded by eight electrons. Each O gets two lone pairs. Draw in the C–H bonds around the indicated atoms. Then count groups to predict shape.

## Solution

Add electron pairs around O atoms:

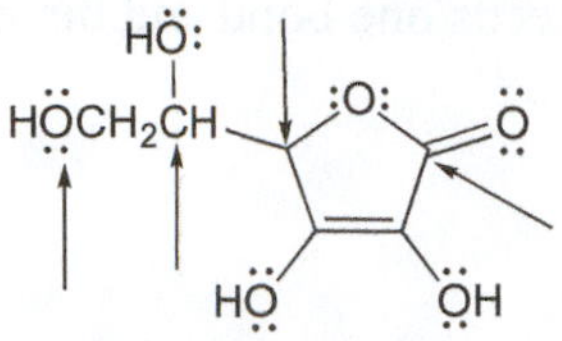

Draw in C–H bonds:

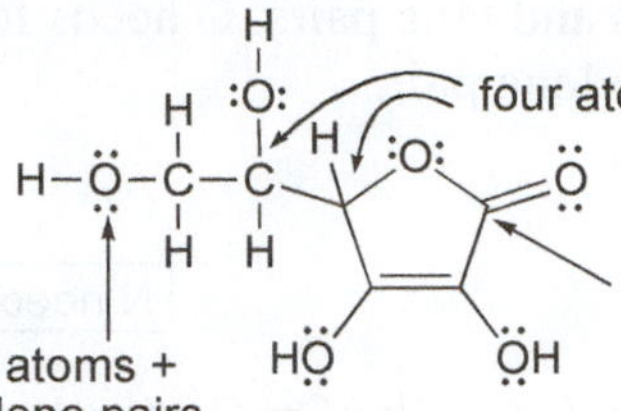

four atoms → **tetrahedral**

three atoms → **trigonal planar**

two atoms + two lone pairs **bent**

---

## [3] Drawing Organic Molecules (11.4)

**Example 11.4** Convert each compound into a condensed structure.

a.

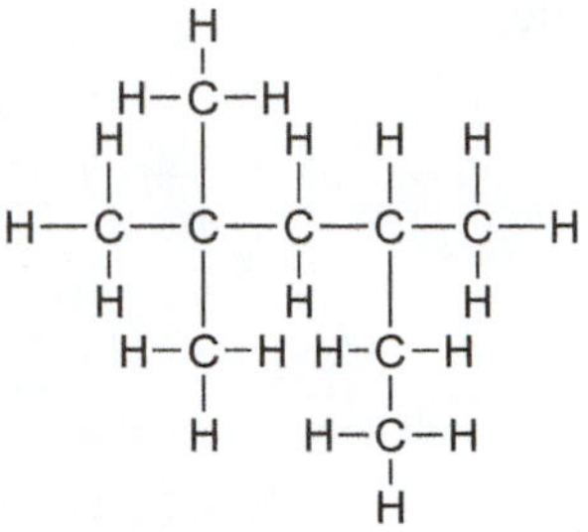

b.

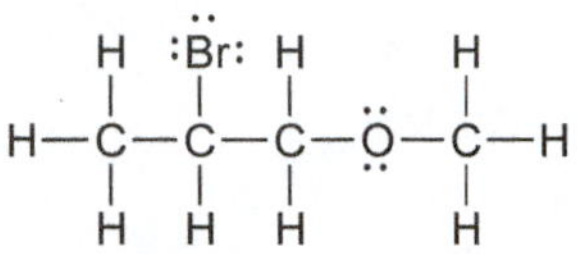

## Analysis

Start at the left and proceed to the right, making sure that each carbon has four bonds. Omit lone pairs on the heteroatoms O and Br. When like groups are bonded together or bonded to the same atom, use parentheses to further simplify the structure.

## Solution

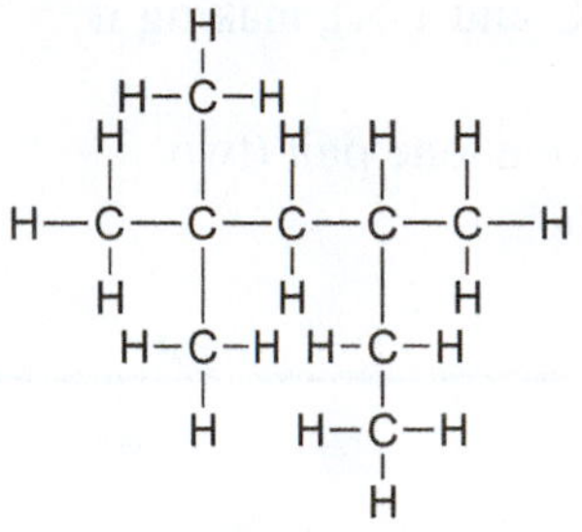

**Condensed structure**

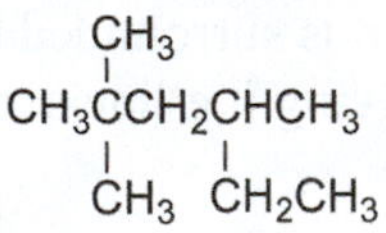

**Further simplified**

a.  =  $CH_3CCH_2CHCH_3$  =  $(CH_3)_3CCH_2CH(CH_2CH_3)CH_3$

$CH_3$ (top)

$CH_3$ $CH_2CH_3$ (bottom)

b. 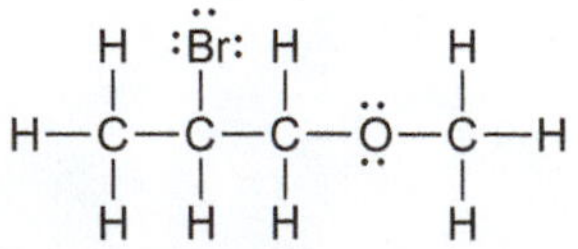  =  $CH_3CHCH_2OCH_3$

Br

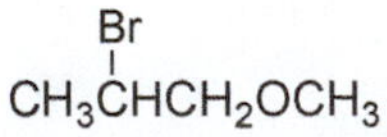

**Example 11.5** Convert each skeletal structure to a complete structure with all C's, H's, and lone pairs drawn in.

a.  —CH₃

b.

**Analysis**

To draw each complete structure, place a C atom at the corner of each polygon and add H's to give each carbon four bonds. The O atom has two bonds so it needs two lone pairs to have eight electrons around it.

**Solution**

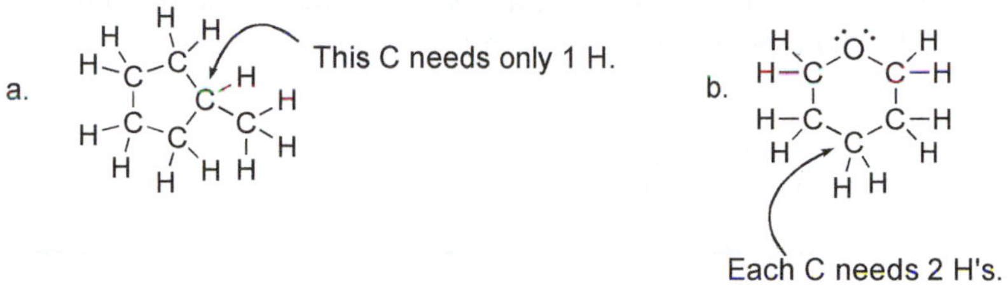

a. This C needs only 1 H.

b.

Each C needs 2 H's.

---

## [4] Functional Groups (11.5)

**Example 11.6** Identify the functional groups in each compound.

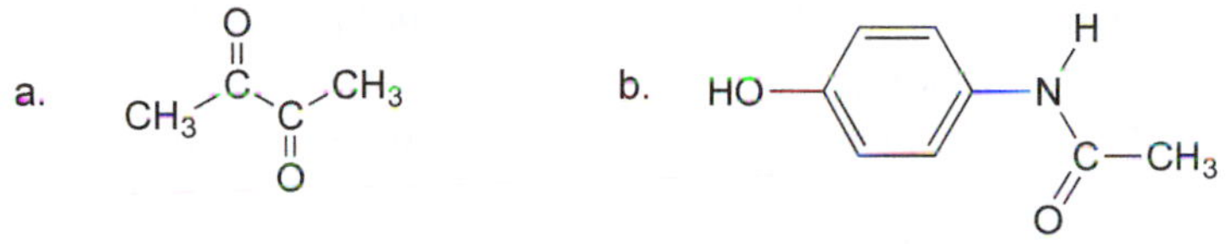

a.

**A**
butanedione
(component of butter flavor)

b.

**B**
acetaminophen
(analgesic in Tylenol)

**Analysis**

Concentrate on the multiple bonds and heteroatoms and refer to Tables 11.2, 11.3, and 11.4.

**Solution**

a.

**A**

b.

**B**

A contains two carbonyl groups. Each carbonyl carbon is bonded to two other carbons, making it a ketone. Thus, **A** contains two ketones.

**B** has a carbon atom bonded to a hydroxyl group (OH), as well as a benzene ring, a six-membered ring with three double bonds. It also has an amide functional group (N bonded to a C=O group). Thus, **B** has three functional groups.

## [5] Properties of Organic Compounds (11.6)

**Example 11.7** Explain why ethanol ($CH_3CH_2OH$) is a polar molecule.

**Analysis**
First, locate the polar bonds in $CH_3CH_2OH$. Then, determine the shape to check whether bond dipoles cancel or reinforce.

**Solution**
The C–C and C–H bonds in $CH_3CH_2OH$ are nonpolar, while the C–O and O–H bonds are polar. Since the O atom is surrounded by two atoms (C and H) and two lone pairs, it has a bent shape and the individual bond dipoles do *not* cancel. This makes $CH_3CH_2OH$ a polar molecule with a net dipole.

$$CH_3CH_2 \overset{\ddot{O}}{\diagdown} H \qquad \Big\Uparrow \text{ net dipole}$$

## Self-Test

### [1] Fill in the blank with one of the terms listed below.

Aldehyde (11.5)  Functional group (11.5)  Organic chemistry (11.1)
Carboxylic acid (11.5)  Heteroatom (11.2)  Polar (11.6)
Ester (11.5)  Hydrocarbon (11.5)  Vitamins (11.7)
Fat-soluble vitamin (11.7)  Nonpolar (11.6)  Water-soluble vitamin (11.7)

1. An _____ has a hydrogen atom bonded directly to the carbonyl carbon.
2. A _____ is an atom or a group of atoms with characteristic chemical and physical properties.
3. If the individual bond dipoles cancel in a molecule, the molecule is _____.
4. A vitamin that dissolves in water and has many polar bonds is classified as a _____.
5. Any atom that is not carbon or hydrogen is called a _____.
6. An _____ contains an OR group bonded directly to the carbonyl carbon.
7. _____ are organic compounds needed in small amounts for normal cell function.
8. A _____ is a compound that contains only the elements of carbon and hydrogen.
9. If the individual bond dipoles do not cancel, a molecule is _____.
10. _____ is the study of compounds that contain the element carbon.
11. A vitamin that dissolves in an organic solvent but is insoluble in water is classified as a _____.
12. A _____ contains an OH group bonded directly to the carbonyl carbon.

### [2] Match the structure with the name of the functional group.

a. $R-\overset{\overset{\displaystyle :O:}{\|}}{C}-H$  b. $R-\overset{\overset{\displaystyle :O:}{\|}}{C}-R$  c. $R-\overset{\overset{\displaystyle :O:}{\|}}{C}-\underset{\underset{\displaystyle H \text{ (or R)}}{|}}{\overset{..}{N}}-H \text{ (or R)}$  d. $R-\ddot{O}H$  e. $R-\ddot{O}-R$

13. Amide
14. Alcohol
15. Ether

16. Ketone
17. Aldehyde

**[3] Fill in the blank with one of the terms listed below.**

a. Alkenes
b. Alkanes

c. Alkynes
d. Aromatic hydrocarbons

18. _____ have only C–C single bonds and no functional group.
19. _____ contain a benzene ring, a six-membered ring with three double bonds.
20. _____ have a C–C triple bond as a functional group.
21. _____ have a C–C double bond as a functional group.

**[4] Identify the functional group present in each compound.**

a. aldehyde
b. ketone

c. amide
d. carboxylic acid

e. amine
f. ester

22. 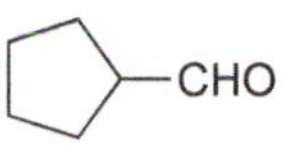 —CHO

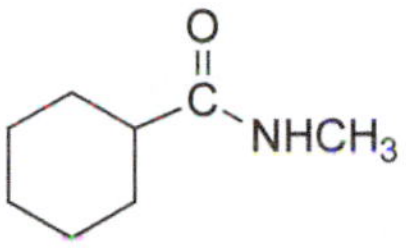

24. 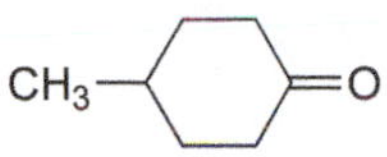 CH$_3$—=O

23.

25. 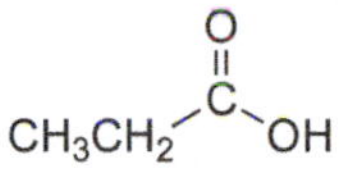 CH$_3$CH$_2$—

## Answers to Self-Test

| | | | | |
|---|---|---|---|---|
| 1. aldehyde | 6. ester | 11. fat-soluble vitamin | 16. b | 21. a |
| 2. functional group | 7. Vitamins | 12. carboxylic acid | 17. a | 22. a |
| 3. nonpolar | 8. hydrocarbon | 13. c | 18. b | 23. c |
| 4. water-soluble vitamin | 9. polar | 14. d | 19. d | 24. b |
| 5. heteroatom | 10. Organic chemistry | 15. e | 20. c | 25. d |

## Solutions to In-Chapter Problems

**11.1**    Organic compounds contain the element carbon.

a. C$_6$H$_{12}$—organic
b. H$_2$O—inorganic

c. KI—inorganic
d. MgSO$_4$—inorganic

e. CH$_4$O—organic
f. NaOH—inorganic

**11.2** Draw in all the H's and lone pairs as in Example 11.1. Each C and heteroatom must be surrounded by eight electrons.

a.
```
    H       H
    |       |
H—C—C=C—C—H
    |   |   |   |
    H   H   H   H
```

c.
```
    H  :O:  H
    |   ||   |
H—C—C—C—H
    |       |
    H       H
```

e.
```
   H   H
    \  /
     C
    / \
   |   :O:
    \ /
     C
    / \
   H   H
```

b.
```
        H
        |
   H  :O:  H
    |   |   |
H—C—C—C—H
    |   |   |
    H   H   H
```

d.
```
    H  :O:  H   H
    |   ||   |   |
H—C—C—N—C—H
    |       |   |
    H       H   H
```

**11.3** To determine the shape around each atom, draw in all the lone pairs on any heteroatom and then count groups as in Example 11.2.

a.
(1) and (2): three groups
**trigonal planar**
```
     (1) (2)
      ↓   ↓
H—C=C—Br
      |   |
      H   H
```

c.
(2): two atoms + two lone pairs
**bent**
```
     O      ..
     ||      |
H—C—O—H
          ..
```
(1): three groups
**trigonal planar**

b.
(1): two groups
**linear**
(3): two atoms + two lone pairs
**bent**
```
            H
            |
H—C≡C—C—O—H
            |  ..
            H
```
(2): four groups
**tetrahedral**

d.
(1): three groups
**trigonal planar**
```
      H   H
       \  /
        C=C       H
       /      \     |
H—C        C—C—C≡N:
       \      /   |
        C—C      H
       /  \
      H    H
```
(2): two groups
**linear**

**11.4** To determine the bond angles, first use the steps in Example 11.2, and then use the values in Table 11.1.

a.
```
  (1) F   H (2)
      \  /
   F—C—C—Br
      |   |
      F   Cl (3)
```
(1), (2), and (3): four groups
tetrahedral
**109.5°**

b.
```
   (1) H (2)
       |
   H—C—C=C—H
       |   |   |
       H   H   H (3)
```
(1): four groups
tetrahedral
**109.5°**

(2) and (3): three groups
trigonal planar
**120°**

c.
```
     H   H (2)
      \  /
       C=C        (3)
      /      \
  H—C        C—O—H
  (1)  \      /
        C—C
       /    \
      H      H
```
(1) and (2): three groups
trigonal planar
**120°**

(3): two atoms + two lone pairs
bent
**109.5°**

**11.5** When drawing three-dimensional structures:
- A solid line is used for bonds in the plane.
- A wedge is used for a bond in front of the plane.
- A dashed line is used for a bond behind the plane.

a.

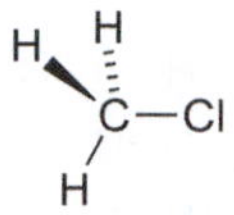

b.

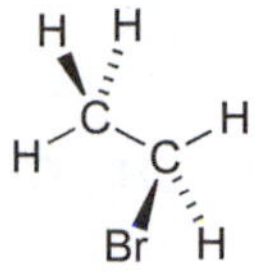

**11.6** Count the groups around each atom to determine the molecular shape as in Example 11.3.

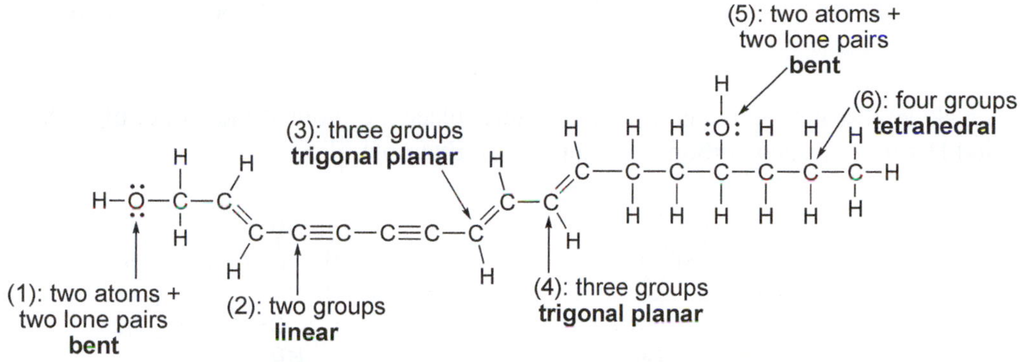

(1): two atoms + two lone pairs **bent**

(2): two groups **linear**

(3): three groups **trigonal planar**

(4): three groups **trigonal planar**

(5): two atoms + two lone pairs **bent**

(6): four groups **tetrahedral**

**11.7** Convert each compound to a condensed formula as in Example 11.4.

a. 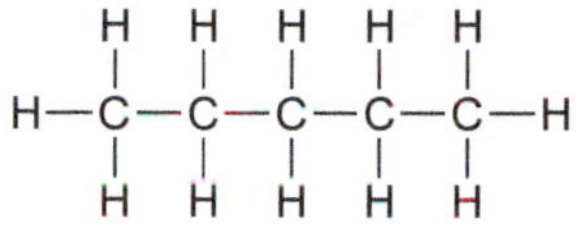 $H-C-C-C-C-C-H$ = $CH_3CH_2CH_2CH_2CH_3$ = $CH_3(CH_2)_3CH_3$

b. 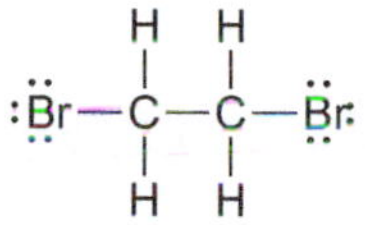 $:Br-C-C-Br:$ = $BrCH_2CH_2Br$ = $Br(CH_2)_2Br$

c. 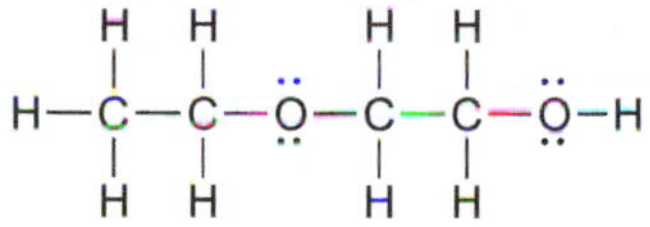 $H-C-C-O-C-C-O-H$ = $CH_3CH_2OCH_2CH_2OH$ = $CH_3CH_2O(CH_2)_2OH$

d. 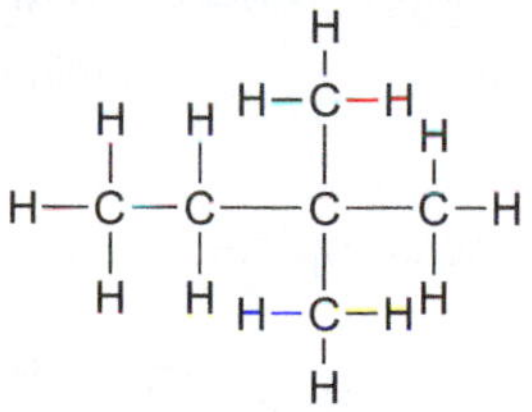 = $CH_3CH_2CCH_3$ with $CH_3$ above and $CH_3$ below = $CH_3CH_2C(CH_3)_3$

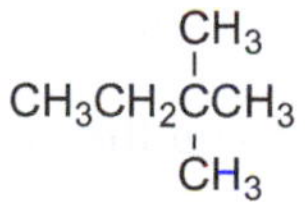

**11.8** Work backwards to convert each condensed formula to a complete structure.

a. $CH_3(CH_2)_8CH_3$ =

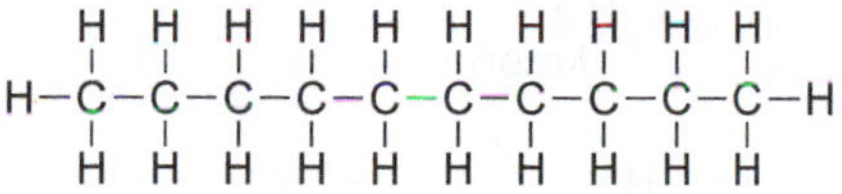

b. $CH_3(CH_2)_4OH$ =

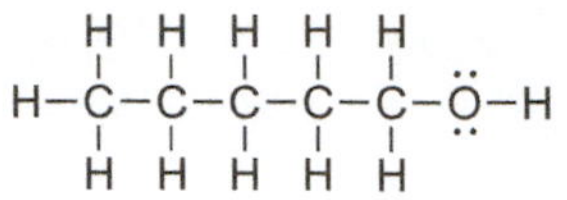

[structure: H–C–C–C–C–C–Ö–H with H's on each carbon]

d. $CH_3(CH_2)_4CH(CH_3)_2$ =

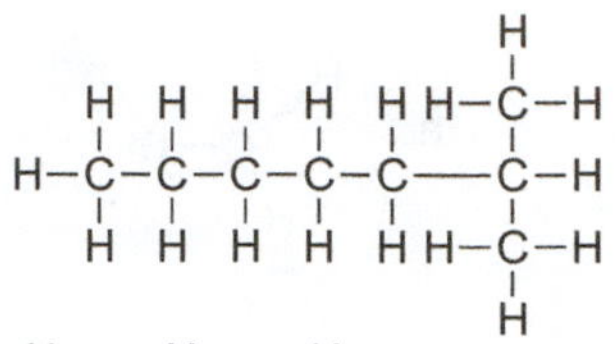

[structure: branched chain]

c. $CH_3CCl_3$ =

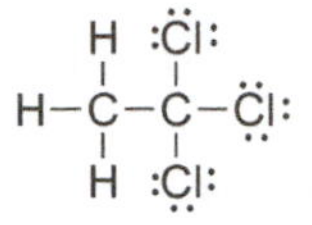

[structure: H–C–C–Cl: with Cl atoms]

e. $(CH_3)_2CHCH_2NH_2$ =

[structure: H–C–C–C–N–H]

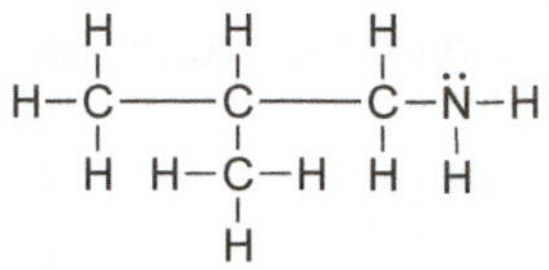

**11.9** To convert each skeletal structure to a complete structure, place a C atom at the corner of each polygon and add H's to give each carbon four bonds as in Example 11.5.

a. [cycloheptane structure with all H's]

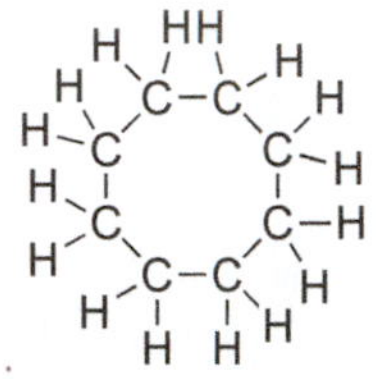

b. [aromatic ring with Cl substituents]

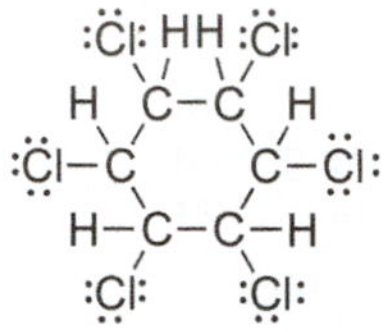

c. [structure with OH and branched carbons]

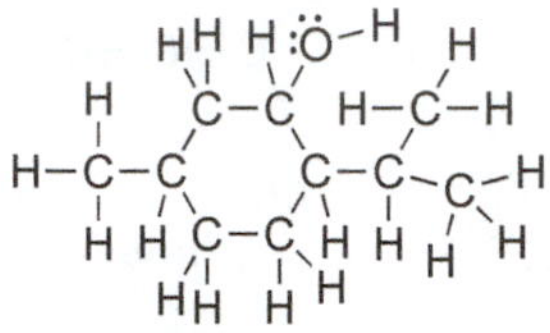

**11.10** Each carbon must have four bonds. To determine the number of H's bonded to each C, count the number of bonds, and then add H's to equal four.

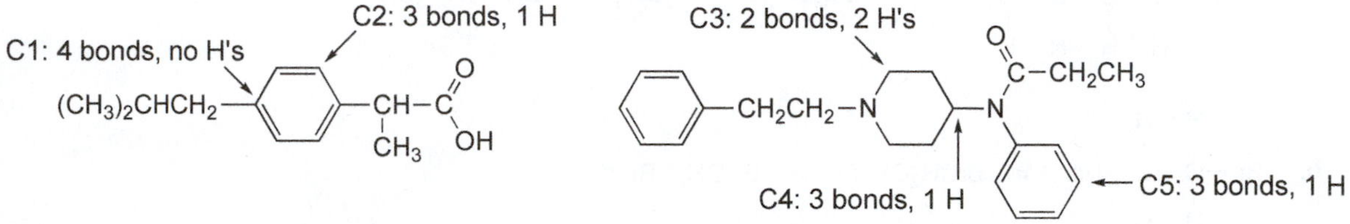

C1: 4 bonds, no H's    C2: 3 bonds, 1 H    C3: 2 bonds, 2 H's

$(CH_3)_2CHCH_2$—[benzene ring]—$CH$–$C$ with $CH_3$, O, OH [aldehyde/carboxylic structure]

[piperidine structure] —$CH_2CH_2$–N ... N ... $C$–$CH_2CH_3$, O

C4: 3 bonds, 1 H        C5: 3 bonds, 1 H

**11.11** Identify the functional groups in each compound. Concentrate on the heteroatoms and multiple bonds.

a. $CH_3CHCH_3$  [OH] hydroxyl

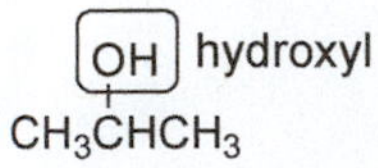

b. [benzene ring] [CH=CH_2] alkene, aromatic ring

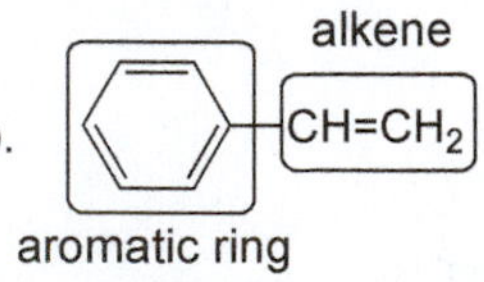

c. [H_2N]$CH_2CH_2CH_2CH_2$[NH_2]  amine, amine

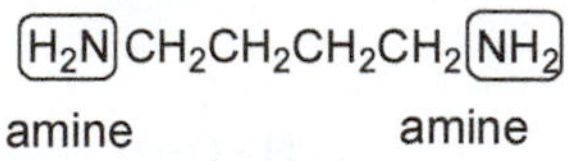

**11.12** Identify the functional groups as in Example 11.6 and then draw out the complete structures.

a. [ring structure] C–C=Ö  aldehyde

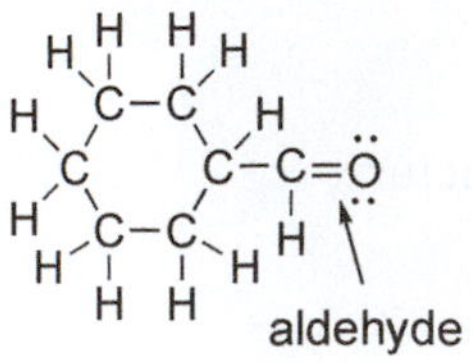

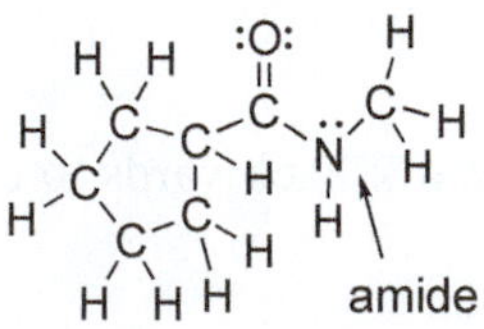

b. H–C–C–C–C–Ö–H with :O:  carboxylic acid

c. [ring structure] C=Ö  ketone

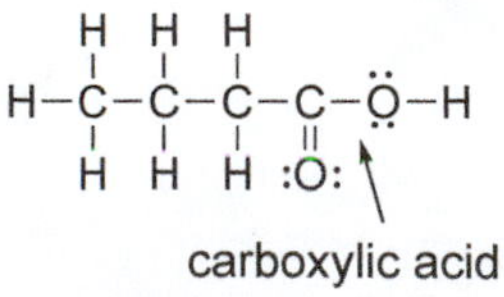

d. H–C–C–C–Ö–C–C–H with :O:  ester

e. [ring structure] C–C, :O:, N–H  amide

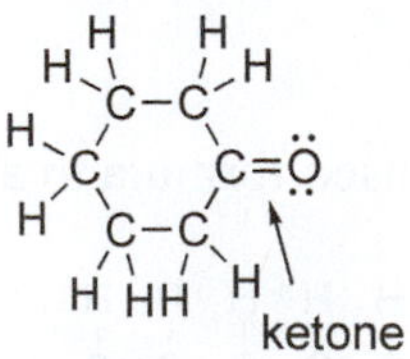

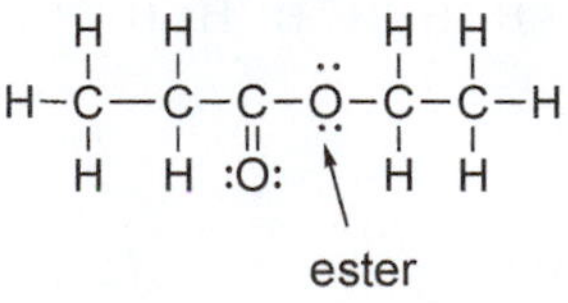

**11.13** Identify the functional groups.

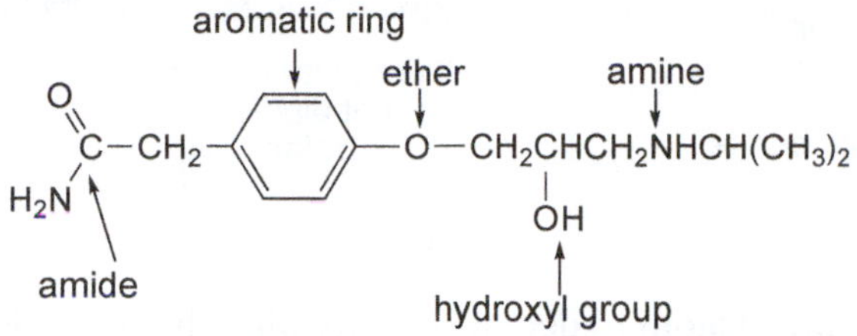

**11.14** Use the common element colors shown on the inside back cover to convert the ball-and-stick model to a shorthand representation, and then identify the functional groups.

a. 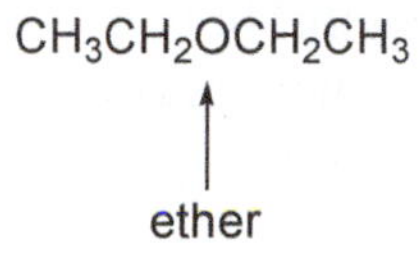 $CH_3CH_2OCH_2CH_3$
↑
ether

b.

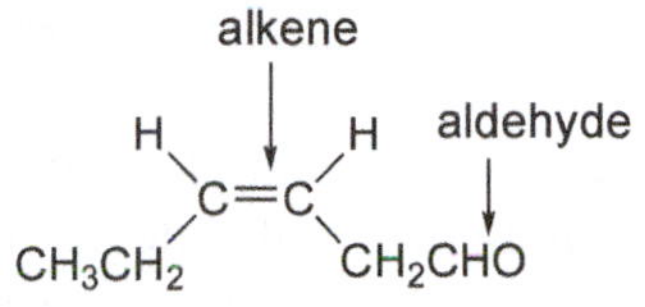

**11.15** Draw the skeletal structure as in Sample Problem 11.9.

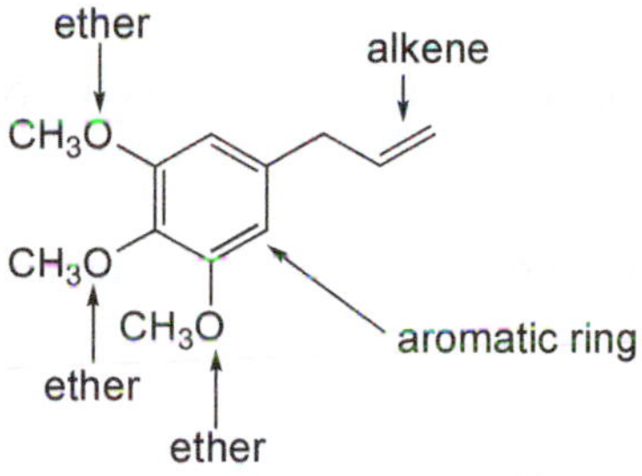

**11.16** All of the bonds except C–C and C–H bonds are polar. The symbol $\delta^+$ is given to the less electronegative atom, and the symbol $\delta^-$ is given to the more electronegative atom.

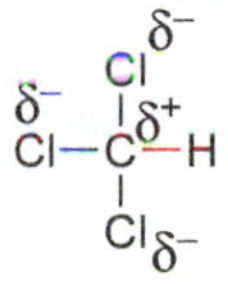

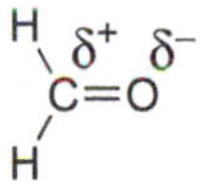

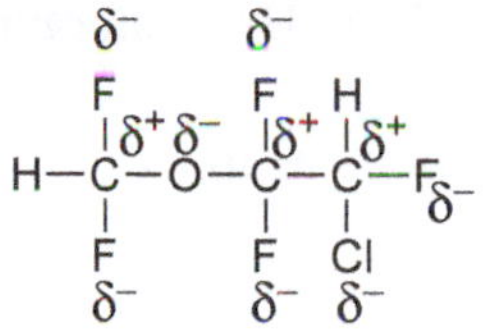

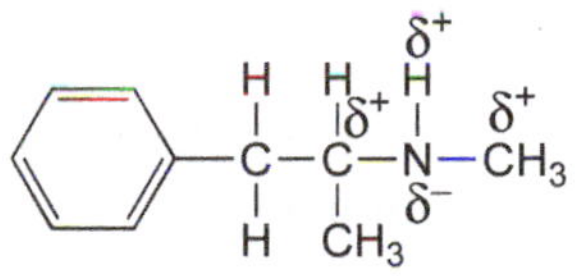

**11.17** With organic compounds that have more than one polar bond, the shape of the molecule determines the overall polarity.
- If the individual bond dipoles cancel in a molecule, the molecule is nonpolar.
- If the individual bond dipoles do not cancel, the molecule is polar.

a.

all nonpolar bonds
**nonpolar**

b.

$CH_3$
$C=O$
$CH_3$

polar C=O
**polar**

c.

Cl
C
Cl
Cl
Cl

four polar bonds
Dipoles cancel.
**nonpolar**

d.  $CH_3CH_2CH_2$

N
H
H

↑ net dipole

three polar bonds
net dipole
**polar**

**11.18** Draw the molecule in three dimensions around the O atom to determine why dimethyl ether is polar.

net dipole ↕  $CH_3$ O $CH_3$

The two C–O bonds are polar.
The molecule is bent, so there is a net dipole.
Therefore, the molecule is polar.

**11.19** Use the rule "like dissolves like" to determine if the compounds are soluble in water.

a.  Octane is a hydrocarbon, and therefore nonpolar. This means it is **insoluble** in water.
b.  Acetone is a small polar molecule due to the C=O. This means it is water **soluble**.
c.  Stearic acid is a fatty acid with a polar functional group, but since it has a very long hydrocarbon chain, it is **insoluble** in water.

**11.20** DDT is a nonpolar, fat-soluble compound that is soluble in tissues. Therefore, once it is ingested by birds, it stays in their tissues for long periods of time, making it detectable.

**11.21** Niacin is soluble in water because it is a small molecule that contains many polar bonds (C–N, C=O, C–O, and O–H).

**11.22** Vitamin $K_1$ is not water soluble because it has relatively few polar bonds compared to the number of nonpolar bonds. This makes it soluble in organic solvents.

## Solutions to Odd-Numbered End-of-Chapter Problems

**11.23** Organic compounds contain the element carbon.

a. $H_2SO_4$—inorganic          b. $Br_2$—inorganic          c. $C_5H_{12}$—organic

**11.25** Draw in all the H's and lone pairs as in Example 11.1. Each C and heteroatom must be surrounded by eight electrons.

a.

H H H
H–C–C=C–C≡C–H
H

c.

H H
H–C≡C–C–C–H
:Cl: H

b.

H H
H–C=C–C–O–H
H H

d.

H H :O:
H–C=C–C–C–N–H
H H H

**11.27** To determine the shape around each atom, draw in all the lone pairs on any heteroatom and then count groups as in Example 11.2.

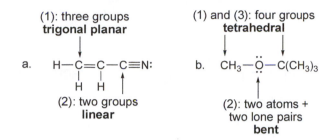

a.  (1): three groups **trigonal planar**

H—C=C—C≡N:
    |  |
    H  H

(2): two groups **linear**

b.  (1) and (3): four groups **tetrahedral**

CH₃—Ö—C(CH₃)₃

(2): two atoms + two lone pairs **bent**

**11.29** To determine the bond angles, first use the steps in Example 11.2 and then use the values in Table 11.1.

a.  (1)
CH₃—C≡C—Cl
(2)

(1) and (2): two groups
linear
**180°**

b.  H (2)
CH₂=C—Cl
(1)

(1) and (2): three groups
trigonal planar
**120°**

c.  (1) H (2)
CH₃—C—Cl
H

(1) and (2): four groups
tetrahedral
**109.5°**

**11.31** To determine the shape around each atom, draw in all the lone pairs on any heteroatom and then count groups as in Example 11.3.

(2): three groups **trigonal planar**

(3): four atoms **tetrahedral**

(1): two atoms + two lone pairs **bent** →:O—[benzene ring with O:]—CH₂CHNCH₃
                                          H
                                        CH₃

(4): three atoms + one lone pair **trigonal pyramidal**

**11.33** Each C in benzene is surrounded by three groups, so each carbon is trigonal planar, giving 120° bond angles.

**11.35** Convert each compound to a condensed structure as in Example 11.4.

a.
$$CH_3CH_2CH_2CCH_3 \quad = \quad CH_3(CH_2)_2C(CH_3)_3$$
with CH₃ above and CH₃ below the fourth carbon.

b. 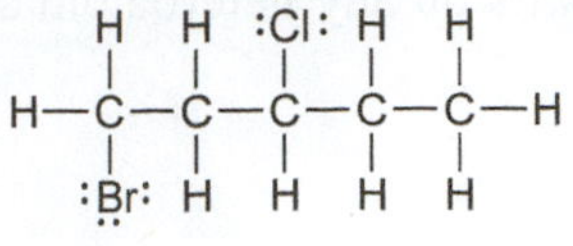 $H-\overset{\overset{\displaystyle H}{|}}{\underset{\underset{\displaystyle :Br:}{|}}{C}}-\overset{\overset{\displaystyle H}{|}}{\underset{\underset{\displaystyle H}{|}}{C}}-\overset{\overset{\displaystyle :\ddot{C}l:}{|}}{\underset{\underset{\displaystyle H}{|}}{C}}-\overset{\overset{\displaystyle H}{|}}{\underset{\underset{\displaystyle H}{|}}{C}}-\overset{\overset{\displaystyle H}{|}}{\underset{\underset{\displaystyle H}{|}}{C}}-H$   =   $BrCH_2CH_2CHCH_2CH_3$   =   $Br(CH_2)_2CHCH_2CH_3$
$\hspace{9.5cm} | \hspace{4.2cm} |$
$\hspace{9.4cm} Cl \hspace{4cm} Cl$

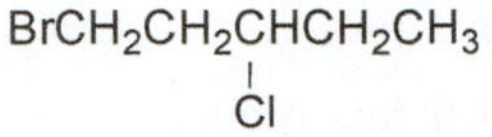

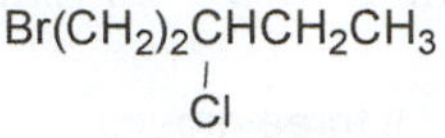

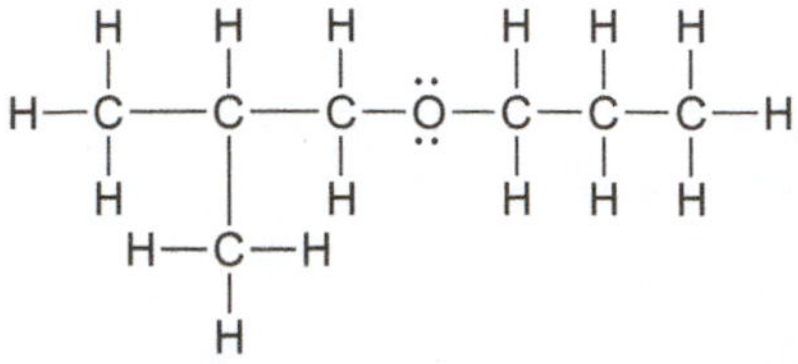

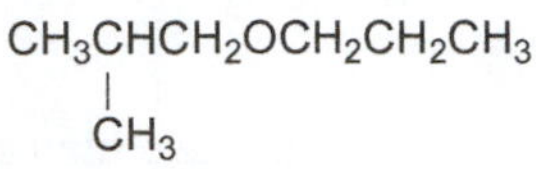

c.   =   $CH_3CHCH_2OCH_2CH_2CH_3$   =   $(CH_3)_2CHCH_2O(CH_2)_2CH_3$
$\hspace{7.6cm} |$
$\hspace{7.5cm} CH_3$

d. 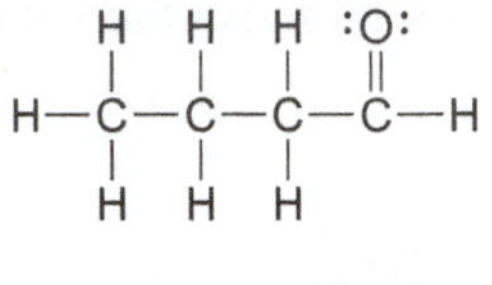   =   $CH_3CH_2CH_2CHO$   =   $CH_3(CH_2)_2CHO$

**11.37** To convert each compound to a skeletal structure, remove all C's and H's bonded to C's. Remove the lone pairs.

a. 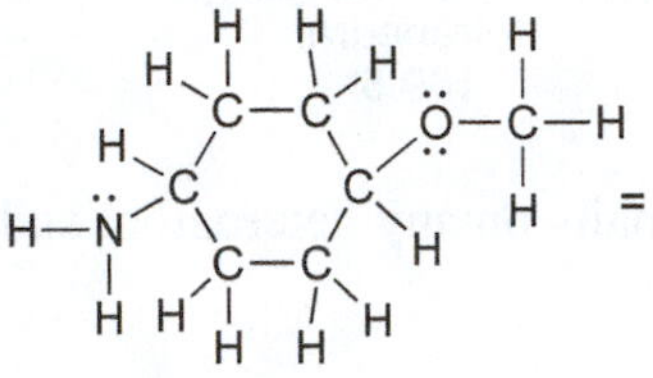   =   $H_2N-\!\!\!\!\bigcirc\!\!\!\!-OCH_3$

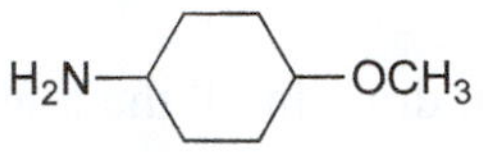

b. 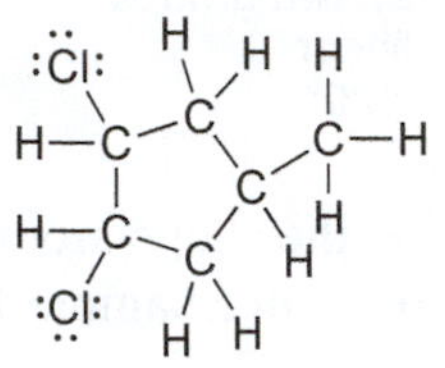   =   (cyclopentane with Cl, Cl, and $CH_3$ substituents)

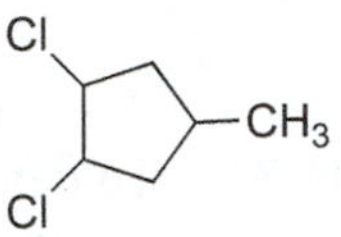

**11.39** Work backwards to convert each shorthand structure to a complete structure.

a. $(CH_3)_2CH(CH_2)_6CH_3$   =

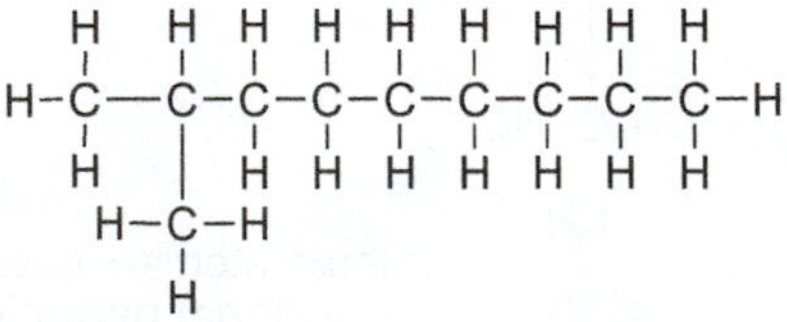

b. $(CH_3)_3COH$   =

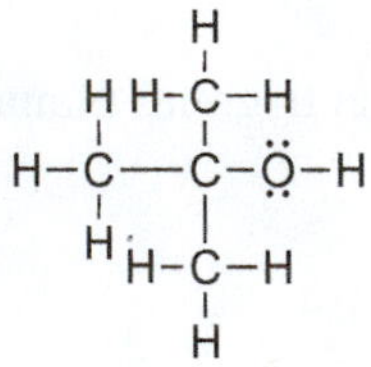

c. $CH_3CO_2(CH_2)_3CH_3$   =

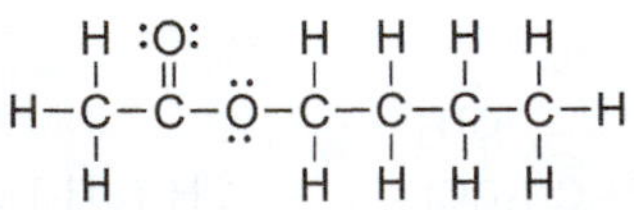

d. 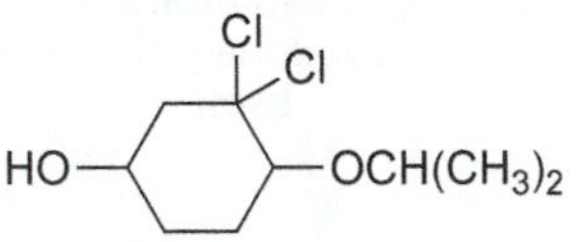 $HO-\!\!\!\!\bigcirc\!\!\!\!-OCH(CH_3)_2$   =

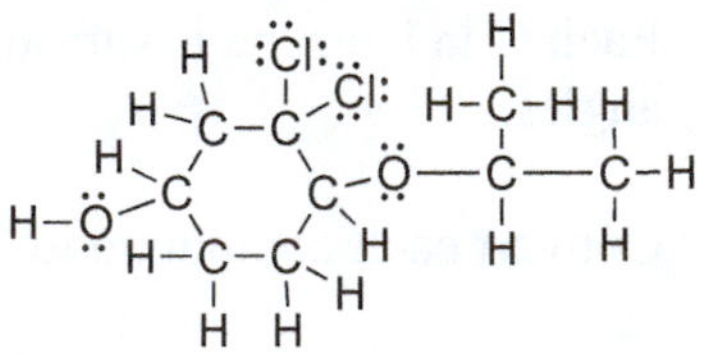

**11.41** Remember that C can have only four bonds.

a. (CH$_3$)$_3$CHCH$_2$CH$_3$   ⌐Five bonds to C

d. 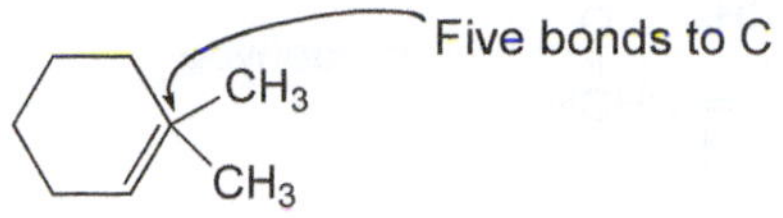 ⌐Five bonds to C

b. CH$_3$(CH$_2$)$_4$CH$_3$OH   ⌐Five bonds to C

e. Five bonds to C

c. (CH$_3$)$_2$C=CH$_3$   ⌐Five bonds to C

**11.43** Convert the shorthand structures to complete structures by drawing in C's, H's, and lone pairs.

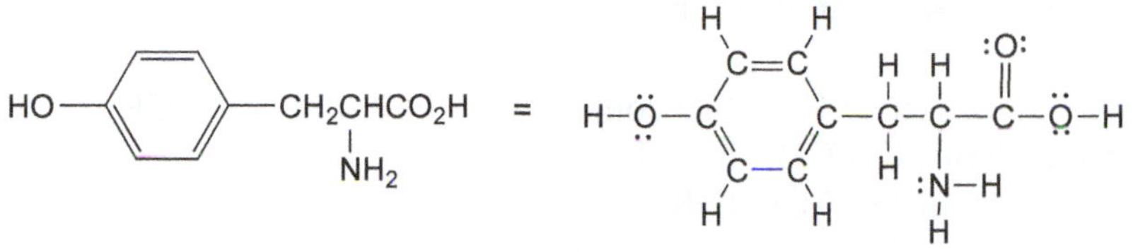

HO— ⬡ —CH$_2$CHCO$_2$H  =  [expanded structure]
           |
         NH$_2$

**11.45** Identify the functional groups in each molecule.

a. 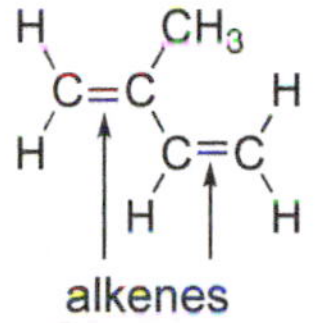 [structure]
    alkenes

b. 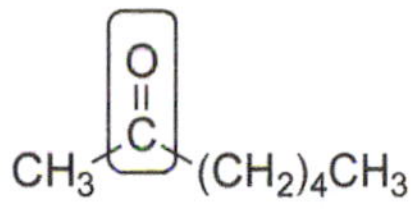 CH$_3$–C(=O)–(CH$_2$)$_4$CH$_3$
    ketone

c. 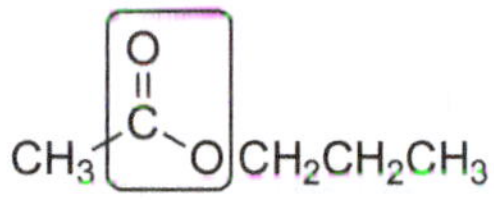 CH$_3$–C(=O)–O–CH$_2$CH$_2$CH$_3$
    ester

**11.47** Identify the functional groups in each molecule.

a. 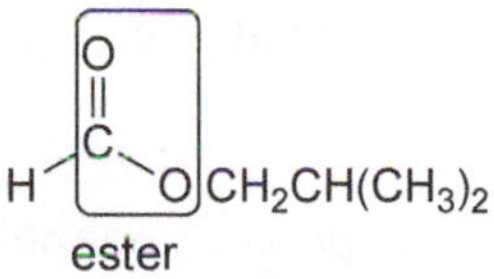 H–C(=O)–O–CH$_2$CH(CH$_3$)$_2$
    ester

c. ether   CH$_3$O
    HO— aromatic ring —C(=O)–H   aldehyde
    hydroxyl group
    aromatic ring

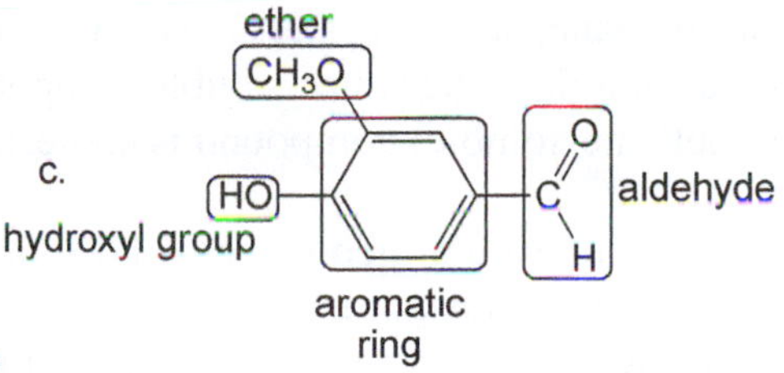

b. 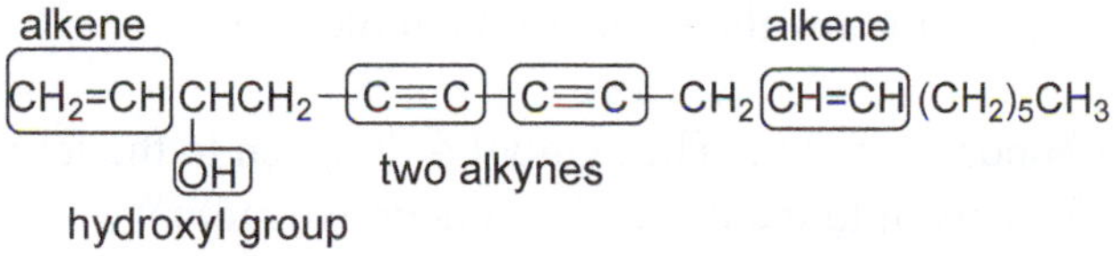    alkene            alkene
CH$_2$=CH CHCH$_2$–C≡C–C≡C–CH$_2$ CH=CH (CH$_2$)$_5$CH$_3$
    |
   OH     two alkynes
  hydroxyl group

**11.49** **Functional groups in dopamine:** two hydroxyl groups (OH groups with red spheres), aromatic ring (six-membered ring with three C=C's), and amine (NH$_2$ group with a blue sphere).

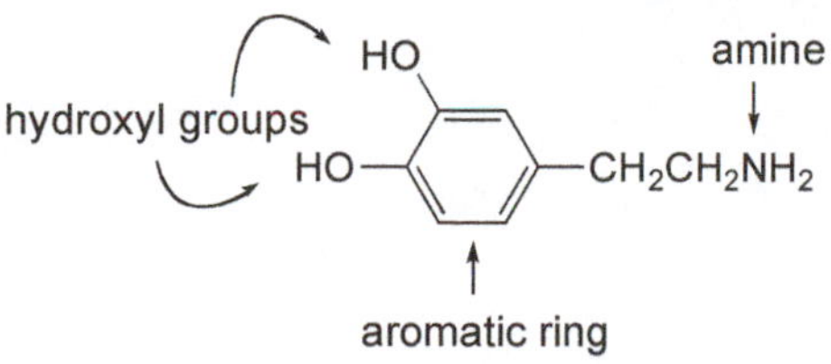

           HO         amine
hydroxyl groups
      HO— ⬡ —CH$_2$CH$_2$NH$_2$

          aromatic ring

**11.51**

**11.53** In a carboxylic acid, the C=O is bonded to a hydroxyl (OH) group. In an ester, the C=O is bonded to an OR group.

Examples: $CH_3CH_2COOH$ (carboxylic acid) and $CH_3CO_2CH_3$ (ester)

**11.55** Draw an organic compound that fits each set of criteria.

a. a hydrocarbon having molecular formula $C_3H_4$ that contains a triple bond: $HC \equiv CCH_3$
b. an alcohol containing three carbons: $CH_3CH_2CH_2OH$
c. an aldehyde containing three carbons: $CH_3CH_2CHO$

d. a ketone having molecular formula $C_4H_8O$:

**11.57** Draw the three five-carbon structures.

**11.59** Solubility properties can help you differentiate unknown compounds. NaCl dissolves in water but not in dichloromethane, and cholesterol dissolves in dichloromethane but not in water. Therefore, when you test the solubility, the water-soluble compound is NaCl, and the water-insoluble (dichloromethane-soluble) compound is cholesterol.

**11.61** Net dipoles and molecular geometry will determine whether a molecule is polar or nonpolar. When a molecule has polar bonds, one must determine whether the bond dipoles cancel or reinforce. To determine if they cancel or reinforce, one must know the shape of the molecule.

**11.63** All of the bonds except C–C and C–H bonds are polar. The symbol $\delta^+$ is given to the less electronegative atom, and the symbol $\delta^-$ is given to the more electronegative atom.

**11.65** With organic compounds that have more than one polar bond, the shape of the molecule determines the overall polarity.

- If the individual bond dipoles cancel in a molecule, the molecule is nonpolar.
- If the individual bond dipoles do not cancel, the molecule is polar.

a.

all nonpolar bonds
**nonpolar**

b. $CH_3CH_2$ —C— $CH_2CH_3$ (C=O)

polar C=O
**polar**

c. (structure with C, H, Br, Br) net dipole

two polar bonds
dipoles reinforce
**polar**

**11.67** Ethylene glycol is more water soluble than butane because it contains two polar hydroxyl groups capable of hydrogen bonding. Butane has no polar groups, and cannot hydrogen bond, and is therefore less water soluble.

**11.69** Sucrose has multiple OH groups and this makes it water soluble. 1-Dodecanol has a long hydrocarbon chain and a single OH group so it is insoluble in water.

**11.71** a. Spermaceti wax is an ester since it contains two R groups bonded to $-CO_2-$;
$CH_3(CH_2)_{14}CO_2(CH_2)_{15}CH_3$.
b. The very long nonpolar hydrocarbon chains make it insoluble in water but soluble in organic solvents.

**11.73** Fat-soluble vitamins will persist in tissues and accumulate, whereas any excess of water-soluble vitamins will be excreted in the urine. Therefore, if large quantities of fat-soluble vitamins are ingested, they can build up to toxic levels because they are not excreted readily in the urine, whereas large quantities of water-soluble vitamins will be eliminated much more readily.

**11.75** Vitamin E is a fat-soluble vitamin, making it soluble in organic solvents and insoluble in water.

**11.77** a. $C_3H_8$: Yes, there are just enough H's to give each C four bonds.
b. $C_3H_9$: No, there are too many H's. The maximum number of H's for 3 C's is 8 H's.
c. $C_3H_6$: Yes, if you put the three C's in a ring, each C gets two H's.

Examples:

a. H—C—C—C—H
$C_3H_8$

b. H—C—C—C—H
$C_3H_9$
One C has 5 bonds.
invalid structure

c. (ring structure)
$C_3H_6$

**11.79**

a. $CH_3CH_2CH_2NHCH_3$

b. $CH_3CH_2CH_2$ —N— $CH_3$ (with $\delta^+$ labels)

c. Four groups around N (three atoms and one lone pair); **trigonal pyramidal** around N.

d. This is a **polar molecule**. The bond dipoles don't cancel.

**11.81**

    a. $C_9H_{11}NO_2$
    b. Five lone pairs are drawn in on benzocaine.
    c. There are seven trigonal planar carbons (*).
    d. Benzocaine is water soluble due to the multiple polar bonds.
    e. Polar bonds are drawn in bold.

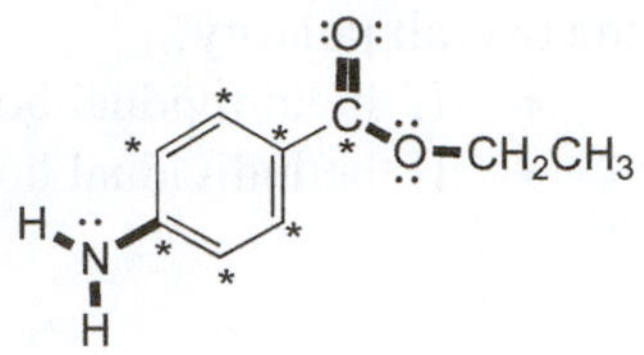

**11.83**    Answer each question about aldosterone.

    a.

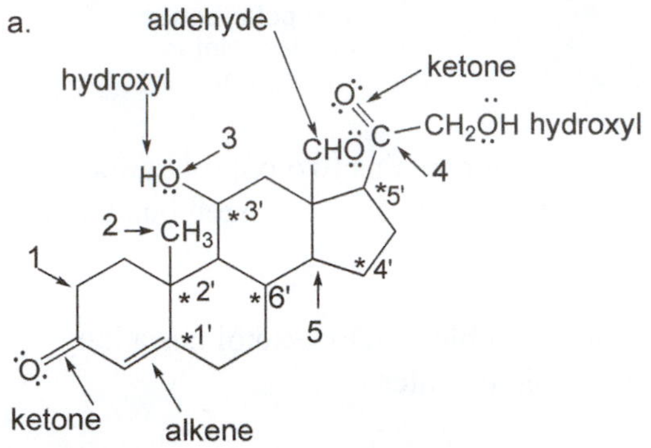

    e. Shape at each carbon
        1: Tetrahedral – 4 atoms
        2: Tetrahedral – 4 atoms
        3: Bent – 2 atoms, 2 lone pairs
        4: Trigonal planar – 3 atoms
        5: Tetrahedral – 4 atoms

    b. Each O needs two lone pairs.
    c. There are 21 C's.
    d. Number of H's at carbon:
        1': 0 H's
        2': 0 H's
        3': 1 H
        4': 2 H's
        5': 1 H
        6': 1 H

    f.

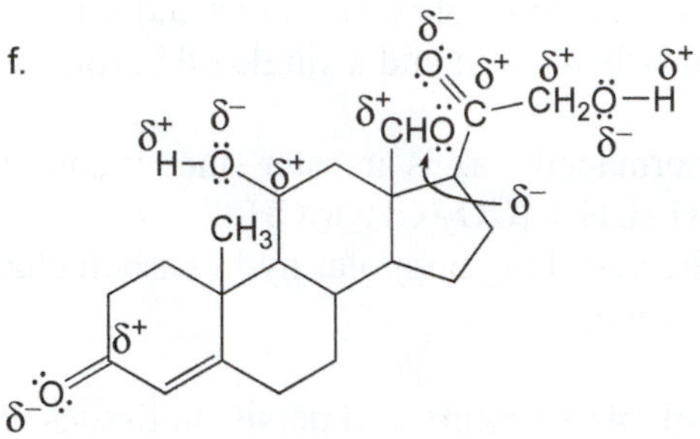

**11.85**    The waxy coating on cabbage leaves will prevent loss of water and keep leaves crisp.

**11.87**    THC is insoluble in water. THC is fat soluble and will therefore persist in tissues for an extended period of time. Ethanol is water soluble and will be quickly excreted in the urine. Therefore, THC is detectable many weeks after exposure, but ethanol is not.

# Chapter 12 Alkanes

**[1] What are the characteristics of an alkane? (12.1, 12.2)**

- Alkanes are hydrocarbons having only nonpolar C–C and C–H single bonds.
- There are two types of alkanes: Acyclic alkanes ($C_nH_{2n+2}$) have no rings. Cycloalkanes ($C_nH_{2n}$) have one or more rings.

- A carbon is classified as $1°$, $2°$, $3°$, or $4°$ by the number of carbons bonded to it. A $1°$ C is bonded to one carbon; a $2°$ C is bonded to two carbons, and so forth.

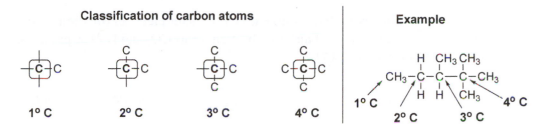

**[2] What are constitutional isomers? (12.2)**

- Isomers are different compounds with the same molecular formula.
- Constitutional isomers differ in the way the atoms are connected to each other. $CH_3CH_2CH_2CH_3$ and $HC(CH_3)_3$ are constitutional isomers because they both have molecular formula $C_4H_{10}$, but one compound has a chain of four carbons in a row and the other does not.

**Constitutional isomers**

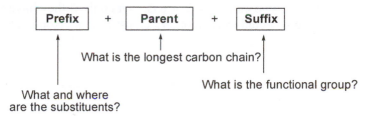

**[3] How are alkanes named? (12.4, 12.5)**

- Alkanes are named using the IUPAC system of nomenclature. A name has three parts: the parent indicates the number of carbons in the longest chain or the ring; the suffix indicates the functional group (*-ane* = alkane); and the prefix tells the number and location of substituents coming off the chain or ring.

| Prefix | + | Parent | + | Suffix |
|--------|---|--------|---|--------|

What is the longest carbon chain?

What is the functional group?

What and where are the substituents?

- Alkyl groups are formed by removing one hydrogen from an alkane. Alkyl groups are named by changing the *-ane* ending of the parent alkane to the suffix *-yl*.

## [4] Characterize the physical properties of alkanes. (12.7)

- Alkanes are nonpolar, so they have weak intermolecular forces, low melting points, and low boiling points.
- The melting points and boiling points of alkanes increase as the number of carbons increases due to increased surface area.

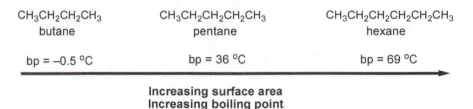

$CH_3CH_2CH_2CH_3$  
butane  

$CH_3CH_2CH_2CH_2CH_3$  
pentane  

$CH_3CH_2CH_2CH_2CH_2CH_3$  
hexane  

bp = –0.5 °C  

bp = 36 °C  

bp = 69 °C  

**Increasing surface area**  
**Increasing boiling point**

- Alkanes are insoluble in water.

## [5] What are the products of the complete combustion and incomplete combustion of an alkane? (12.8)

- Alkanes burn in the presence of air. Combustion forms $CO_2$ and $H_2O$ as products. Incomplete combustion forms CO and $H_2O$.

**Complete combustion**  
$2\ (CH_3)_3CCH_2CH(CH_3)_2\ +\ 25\ O_2\ \xrightarrow{\text{flame}}\ 16\ CO_2\ +\ 18\ H_2O\ +\ \text{energy}$

**Incomplete combustion**  
$2\ CH_4\ +\ 3\ O_2\ \xrightarrow{\text{flame}}\ 2\ CO\ +\ 4\ H_2O\ +\ \text{energy}$

- Alkane combustion has increased the level of $CO_2$, a greenhouse gas, in the atmosphere.
- The incomplete combustion forms carbon monoxide (CO), a toxin and air pollutant.

## Problem Solving

## [1] Simple Alkanes (12.2)

**Example 12.1** Are the compounds in each pair constitutional isomers or are they not isomers of each other?

a.  $CH_3CH_2CH_2CH_3$  and  $CH_3CH_2CH_2CH_2CH_3$

b.  $CH_3CH_2CH_2CH_2CH_3$  and

## Analysis

First compare molecular formulas; two compounds are isomers only if they have the same molecular formula. Then check how the atoms are connected to each other. Constitutional isomers have atoms bonded to different atoms.

**Solution**

a.   $CH_3CH_2CH_2CH_3$   and   $CH_3CH_2CH_2CH_2CH_3$

     molecular formula        molecular formula
        $C_4H_{10}$                 $C_5H_{12}$

The two compounds have a different number of C's and H's, so they have different molecular formulas. Thus, they are *not* isomers of each other.

b.   $CH_3CH_2CH_2CH_2CH_3$   and   (pentagon)

    molecular formula        molecular formula
      $C_5H_{12}$              $C_5H_{10}$

The two compounds have the same number of C's but a different number of H's, so they have different molecular formulas. Thus, they are *not* isomers of each other.

---

**Example 12.2** Draw two isomers with molecular formula $C_4H_{10}$.

**Analysis**
Since isomers are different compounds with the same molecular formula, we can draw one molecule with four carbons in a straight chain, but the other must have a one-carbon branch. Then add enough H's to give each C four bonds.

**Solution**

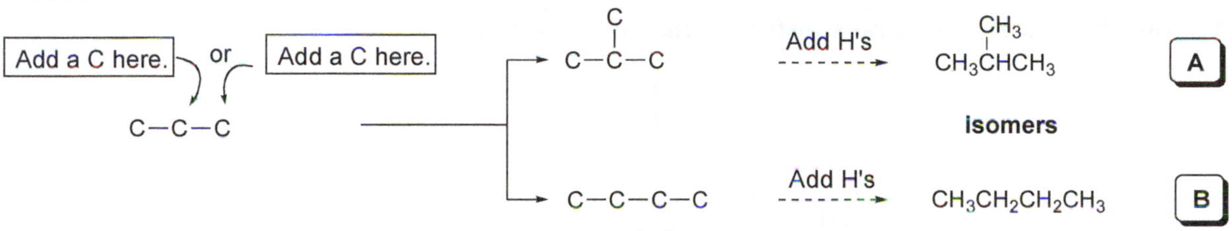

Compounds **A** and **B** are isomers because the atoms are connected together differently.

---

**Example 12.3** Classify each carbon atom in the following molecule as 1°, 2°, 3°, or 4°.

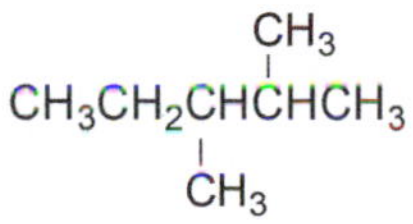

$$CH_3CH_2CH\overset{\overset{\textstyle CH_3}{|}}{C}HCH_3$$
$$\underset{\underset{\textstyle CH_3}{|}}{}$$

**Analysis**
To classify a carbon atom, count the number of C's bonded to it. Draw a complete structure with all bonds and atoms to clarify the structure if necessary.

**Solution**

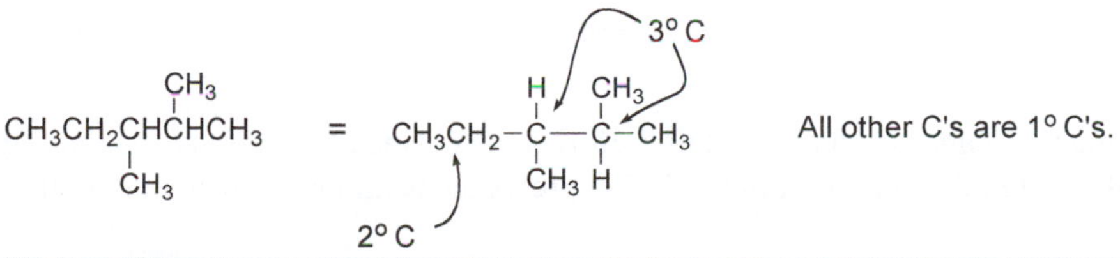

All other C's are 1° C's.

## [2] Alkane Nomenclature (12.4)

**Example 12.4** Give the IUPAC name for the following compound.

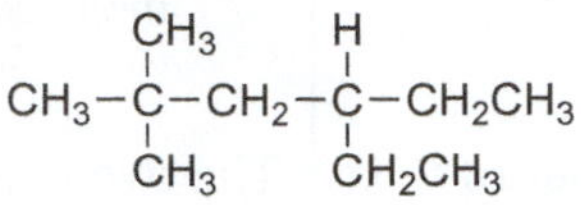

$$CH_3-\underset{\underset{CH_3}{|}}{\overset{\overset{CH_3}{|}}{C}}-CH_2-\underset{\underset{CH_2CH_3}{|}}{\overset{\overset{H}{|}}{C}}-CH_2CH_3$$

**Analysis and Solution**

**[1] Name the parent and use the suffix *-ane* since the molecule is an alkane.**

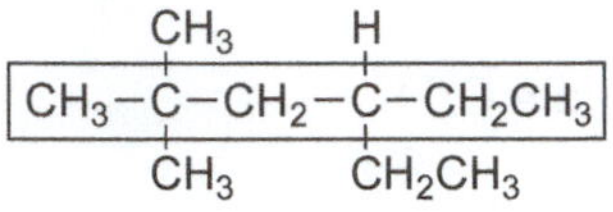

$$\boxed{CH_3-\underset{\underset{CH_3}{|}}{\overset{\overset{CH_3}{|}}{C}}-CH_2-\underset{\underset{CH_2CH_3}{|}}{\overset{\overset{H}{|}}{C}}-CH_2CH_3}$$

6 C's in the longest chain $\cdots\cdots\rightarrow$ **hexane**

- Box in the atoms of the longest chain to clearly show which carbons are part of the longest chain and which carbons are substituents.

**[2] Number the chain to give the first substituent the lower number.**

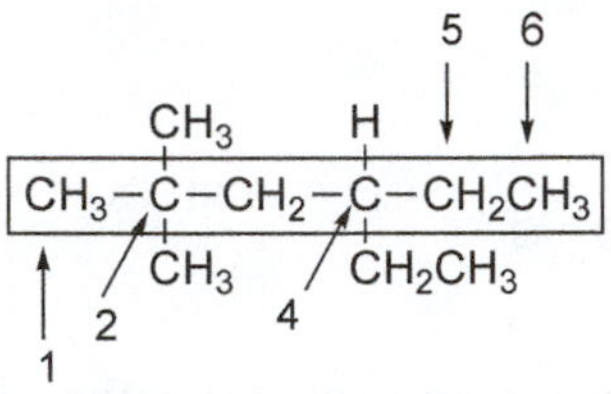

- Numbering from left to right puts the first substituent at C2.

**[3] Name and number the substituents.**

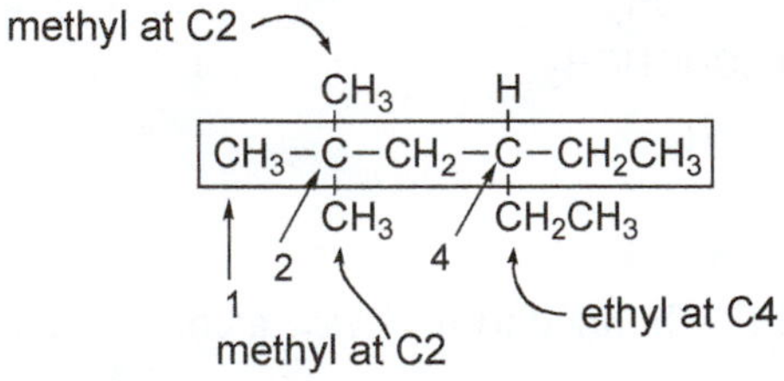

- This compound has three substituents: two methyl groups at C2 and an ethyl group at C4.

**[4] Combine the parts.**

- Write the name as one word and use the prefix di- before methyl since there are two methyl groups.
- Alphabetize the *e* of **e**thyl before the *m* of **m**ethyl. The prefix di- is ignored when alphabetizing.

methyl at C2

CH₃        H
$CH_3-C-CH_2-C-CH_2CH_3$ ◄— **hexane**
CH₃     CH₂CH₃
2      4
1
ethyl at C4
methyl at C2

**Answer: 4-ethyl-2,2-dimethylhexane**

---

**Example 12.5** Give the structure with the following IUPAC name: 2,2,3,4-tetramethylhexane.

**Analysis**
To derive a structure from a name, first look at the end of the name to find the parent name and suffix. From the parent we know the number of C's in the longest chain, and the suffix tells us the functional group; the suffix -*ane* = an alkane. Then, number the carbon chain from either end and add the substituents. Finally, add enough H's to give each C four bonds.

**Solution**
2,2,3,4-Tetramethyl**hexane** has **hexane** (6 C's) as the longest chain and four methyl groups at carbons 2, 2, 3, and 4.

Draw 6 C's and                                                    **Answer:**
number the chain:

$C-C-C-C-C-C$        ┈┈┈┈►        $C-C-C-C-C$        ┈┈┈┈►
1  2  3  4  5  6                               C   3   4
| Add 4 CH₃ |                               2              | Add H's |
| groups |

CH₃ CH₃  CH₃
$CH_3-C-C-CHCH_2CH_3$
CH₃ H

**2,2,3,4-tetramethylhexane**

---

## [3] Cycloalkanes (12.5)

**Example 12.6** Give the IUPAC name for the following compound.

CH₂CH₂CH₃

CH₂CH₂CH₂CH₃

**Analysis and Solution**
**[1] Name the ring.** The ring has 6 C's so the molecule is named as a cyclohexane.

---

**[2] Name and number the substituents.**
- There are two substituents: $CH_3CH_2CH_2-$ is a propyl group and $CH_3CH_2CH_2CH_2-$ is a butyl group.
- Number to put the two groups at C1 and C2 (not C1 and C6).
- Place the butyl group at C1 since the *b* of **butyl** comes before the *p* of **propyl** in the alphabet.

2 $CH_2CH_2CH_3$   propyl at C2

6 C's ···→ cyclohexane   1

$CH_2CH_2CH_2CH_3$   butyl at C1

Start numbering here.

**Answer: 1-butyl-2-propylcyclohexane**

---

## Self-Test

**[1] Fill in the blank with one of the terms listed below.**

| | | |
|---|---|---|
| Acyclic alkanes (12.1) | Constitutional isomers (12.2) | Oxidation (12.8) |
| Alkanes (12.1) | Cycloalkanes (12.1) | Pheromone (12.1) |
| Branched-chain alkane (12.2) | Greenhouse gas (12.8) | Reduction (12.8) |
| Combustion (12.8) | Isomers (12.2) | Saturated hydrocarbons (12.1) |

1. A _____ is a chemical substance used for communication in a specific animal species, most commonly an insect population.
2. _____ contain carbons joined in one or more rings.
3. $CO_2$ is a _____ because it absorbs thermal energy that normally radiates from the earth's surface, and redirects it back to the surface.
4. _____ results in an increase in the number of C–O bonds or a decrease in the number of C–H bonds.
5. Alkanes that contain chains of carbon atoms but no rings are called _____.
6. _____ differ in the way the atoms are connected to each other.
7. A _____ is an alkane that contains one or more carbon substituents bonded to a carbon chain.
8. _____ results in a decrease in the number of C–O bonds or an increase in the number of C–H bonds.
9. Acyclic alkanes are also called _____ because they have the maximum number of hydrogen atoms per carbon.
10. _____ are hydrocarbons having only C–C and C–H single bonds.
11. _____ are two different compounds with the same molecular formula.
12. Alkanes undergo _____—that is, they burn in the presence of oxygen to form carbon dioxide ($CO_2$) and water.

**[2] Label each carbon as 1°, 2°, 3°, or 4°.**

$CH_3$

$CH_3CH_2CHCHCH_2CH_3$
$CH_3$

13.

$CH_3$

$CH_3$

14. 15.

$(CH_3)_3CC(CH_3)_3$

16.

$CH_2CH_3$

17. 18.

[3] **Label each pair of compounds as constitutional isomers, identical molecules, or not isomers of each other.**

|   |   |
|---|---|
| a. constitutional isomers | c. not isomers |
| b. identical molecules |   |

19.
$$
\begin{array}{c}
\text{CH}_2-\text{CH}_2 \\
| \quad\quad | \\
\text{CH}_3\text{CH}_2 \quad \text{CH}_2\text{CH}_3
\end{array}
\quad \text{and} \quad \text{CH}_3(\text{CH}_2)_4\text{CH}_3
$$

21. $\text{CH}_3\text{CH}_2\text{CH}_2\text{CH}_3$ and ▷–CH₃

20. $\text{CH}_3-\overset{\overset{\displaystyle O}{\|}}{\underset{\text{OCH}_3}{C}}$ and $\text{CH}_3\text{CH}_2-\overset{\overset{\displaystyle O}{\|}}{\underset{\text{OH}}{C}}$

[4] **Pick the correct IUPAC name from each pair.**

22. (a) 4-methylpentane   or   (b) 2-methylpentane
23. (a) 1,2-diethylcyclohexane   or   (b) 1,6-diethylcyclohexane
24. (a) 3-methylhexane   or   (b) 1,3-dimethylpentane
25. (a) 2-ethyl-2-methylcycloheptane   or   (b) 1-ethyl-1-methylcycloheptane

## Answers to Self-Test

| | | | | |
|---|---|---|---|---|
| 1. pheromone | 6.  Constitutional isomers | 11. Isomers | 16. 4° | 21. c |
| 2. Cycloalkanes | 7.  branched-chain alkane | 12. combustion | 17. 3° | 22. b |
| 3. greenhouse gas | 8.  Reduction | 13. 3° | 18. 1° | 23. a |
| 4. Oxidation | 9.  saturated hydrocarbons | 14. 2° | 19. b | 24. a |
| 5. acyclic alkanes | 10. Alkanes | 15. 3° | 20. a | 25. b |

## Solutions to In-Chapter Problems

**12.1**
- An acyclic alkane has the molecular formula $C_nH_{2n+2}$, where $n$ is the number of carbons it contains.
- A cycloalkane has two fewer H's than an acyclic alkane with the same number of carbons, and its general formula is $C_nH_{2n}$.

   a. An acyclic alkane with three carbons has eight hydrogen atoms; $C_3H_8$.
   b. A cycloalkane with four carbons has eight hydrogen atoms; $C_4H_8$.
   c. A cycloalkane with nine carbons has 18 hydrogen atoms; $C_9H_{18}$.
   d. An acyclic alkane with seven carbons has 16 hydrogen atoms; $C_7H_{16}$.

**12.2**   An acyclic alkane has the molecular formula $C_nH_{2n+2}$, whereas a cycloalkane has the molecular formula $C_nH_{2n}$.

   a. $C_5H_{12}$, $5 \times 2 + 2 = 12$: acyclic
   b. $C_4H_8$, $4 \times 2 = 8$: cyclic
   c. $C_{12}H_{24}$, $12 \times 2 = 24$: cyclic
   d. $C_{10}H_{22}$, $10 \times 2 + 2 = 22$: acyclic

**12.3**    To determine if the compounds are isomers, first compare molecular formulas as in Example 12.1; two compounds are isomers only if they have the same molecular formula. Then check how the atoms are connected to each other. Constitutional isomers have atoms bonded to different atoms.

a. $CH_3CH_2CH_2CH_3$ and $CH_3CH_2CH_3$

      $C_4H_{10}$             $C_3H_8$

    different molecular formulas
        **not isomers**

c. and

    $C_6H_{12}$           $C_6H_{12}$

    same molecular formula
  different arrangement of atoms
        **isomers**

b. $CH_3CH_2CH_2OH$ and $CH_3OCH_2CH_3$

    $C_3H_8O$            $C_3H_8O$

    same molecular formula
  different arrangement of atoms
        **isomers**

**12.4**    Draw two isomers as in Example 12.2. Since isomers are different compounds with the same molecular formula, add a one-carbon branch to two different carbons to form two different molecules. Then add enough H's to give each C four bonds.

**12.5**    Isopentane has 4 C's in a row with a one-carbon branch.

**12.6**    To classify a carbon atom, count the number of C's bonded to it as in Example 12.3. Draw a complete structure with all bonds and atoms to clarify the structure if necessary.

**12.7** Draw a structure to fit each description.

a. 
$$1° \quad 1°$$
$$\downarrow \quad \downarrow$$
$$CH_3CH_2CH_3$$
$$\uparrow$$
$$2°$$

b. 
$$CH_3CHCH_2CH_3$$
$$|$$
$$CH_3 \quad 3°$$
$$C_5H_{12}$$

c. 
$$CH_3 \quad CH_3$$
$$| \quad |$$
$$CH_3CCH_2CCH_3$$
$$| \quad |$$
$$CH_3 \quad CH_3$$
$$4° \qquad\qquad 4°$$

**12.8** Draw a skeletal structure for each compound.

a. $CH_3CH_2CH_2CH_2CH_2CH_3$ =  /\/\/

b. $CH_3(CH_2)_5CH_3$ = /\/\/\

**12.9** Convert each skeletal structure to a complete structure with all atoms and bond lines.

a. /\/\

$$\begin{array}{ccccc} H & H & H & H & H \\ | & | & | & | & | \\ H-C-C-C-C-C-H \\ | & | & | & | & | \\ H & H & H & H & H \end{array}$$

c.  (cyclopentane with propyl chain)

b. /\/\|

$$\begin{array}{ccccc} H & H & H & H & H \\ | & | & | & | & | \\ H-C-C-C-C-C-H \\ | & | & | & | & | \\ H & H & | & H & H \\ & & H-C-H \\ & & | \\ & & H \end{array}$$

**12.10** To give the IUPAC name for each compound, follow the steps in Example 12.4:
[1] Name the parent and use the suffix -*ane* since the molecule is an alkane.
[2] Number the chain to give the first substituent the lower number.
[3] Name and number the substituents.
[4] Combine the parts.

a.
$$\boxed{CH_3CH_2CHCH_2CH_3}$$
$$|$$
$$CH_3$$
5 C's in the longest chain
**pentane**

- - - - - - - →

$$CH_3CH_2CHCH_2CH_3$$
$$\uparrow \quad \uparrow \quad | \quad \uparrow$$
$$\quad\quad CH_3$$
$$1 \quad 2 \quad 3 \quad 5$$

- - - - - - - →

$$CH_3CH_2CHCH_2CH_3$$
$$|$$
$$CH_3 \quad 3$$
methyl at C3

- - - - - - - → **3-methylpentane**

b.
$$\boxed{\begin{array}{c} CH_3 \; CH_2CH_2CH_3 \\ | \quad | \\ CH_3CH_2CH_2-C-C-CH_2CH_2CH_3 \\ | \quad | \\ H \quad H \end{array}}$$
9 C's in the longest chain
**nonane**

- - - →

$$\begin{array}{c} CH_3 \; CH_2CH_2CH_3 \\ | \quad | \\ CH_3CH_2CH_2-C-C-CH_2CH_2CH_3 \\ \uparrow \quad \uparrow \quad H \quad H \\ 1 \quad 2 \quad 4 \quad 5 \end{array}$$

propyl at C5
methyl at C4

$$\begin{array}{c} CH_3 \; CH_2CH_2CH_3 \\ | \quad | \\ CH_3CH_2CH_2-C-C-CH_2CH_2CH_3 \\ | \quad | \\ H \quad H \end{array}$$

- - - → **4-methyl-5-propylnonane**

c.   H–C–CH$_2$–CHCH$_3$  with CH$_2$CH$_3$ and CH$_3$, CH$_3$ groups  - - - - →  H–C–CH$_2$–CHCH$_3$ (CH$_2$CH$_3$; CH$_3$ at 4, CH$_3$ at 2, position 1) - - - - →  H–C–CH$_2$–CHCH$_3$ (CH$_2$CH$_3$; CH$_3$ at 4, CH$_3$ at 2) - - - - → **2,4-dimethylhexane**

6 C's in the longest chain
**hexane**

methyls at C2 and C4

**12.11**   To give the IUPAC name for each compound follow the steps in Example 12.4.

a.   (CH$_3$)$_2$CHCH(CH$_3$)$_2$  - - →  CH$_3$–C–C–CH$_3$ (H H; CH$_3$ CH$_3$)  - - →  CH$_3$–C–C–CH$_3$ (H H; CH$_3$ CH$_3$; positions 1, 2, 4)  - - → **2,3-dimethylbutane**

4 C's in the longest chain
**butane**

methyls at C2 and C3

b.   CH$_3$CH$_2$CH$_2$CHCH$_2$C–H (CH$_2$CH$_2$CH$_2$CH$_3$; CH$_3$CH$_2$, H)  - - - - - →  CH$_3$CH$_2$CH$_2$CHCH$_2$C–H (4; CH$_2$CH$_2$CH$_2$CH$_3$; CH$_3$CH$_2$, H; positions 1, 2, 6)  - - - - - → **4-ethyldecane**

10 C's in the longest chain
**decane**

ethyl at C4

c.   CH$_3$CH$_2$CH$_2$CH$_2$–C–C–CH$_2$CH$_3$ (CH$_3$ H; CH$_2$ CH$_3$; CH$_3$)  - - - →  CH$_3$CH$_2$CH$_2$CH$_2$–C–C–CH$_2$CH$_3$ (4; CH$_3$ H 3; CH$_2$ CH$_3$; CH$_3$ 2 1)  - - - - → **4-ethyl-3,4-dimethyloctane**

methyl at C4

8 C's in the longest chain
**octane**

ethyl at C4       methyl at C3

**12.12**   To draw the structure corresponding to each IUPAC name, follow the steps in Example 12.5.

a. 3-methyl**hexane**

Draw 6 C's and
number the chain:

C–C–C–C–C–C
1  2  3  4  5  6

- - - - - - - - - →   | Add 1 CH$_3$ group |

C–C–C–C–C–C (with C at position 3)
2     3

- - - - - →   | Add H's |

CH$_3$CH$_2$CHCH$_2$CH$_2$CH$_3$
CH$_3$

b. 3,3-dimethyl**pentane**

Draw 5 C's and
number the chain:

C–C–C–C–C
1  2  3  4  5

- - - - - - - - - →   | Add 2 CH$_3$ groups |

C–C–C–C–C (with C above and below position 3)
2     3

- - - - - →   | Add H's |

CH$_3$CH$_2$CCH$_2$CH$_3$ (with CH$_3$ above and CH$_3$ below)

c. 3,5,5-trimethyl**octane**

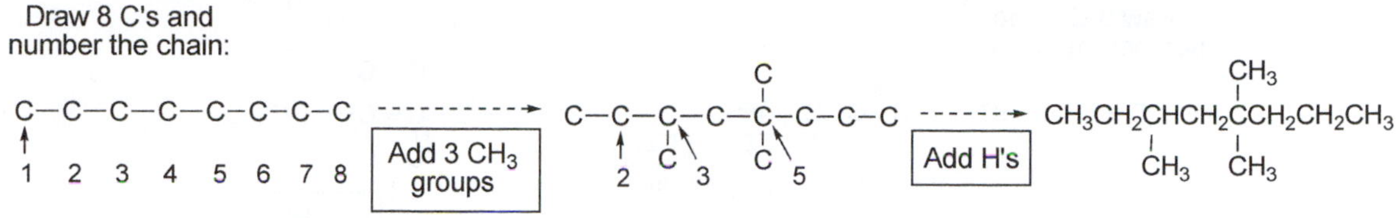

Draw 8 C's and
number the chain:

C—C—C—C—C—C—C—C   - - - - - - - →   Add 3 CH$_3$ groups   →   C—C—C—C—C—C—C—C   - - - - →   Add H's   →   $CH_3CH_2CHCH_2CCH_2CH_2CH_3$

1  2  3  4  5  6  7  8

d. 3-ethyl-4-methyl**hexane**

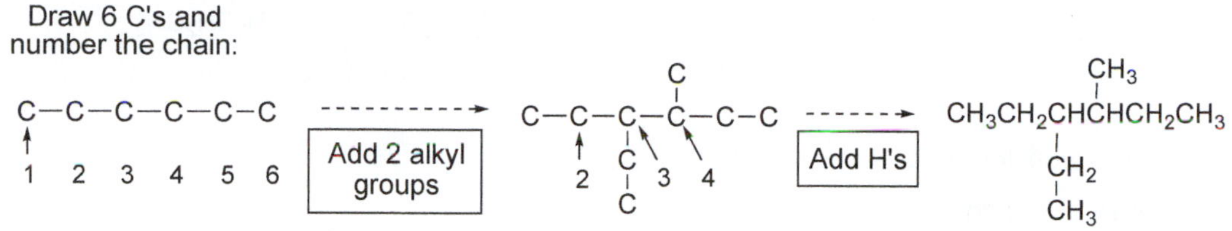

Draw 6 C's and
number the chain:

C—C—C—C—C—C   - - - - - - →   Add 2 alkyl groups   →   C—C—C—C—C—C   - - - - →   Add H's   →   $CH_3CH_2CHCHCH_2CH_3$

1  2  3  4  5  6

**12.13**   To draw the structure corresponding to each IUPAC name, follow the steps in Example 12.5.

a. 2,2-dimethyl**butane**

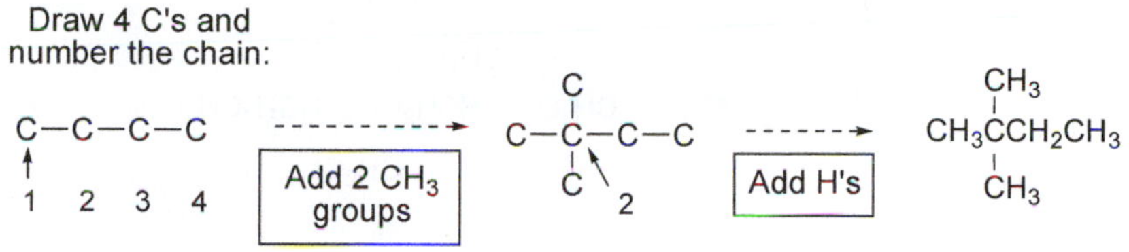

Draw 4 C's and
number the chain:

C—C—C—C   - - - - - - →   Add 2 CH$_3$ groups   →   C—C—C—C   - - - - →   Add H's   →   $CH_3CCH_2CH_3$

1  2  3  4

b. 6-butyl-3-methyl**decane**

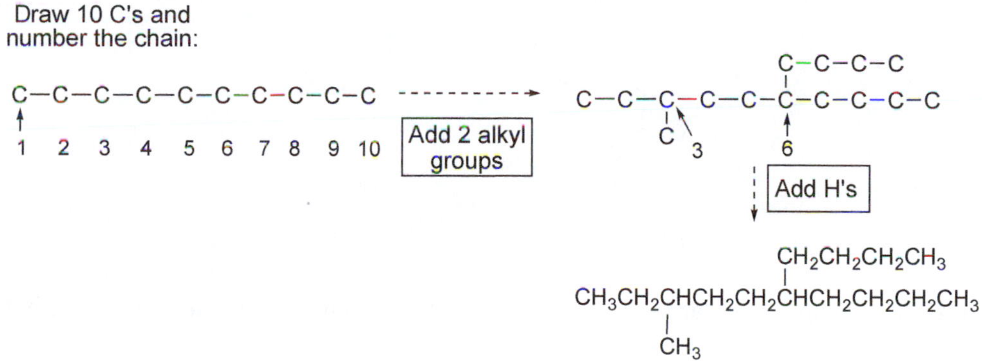

Draw 10 C's and
number the chain:

C—C—C—C—C—C—C—C—C—C   - - - - - - →   Add 2 alkyl groups   →   C—C—C—C—C—C—C—C—C—C

1  2  3  4  5  6  7  8  9  10

Add H's

$CH_3CH_2CHCH_2CH_2CHCH_2CH_2CH_2CH_3$
with substituents $CH_3$ and $CH_2CH_2CH_2CH_3$

c. 4,4,5,5-tetramethyl**nonane**

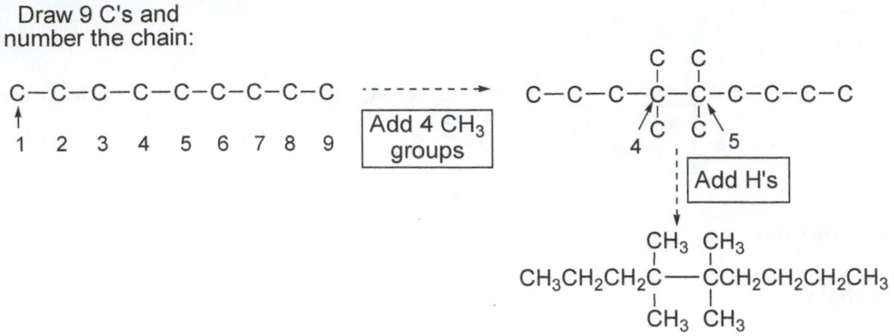

Draw 9 C's and
number the chain:

C—C—C—C—C—C—C—C—C

1 2 3 4 5 6 7 8 9

Add 4 CH₃ groups

C—C—C—C—C—C—C—C—C

Add H's

CH₃CH₂CH₂C—CCH₂CH₂CH₂CH₃

d. 3-ethyl-5-propyl**nonane**

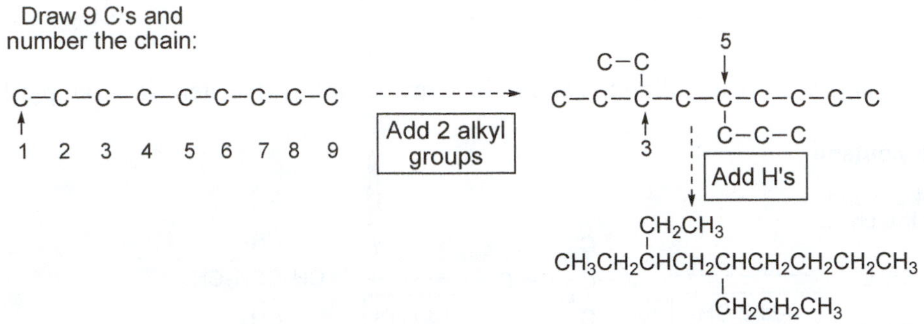

Draw 9 C's and
number the chain:

C—C—C—C—C—C—C—C—C

1 2 3 4 5 6 7 8 9

Add 2 alkyl groups

C—C—C—C—C—C—C—C—C

Add H's

CH₃CH₂CHCH₂CHCH₂CH₂CH₂CH₃

**12.14** Label the functional groups.

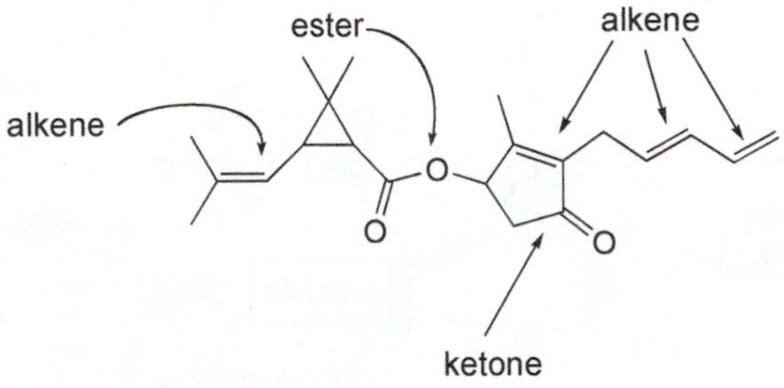

ester

alkene

alkene

ketone

**12.15** Vitamin D₃ has many nonpolar C–C and C–H bonds, which makes it water insoluble but fat soluble.

**12.16** To give the IUPAC name for each compound, follow the steps in Example 12.6:
[1] Name the ring.
[2] Name and number the substituents.

Start numbering here.

a.

1 CH₃

methyl at C1

4 C's ··→ cyclobutane

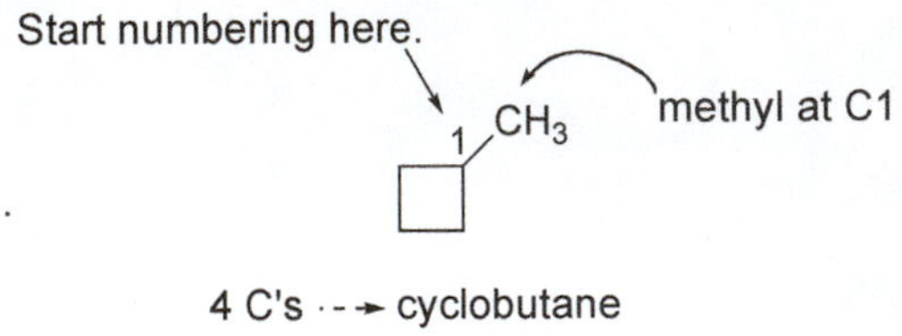

**Answer: methylcyclobutane**

Start numbering here.

b.

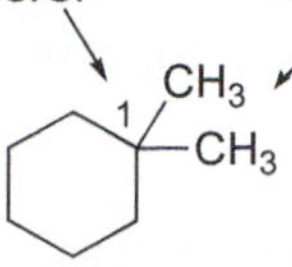

two methyls at C1

**Answer: 1,1-dimethylcyclohexane**

6 C's ·· → cyclohexane

Start numbering here.

c.

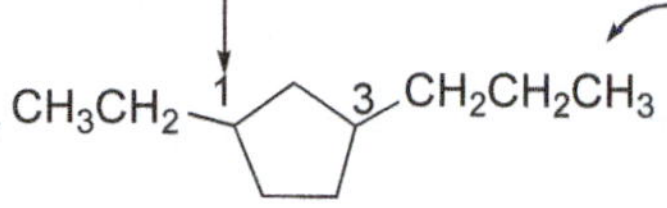

propyl at C3

**Answer: 1-ethyl-3-propylcyclopentane**

5 C's ·· → cyclopentane

ethyl at C1

Start numbering here.

d.

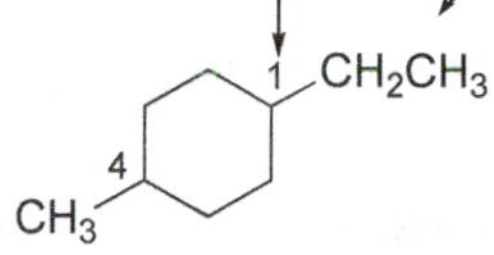

ethyl at C1

**Answer: 1-ethyl-4-methylcyclohexane**

6 C's ·· → cyclohexane

methyl at C4

**12.17**  Give the structure corresponding to each IUPAC name.

a. propylcyclopentane

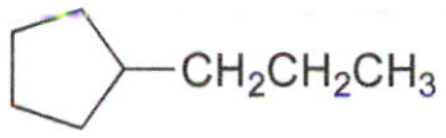

—CH₂CH₂CH₃

c. 1,1,2-trimethylcyclopropane

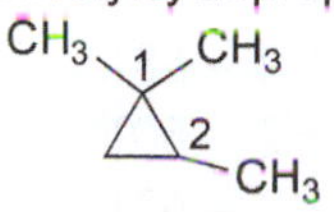

b. 1,2-dimethylcyclobutane

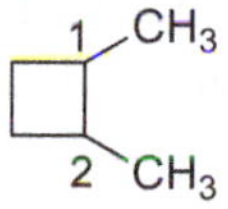

d. 4-ethyl-1,2-dimethylcyclohexane

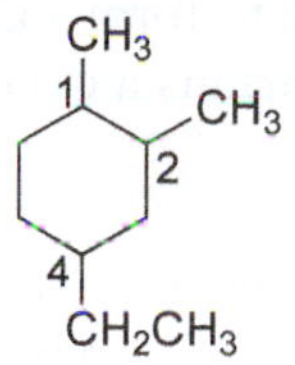

**12.18**  The melting points and boiling points of alkanes increase as the number of carbons increases.

a.  Highest boiling point, most carbons: decane
b.  Lowest boiling point, fewest carbons: pentane
c.  Highest melting point, most carbons: decane
d.  Lowest melting point, fewest carbons: pentane

**12.19**  Vaseline is a complex mixture of hydrocarbons, and is nonpolar, so it is insoluble in a polar solvent like water.  In a weakly polar solvent like dichloromethane, it is soluble.

**12.20** Elastol is a high molecular weight alkane, making it nonpolar. By the solubility rule, "Like dissolves like," nonpolar oil has similar intermolecular forces to nonpolar polyisobutylene, so it is soluble in Elastol.

**12.21** Write a balanced equation for each reaction. Combustion reactions release $CO_2$ and $H_2O$.

a. $CH_3CH_2CH_3 + 5 O_2 \xrightarrow{\text{flame}} 3 CO_2 + 4 H_2O$

b. $2 CH_3CH_2CH_2CH_3 + 13 O_2 \xrightarrow{\text{flame}} 8 CO_2 + 10 H_2O$

**12.22** Write a balanced equation for the reaction. Incomplete combustion reactions release $CO$ and $H_2O$.

$2 CH_3CH_3 + 5 O_2 \longrightarrow 4 CO + 6 H_2O$

## Solutions to Odd-Numbered End-of-Chapter Problems

**12.23** Use the formula $C_nH_{2n+2}$ to determine the number of hydrogen atoms.

$(31 \times 2) + 2 = 64$ H's

**12.25** Use the formula $C_nH_{2n+2}$ to determine the number of hydrogen atoms.

A $C_{26}$ alkane has 54 H's $[(26 \times 2) + 2 = 54]$.
A $C_{27}$ alkane has 56 H's $[(27 \times 2) + 2 = 56]$.
A $C_{28}$ alkane has 58 H's $[(28 \times 2) + 2 = 58]$.
A $C_{29}$ alkane has 60 H's $[(29 \times 2) + 2 = 60]$.
A $C_{30}$ alkane has 62 H's $[(30 \times 2) + 2 = 62]$.

**12.27** To form a cycloalkane, you need to form a C–C bond between 2 C's in a chain. To do this, 1 H from each C must be removed. This means a cycloalkane always has two fewer H's than an acyclic alkane of the same number of C's.

**12.29** To classify a carbon atom, count the number of C's bonded to it as in Example 12.3. Draw a complete structure with all bonds and atoms to clarify the structure if necessary.

a. $CH_3(CH_2)_3CH_3$

c. $(CH_3)_3CC(CH_3)_3$

b. $CH_3CH_2CHCHCH_3$

d.

**12.31** Draw a structure to satisfy the requirements.

a.

$$CH_3CH_2CH_2\overset{\overset{\displaystyle CH_3}{|}}{\underset{\underset{\displaystyle CH_3}{|}}{C}}-CH_3 \quad 4°$$

b.

$$\overset{1°\downarrow}{\quad}\overset{2°}{\overline{\quad\quad\quad\quad}}\overset{1°\downarrow}{\quad}$$
$$CH_3CH_2CH_2CH_2CH_2CH_2CH_3$$

only 1° and 2° C's

**12.33** To determine if the compounds are isomers, first compare molecular formulas as in Example 12.1; two compounds are isomers only if they have the same molecular formula. Then check how the atoms are connected to each other. Constitutional isomers have atoms bonded to different atoms.

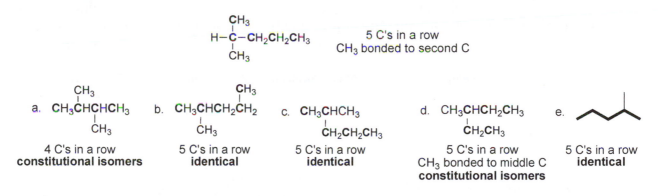

a.
$$\overset{\overset{\displaystyle CH_2CH_3}{|}}{CH_3CHCHCH_3} \quad and \quad CH_3CH_2CHCH_2CH(CH_3)_2$$
(with $\overset{|}{CH_2CH_3}$ and $\overset{|}{CH_3}$)

$C_8H_{18}$  ·  $C_8H_{18}$

same molecular formula
**constitutional isomers**

c.

$C_5H_{10}$    $C_5H_{10}$

same molecular formula
**constitutional isomers**

b.
$$\overset{\overset{\displaystyle CH_3}{|}}{CH_3CH_2CHCHCH_2CH_3} \quad and \quad CH_3CHCHCH_3$$

$C_8H_{18}$    $C_8H_{18}$

same molecular formula
identical connectivity
**identical**

d.

$C_8H_{16}$    $C_8H_{16}$

same molecular formula
**constitutional isomers**

**12.35** To determine if the molecules are constitutional isomers or identical, determine if the atoms are connected to each other in the same way.

$$H-\overset{\overset{\displaystyle CH_3}{|}}{\underset{\underset{\displaystyle CH_3}{|}}{C}}-CH_2CH_2CH_3$$

5 C's in a row
$CH_3$ bonded to second C

a. $\overset{\overset{\displaystyle CH_3}{|}}{\underset{\underset{\displaystyle CH_3}{|}}{CH_3CHCHCH_3}}$

4 C's in a row
**constitutional isomers**

b. $\overset{\overset{\displaystyle CH_3}{|}}{\underset{\underset{\displaystyle CH_3}{|}}{CH_3CHCH_2CH_2}}$

5 C's in a row
**identical**

c. $\underset{\underset{\displaystyle CH_2CH_2CH_3}{|}}{CH_3CHCH_3}$

5 C's in a row
**identical**

d. $\underset{\underset{\displaystyle CH_2CH_3}{|}}{CH_3CHCH_2CH_3}$

5 C's in a row
$CH_3$ bonded to middle C
**constitutional isomers**

e.

5 C's in a row
**identical**

**12.37** Draw structures that fit each description.

a.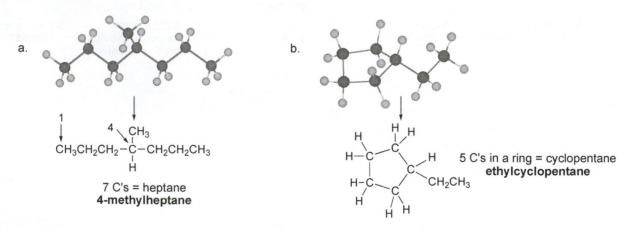
cycloalkanes
constitutional isomers
$C_7H_{14}$

c. $CH_3CH_2CH_2Cl$    $CH_3CHCH_3$
                              |
                              Cl
constitutional isomers
$C_3H_7Cl$

b. $CH_3CH_2CH_2CH_2CH_2OH$        $CH_3OCH_2CH_2CH_2CH_3$
alcohol        constitutional isomers        ether
$C_5H_{12}O$

**12.39** Draw all of the structures that satisfy the criteria.

$CH_3$
|
$CH_3CHCH_2CH_2CH_2CH_2CH_3$

$CH_3$
|
$CH_3CH_2CHCH_2CH_2CH_2CH_3$

$CH_3$
|
$CH_3CH_2CH_2CHCH_2CH_2CH_3$

molecular formula $C_8H_{18}$
7 C's in the longest chain
one $CH_3$ group bonded to each chain

**12.41** Draw all of the structures that satisfy the criteria.

molecular formula $C_3H_6O$:

△—OH

alcohol

$\overset{O}{\overset{||}{CH_3CCH_3}}$

ketone

▷—$CH_3$ (with O)

cyclic ether

**12.43** Give the IUPAC name for each molecule.

$CH_3CH_2CH_2CH_2CH_2CH_3$

hexane

$CH_3$
|
$CH_3CH_2CH_2-C-CH_3$
|
H

2-methylpentane

$CH_3$
|
$CH_3CH_2-C-CH_2CH_2$
|
H

3-methylpentane

$CH_3$
|
$CH_3CH_2-C-CH_3$
|
$CH_3$

2,2-dimethylbutane

$CH_3\ CH_3$
|      |
$CH_3-C-C-CH_3$
|    |
H   H

2,3-dimethylbutane

**12.45** Give the IUPAC name for each molecule.

a.

1 ↓
4 ↓ $CH_3$
        |
$CH_3CH_2CH_2-C-CH_2CH_2CH_3$
        |
        H

7 C's = heptane
**4-methylheptane**

b.

5 C's in a ring = cyclopentane
**ethylcyclopentane**

**12.47**  To give the IUPAC name for each compound, follow the steps in Example 12.4.

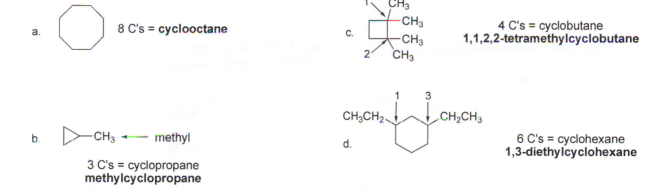

a.  ┌─────────────────────────┐
    │CH₃CH₂CHCH₂CH₂CH₂CH₃│
    └─────────────────────────┘
         1      3
              CH₃ ◄── methyl at C3
    7 C's = heptane
    **3-methylheptane**

b.  ┌──────────────────────────┐
    │CH₃CH₂CHCH₂CHCH₂CH₂CH₃│
    └──────────────────────────┘
         1     3     5
             CH₃   CH₃ ◄── methyls at C3 and C5
    8 C's = octane
    **3,5-dimethyloctane**

c.  CH₃CH₂CH₂C(CH₂CH₃)₃
                    CH₂CH₃ ◄── two ethyls at C3
    CH₃CH₂CH₂C─CH₂CH₃
         3      CH₂CH₃   1
    6 C's = hexane
    **3,3-diethylhexane**

methyls at C2 and C6

d.  CH₃CHCH₂C HCH₂CH₃ ◄── ethyl at C4
         CH₂CH₃
    8 C's = octane
    **4-ethyl-2,6-dimethyloctane**

e.  (CH₃CH₂)₂CHCH₂CH₂CH₂CH(CH₃)₂
    ethyl at C6 ──► CH₃CH₂
    CH₃CH₂─CCH₂CH₂CH₂CCH₃
         H              CH₃ ◄── methyl at C2
    8 C's = octane
    **6-ethyl-2-methyloctane**

f.  CH₃
    CH₃CH₂─C─CH₂CH₂            CH₃
         8    CH₃   CH₂CH₂CH₂─C─CH₃
                              CH₃ ◄── two methyls at C2
                                     two methyls at C8
    10 C's = decane
    **2,2,8,8-tetramethyldecane**

**12.49**  To give the IUPAC name for each compound, follow the steps in Example 12.6.

a.  (octagon)   8 C's = **cyclooctane**

b.  (triangle)─CH₃ ◄── methyl
    3 C's = cyclopropane
    **methylcyclopropane**

c.  (square) with CH₃ groups
    4 C's = cyclobutane
    **1,1,2,2-tetramethylcyclobutane**

d.  CH₃CH₂ (cyclohexane ring) CH₂CH₃
         1          3
    6 C's = cyclohexane
    **1,3-diethylcyclohexane**

**12.51**  Give the structure corresponding to each IUPAC name.

a. 3-ethyl**hexane**

CH₃CH₂CHCH₂CH₂CH₃
        CH₂CH₃ ◄── ethyl at C3

b. 3-ethyl-3-methyl**octane**

CH₃ ◄── methyl at C3
CH₃CH₂CCH₂CH₂CH₂CH₂CH₃
        CH₂CH₃ ◄── ethyl at C3

c. 2,3,4,5-tetramethyl**decane**

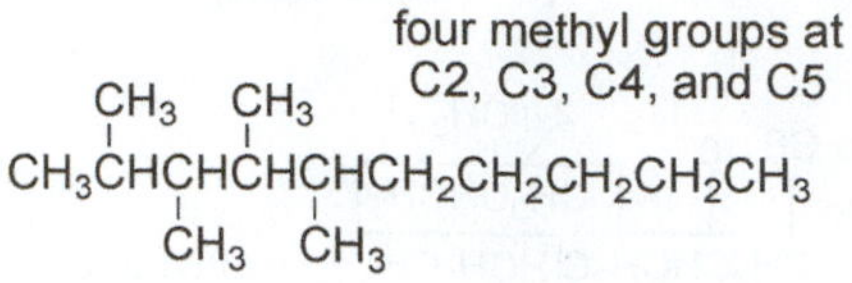

four methyl groups at
C2, C3, C4, and C5

$$CH_3CHCHCHCHCH_2CH_2CH_2CH_2CH_3$$

with $CH_3$ $CH_3$ above (at C2, C3) and $CH_3$ $CH_3$ below (at C4, C5)

d. cyclo**nonane**

e. 1,1,3-trimethyl**cyclohexane**

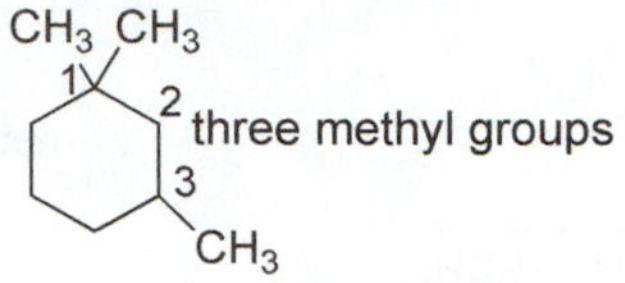

$CH_3$  $CH_3$

three methyl groups

$CH_3$

f. 1-ethyl-2,3-dimethyl**cyclopentane**

$CH_2CH_3$ ⟵ ethyl at C1

$CH_3$ ⟵

$CH_3$ ⟵ two methyl groups

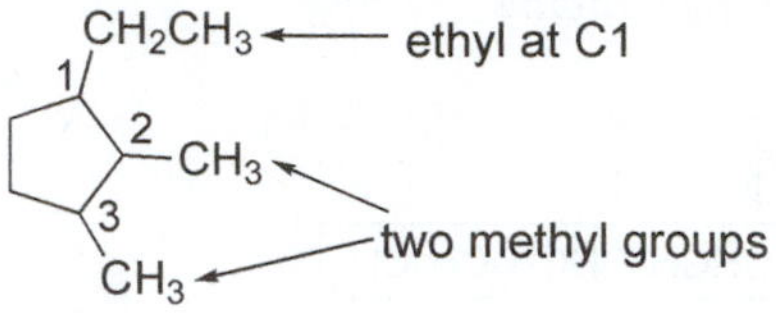

**12.53**  Explain why each IUPAC name is incorrect.

a. 2-methylbutane: Number to give $CH_3$ the lower number, 2 not 3.

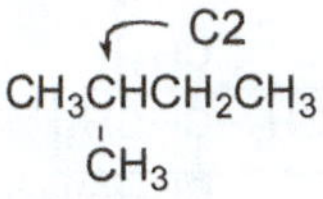

⌐ C2

$CH_3CHCH_2CH_3$
$CH_3$

b. methylcyclopentane: no number assigned if only one substituent

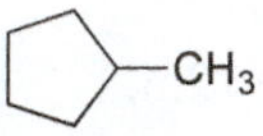

⬠—$CH_3$

c. 2-methylpentane: five-carbon chain

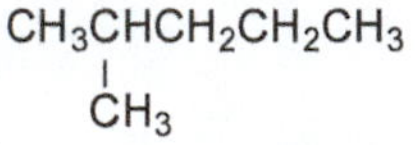

$CH_3CHCH_2CH_2CH_3$
$CH_3$

d. 2,5-dimethylheptane: longest chain not chosen

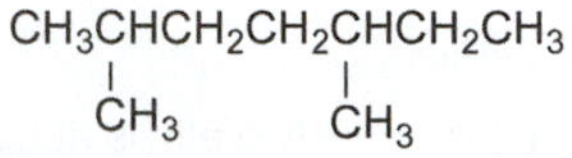

$CH_3CHCH_2CH_2CHCH_2CH_3$
$CH_3$   $CH_3$

e. 1,3-dimethylcyclohexane: Number to give the second substituent the lower number.

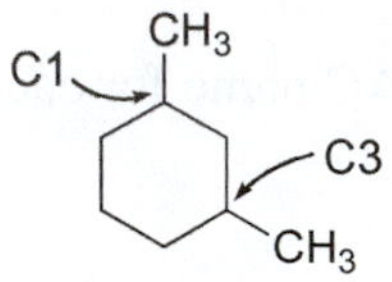

C1 ⟵ $CH_3$

—C3

$CH_3$

f. 1-ethyl-2-propylcyclopentane: lower number assigned alphabetically

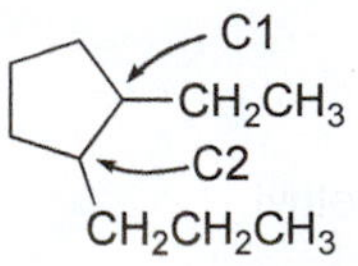

⟋ C1

—$CH_2CH_3$

⟵C2

$CH_2CH_2CH_3$

**12.55**  Draw the isomers and then give the IUPAC name.

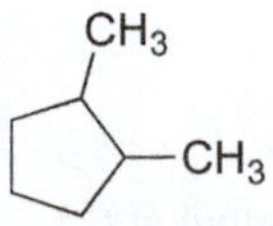

$CH_3$

—$CH_3$

1,2-dimethylcyclopentane

$CH_3$

$CH_3$

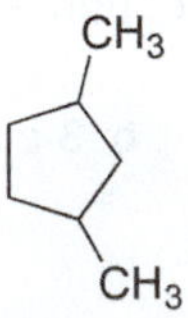

1,3-dimethylcyclopentane

$CH_3$

$CH_3$

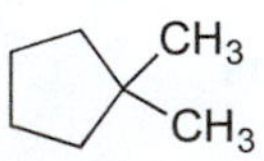

1,1-dimethylcyclopentane

**12.57** Draw a skeletal structure for each compound.

a. octane     b. 1,2-dimethylcyclopentane     c. $CH_3CHCH_2CH_2CH_2CH_2CH_3$
                                                          |
                                                          $CH_3$

**12.59** Convert each structure to a complete structure with all atoms drawn in.

a.     b.     c.

**12.61** The melting points and boiling points of alkanes increase as the number of carbons increases.

a.      or

**more carbon atoms
higher melting point**

b.     or

**more carbon atoms
higher melting point**

**12.63** Branched alkanes have lower boiling points than linear alkanes.

a. increasing boiling point: $(CH_3)_4C$  <  $(CH_3)_2CHCH_2CH_3$  <  $CH_3CH_2CH_2CH_2CH_3$
                                most branching                        no branching

b. increasing boiling point: $(CH_3)_2CHCH(CH_3)_2$  <  $CH_3CH_2CH_2CH(CH_3)_2$  <  $CH_3(CH_2)_4CH_3$
                                most branching                                    no branching

**12.65** Hexane is a nonpolar hydrocarbon, making it soluble in organic solvents like dichloromethane, but insoluble in water.

**12.67** Write a balanced equation for each reaction. Combustion reactions form $CO_2$ and $H_2O$.

a.  $2\ CH_3CH_3 + 7\ O_2 \longrightarrow 4\ CO_2 + 6\ H_2O$          b.  $(CH_3)_2CHCH_2CH_3 + 8\ O_2 \longrightarrow 5\ CO_2 + 6\ H_2O$

**12.69** Write a balanced equation for each reaction. Incomplete combustion reactions form CO and $H_2O$.

a. $2 \ CH_3CH_2CH_3 + 7 \ O_2 \longrightarrow 6 \ CO + 8 \ H_2O$     b. $2 \ CH_3CH_2CH_2CH_3 + 9 \ O_2 \longrightarrow 8 \ CO + 10 \ H_2O$

**12.71** Write a balanced equation for the oxidation of glucose to form $CO_2$ and $H_2O$.

$$C_6H_{12}O_6 \ + \ 6 \ O_2 \longrightarrow 6 \ CO_2 \ + \ 6 \ H_2O$$

**12.73** Higher molecular weight alkanes in warmer weather means less evaporation. Lower molecular weight alkanes in colder weather means the gasoline won't freeze.

**12.75** The mineral oil can prevent the body's absorption of important fat-soluble vitamins. The vitamins dissolve in the mineral oil, and are thus not absorbed. Instead, they are expelled with the mineral oil.

**12.77** c. The nonpolar asphalt will be most soluble in the paint thinner because "like dissolves like." The liquid alkanes of the paint thinner dissolve the high molecular weight hydrocarbons of the asphalt.

**12.79** Answer each question about the compound.

a. 
$$\begin{array}{c} \quad\quad\ \ CH_3 \ \ CH_2CH_3 \\ \quad\quad\quad | \quad\quad | \\ CH_3CH_2CH_2-C-\!-\!C-CH_2CH_3 \\ \quad\quad\quad | \quad\quad | \\ \quad\quad\ \ CH_3 \ \ \ H \end{array}$$

7 C's = heptane
**3-ethyl-4,4-dimethylheptane**

b. 
$$\begin{array}{c} \quad\quad\ \ CH_3 \ CH_3 \ CH_2CH_3 \\ \quad\quad\quad\ | \quad\ \ | \quad\quad | \\ CH_3CH_2CH-C-\!-\!C-CH_2CH_3 \\ \quad\quad\quad\ | \quad\ \ | \\ \quad\quad\ \ \ H \quad\ \ H \end{array}$$

constitutional isomer

c. not water soluble
d. soluble in organic solvents

e. $\ C_{11}H_{24} + 17 \ O_2 \longrightarrow 11 \ CO_2 + 12 \ H_2O$

f.

**12.81**

a. 

5 C's in a ring = cyclopentane
**propylcyclopentane**

b. 
$$\begin{array}{c} CH_3 \\ \diagdown \\ \\ CH_3CH_2 \diagup \end{array}$$

constitutional isomer

c. not water soluble
d. soluble in organic solvents

e. $\ C_8H_{16} + 12 \ O_2 \longrightarrow 8 \ CO_2 + 8 \ H_2O$

f.

**12.83** A compound with 10 carbons and two rings will have $2n - 2$ H's.
$(10 \times 2) - 2 = 18$
$C_{10}H_{18}$

**12.85** Cyclopentane has a more rigid structure. The rings can get closer together since they are not floppy, resulting in an increased force of attraction. Therefore, the boiling point is higher.

# Chapter 13 Unsaturated Hydrocarbons

## Chapter Review

**[1] What are the characteristics of alkenes, alkynes, and aromatic compounds?**
- Alkenes are unsaturated hydrocarbons that contain a carbon–carbon double bond and have molecular formula $C_nH_{2n}$. Each carbon of the double bond is trigonal planar. (13.1)

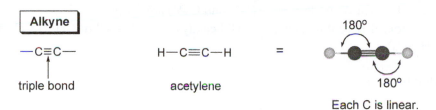

- Alkynes are unsaturated hydrocarbons that contain a carbon–carbon triple bond and have molecular formula $C_nH_{2n-2}$. Each carbon of the triple bond is linear. (13.1)

- Benzene, molecular formula $C_6H_6$, is the most common aromatic hydrocarbon. Benzene is a stable hybrid of two resonance structures, each containing a six-membered ring and three double bonds. Each carbon of benzene is trigonal planar. (13.9)

**[2] How are alkenes, alkynes, and substituted benzenes named?**
- An alkene is identified by the suffix -*ene*, and the carbon chain is numbered to give the C=C the lower number. (13.2)
- An alkyne is identified by the suffix -*yne*, and the carbon chain is numbered to give the C≡C the lower number. (13.2)
- Substituted benzenes are named by naming the substituent and adding the word *benzene*. When two substituents are bonded to the ring, the prefixes ortho, meta, and para are used to show the relative position of the two groups: 1,2-, 1,3-, or 1,4-, respectively. With three substituents on a benzene ring, number to give the lowest possible numbers. (13.10)

**[3] What is the difference between constitutional isomers and stereoisomers? How are cis and trans isomers different? (13.3)**
- Constitutional isomers differ in the way the atoms are bonded to each other.
- Stereoisomers differ only in the three-dimensional arrangement of the atoms.
- Cis and trans isomers are one type of stereoisomer. A cis alkene has two alkyl groups on the same side of the double bond. A trans alkene has two alkyl groups on opposite sides of the double bond.

| General structure | Two possible arrangements |
|---|---|

$CH_3CH=CHCH_3$

2-butene

two $CH_3$ groups on the **same** side

**cis** isomer

two $CH_3$ groups on **opposite** sides

**trans** isomer

## [4] How do saturated and unsaturated fatty acids differ? (13.3B)

- Fatty acids are carboxylic acids (RCOOH) with long carbon chains. Saturated fatty acids have no double bonds in the carbon chain and unsaturated fatty acids have one or more double bonds in their long carbon chains.
- Generally, double bonds in naturally occurring fatty acids are cis.
- As the number of double bonds in the fatty acid increases, the melting point decreases.

## [5] What types of reactions do alkenes undergo? (13.6)

- Alkenes undergo addition reactions with reagents X–Y. One bond of the double bond and the X–Y bond break and two new single bonds (C–X and C–Y) are formed.
- Alkenes react with four different reagents—$H_2$ (Pd catalyst), $X_2$ (X = Cl or Br), HX (X = Cl or Br), and $H_2O$ (with $H_2SO_4$).

General addition reaction:

$$\underset{\text{One bond is broken.}}{\phantom{x}C=C\phantom{x}} + \quad X-Y \quad \longrightarrow \quad \underset{\text{Two single bonds are formed.}}{-\overset{|}{\underset{X}{C}}-\overset{|}{\underset{Y}{C}}-}$$

Hydrogenation (13.6A):     $CH_2=CH_2$  +  $H_2$  $\xrightarrow{\text{Pd}}$  $\underset{H \quad H}{CH_2-CH_2}$

Halogenation (13.6B):     $CH_2=CH_2$  +  $X_2$  $\longrightarrow$  $\underset{X \quad X}{CH_2-CH_2}$     (X = Cl or Br)

Hydrohalogenation (13.6C):     $CH_2=CH_2$  +  HX  $\longrightarrow$  $\underset{H \quad X}{CH_2-CH_2}$     (X = Cl or Br)

Hydration (13.6D):     $CH_2=CH_2$  +  $H_2O$  $\xrightarrow{H_2SO_4}$  $\underset{H \quad OH}{CH_2-CH_2}$

## [6] What is Markovnikov's rule? (13.6)

- Markovnikov's rule explains the selectivity observed when an unsymmetrical reagent HZ (Z = Cl, Br, or OH) adds to an unsymmetrical alkene like propene ($CH_3CH=CH_2$). The H of HZ is added to the end of the C=C that has more H's to begin with, forming $CH_3CH(Z)CH_3$.

This C has no H's so the OH bonds here.

| Example |

This C has 1 H so the H bonds here.

**[7] What products are formed when a vegetable oil is partially hydrogenated? (13.7)**
- When an unsaturated oil is partially hydrogenated, some but not all of the cis C=C's add $H_2$, reducing the number of double bonds and increasing the melting point.
- Some of the cis double bonds are converted to trans double bonds, thus forming trans fats, whose shape and properties closely resemble those of saturated fats.

**[8] What are polymers, and how are they formed from alkene monomers? (13.8)**
- Polymers are large molecules made up of repeating smaller molecules called monomers covalently bonded together. When alkenes are polymerized, one bond of the double bond breaks, and two new single bonds join the alkene monomers together in long carbon chains.

| Three monomer units joined together |

$CH_2=CHZ$ + $CH_2=CHZ$ + $CH_2=CHZ$ ⟶

**Repeating unit:**

shorthand structure

Many monomers are joined together.

**[9] What types of reactions does benzene undergo? (13.13)**
- To keep the stable aromatic ring intact, benzene undergoes substitution, not addition, reactions. One H atom on the ring is replaced by another atom or group of atoms. Reactions include chlorination (substitution by Cl), nitration (substitution by $NO_2$), and sulfonation (substitution by $SO_3H$).

Chlorination (13.13A): + $Cl_2$ →($FeCl_3$) Cl + HCl

Nitration (13.13B): + $HNO_3$ →($H_2SO_4$) $NO_2$ + $H_2O$

Sulfonation (13.13C): + $SO_3$ →($H_2SO_4$) $SO_3H$

## Problem Solving

## [1] Alkenes and Alkynes (13.1)

**Example 13.1** Draw a complete structure for each alkene or alkyne.

a. $(CH_3)_2C=CHCH_2CH_2CH_3$      b.   $CH_3CH_2CH_2C\equiv CCH_3$

**Analysis**

First, draw the multiple bond in each structure. Draw an alkene so that each C of the double bond has three atoms around it. Draw an alkyne so that each C of the triple bond has two atoms around it. All other C's have four single bonds.

**Solution**

a. One C has bonds to two $CH_3$ groups. The other C has one bond to 1 H and one bond to a $CH_2CH_2CH_3$ group.

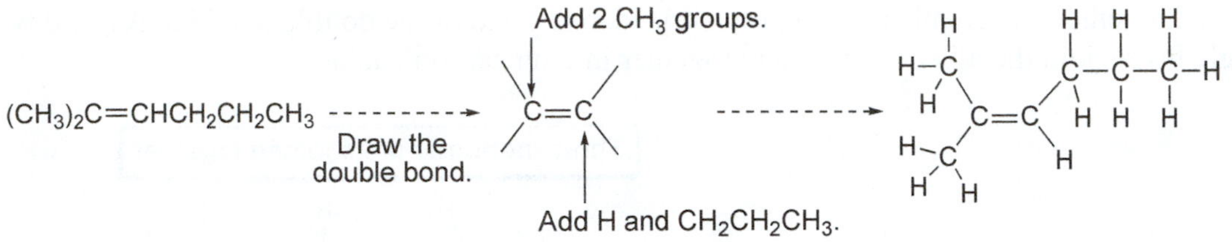

b. One C has a single bond to a $CH_2CH_2CH_3$ group. The other C has a single bond to a $CH_3$ group.

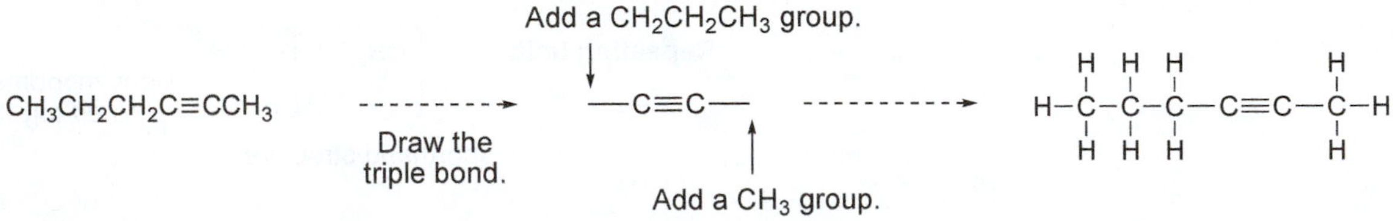

## [2] Nomenclature of Alkenes and Alkynes (13.2)

**Example 13.2** Give the IUPAC name for the following compound.

$$CH_3C\equiv CCHCH_3$$
$$|$$
$$CH_2CH_2CH_3$$

**Analysis and Solution**

**[1] Find the longest chain containing both carbon atoms of the multiple bond.**

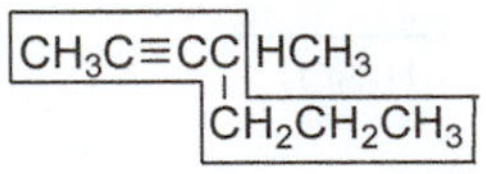

7 C's in the longest chain $\cdots\cdots\rightarrow$ **heptyne**

**[2] Number the chain to give the triple bond the lower number.**

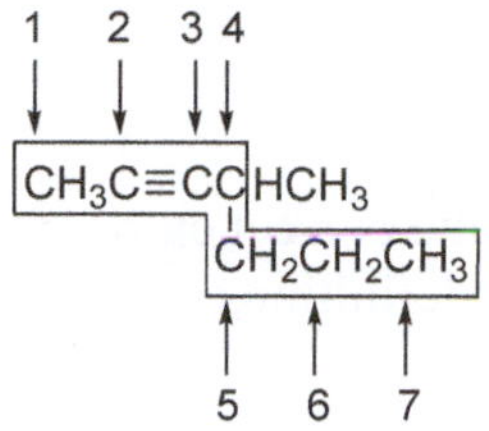

$$\overset{1\ \ \ 2\ \ \ 3\ 4}{\downarrow\ \ \downarrow\ \ \downarrow\downarrow}$$

$CH_3C \equiv CCHCH_3$
$\quad\quad\quad CH_2CH_2CH_3$

$$\overset{\ }{\underset{5\ \ \ 6\ \ \ 7}{\uparrow\ \ \uparrow\ \ \uparrow}}$$

- Numbering from left to right is preferred since the triple bond begins at C2 (not C5). The molecule is named as a **2-heptyne**.

**[3] Name and number the substituents and write the complete name.**

- The alkyne has one methyl group located at C4.

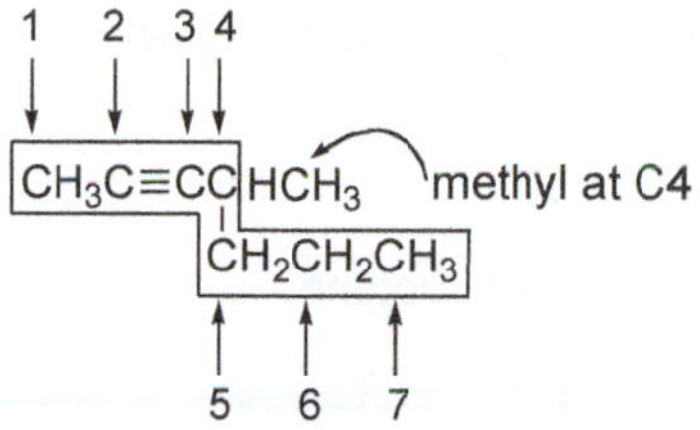

$$\overset{1\ \ \ 2\ \ \ 3\ 4}{\downarrow\ \ \downarrow\ \ \downarrow\downarrow}$$

$CH_3C \equiv CCHCH_3$ ⟍methyl at C4
$\quad\quad\quad CH_2CH_2CH_3$

$$\overset{\ }{\underset{5\ \ \ 6\ \ \ 7}{\uparrow\ \ \uparrow\ \ \uparrow}}$$

**Answer: 4-methyl-2-heptyne**

**Example 13.3** Draw the structure corresponding to the IUPAC name: 1,3-dimethylcyclopentene.

**Analysis**
First identify the parent name to find the longest carbon chain or ring, and then use the suffix to determine the functional group; the suffix *-ene* = an alkene and *-yne* = an alkyne. Then number the carbon chain or ring and place the functional group at the indicated carbon. For a cycloalkene, the double bond is located between C1 and C2. Add the substituents and enough hydrogens to give each carbon four bonds.

**Solution**
1,3-Dimethylcyclopentene has 5 C's in a ring (cyclopent-) and a double bond that begins at C1. Two methyl groups are bonded to C1 and C3.

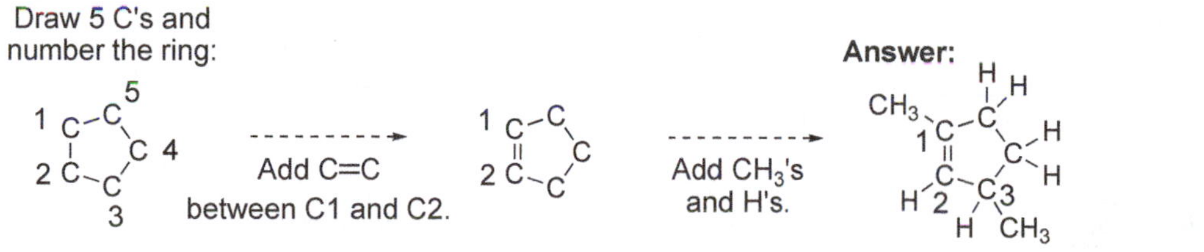

Draw 5 C's and
number the ring:

Add C=C
between C1 and C2.

Add CH$_3$'s
and H's.

**Answer:**

**1,3-dimethylcyclopentene**

## [3] Cis–Trans Isomers (13.3)

**Example 13.4** Draw *cis-* and *trans*-3-heptene.

**Analysis**
First, use the parent name to draw the carbon skeleton, and place the double bond at the correct carbon; 3-heptene indicates a 7 C chain with the double bond beginning at C3. Then use the definitions of cis and trans to draw the isomers.

**Solution**
Each C of the double bond is bonded to an alkyl group and a hydrogen. A cis isomer has the two alkyl groups bonded to the same side of the double bond. A trans isomer has the two alkyl groups bonded to the opposite sides of the double bond.

$CH_3CH_2CH=CHCH_2CH_2CH_3$  - - - - - - →

1  2  3   4  5  6  7

3-heptene          cis-3-heptene 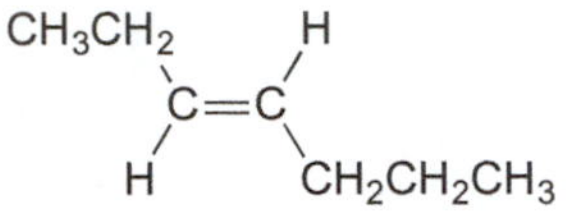          trans-3-heptene

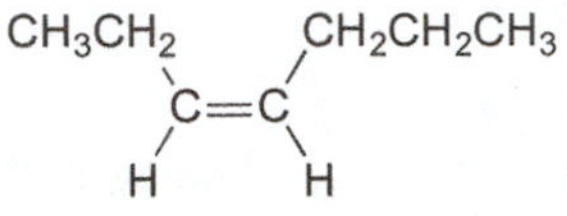

---

## [4] Reactions of Alkenes (13.6)

**Example 13.5** Draw the product of the following reaction.

**Analysis**
To draw the product of a hydrogenation reaction:
- Locate the C=C and mentally break one bond in the double bond.
- Mentally break the H–H bond of the reagent.
- Add one H atom to each C of the C=C, thereby forming two new C–H single bonds.

**Solution**

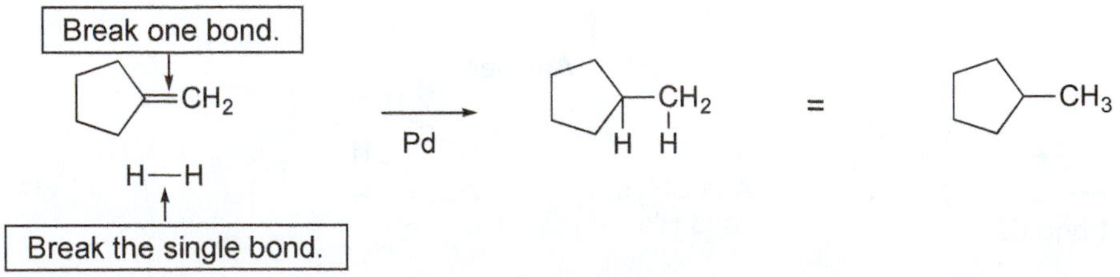

---

**Example 13.6** Draw the product of the following reaction.

$(CH_3)_2C=CHCH_2CH_3$   +   HBr   ⟶

**Analysis**
Alkenes undergo addition reactions, so the elements of H and Br must be added to the double bond. Since the alkene is unsymmetrical, the H atom of HBr bonds to the carbon that has more H's to begin with.

**Solution**

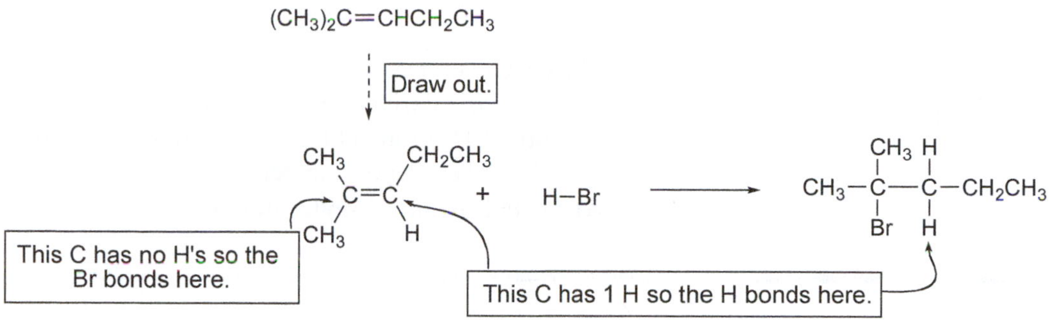
(CH3)2C=CHCH2CH3
Draw out.
CH3
CH2CH3
CH3
H
+
H−Br
CH3 H
CH3−C—C−CH2CH3
Br H
This C has no H's so the Br bonds here.
This C has 1 H so the H bonds here.

## [5] Polymers—The Fabric of Modern Society (13.8)

**Example 13.7** What polymer is formed when $CH_2$=$CHNHCOCH_3$ is polymerized?

**Analysis**
Draw three or more alkene molecules and arrange the carbons of the double bonds next to each other. Break one bond of each double bond, and join the alkenes together with single bonds. With unsymmetrical alkenes, substituents are bonded to every other carbon.

**Solution**

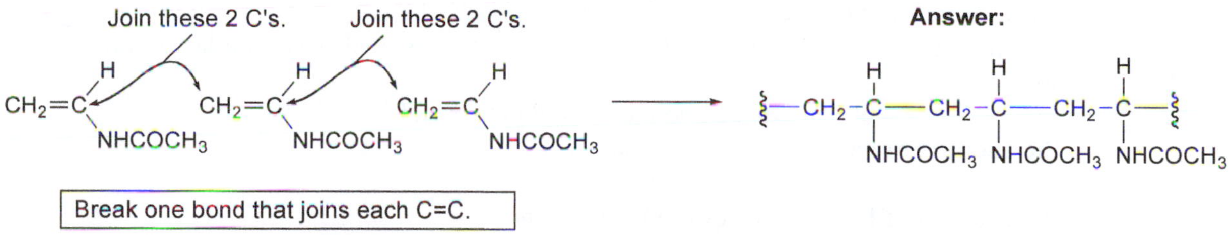
Join these 2 C's.
Join these 2 C's.
Answer:
CH2=C
NHCOCH3
H
CH2−C—CH2−C—CH2−C
NHCOCH3 NHCOCH3 NHCOCH3
Break one bond that joins each C=C.

## [6] Nomenclature of Benzene Derivatives (13.10)

**Example 13.8** Name each of the following aromatic compounds.

a. 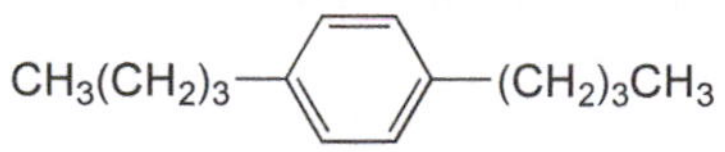
CH3(CH2)3
(CH2)3CH3

b. 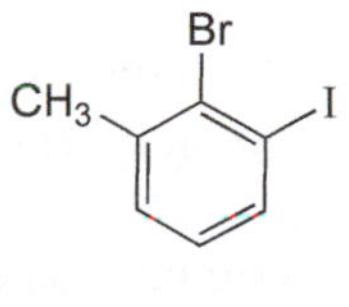
Br
CH3
I

**Analysis**
Name the substituents on the benzene ring. With two groups, alphabetize the substituent names and use the prefix ortho, meta, or para to indicate their location. With three substituents, alphabetize the substituent names, and number to give the lowest set of numbers.

## Solution

a.  $CH_3(CH_2)_3$ —⟨ ⟩— $(CH_2)_3CH_3$

butyl
group

butyl
group

- The two substituents are located 1,4- or **para** to each other.
- **Answer:** *p*-dibutylbenzene

---

b.

Br

$CH_3$        I

1        3

toluene

- Since a $CH_3$– group is bonded to the ring, name the molecule as a derivative of toluene.
- Place the $CH_3$ group at the "1" position, and number to give the lowest set of numbers.
- **Answer: 2-bromo-3-iodotoluene**

## Self-Test

**[1] Fill in the blank with one of the terms listed below.**

| | | |
|---|---|---|
| Addition reaction (13.6) | Hydration (13.6) | Stereoisomers (13.3) |
| Alkenes (13.1) | Halogenation (13.6) | Substitution reaction (13.13) |
| Alkynes (13.1) | Hydrogenation (13.6) | Trans isomer (13.3) |
| Cis isomer (13.3) | Hydrohalogenation (13.6) | Unsaturated hydrocarbons (13.1) |
| Fats (13.3) | Oils (13.3) | |

1. _____ are compounds that contain a carbon–carbon triple bond.
2. When two alkyl groups are on the *same side* of a double bond, the compound is called the _____.
3. _____ are liquids at room temperature and are formed from fatty acids having a large number of double bonds.
4. _____ are compounds that contain a carbon–carbon double bond.
5. _____ is the addition of halogen ($X_2$) to an alkene.
6. _____ is the addition of hydrogen ($H_2$) to an alkene.
7. _____ is the addition of HX (X = Cl or Br) to an alkene.
8. _____ are isomers that differ *only* in the three-dimensional arrangement of atoms.
9. _____ are solids at room temperature and are generally formed from fatty acids having few double bonds.
10. _____ are compounds that contain fewer than the maximum number of hydrogen atoms per carbon.
11. When two alkyl groups are on *opposite sides* of a double bond, the compound is called the _____.
12. A _____ is a reaction in which an atom is *replaced* by another atom or a group of atoms.
13. An _____ is a reaction in which elements are added to a compound.
14. _____ is the addition of water to an alkene.

**[2] Label the substituents on each benzene ring as ortho, meta, or para.**

a. ortho                    b. meta                    c. para

15.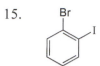

16. 

17. 

**[3] Label each statement as true (T) or false (F).**

18. Cis alkenes have two alkyl groups on the same side of the double bond.
19. Stereoisomers have atoms bonded to different atoms.
20. Trans alkenes have two alkyl groups on the same side of the double bond.
21. Two compounds with the same molecular formula can be stereoisomers.

**[4] For each reaction, pick the missing reagent needed to produce the products.**

a. $HNO_3$      b. $Br_2$      c. $Cl_2$      d. $H_2O$

22. $CH_2{=}CHCH_3$ + ? $\longrightarrow$ $BrCH_2\overset{Br}{\underset{|}{C}}HCH_3$

23. ⬡ + ? $\xrightarrow{H_2SO_4}$ (benzene-$NO_2$) + $H_2O$

24. ⬡ + ? $\xrightarrow{FeCl_3}$ (benzene-$Cl$) + HCl

25. $CH_3CH{=}CHCH_3$ + ? $\xrightarrow{H_2SO_4}$ $CH_3CH_2\overset{HO}{\underset{|}{C}}HCH_3$

## Answers to Self-Test

| | | | | |
|---|---|---|---|---|
| 1. Alkynes | 6. Hydrogenation | 11. trans isomer | 16. c | 21. T |
| 2. cis isomer | 7. Hydrohalogenation | 12. substitution reaction | 17. b | 22. b |
| 3. Oils | 8. Stereoisomers | 13. addition reaction | 18. T | 23. a |
| 4. Alkenes | 9. Fats | 14. Hydration | 19. F | 24. c |
| 5. Halogenation | 10. Unsaturated hydrocarbons | 15. a | 20. F | 25. d |

## Solutions to In-Chapter Problems

**13.1** To draw a complete structure for each condensed structure, first draw in the multiple bonds. Then draw in all the other C's and H's, as in Example 13.1.

a. $CH_2{=}CHCH_2OH$ = 

b. $(CH_3)_2C{=}CH(CH_2)_2CH_3$ =

c. $(CH_3)_2CHC{\equiv}CCH_2C(CH_3)_3$ =

**13.2** To determine whether each molecular formula corresponds to a saturated hydrocarbon, alkene, or alkyne, recall that the formula for a saturated hydrocarbon is $C_nH_{2n+2}$, the formula for an alkene is $C_nH_{2n}$, and the formula for an alkyne is $C_nH_{2n-2}$.

a. $C_3H_6$ = $C_nH_{2n}$ = alkene

b. $C_5H_{12}$ = $C_nH_{2n+2}$ = saturated hydrocarbon

c. $C_8H_{14}$ = $C_nH_{2n-2}$ = alkyne

d. $C_6H_{12}$ = $C_nH_{2n}$ = alkene

**13.3** Use the general formulas [saturated hydrocarbon ($C_nH_{2n+2}$), alkene ($C_nH_{2n}$), and alkyne ($C_nH_{2n-2}$)] to determine the molecular formula for each compound.

a. alkene = $C_nH_{2n}$, $4 \times 2 = 8$, $C_4H_8$

b. saturated hydrocarbon = $C_nH_{2n+2}$, $(6 \times 2) + 2 = 14$, $C_6H_{14}$

c. alkyne = $C_nH_{2n-2}$, $(7 \times 2) - 2 = 12$, $C_7H_{12}$

d. alkene = $C_nH_{2n}$, $5 \times 2 = 10$, $C_5H_{10}$

**13.4** Give the IUPAC name for each compound using the following steps, as in Example 13.2:

[1] Find the longest chain containing both carbon atoms of the multiple bond.

[2] Number the chain to give the double bond the lower number.

[3] Name and number the substituents and write the complete name.

a.

```
        1    2 3 4 5
CH2=CHCHCH2CH3  --------→  CH2=CHCHCH2CH3  ---------→  Answer: 3-methyl-1-pentene
       |                          |
      CH3                        CH3
```

5 C's in the longest chain **pentene**

double bond at C1

methyl at C3

b. $(CH_3CH_2)_2C{=}CHCH_2CH_2CH_3$

ethyl at C3

```
         CH2CH3                    3
         |                 1    2 \ CH2CH3
CH3CH2C=CHCH2CH2CH3  --------→  CH3CH2C=CHCH2CH2CH3  -------→  Answer: 3-ethyl-3-heptene
                                       ↑ 4  5  6  7
```

7 C's in the longest chain **heptene**

double bond at C3

c.

```
                                7  6  5   4  3  2  1
CH3CH2CH=CHCH=CHCH3  --------→  CH3CH2CH=CHCH=CHCH3  ---------→  Answer: 2,4-heptadiene
```

7 C's in the longest chain **heptadiene**

double bonds at C2 and C4

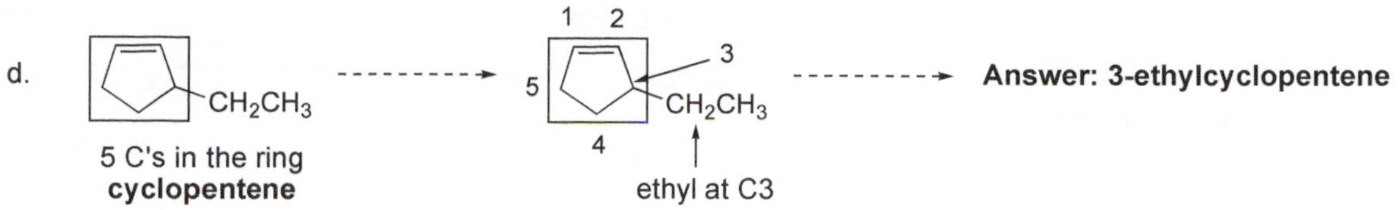

d.     --------→     --------→   **Answer: 3-ethylcyclopentene**

5 C's in the ring
**cyclopentene**

ethyl at C3

**13.5**   Give the IUPAC name for each compound using the following steps, as in Example 13.2:
[1] Find the longest chain containing both carbon atoms of the multiple bond.
[2] Number the chain to give the multiple bond the lower number.
[3] Name and number the substituents and write the complete name.

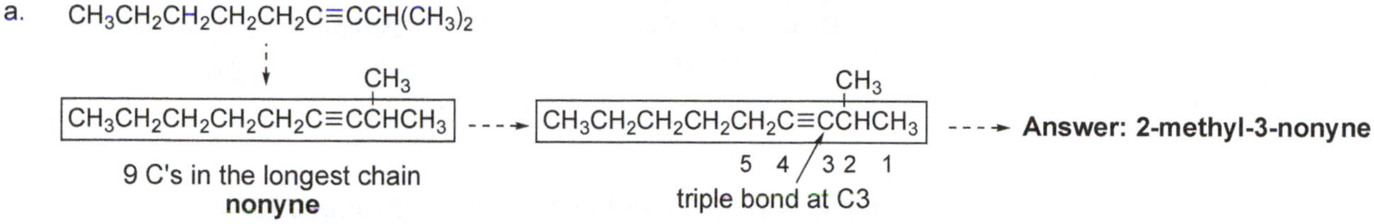

a.   $CH_3CH_2CH_2CH_2CH_2C{\equiv}CCH(CH_3)_2$

$$\boxed{CH_3CH_2CH_2CH_2CH_2C{\equiv}CCHCH_3} \overset{CH_3}{}$$

----→   $\boxed{CH_3CH_2CH_2CH_2CH_2C{\equiv}CCHCH_3}$ ----→   **Answer: 2-methyl-3-nonyne**

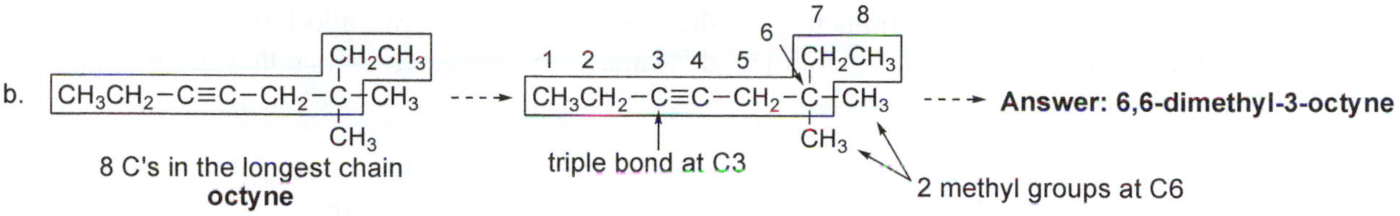

9 C's in the longest chain
**nonyne**

5  4 / 3 2  1
triple bond at C3

b.   $\boxed{CH_3CH_2-C{\equiv}C-CH_2-C{-}CH_3}$

with CH₂CH₃ above and CH₃ below the C

8 C's in the longest chain
**octyne**

----→   $\boxed{CH_3CH_2-C{\equiv}C-CH_2-C{-}CH_3}$

1  2   3  4   5   6   7  8

triple bond at C3

2 methyl groups at C6

----→   **Answer: 6,6-dimethyl-3-octyne**

**13.6**   To draw the structure corresponding to each name, follow the steps in Example 13.3.
• Identify the parent name to find the longest carbon chain or ring, and then use the suffix to determine the functional group; the suffix -*ene* = alkene and -*yne* = alkyne.
• Number the carbon chain or ring and place the functional group at the indicated carbon. Add the substituents and enough hydrogens to give each carbon four bonds.

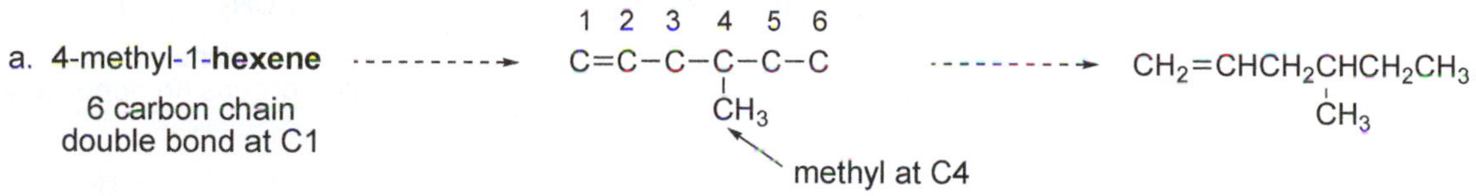

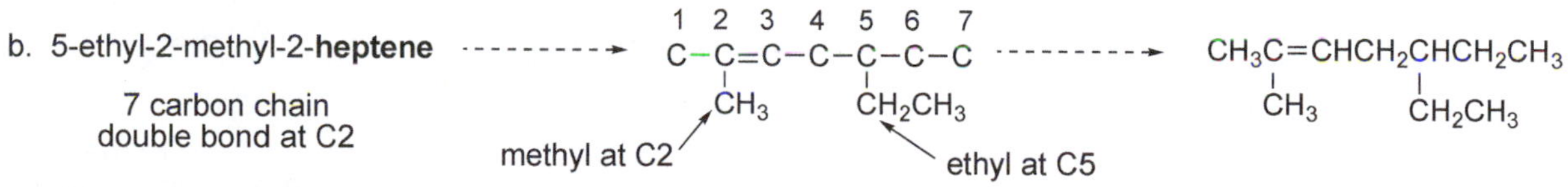

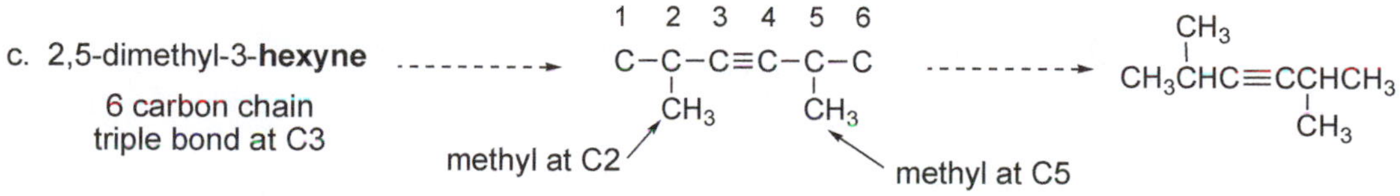

a. 4-methyl-1-**hexene**   --------→   1 2 3  4  5  6
C=C–C–C–C–C
          |
          CH₃

6 carbon chain
double bond at C1

methyl at C4

--------→   $CH_2{=}CHCH_2CHCH_2CH_3$
with CH₃ below

b. 5-ethyl-2-methyl-2-**heptene**   --------→   1 2 3  4  5  6  7
C–C=C–C–C–C–C
      |        |
      CH₃    CH₂CH₃

7 carbon chain
double bond at C2

methyl at C2

ethyl at C5

--------→   $CH_3C{=}CHCH_2CHCH_2CH_3$
with CH₃ and CH₂CH₃ below

c. 2,5-dimethyl-3-**hexyne**   ---------→   1  2  3  4  5  6
C–C–C≡C–C–C
     |        |
     CH₃    CH₃

6 carbon chain
triple bond at C3

methyl at C2

methyl at C5

---------→   $CH_3CHC{\equiv}CCHCH_3$
with CH₃ above and CH₃ below

d. 1-propyl**cyclobutene** - - - - - - - - →

$$
\begin{array}{c}
4C - C^{1}\hspace{-0.4em} \diagup CH_2CH_2CH_3 \\
\| \hspace{1em} \| \\
3C - C^{2}
\end{array}
$$
← propyl at C1 - - - - - - - - → 

4 carbon ring
double bond at C1

e. 1,3-**cyclohexadiene** - - - - - - - - →

$$
\begin{array}{c}
1 \\
6 \hspace{2em} 2 \\
5 \hspace{2em} 3 \\
4
\end{array}
$$

6 carbon ring
2 double bonds (C1 and C3)

f. 4-ethyl-1-**decyne** - - → 

10  9  8  7  6  5  4  3  2  1
$$C-C-C-C-C-C-C-C-C\equiv C$$
$$\underset{CH_2CH_3}{\overset{|}{\phantom{C}}}$$
- - → $CH_3CH_2CH_2CH_2CH_2CH_2CHCH_2C\equiv CH$
$$\underset{CH_2CH_3}{\overset{|}{\phantom{C}}}$$

10 carbon chain
triple bond at C1

↖ ethyl at C4

**13.7** To draw the structures of the cis and trans isomers, follow the steps in Example 13.4.

- Use the parent name to draw the carbon skeleton and place the double bond at the correct carbon.
- Use the definitions of cis and trans to draw the isomers. When the two alkyl groups are on the same side of the double bond, the compound is called the cis isomer. When they are on opposite sides, it is called the trans isomer.

a. *cis*-2-**octene** - - - - - - → $CH_3CH=CHCH_2CH_2CH_2CH_3$ - - - - - - →

8 carbon chain

1  2  3 4  5  6  7  8

$$
\begin{array}{c}
H \hspace{2.5em} H \\
\diagdown \hspace{1em} \diagup \\
C = C \\
\diagup \hspace{1em} \diagdown \\
CH_3 \hspace{2em} CH_2CH_2CH_2CH_2CH_3
\end{array}
$$

cis isomer
both alkyl groups on the same side

b. *trans*-3-**heptene** - - - - - - → $CH_3CH_2CH=CHCH_2CH_2CH_3$ - - - - - - →

7 carbon chain

1  2  3  4  5  6  7

$$
\begin{array}{c}
H \hspace{2.5em} CH_2CH_2CH_3 \\
\diagdown \hspace{1em} \diagup \\
C = C \\
\diagup \hspace{1em} \diagdown \\
CH_3CH_2 \hspace{2em} H
\end{array}
$$

trans isomer
alkyl groups on opposite sides

c. *trans*-4-methyl-2-**pentene** - - - - - - → $CH_3CH=CHCH(CH_3)CH_3$ - - - - - - →

5 carbon chain

1  2  3 4  5

$$
\begin{array}{c}
\hspace{5em} CH_3 \\
\hspace{5em} | \\
H \hspace{2em} CHCH_3 \\
\diagdown \hspace{1em} \diagup \\
C = C \\
\diagup \hspace{1em} \diagdown \\
CH_3 \hspace{2em} H
\end{array}
$$

trans isomer
alkyl groups on opposite sides

**13.8** Whenever the two groups on *each* end of a C=C are *different from each other*, two isomers are possible.

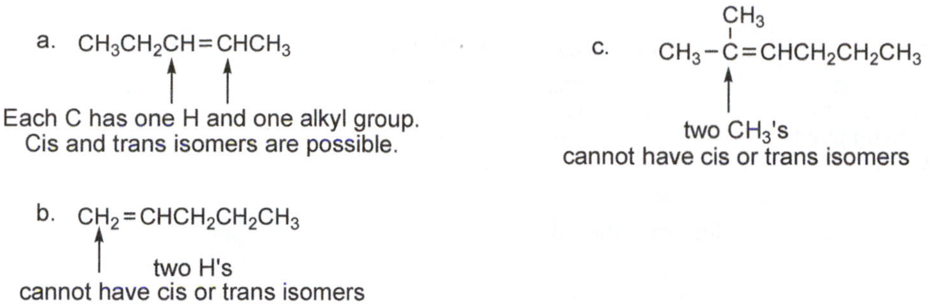
a. CH3CH2CH=CHCH3
Each C has one H and one alkyl group.
Cis and trans isomers are possible.
c. CH3–C(CH3)=CHCH2CH2CH3
two CH3's
cannot have cis or trans isomers
b. CH2=CHCH2CH2CH3
two H's
cannot have cis or trans isomers

**13.9** When the two alkyl groups are on the same side of the double bond, the compound is called the cis isomer. When they are on opposite sides, it is called the trans isomer.

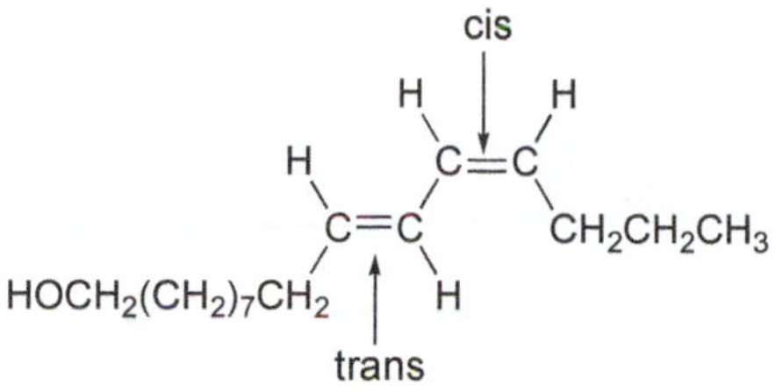
cis
HOCH2(CH2)7CH2
CH2CH2CH3
trans

**13.10** Stereoisomers differ only in the three-dimensional arrangement of atoms. Constitutional isomers differ in the way the atoms are bonded to each other.

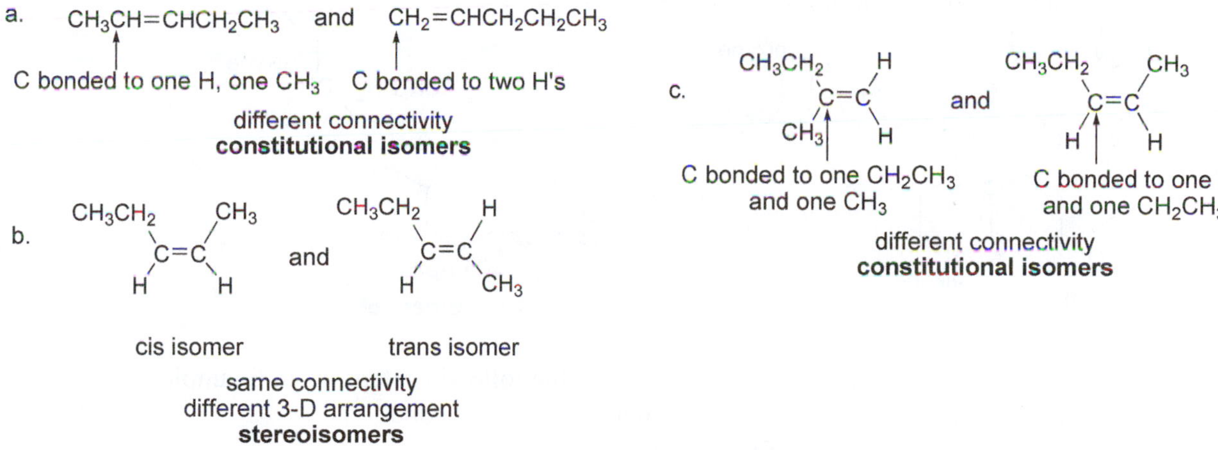
a. CH3CH=CHCH2CH3 and CH2=CHCH2CH2CH3
C bonded to one H, one CH3
C bonded to two H's
different connectivity
constitutional isomers
b.
and
cis isomer
trans isomer
same connectivity
different 3-D arrangement
stereoisomers
c.
and
C bonded to one CH2CH3 and one CH3
C bonded to one H and one CH2CH3
different connectivity
constitutional isomers

**13.11** Double bonds in naturally occurring fatty acids are cis.

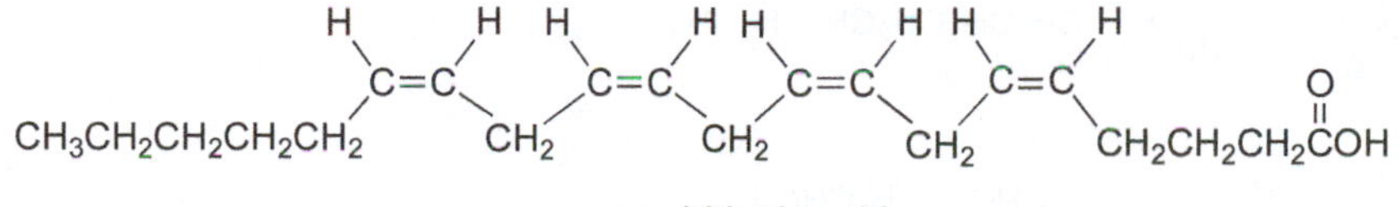

arachidonic acid

**13.12** Fats are solids at room temperature because of their higher melting point. They are formed from fatty acids with few double bonds. Oils are liquids at room temperature because of their lower melting points. They are also formed from fatty acids, but have more double bonds.

$CH_3(CH_2)_{14}COOH$          $CH_3(CH_2)_5CH=CH(CH_2)_7COOH$

palmitic acid                     palmitoleic acid
no double bonds                   one double bond
higher melting point              lower melting point
63 °C                             1 °C

**13.13** The functional groups in tamoxifen are labeled.

**13.14** The functional groups in RU 486 and levonorgestrel are labeled.

**13.15** To draw the product of a hydrogenation reaction, use the following steps, as in Example 13.5:
- Locate the C=C and mentally break one bond in the double bond.
- Mentally break the H–H bond of the reagent.
- Add one H atom to each C of the C=C, thereby forming two new C–H single bonds.

a.   $CH_3CH_2CH=CHCH_2CH_3$  $\xrightarrow[Pd]{H_2}$  $CH_3CH_2CH_2CH_2CH_2CH_3$

b.

c. $\xrightarrow[\text{Pd}]{\text{H}_2}$

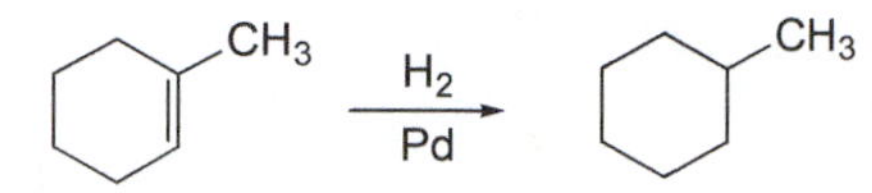

**13.16** To draw the product of each halogenation reaction, add a halogen to both carbons of the double bond.

a. $CH_3CH_2CH=CH_2$ + $Cl_2$ $\longrightarrow$ $CH_3CH_2\underset{\underset{H}{|}}{\overset{\overset{Cl}{|}}{C}}-\underset{\underset{H}{|}}{\overset{\overset{Cl}{|}}{C}}-H$

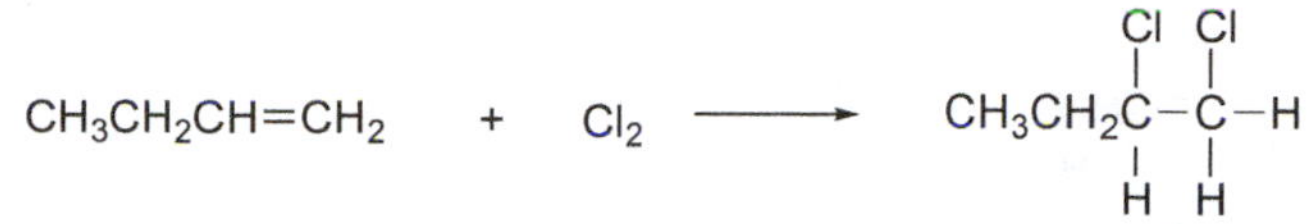

b. + $Br_2$ $\longrightarrow$

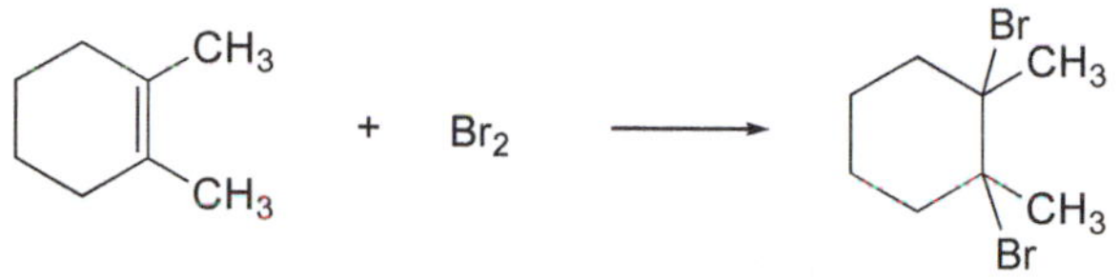

**13.17** In hydrohalogenation reactions, the elements of H and Br (or H and Cl) must be added to the double bond. When the alkene is unsymmetrical, the H atom of HX bonds to the carbon that has more H's to begin with.

a. $CH_3CH=CHCH_3$ + HBr $\longrightarrow$ $CH_3\underset{\underset{Br}{|}}{\overset{\overset{H}{|}}{C}}-\underset{\underset{H}{|}}{\overset{\overset{H}{|}}{C}}CH_3$

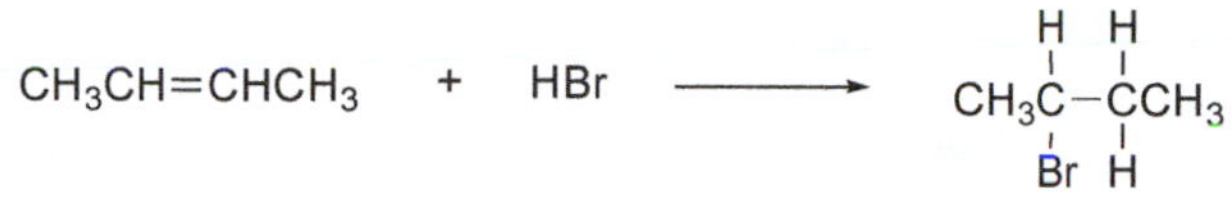

b. This C does not have any H's.
Add Br here.
+ HBr $\longrightarrow$
This C has more H's.
Add H here.

c. This C does not have any H's.
Add Cl here.
$(CH_3)_2C=CHCH_3$ + HCl $\longrightarrow$ $CH_3-\underset{\underset{Cl}{|}}{\overset{\overset{CH_3}{|}}{C}}-\underset{\underset{H}{|}}{\overset{\overset{H}{|}}{C}}CH_3$
This C has more H's.
Add H here.

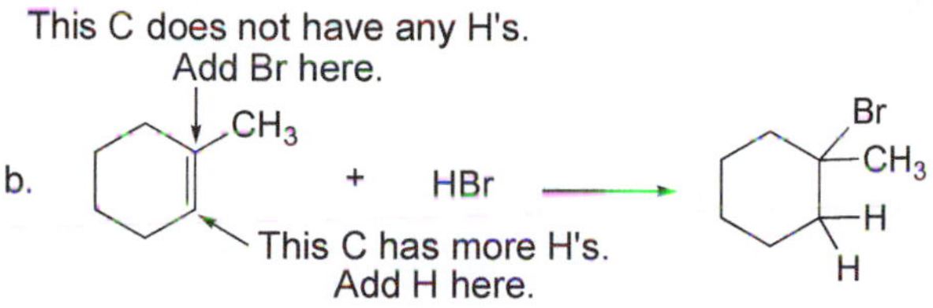

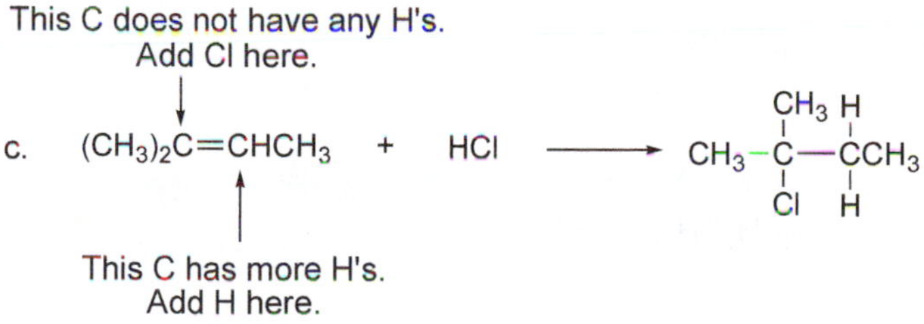

d. + HCl $\longrightarrow$

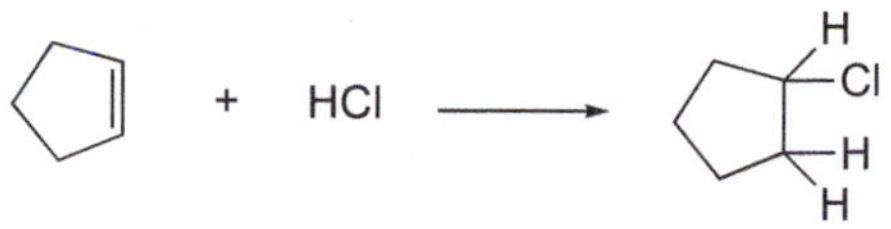

**13.18** In hydration reactions, the elements of H and OH are added to the double bond. In unsymmetrical alkenes, the H atom bonds to the less substituted carbon.

a. $CH_3CH=CHCH_3$ $\xrightarrow[\text{H}_2\text{SO}_4]{\text{H—OH}}$ $CH_3\underset{\underset{H}{|}}{\overset{\overset{HO}{|}}{C}}-\underset{\underset{H}{|}}{\overset{\overset{H}{|}}{C}}CH_3$

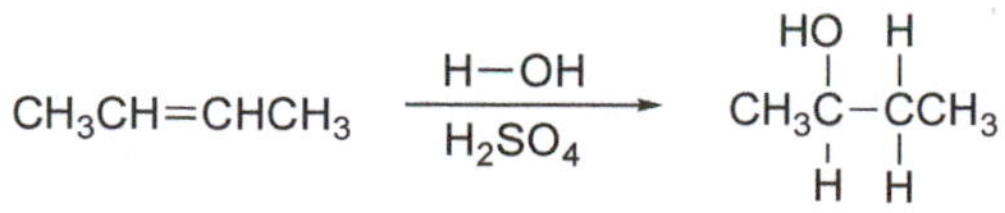

This C has one H so the OH bonds here.

b.   $CH_3CH_2CH=CH_2$ $\xrightarrow[\text{H}_2\text{SO}_4]{\text{H-OH}}$  $CH_3CH_2\overset{\text{HO}}{\underset{\text{H}}{C}}-\overset{\text{H}}{\underset{\text{H}}{C}}-H$

This C has 2 H's so the H bonds here.

c.  $\xrightarrow[\text{H}_2\text{SO}_4]{\text{H-OH}}$

**13.19**   Draw the products of each reaction.

a.   $CH_3CH_2CH_2CH=CH_2$ $\xrightarrow[\text{Pd}]{\text{H}_2}$  $CH_3CH_2CH_2\overset{\text{H}}{\underset{\text{H}}{C}}-\overset{\text{H}}{\underset{\text{H}}{C}}-H$

b.   $CH_3CH_2CH_2CH=CH_2$ $\xrightarrow{\text{Cl}_2}$  $CH_3CH_2CH_2\overset{\text{H}}{\underset{\text{Cl}}{C}}-\overset{\text{H}}{\underset{\text{Cl}}{C}}-H$

c.   $CH_3CH_2CH_2CH=CH_2$ $\xrightarrow{\text{Br}_2}$  $CH_3CH_2CH_2\overset{\text{Br}}{\underset{\text{H}}{C}}-\overset{\text{Br}}{\underset{\text{H}}{C}}-H$

d.   $CH_3CH_2CH_2CH=CH_2$ $\xrightarrow{\text{H-Br}}$  $CH_3CH_2CH_2\overset{\text{H}}{\underset{\text{Br}}{C}}-\overset{\text{H}}{\underset{\text{H}}{C}}-H$

e.   $CH_3CH_2CH_2CH=CH_2$ $\xrightarrow{\text{H-Cl}}$  $CH_3CH_2CH_2\overset{\text{H}}{\underset{\text{Cl}}{C}}-\overset{\text{H}}{\underset{\text{H}}{C}}-H$

f.   $CH_3CH_2CH_2CH=CH_2$ $\xrightarrow[\text{H}_2\text{SO}_4]{\text{H-OH}}$  $CH_3CH_2CH_2\overset{\text{HO}}{\underset{\text{H}}{C}}-\overset{\text{H}}{\underset{\text{H}}{C}}-H$

**13.20** Draw the products of the hydrogenation reactions.

a.

b. $CH_3CH_2CH_2CH_2CH_2CH_2CH_2$ $CH_2CH_2CH_2CH_2CH_2CH_2CH_2CH_2CH_2CH_2COOH$

**13.21** To draw the polymers, draw three or more alkene molecules and arrange the carbons of the double bonds next to each other. Break one bond of each double bond, and join the alkenes together with single bonds. With unsymmetrical alkenes, substituents are bonded to every other carbon. Use Example 13.7 as a guide.

a.

b.

c.

**13.22** Work backwards to determine what monomer is used to form the polymer.

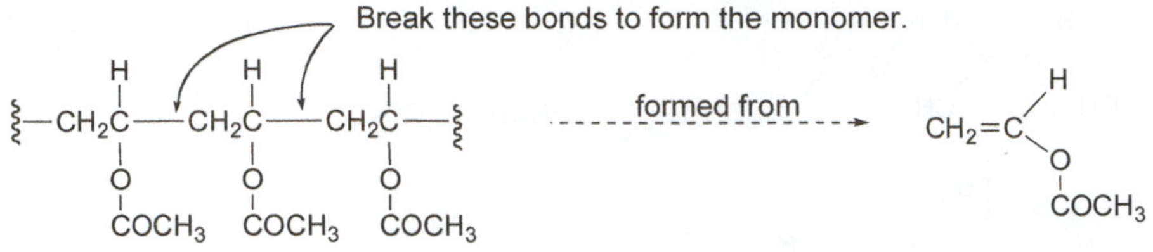
Break these bonds to form the monomer.
H H H
CH2C CH2C CH2C
O O O
COCH3 COCH3 COCH3
formed from
H
CH2=C
O
COCH3

poly(vinyl acetate)

**13.23** Name each aromatic compound as in Example 13.8. Name the substituents on the benzene ring. With two groups, alphabetize the substituent names and use the prefix ortho, meta, or para to indicate their location. With three substituents, alphabetize the substituent names, and number to give the lowest set of numbers.

a. 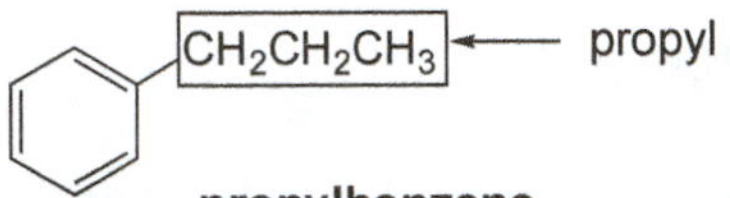
CH2CH2CH3
propyl

**propylbenzene**

c. 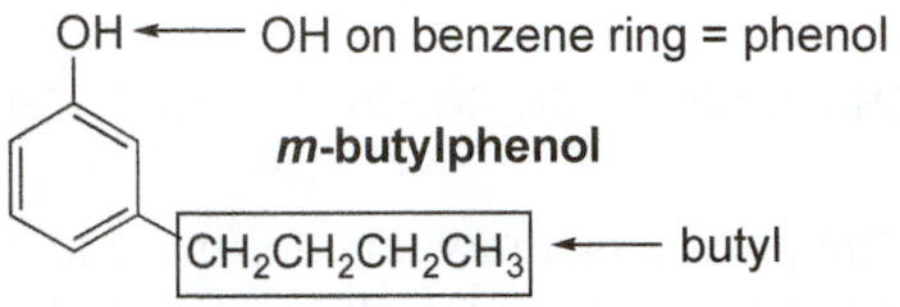
OH
OH on benzene ring = phenol
CH2CH2CH2CH3
butyl

***m*-butylphenol**

b. 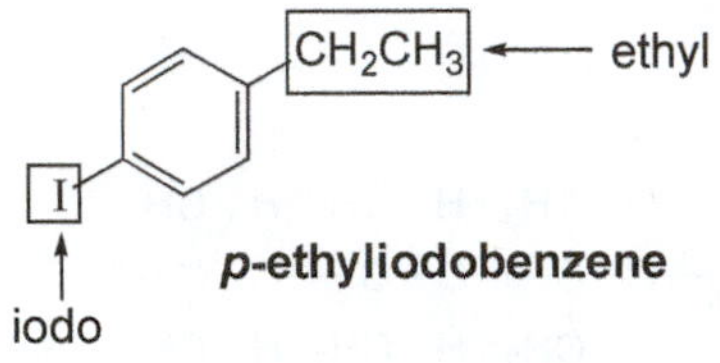
CH2CH3
ethyl
I
iodo

***p*-ethyliodobenzene**

d. 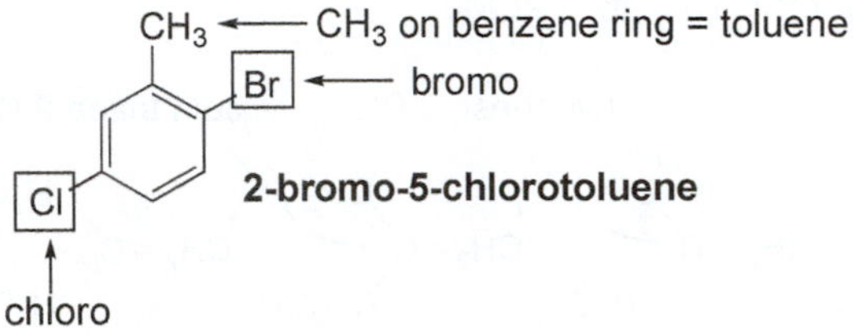
CH3
CH3 on benzene ring = toluene
Br
bromo
Cl
chloro

**2-bromo-5-chlorotoluene**

**13.24** Draw the structure corresponding to each name.

a. pentylbenzene

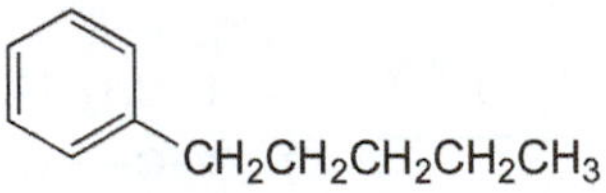
CH2CH2CH2CH2CH3

c. *m*-bromoaniline

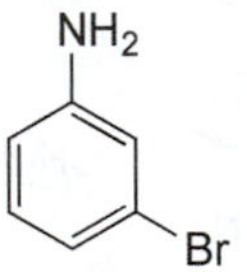
NH2
Br

b. *o*-dichlorobenzene

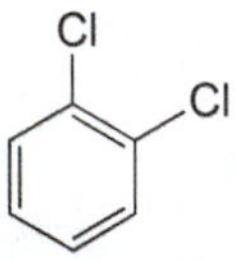
Cl
Cl

d. 4-chloro-1,2-diethylbenzene

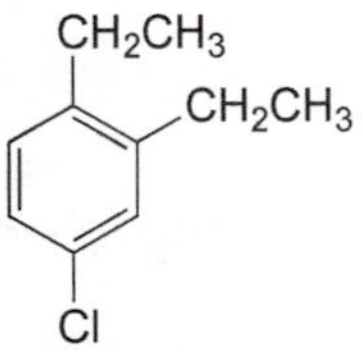
CH2CH3
CH2CH3
Cl

**13.25** Commercially available sunscreens contain a benzene ring. Therefore, compound (a) might be found in a sunscreen since it contains two aromatic rings. Compound (b) does not contain any aromatic rings.

**13.26** Phenols are antioxidants because the OH group on the benzene ring prevents unwanted oxidation reactions from occurring. Of the compounds listed, only curcumin (b) contains a phenol group (OH on a benzene ring), making it an antioxidant.

**13.27** Draw the products of each substitution reaction.
- Chlorination replaces one of the H's on the benzene ring with Cl.
- Nitration replaces one of the H's on the benzene ring with $NO_2$.
- Sulfonation replaces one of the H's on the benzene ring with $SO_3H$.

a.

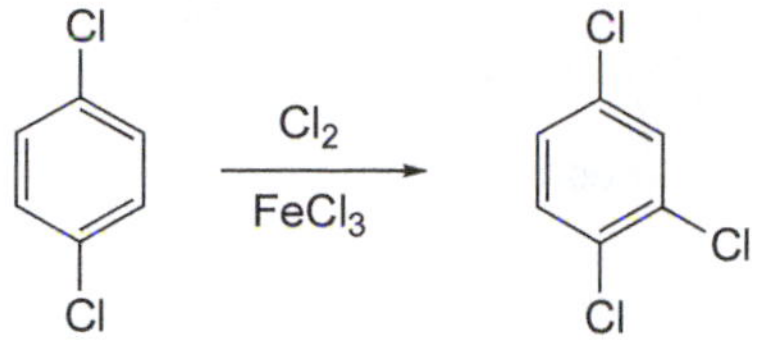

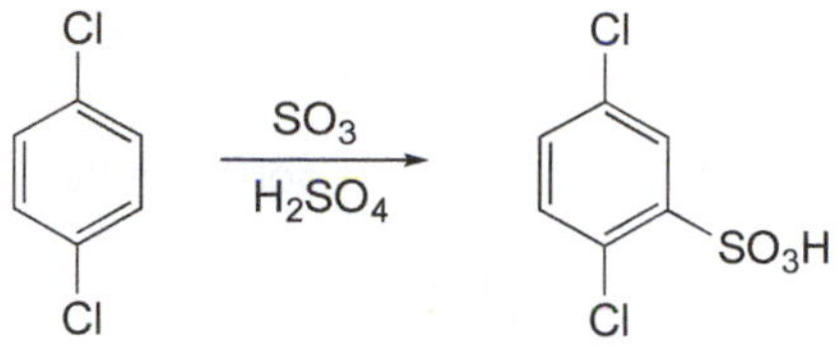

b.

c.

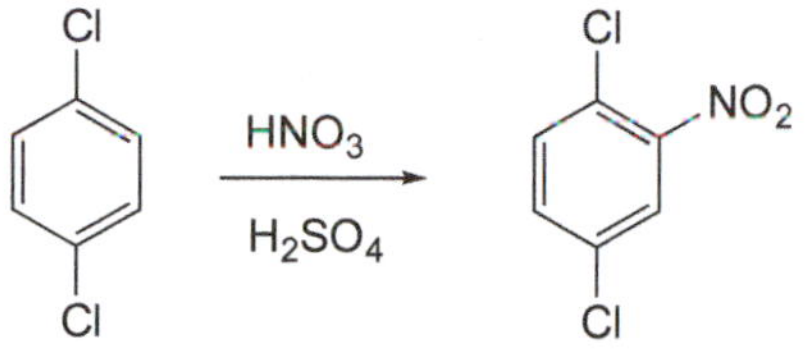

**13.28** Draw the products of the substitution reaction. The Cl can replace any of the H's on the benzene ring, giving three different products.

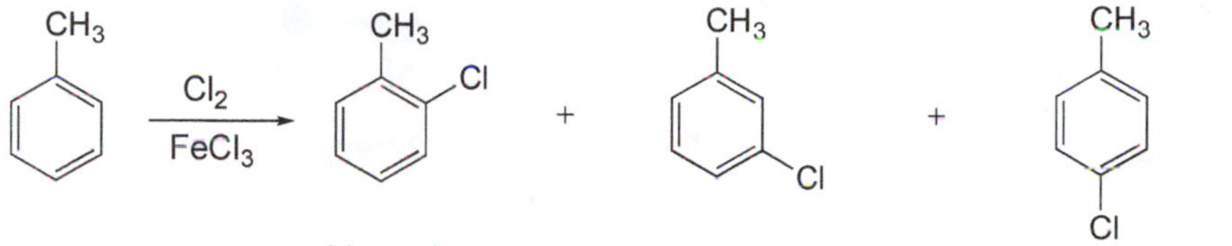

o-chlorotoluene    m-chlorotoluene    p-chlorotoluene

## Solutions to Odd-Numbered End-of-Chapter Problems

**13.29**

a. molecular formula: $C_{10}H_{12}O$
b. aromatic ring, alkene, ether
c. trans
d. Tetrahedral C's are indicated. All other C's are trigonal planar.

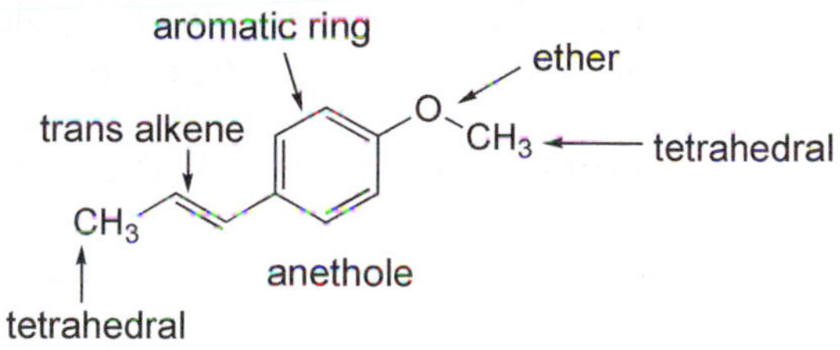

aromatic ring
ether
trans alkene
tetrahedral
anethole
tetrahedral

**13.31** Use the general formulas [saturated hydrocarbon ($C_nH_{2n+2}$), alkene ($C_nH_{2n}$), and alkyne ($C_nH_{2n-2}$)] to determine the molecular formula for each compound with 10 C's.

a. $(10 \times 2) + 2 = 22$: molecular formula $C_{10}H_{22}$    c. $(10 \times 2) - 2 = 18$: molecular formula $C_{10}H_{18}$
b. $10 \times 2 = 20$: molecular formula $C_{10}H_{20}$

**13.33** Draw three alkynes with molecular formula $C_5H_8$.

$HC{\equiv}CCH_2CH_2CH_3$      $CH_3C{\equiv}CCH_2CH_3$

**13.35** Label each carbon as tetrahedral, trigonal planar, or linear by counting groups.

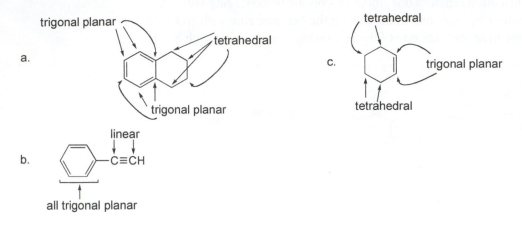

a.

trigonal planar

tetrahedral

trigonal planar

c.

tetrahedral

trigonal planar

tetrahedral

b.

linear

—C≡CH

all trigonal planar

**13.37** Give the IUPAC name for each compound.

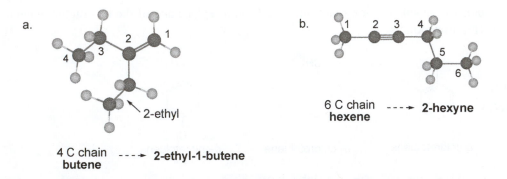

a.

2-ethyl

4 C chain ----> **2-ethyl-1-butene**
butene

b.

6 C chain ----> **2-hexyne**
hexene

**13.39** Give the IUPAC name for each compound using the following steps, as in Example 13.2:
[1] Find the longest chain containing both carbon atoms of the multiple bond.
[2] Number the chain to give the multiple bond the lower number.
[3] Name and number the substituents and write the complete name.

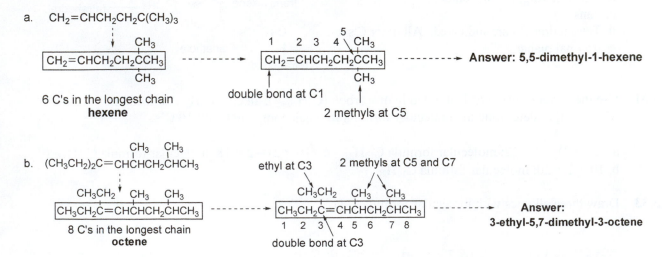

a.   CH₂=CHCH₂CH₂C(CH₃)₃

$CH_2=CHCH_2CH_2\overset{\underset{|}{CH_3}}{\underset{|}{C}}CH_3$

6 C's in the longest chain
**hexene**

- - - - - - ->

$\overset{5}{\underset{1\quad 2\quad 3\quad 4}{CH_2=CHCH_2CH_2\overset{CH_3}{\underset{CH_3}{C}}CH_3}}$

double bond at C1

2 methyls at C5

- - - - - - -> **Answer: 5,5-dimethyl-1-hexene**

b.   (CH₃CH₂)₂C=CHCHCH₂CHCH₃
         CH₃    CH₃

$CH_3CH_2\overset{CH_3CH_2}{C}=CH\overset{CH_3}{CH}CH_2\overset{CH_3}{CH}CH_3$

8 C's in the longest chain
**octene**

- - - - - - ->

ethyl at C3      2 methyls at C5 and C7

$\underset{1\quad 2\quad 3\quad 4\; 5\; 6\quad 7\; 8}{CH_3CH_2C=CHCHCH_2CHCH_3}$

double bond at C3

- - - - - - -> **Answer:**
**3-ethyl-5,7-dimethyl-3-octene**

c.  
$CH_2=C\,CH_2CH_3$  
$CH_2CH_2CH_2CH_2CH_3$  

7 C's in the longest chain  
**heptene**

$\longrightarrow$

ethyl at C2

$\overset{1\quad 2}{CH_2=C\,CH_2CH_3}$  
$\underset{3\quad 4\quad 5\quad 6\quad 7}{CH_2CH_2CH_2CH_2CH_3}$

double bond at C1

$\longrightarrow$  **Answer: 2-ethyl-1-heptene**

---

d.  $CH_3C{\equiv}CCH_2C(CH_3)_3$

$CH_3$  
$CH_3C{\equiv}CCH_2CCH_3$  
$CH_3$

6 C's in the longest chain  
**hexyne**

$\longrightarrow$

$\overset{1\quad 2\quad 3}{CH_3C{\equiv}CCH_2}\overset{5}{\underset{6}{C}}CH_3$  
$CH_3$

triple bond at C2

2 methyls at C5

$\longrightarrow$  **Answer: 5,5-dimethyl-2-hexyne**

---

e.  
$CH_3\ \ CH_3$  
$CH_3C{\equiv}C-CH_2-CH-C-CH_2CH_3$  
$CH_2CH_3$

8 C's in the longest chain  
**octyne**

$\longrightarrow$

2 methyls at C5 and C6

$\overset{1\quad 2\quad 3\quad 4}{CH_3C{\equiv}C-CH_2-}\overset{5}{CH}\overset{6}{-C-}\overset{8}{CH_2CH_3}$  
$CH_3\ CH_3$  
$CH_2CH_3$

triple bond at C2

ethyl at C6

$\longrightarrow$  **Answer:**  
**6-ethyl-5,6-dimethyl-2-octyne**

---

f.  
$CH_3$  
$CH_2=CHCH_2-C-CH=CH_2$  
$CH_3$

6 C's in the longest chain  
**hexadiene**

$\longrightarrow$

$\overset{6\quad\quad 5\quad 4}{CH_2=CHCH_2-}\overset{3}{C}\overset{1}{-CH=CH_2}$  
$CH_3$

2 methyls at C3  
double bonds at C1 and C5

$\longrightarrow$  **Answer:**  
**3,3-dimethyl-1,5-hexadiene**

---

**13.41**  Give the IUPAC name for each compound using the steps in Answer 13.39 and Example 13.2.

a.  
double bond at C1  

$\overset{2\quad\quad 3}{\underset{1\quad\quad\quad 4}{\hexagon}}-CH_3$

methyl at C4  
6 carbon ring  
**cyclohexene**

$\dashrightarrow$  **4-methylcyclohexene**

b.  
$CH_2CH_3$  2 ethyl groups at C3  

$\overset{2\quad 3}{\underset{1\quad 4}{\square}}-CH_2CH_3$

4 carbon ring  
**cyclobutene**

$\dashrightarrow$  **3,3-diethylcyclobutene**

---

**13.43**  To draw the structure corresponding to each name, follow the steps in Example 13.3.
- Identify the parent name to find the longest carbon chain or ring, and then use the suffix to determine the functional group; the suffix -*ene* = alkene and -*yne* = alkyne.
- Number the carbon chain or ring and place the functional group at the indicated carbon. Add the substituents and enough hydrogens to give each carbon four bonds.

a. 3-methyl-1-**octene**  
8 carbon chain  
double bond at C1

$\longrightarrow$

$\overset{1\quad 2\quad 3\quad 4\quad 5\quad 6\quad 7\quad 8}{C=C-C-C-C-C-C-C}$  
$CH_3$

methyl at C3

$\longrightarrow$

$CH_2=CHCHCH_2CH_2CH_2CH_3$  
$CH_3$

b. 1-ethyl**cyclobutene**   - - - - - →   $C=C$ ─ $CH_2CH_3$   ethyl at C1   - - - - - →   [cyclobutene with $CH_2CH_3$]
   4 carbon ring
   double bond at C1

c. 2-methyl-3-**hexyne**   - - - - - - →   
1 2 3 4 5 6
C─C─C≡C─C─C
       |
       $CH_3$   methyl at C2
   6 carbon chain
   triple bond at C3
   - - - - - - →   $CH_3$─$\overset{H}{\underset{CH_3}{C}}$─C≡CCH$_2$CH$_3$

d. 3,5-diethyl-2-methyl-3-**heptene**   - - - - - - →   
        $CH_2CH_3$
1 2  |  4 5 6 7
C─C─C=C─C─C─C
   |        |
   $CH_3$3   $CH_2CH_3$
methyl at C2      2 ethyl groups at C3 and C5
   7 carbon chain
   double bond at C3
   - - - - - - →   $CH_3$─$\overset{H}{\underset{CH_3}{C}}$─$\overset{CH_2CH_3}{C}$=$\overset{H}{C}$─$\overset{H}{\underset{CH_2CH_3}{C}}CH_2CH_3$

e. 1,3-**heptadiene**   - - - - - →   
1 2 3 4 5 6 7
C=C─C=C─C─C─C
   - - - - - - →   $CH_2$=$\overset{}{\underset{H}{C}}$─$\overset{}{\underset{H}{C}}$=CHCH$_2$CH$_2$CH$_3$
   7 carbon chain
   double bonds at
   C1 and C3

f. *cis*-7-methyl-2-**octene**   - - - - - →   
1 2 3 4 5 6 7 8
C─C=C─C─C─C─C─C
            |
            $CH_3$
            methyl at C7
   8 carbon chain
   double bond at C2
   - - - - - - →   $\overset{CH_3}{\underset{H}{}}C=C\overset{CH_2CH_2CH_2CHCH_3}{\underset{H}{}}\overset{}{\underset{CH_3}{}}$   cis

**13.45**   Correct each of the incorrect IUPAC names.

a. The name 5-methyl-4-hexene places the double bond at C4
   instead of C2. Assign the lower number to the alkene:
   2-methyl-2-hexene.

   $\overset{CH_3}{\underset{}{}}$
   $CH_3C=CHCH_2CH_2CH_3$
   2-methyl-2-hexene

b. The name 1-methylbutene makes the last carbon in the chain
   a substituent. In addition, the location of the double bond is
   not specified. There are five carbons in the chain (not a
   methyl substituent): 2-pentene.

   $\overset{CH_3}{\underset{}{}}$
   $CH=CHCH_2CH_3$
   2-pentene

c. The name 2,3-dimethylcyclohexene starts numbering
   substituents at C2 instead of C1. Number to put the C=C
   between C1 and C2, and then give the first substituent the
   lower number: 1,6-dimethylcyclohexene.

   [cyclohexene ring with $CH_3$ at 1, numbering 1 2 3 4 5 6, $CH_3$ at 6]
   1,6-dimethylcyclohexene

d.  The name 3-butyl-1-butyne does not name the longest chain. Name the seven carbon chain: 3-methyl-1-heptyne.

$$HC\equiv\overset{\underset{\displaystyle CH_2CH_2CH_2CH_3}{|}}{\underset{2}{C}}\overset{3}{\overset{H}{\diagup}}CH_3$$

3-methyl-1-heptyne

**13.47**  When the two alkyl groups are on the same side of the double bond, the compound is called the cis isomer. When they are on opposite sides it is called the trans isomer.

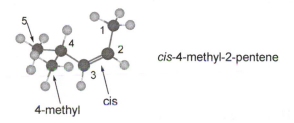

**13.49**  Give the IUPAC name for the alkene. Use the definition in Answer 13.47 to determine if it is the cis or trans isomer.

*cis*-4-methyl-2-pentene

**13.51**  Draw the cis and trans isomers for each compound, as in Example 13.4.

a.
$$\underset{CH_3}{\overset{H}{\diagdown}}C=C\underset{CH_2CH_2CH_2CH_2CH_2CH_3}{\overset{H}{\diagup}}$$

**cis-2-nonene**

b.  $$\underset{\underset{\displaystyle CH_3}{|}}{CH_3CH}\overset{H}{\diagdown}C=C\underset{CH_2CH_2CH_3}{\overset{H}{\diagup}}$$

**cis-2-methyl-3-heptene**

$$\underset{CH_3}{\overset{H}{\diagdown}}C=C\underset{H}{\overset{CH_2CH_2CH_2CH_2CH_2CH_3}{\diagup}}$$

**trans-2-nonene**

$$\underset{\underset{\displaystyle CH_3}{|}}{CH_3CH}\overset{H}{\diagdown}C=C\underset{H}{\overset{CH_2CH_2CH_3}{\diagup}}$$

**trans-2-methyl-3-heptene**

**13.53**  Constitutional isomers have the same molecular formula, but have the atoms bonded to different atoms. Stereoisomers have atoms bonded to the same atoms but in a different three-dimensional arrangement.

**13.55** Determine if the molecules are constitutional isomers, stereoisomers, or identical.

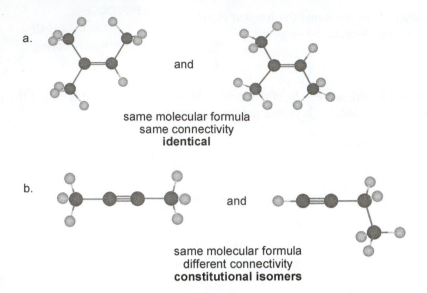

a.                    and

same molecular formula
same connectivity
**identical**

b.                    and

same molecular formula
different connectivity
**constitutional isomers**

**13.57** Draw the products of each reaction by adding $H_2$ to the double bond.

a.   $CH_2=CHCH_2CH_2CH_2CH_3$   $\xrightarrow[\text{Pd}]{H_2}$   $CH_3CH_2CH_2CH_2CH_2CH_3$

c.   (cyclohexene with $CH_3$)   $\xrightarrow[\text{Pd}]{H_2}$   (cyclohexane with $CH_3$)

b.   $(CH_3)_2C=CHCH_2CH_2CH_3$   $\xrightarrow[\text{Pd}]{H_2}$   $(CH_3)_2CHCH_2CH_2CH_2CH_3$

d.   (cyclohexane with $=CH_2$)   $\xrightarrow[\text{Pd}]{H_2}$   (cyclohexane with H and $CH_3$)

**13.59** Draw the products of each reaction by adding HCl to the double bond.

a.   (cyclobutene)   $\xrightarrow{\text{HCl}}$   (cyclobutane with Cl)

c.   $CH_2=CHCH_2CH(CH_3)_2$   $\xrightarrow{\text{HCl}}$   $CH_3\overset{\text{Cl}}{C}HCH_2CH(CH_3)_2$

b.   $(CH_3)_2C=C(CH_3)_2$   $\xrightarrow{\text{HCl}}$   $(CH_3)_2\overset{\text{Cl}}{C}-\overset{\text{H}}{C}(CH_3)_2$

d.   (cyclohexane with $=CH_2$)   $\xrightarrow{\text{HCl}}$   (cyclohexane with $CH_3$ and Cl)

**13.61**  Draw the products of each reaction by adding the specified reagent to the double bond.

a.   $\xrightarrow[\text{Pd}]{\text{H}_2}$

d.   $\xrightarrow{\text{HCl}}$

b.   $\xrightarrow{\text{Cl}_2}$

e.   $\xrightarrow{\text{HBr}}$

c.   $\xrightarrow{\text{Br}_2}$

f.   $\xrightarrow[\text{H}_2\text{SO}_4]{\text{H}_2\text{O}}$

**13.63**  Work backwards to determine what alkene is needed as a starting material to prepare each of the alkyl halides or dihalides.

a.  $CH_2=CH_2$  $\xrightarrow{\text{HBr}}$  $CH_3CH_2Br$

c.  $-CH_3$  $\xrightarrow{\text{Cl}_2}$  $-CH_3$

b.   $\xrightarrow{\text{HCl}}$

d.  $CH_2=CHCH_2CH(CH_3)_2$  $\xrightarrow{\text{Br}_2}$  $BrCH_2CHCH_2CH(CH_3)_2$
$\quad\quad\quad\quad\quad\quad\quad\quad\quad\quad\quad\quad\quad\quad\quad\quad\quad\quad\quad\quad\quad\quad\quad\overset{|}{Br}$

**13.65**  Work backwards to determine what reagent is needed to convert 2-methylpropene to each product.

a.  $(CH_3)_2C=CH_2$  $\xrightarrow{\text{HCl}}$  $(CH_3)_3CCl$

d.  $(CH_3)_2C=CH_2$  $\xrightarrow{\text{HBr}}$  $(CH_3)_3CBr$

b.  $(CH_3)_2C=CH_2$  $\xrightarrow[\text{Pd}]{\text{H}_2}$  $(CH_3)_3CH$

e.  $(CH_3)_2C=CH_2$  $\xrightarrow{\text{Br}_2}$  $(CH_3)_2CCH_2Br$
$\quad\quad\quad\quad\quad\quad\quad\quad\quad\quad\quad\quad\quad\quad\quad\quad\quad\overset{|}{Br}$

c.  $(CH_3)_2C=CH_2$  $\xrightarrow[\text{H}_2\text{SO}_4]{\text{H}_2\text{O}}$  $(CH_3)_3COH$

f.  $(CH_3)_2C=CH_2$  $\xrightarrow{\text{Cl}_2}$  $(CH_3)_2CCH_2Cl$
$\quad\quad\quad\quad\quad\quad\quad\quad\quad\quad\quad\quad\quad\quad\quad\quad\quad\overset{|}{Cl}$

**13.67**  To draw the polymer, draw three or more alkene molecules and arrange the carbons of the double bonds next to each other.  Break one bond of each double bond, and join the alkenes together with single bonds.  With unsymmetrical alkenes, substituents are bonded to every other carbon. Use Example 13.7 as a guide.

Join these 2 C's.     Join these 2 C's.

**13.69** Draw the polymers using the steps in Example 13.7.

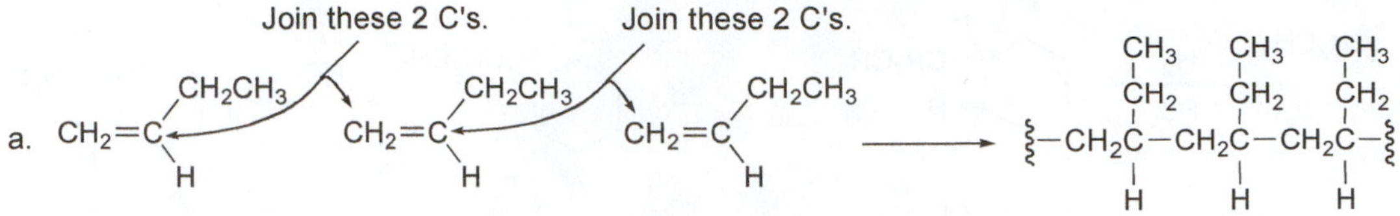

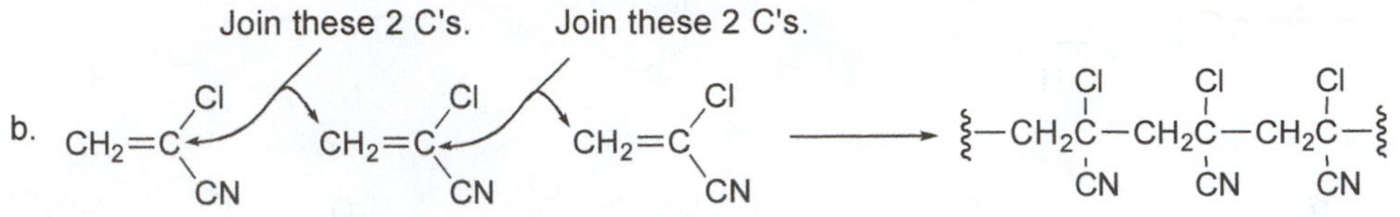

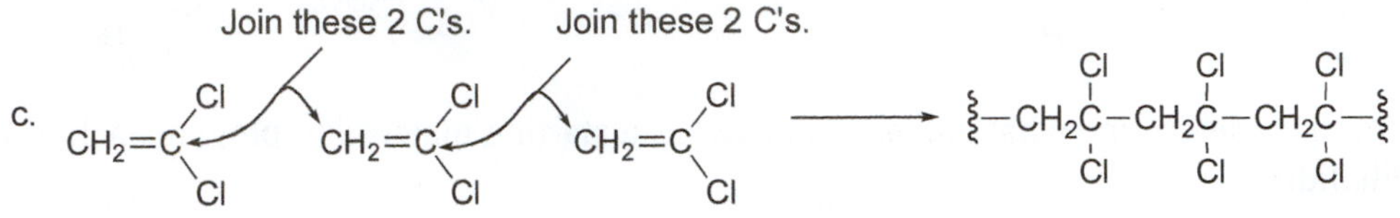

a.

Join these 2 C's.    Join these 2 C's.

$CH_2=C$(CH$_2$CH$_3$)(H) → $CH_2=C$(CH$_2$CH$_3$)(H) → $CH_2=C$(CH$_2$CH$_3$)(H) ⟶

$$\{-CH_2C(CH_3)(H)-CH_2C(CH_3)(H)-CH_2C(CH_3)(H)-\}$$

b.

Join these 2 C's.    Join these 2 C's.

$CH_2=C$(Cl)(CN) → $CH_2=C$(Cl)(CN) → $CH_2=C$(Cl)(CN) ⟶

$$\{-CH_2C(Cl)(CN)-CH_2C(Cl)(CN)-CH_2C(Cl)(CN)-\}$$

c.

Join these 2 C's.    Join these 2 C's.

$CH_2=C$(Cl)(Cl) → $CH_2=C$(Cl)(Cl) → $CH_2=C$(Cl)(Cl) ⟶

$$\{-CH_2C(Cl)(Cl)-CH_2C(Cl)(Cl)-CH_2C(Cl)(Cl)-\}$$

**13.71** Work backwards to determine what monomer was used to form the polymer.

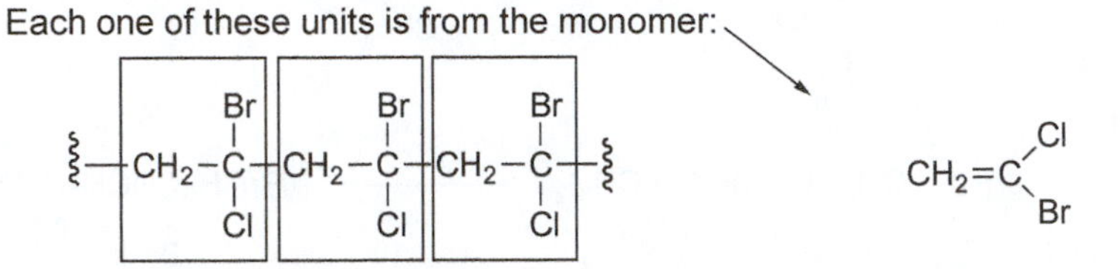

Each one of these units is from the monomer:

$$\{-CH_2-C(Br)(Cl)-CH_2-C(Br)(Cl)-CH_2-C(Br)(Cl)-\}$$

$CH_2=C$(Cl)(Br)

**13.73** Draw two resonance structures by moving the double bonds.

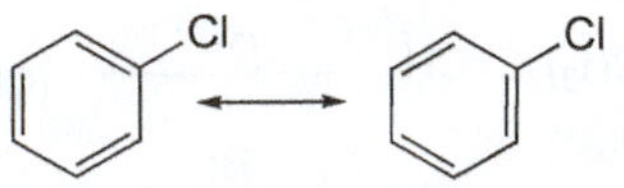

**13.75** Name each aromatic compound as in Example 13.8. Name the substituents on the benzene ring. With two groups, alphabetize the substituent names and use the prefix ortho, meta, or para to indicate their location. With three substituents, alphabetize the substituent names, and number to give the lowest set of numbers.

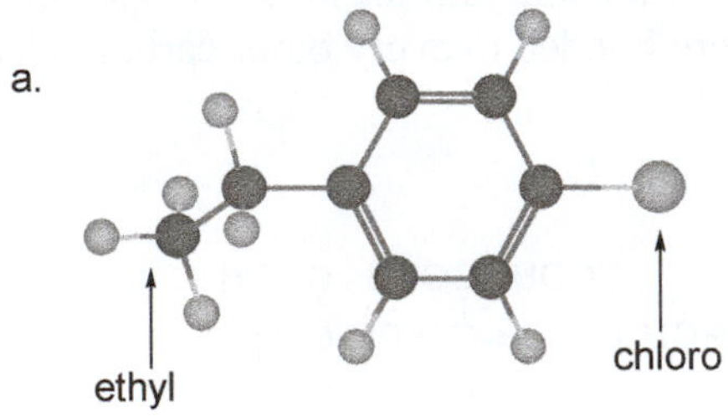

a.

ethyl    chloro

**p-chloroethylbenzene**

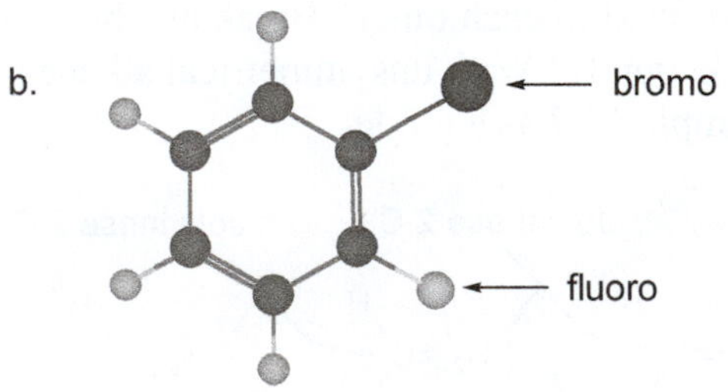

b.

← bromo

← fluoro

**o-bromofluorobenzene**

**13.77** Name each aromatic compound as in Example 13.8 and Answer 13.75.

a.
Cl ← chloro
NO$_2$ ← nitro
***m*-chloronitrobenzene**

b. H$_2$N—⟨ ⟩—NO$_2$ ← nitro
***p*-nitroaniline**
NH$_2$ on benzene ring = aniline

c.
← butyl
← ethyl
***o*-butylethylbenzene**

d.
Cl
OH ← OH on benzene ring = phenol
Cl ← 2,5-dichloro
**2,5-dichlorophenol**

**13.79** Draw and name the three isomers with Cl and NH$_2$ as substituents. Recall that a benzene ring with an NH$_2$ group is named aniline.

*o*-chloroaniline          *m*-chloroaniline          *p*-chloroaniline

**13.81** Work backwards to draw the structure from the IUPAC name.

a.
NO$_2$ ← nitro
CH$_2$CH$_2$CH$_3$ ← propyl
***p*-nitropropylbenzene**

b.
CH$_2$CH$_2$CH$_2$CH$_3$ ← butyl
CH$_2$CH$_2$CH$_2$CH$_3$ ← butyl
***m*-dibutylbenzene**

c.
OH ← OH on benzene ring = phenol
I ← iodo
***o*-iodophenol**

d.
CH$_3$ ← CH$_3$ on benzene ring = toluene
Br ← bromo
Cl ← chloro
**2-bromo-4-chlorotoluene**

e.
NH$_2$ ← NH$_2$ on benzene ring = aniline
Cl
**2-chloro-6-iodoaniline**
iodo          chloro

**13.83** Draw the products of each reaction.

a. CH$_3$—⟨ ⟩—CH$_3$  $\xrightarrow[\text{FeCl}_3]{\text{Cl}_2}$  CH$_3$—⟨ ⟩—CH$_3$ (Cl)

b. CH$_3$—⟨ ⟩—CH$_3$  $\xrightarrow[\text{H}_2\text{SO}_4]{\text{HNO}_3}$  CH$_3$—⟨ ⟩—CH$_3$ (NO$_2$)

c. CH$_3$—⟨ ⟩—CH$_3$  $\xrightarrow[\text{H}_2\text{SO}_4]{\text{SO}_3}$  CH$_3$—⟨ ⟩—CH$_3$ (SO$_3$H)

**13.85** Draw the three products formed in the reaction of bromobenzene.

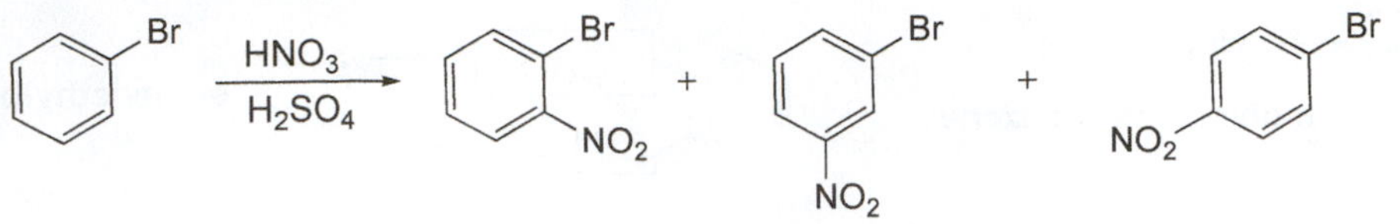

**13.87** Vitamin E is an antioxidant because of the phenol, which has an OH bonded to the benzene ring.

**13.89** Methoxychlor is more water soluble than DDT. The $OCH_3$ groups can hydrogen bond to water. This increase in water solubility makes methoxychlor more biodegradable.

**13.91**

a.

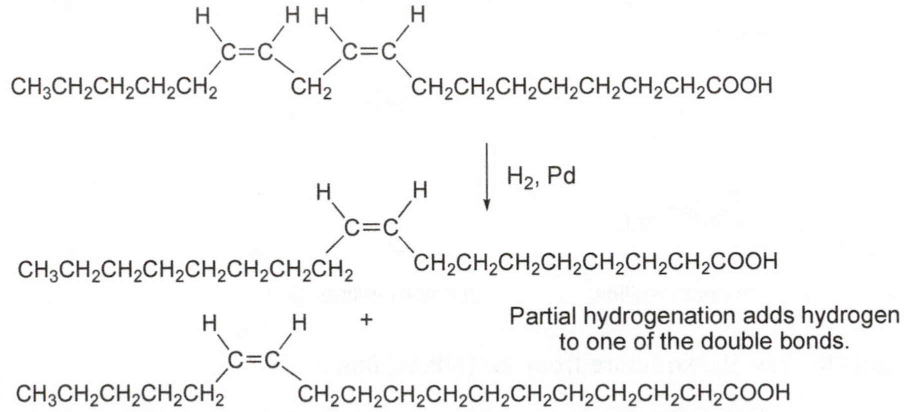

$H_2$, Pd

Partial hydrogenation adds hydrogen to one of the double bonds.

b. Complete hydrogenation adds hydrogen to both of the double bonds, forming:

$$CH_3CH_2CH_2CH_2CH_2CH_2CH_2CH_2CH_2CH_2CH_2CH_2CH_2CH_2CH_2CH_2CH_2COOH$$

one possibility:

c.

trans

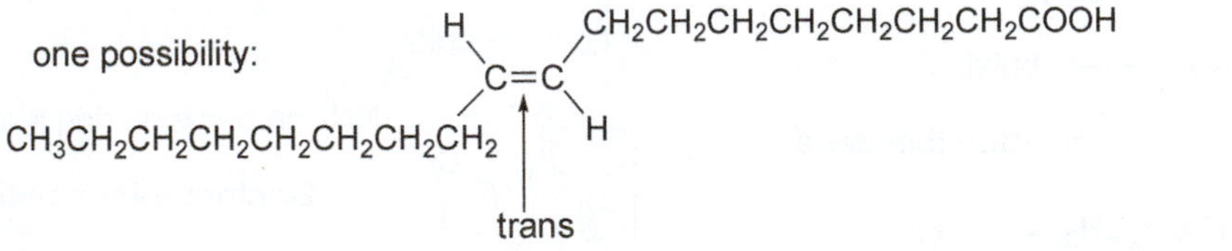

**13.93** Recall from Section 13.12 that many phenols are antioxidants.

a. 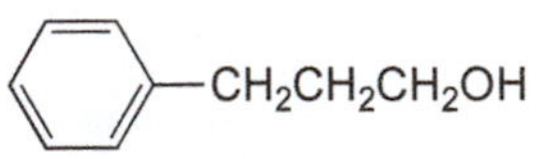 —$CH_2CH_2CH_2OH$

an alcohol
not an antioxidant

b. 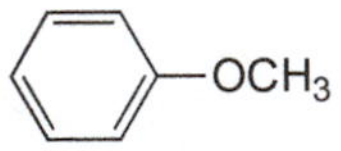 —$OCH_3$

an ether
not an antioxidant

c. 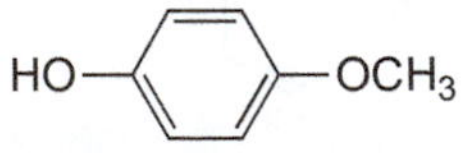 $HO$——$OCH_3$

This compound could be an
antioxidant because it has an OH
group bonded to the aromatic ring.

**13.95** When benzene is oxidized to phenol, it is converted to a more water-soluble compound that can then be excreted in the urine.

**13.97**

a.

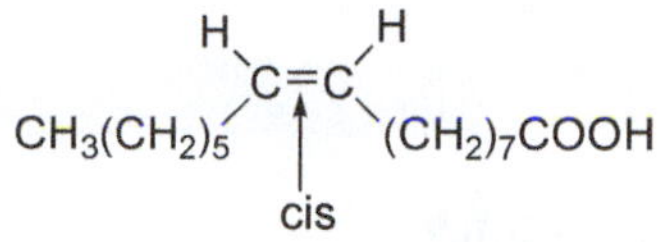

cis
**palmitoleic acid**

b.

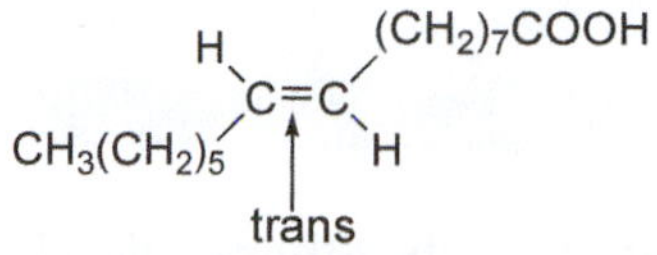

trans
**stereoisomer**

c.

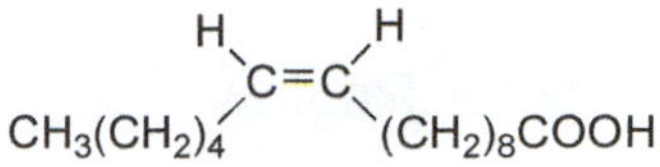

**constitutional isomer**
one possibility

**13.99**  All the carbons in benzene are trigonal **planar** with $120°$ bond angles, resulting in a flat ring. In cyclohexane, all the carbons are tetrahedral, so the ring is puckered.

**13.101**

a. $CH_2=CH(CH_2)_4CH_3$   7 C chain
**1-heptene**

b. $CH_2=CH(CH_2)_4CH_3$ $\xrightarrow{H_2}$ $CH_3(CH_2)_5CH_3$

c. $CH_2=CH(CH_2)_4CH_3$ $\xrightarrow{H_2O}$ $CH_3CH(OH)CH_2CH_2CH_2CH_2CH_3$

d. polymerization:

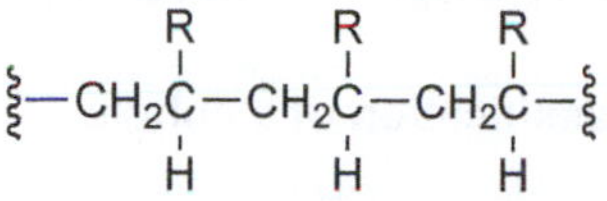

$R = (CH_2)_4CH_3$

**13.103** *cis*-2-Hexene and *trans*-3-hexene are constitutional isomers because the double bond is located in a different place on the carbon chain (C2 vs. C3).

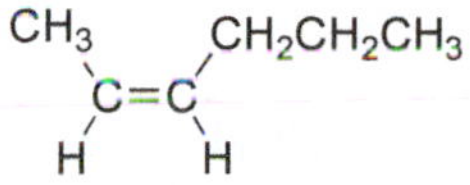

*cis*-2-hexene 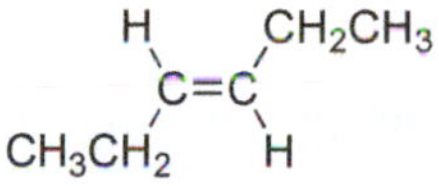          *trans*-3-hexene

# Chapter 14 Organic Compounds That Contain Oxygen, Halogen, or Sulfur

## Chapter Review

**[1] What are the characteristics of alcohols, ethers, alkyl halides, and thiols?**
- Alcohols contain a hydroxyl group (OH group) bonded to a tetrahedral carbon. Since the O atom is surrounded by two atoms and two lone pairs, alcohols have a bent shape around O. (14.2)
- Ethers have two alkyl groups bonded to an oxygen atom. Since the O atom is surrounded by two atoms and two lone pairs, ethers have a bent shape around O. (14.7)
- Alkyl halides contain a halogen atom (X = F, Cl, Br, or I) singly bonded to a tetrahedral carbon. (14.9)
- Thiols contain a sulfhydryl group (SH group) bonded to a tetrahedral carbon. Since the S atom is surrounded by two atoms and two lone pairs, thiols have a bent shape around S. (14.10)

**[2] How are alcohols and alkyl halides classified?**
- Alcohols and alkyl halides are classified by the number of C's bonded to the C with the functional group.
- $RCH_2OH = 1°$ alcohol; $R_2CHOH = 2°$ alcohol; $R_3COH = 3°$ alcohol. (14.2)

**Alcohol**

**Classification of alcohols**

- $RCH_2X = 1°$ alkyl halide; $R_2CHX = 2°$ alkyl halide; $R_3CX = 3°$ alkyl halide. (14.9)

**Alkyl halide**

**Classification of alkyl halides**

X = F, Cl, Br, I

**[3] What are the properties of alcohols, ethers, alkyl halides, and thiols?**
- Alcohols have a bent shape and polar C–O and O–H bonds, so they have a net dipole. Their OH bond allows for intermolecular hydrogen bonding between two alcohol molecules or between an alcohol molecule and water. As a result, alcohols have the strongest intermolecular forces of the four families of molecules in this chapter. (14.2)

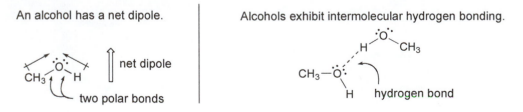

An alcohol has a net dipole.

Alcohols exhibit intermolecular hydrogen bonding.

- Ethers have a bent shape and two polar C–O bonds so they have a net dipole. (14.7)

- Alkyl halides with one halogen have one polar bond and a net dipole. (14.9)

X = F, Cl, Br, I

- Thiols have lower boiling points than alcohols with the same number of carbons. (14.10)

## [4] How are alcohols, ethers, alkyl halides, and thiols named?
- Alcohols are identified by the suffix -ol. (14.3)
- Ethers are named in two ways. Simple ethers are named by naming the alkyl groups bonded to the ether oxygen and adding the word *ether*. More complex ethers are named as *alkoxy alkanes*; that is, the simpler alkyl group is named as an alkoxy group (RO) bonded to an alkane. (14.7B)
- Alkyl halides are named as *halo alkanes*; that is, the halogen is named as a substituent (halo group) bonded to an alkane. (14.9B)
- Thiols are identified by the suffix -*thiol*. (14.10)

## [5] What products are formed when an alcohol undergoes dehydration? (14.5A)
- Alcohols form alkenes on treatment with strong acid. The elements of H and OH are lost from two adjacent atoms and a new carbon–carbon double bond is formed.

- Dehydration follows the Zaitsev rule: When more than one alkene can be formed, the major product of elimination is the alkene that has more alkyl groups bonded to it.

## [6] What products are formed when an alcohol is oxidized? (14.5B)
- Primary alcohols ($RCH_2OH$) are oxidized to aldehydes (RCHO), which are further oxidized to carboxylic acids ($RCO_2H$).

- Secondary alcohols ($R_2CHOH$) are oxidized to ketones.

- Tertiary alcohols have no C–H bond on the carbon with the OH group, so they are not oxidized.

$$R_3COH \xrightarrow{[O]} \text{No reaction}$$

## [7] What product is formed when a thiol is oxidized? (14.10)
- Thiols (RSH) are oxidized to disulfides (RSSR).
- Disulfides are reduced to thiols.

$$2 \; R-S-H \underset{[H]}{\overset{[O]}{\rightleftharpoons}} R-S-S-R$$

$$\text{thiol} \qquad\qquad \text{disulfide}$$

## Problem Solving

## [1] Structure and Properties of Alcohols (14.2)

**Example 14.1** Classify each alcohol as 1°, 2°, or 3°.

a.

b.   $CH_3CH_2CH_2OH$

**Analysis**
To determine whether an alcohol is 1°, 2°, or 3°, locate the C with the OH group and count the number of C's bonded to it. A 1° alcohol has the OH group on a C bonded to one C, and so forth.

**Solution**
Draw out the structure or add H's to the skeletal structure to clearly see how many C's are bonded to the C bearing the OH group.

a.

This C is bonded to 2 C's
in the ring and one $CH_3$.
**3° alcohol**

b.

This C is bonded to 1 C.
**1° alcohol**

## [2] Nomenclature of Alcohols (14.3)

**Example 14.2** Give the IUPAC name of the following alcohol.

$$CH_3CH_2-\underset{\underset{H}{|}}{\overset{\overset{CH_3}{|}}{C}}-CH_2OH$$

**Analysis and Solution**
**[1] Find the longest carbon chain that contains the carbon bonded to the OH group.**

$$CH_3CH_2-\underset{\underset{H}{|}}{\overset{\overset{CH_3}{|}}{C}}-CH_2\text{—OH}$$

4 C's in the longest chain $\text{-----}\rightarrow$ **butanol**

- Change the *-e* ending of the parent alkane to the suffix *-ol*.

**[2] Number the carbon chain to give the OH group the lower number, and apply all other rules of nomenclature.**

a. **Number** the chain.

$$CH_3CH_2-\underset{\underset{2}{\underset{H}{|}}}{\overset{\overset{CH_3}{|}}{C}}-CH_2\text{—OH}$$
4    3         1

1-butanol

b. **Name** and **number** the substituent.

$\overset{\frown}{\phantom{x}}$ methyl at C2

$$CH_3CH_2-\underset{\underset{2}{\underset{H}{|}}}{\overset{\overset{CH_3}{|}}{C}}-CH_2\text{—OH}$$
4    3         1

**Answer: 2-methyl-1-butanol**

## [3] Reactions of Alcohols (14.5)

**Example 14.3** Draw all possible products of dehydration of the following alcohol, and predict which one is the major product.

$$CH_3CH_2-\underset{\underset{OH}{|}}{\overset{\overset{CH_3}{|}}{C}}-CH_2CH_3$$

**Analysis**

To draw the products of dehydration:

- First find the carbon bonded to the OH group, and then identify all carbons with H's bonded to this carbon.
- Remove the elements of H and OH from two adjacent C's, and draw a double bond between these C's in the product.
- When two different alkenes are formed, the major product has more C's bonded to the C=C.

**Solution**

In this example, there are two different C's bonded to the C with the OH. Elimination of H and OH forms two different alkenes. The major product is 3-methyl-2-pentene because it has three C's bonded to the C=C, whereas 2-ethyl-1-butene has only two C's bonded to the C=C.

$$CH_3CH_2-\underset{\underset{OH}{|}}{\overset{\overset{H-C-H}{|}}{C}}-CH_2CH_3 \xrightarrow{H_2SO_4} CH_3CH_2-\underset{}{\overset{\overset{CH_2}{||}}{C}}-CH_2CH_3 \quad + \quad H_2O$$

**2-ethyl-1-butene**
2 C's bonded to the double bond
**minor product**

$$CH_3-\underset{\underset{H}{|}}{\overset{\overset{H}{|}}{C}}-\underset{\underset{OH}{|}}{\overset{\overset{CH_3}{|}}{C}}-CH_2CH_3 \xrightarrow{H_2SO_4} CH_3-\overset{\overset{H}{|}}{C}=\overset{\overset{CH_3}{|}}{C}-CH_2CH_3 \quad + \quad H_2O$$

**3-methyl-2-pentene**
3 C's bonded to the double bond
**major product**

---

**Example 14.4** Draw the carbonyl products formed when each alcohol is oxidized with $K_2Cr_2O_7$.

a. —OH

b. $CH_3CHCH_2OH$
$\qquad\quad\underset{CH_3}{|}$

## Analysis

Classify the alcohol as 1°, 2°, or 3° by drawing in all the H atoms on the C with the OH. Then concentrate on the C with the OH group and replace H atoms by bonds to O. Keep in mind:

- $RCH_2OH$ (1° alcohols) are oxidized to RCHO, which are then oxidized to RCOOH.
- $R_2CHOH$ (2° alcohols) are oxidized to $R_2CO$.
- $R_3COH$ (3° alcohols) are not oxidized since they have no H atom on the C with the OH.

## Solution

a. Since this is a 2° alcohol with only one H atom on the C bonded to the OH group, it is oxidized to a ketone.

only 1 H atom

2° alcohol          ketone

b. $(CH_3)_2CHCH_2OH$ is a 1° alcohol with two H atoms on the C bonded to the OH group. Thus, it is first oxidized to an aldehyde and then to a carboxylic acid.

1° alcohol          aldehyde          carboxylic acid

---

## [4] Ethers (14.7)

**Example 14.5** Rank the following compounds in order of increasing boiling point.

$$CH_3CH_2CH_2-O-CH_3 \qquad CH_3CH_2CH_2CH_2OH \qquad CH_3CH_2CH_2CH_2CH_3$$

**A**                      **B**                   **C**

**Analysis**

Look at the functional groups to determine the strength of the intermolecular forces—**the stronger the forces, the higher the boiling point.**

**Solution**

**C** is an alkane with only nonpolar C–C and C–H bonds, so it has the weakest intermolecular forces and therefore the lowest boiling point. **B** is an alcohol capable of intermolecular hydrogen bonding, so it has the strongest intermolecular forces and highest boiling point. **A** is an ether, so it contains a net dipole but is incapable of intermolecular hydrogen bonding. **A,** therefore, has intermolecular forces of intermediate strength and has a boiling point between the boiling points of **B** and **C.**

$$CH_3CH_2CH_2CH_2CH_3 \qquad CH_3CH_2CH_2-O-CH_3 \qquad CH_3CH_2CH_2CH_2OH$$

**C**                    **A**                   **B**

Increasing intermolecular forces
Increasing boiling point

---

**Example 14.6** Give the IUPAC name for the following ether.

$$CH_3CHCH_2-O-CH_3$$
$$\overset{|}{CH_3}$$

**Analysis and Solution**

[1] Name the longer chain as an alkane and the shorter chain as an alkoxy group.

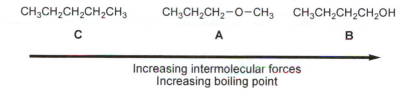

$\boxed{CH_3CHCH_2}-O-CH_3$
$\overset{|}{CH_3}$ ⟶ methoxy group

3 C's ----→ propane

[2] Apply other nomenclature rules to complete the name.

$\overset{3\ \ \ 2\ \ \ 1}{\boxed{CH_3CHCH_2}-O-CH_3}$
$\overset{|}{CH_3}$ ⟵ methyl at C2

**Answer: 1-methoxy-2-methylpropane**

---

## [5] Organic Compounds That Contain Sulfur (14.10)

**Example 14.7** Give the IUPAC name for the following thiol.

$$\overset{\quad CH_3 \quad\ SH}{\underset{\quad |\quad\quad\ |}{CH_3CH_2CHCH_2CHCH_2CH_3}}$$

**Analysis and Solution**

**[1] Find the longest carbon chain that contains the carbon bonded to the SH group.**

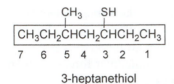

7 C's in the longest chain ----→ **heptane**

- Name the alkane and add the suffix -*thiol*: **heptanethiol**.

---

**[2] Number the carbon chain to give the SH group the lower number and apply all other rules of nomenclature.**

a. **Number** the chain.

$$CH_3CH_2CHCH_2CHCH_2CH_3$$
7  6  5  4  3  2  1

3-heptanethiol

b. **Name** and **number** the substituent.

methyl at C5 → $CH_3$ SH
$$CH_3CH_2CHCH_2CHCH_2CH_3$$
5     3

**Answer: 5-methyl-3-heptanethiol**

---

## Self-Test

**[1] Fill in the blank with one of the terms listed below.**

Alcohols (14.1)
Alkyl halides (14.1)
Chlorofluorocarbons (14.9)
Dehydration (14.5)

Disulfides (14.10)
Elimination (14.5)
Ethers (14.1)
Glycols (14.3)

Heterocycle (14.7)
Oxidation (14.5)
Sulfhydryl group (14.1)
Thiols (14.1)

1. _____ are organic compounds that have two alkyl groups bonded to an oxygen atom.
2. _____ are compounds that contain a sulfur–sulfur bond.
3. _____ results in an increase in the number of C–O bonds or a decrease in the number of C–H bonds.
4. Loss of $H_2O$ from a starting material is called _____.
5. _____ contain a sulfhydryl group (SH group) bonded to a tetrahedral carbon atom.
6. _____ contain a hydroxyl group (OH group) bonded to a tetrahedral carbon atom.
7. _____ are organic molecules containing a halogen atom bonded to a tetrahedral carbon atom.
8. A ring that contains a heteroatom is called a _____.
9. _____ is a reaction in which elements of the starting material are "lost" and a new multiple bond is formed.
10. Compounds with two hydroxyl groups are called diols (using the IUPAC system) or _____.
11. A _____ is the SH group.
12. _____ are simple halogen-containing compounds having the general molecular formula $CF_xCl_{4-x}$.

**[2] Classify each alkyl halide as 1°, 2°, or 3°.**

a. 1° alkyl halide     b. 2° alkyl halide     c. 3° alkyl halide

13. CH$_3$CH$_2$CH$_2$–$\overset{\text{H}}{\underset{\text{CH}_3}{\text{C}}}$–Br

15. CH$_3$CH$_2$$\overset{\text{Br}}{\text{CH}}$CH$_2$CH$_3$

14. H–$\overset{\text{Br}}{\underset{\text{CH}_2\text{CH}_2\text{CH}_3}{\text{C}}}$–H

16. CH$_3$CH$_2$$\overset{\text{Br}}{\underset{\text{CH}_3}{\text{C}}}$CH$_2$CH$_2$CH$_2$CH$_3$

## [3] Match the reactants with the products.

17. CH$_3$CH$_2$CH$_2$–$\overset{\text{H}}{\underset{\text{CH}_3}{\text{C}}}$–OH $\xrightarrow{\text{H}_2\text{SO}_4}$

a. CH$_3$CH$_2$CH$_2$–$\overset{\text{O}}{\overset{\|}{\text{C}}}$–CH$_3$

18. CH$_3$CH$_2$CH$_2$–$\overset{\text{H}}{\underset{\text{CH}_3}{\text{C}}}$–OH $\xrightarrow{\text{[O]}}$

b. CH$_3$CH$_2$CH=CHCH$_3$

19. CH$_3$CH$_2$CH$_2$–$\overset{\text{H}}{\underset{\text{H}}{\text{C}}}$–OH $\xrightarrow[\text{(first step)}]{\text{[O]}}$

c. CH$_3$CH$_2$CH=CH$_2$

20. CH$_3$CH$_2$CH$_2$–$\overset{\text{O}}{\overset{\|}{\text{C}}}$–H $\xrightarrow{\text{[O]}}$

d. CH$_3$CH$_2$CH$_2$–$\overset{\text{O}}{\overset{\|}{\text{C}}}$–CH$_2$CH$_3$

21. CH$_3$CH$_2$CH$_2$–$\overset{\text{H}}{\underset{\text{H}}{\text{C}}}$–OH $\xrightarrow{\text{H}_2\text{SO}_4}$

e. CH$_3$CH$_2$CH$_2$–$\overset{\text{O}}{\overset{\|}{\text{C}}}$–H

22. CH$_3$CH$_2$CH$_2$–$\overset{\text{OH}}{\underset{\text{H}}{\text{C}}}$–CH$_2$CH$_3$ $\xrightarrow{\text{[O]}}$

f. CH$_3$CH$_2$CH$_2$–$\overset{\text{O}}{\overset{\|}{\text{C}}}$–OH

## [4] Rank the compounds in each group in order of increasing boiling point.

23.     CH$_3$CH$_2$CH$_2$CH$_2$OH          CH$_3$CH$_2$CH$_2$OCH$_3$          CH$_3$CH$_2$CH$_2$CH$_2$CH$_3$

        **A**                       **B**                       **C**

24. CH$_3$CH$_2$–$\overset{\text{OCH}_2\text{CH}_3}{\underset{\text{H}}{\text{C}}}$–CH$_2$CH$_2$CH$_3$     CH$_3$CH$_2$–$\overset{\text{CH}_2\text{CH}_2\text{CH}_3}{\underset{\text{H}}{\text{C}}}$–CH$_2$CH$_2$CH$_3$     CH$_3$CH$_2$CH$_2$–$\overset{\text{OH}}{\underset{\text{H}}{\text{C}}}$–CH$_2$CH$_2$CH$_2$CH$_3$

          **A**                       **B**                       **C**

25. CH$_3$CH$_2$CH$_2$CH$_2$CH(CH$_3$)$_2$     CH$_3$CH$_2$CH$_2$CH(CH$_3$)CH$_2$OH     CH$_3$CH$_2$CH(CH$_3$)CH$_2$OCH$_3$

        **A**                       **B**                       **C**

## Answers to Self-Test

| | | | | |
|---|---|---|---|---|
| 1. Ethers | 6. Alcohols | 11. sulfhydryl group | 16. c | 21. c |
| 2. Disulfides | 7. Alkyl halides | 12. Chlorofluorocarbons | 17. b | 22. d |
| 3. Oxidation | 8. heterocycle | 13. b | 18. a | 23. C < B < A |
| 4. dehydration | 9. Elimination | 14. a | 19. e | 24. B < A < C |
| 5. Thiols | 10. glycols | 15. b | 20. f | 25. A < C < B |

## Solutions to In-Chapter Problems

**14.1** Label the –OH groups, –SH groups, halogens, and ether oxygens in each compound.

b. The OH on the benzene ring of salmeterol is part of a phenol, so it is *not* an alcohol.

**14.2** Identify the additional functional groups in taxol.

**14.3** To determine whether an alcohol is 1°, 2°, or 3°, locate the C with the OH group and count the number of C's bonded to it. A 1° alcohol has the OH group on a C bonded to one C, and so forth, as in Example 14.1.

**14.4**   Use the definition in Example 14.1 and Answer 14.3 to label the hydroxyl groups.

C bonded to 1 C
**1° alcohol**

$HOCH_2-\overset{\overset{OH}{|}}{C}-\overset{\overset{H}{|}}{\underset{\underset{OH}{|}}{C}}-\overset{\overset{OH}{|}}{\underset{\underset{H}{|}}{C}}-\overset{\overset{H}{|}}{\underset{\underset{OH}{|}}{C}}-CH_2OH$

C bonded to 1 C
**1° alcohol**

All other hydroxyl groups are on C's bonded to 2 C's.
**2° alcohols**

**14.5**   Alcohols have stronger intermolecular forces and therefore higher boiling points than hydrocarbons of comparable size and shape.

a.   or

alcohol
**higher boiling point**

b.   $(CH_3)_3C-OH$   or   $(CH_3)_4C$

alcohol
**higher boiling point**

**14.6**   Hydrocarbons are insoluble in water.  Low molecular weight alcohols (< 6 C's) are water soluble. Higher molecular weight alcohols (≥ 6 C's) are insoluble in water.

a.

hydrocarbon
**insoluble**

b.   $(CH_3)_3C-OH$

alcohol with 4 C's
**soluble**

c.   $CH_2=CHCH_2CH_3$

hydrocarbon
**insoluble**

d.

alcohol with 10 C's
**insoluble**

**14.7**   To name alcohols using the IUPAC system, follow the steps in Example 14.2:
[1]  Find the longest carbon chain that contains the carbon bonded to the OH group.
[2]  Number the carbon chain to give the OH group the lower number, and apply all other rules of nomenclature.

a.  $CH_3\overset{\overset{OH}{|}}{C}H(CH_2)_4CH_3$

7 C's in the longest chain
**heptanol**

→ 2-heptanol

b.  $(CH_3CH_2)_2\overset{\overset{OH}{|}}{C}HCHCH_2CH_3$

6 C's in the longest chain
**hexanol**

→ **4-ethyl-3-hexanol**

c.

CH3
(cyclohexane ring with CH3 and OH)
OH
6 C's in the ring
**cyclohexanol**

- - - - →

2 CH3 ← methyl group at C2
(cyclohexane ring)
1 OH ← OH group at C1

- - - - → **2-methylcyclohexanol**

d.  $\boxed{CH_3CH_2CH_2CHCHCH_2CHCH_2CH_3}$ with CH3 and OH substituents and CH2CH3 below
CH3   OH
|
CH2CH3

9 C's in the longest chain
**nonanol**

- - → methyl group at C6 → CH3 ⁵ , OH ← OH group at C3

$\boxed{CH_3CH_2CH_2CHCHCH_2CHCH_2CH_3}$
6 CH2CH3 ₃   1
↑
ethyl group at C5

- - → **5-ethyl-6-methyl-3-nonanol**

**14.8**   Work backwards to draw the structure corresponding to each name.

a. 7,7-dimethyl-4-**octanol**  - - - - - →

8 carbon chain
OH at C4

1 2 3 4 5 6 CH3
C–C–C–C–C–C–C–C
|         |
OH      CH3
two methyl groups at C7

- - - - - →

CH3
|
CH3CH2CH2CHCH2CH2CCH3
|              |
OH          CH3

b. 5-methyl-4-propyl-3-**heptanol**  - - →

7 carbon chain
OH at C3

methyl group at C5
1 2 3 4 CH3 7
C–C–C–C–C–C–C
|    |
HO  CH2CH2CH3
↑
propyl group at C4

- - - - - →

CH3
|
CH3CH2CHCHCHCH2CH3
|    |
HO  CH2CH2CH3

c. 2-ethyl-3-methyl**cyclohexanol**  - - - - →

6 carbon ring
OH at C1

(cyclohexane ring)
1 OH
2 CH2CH3 ← ethyl group at C2
3 CH3
↑
methyl group at C3

d. 1,3-**cyclohexanediol**  - - - - →

6 carbon ring
OH at C1 and C3

OH
1
(cyclohexane ring) 2
3 OH

**14.9**   To draw the products of the dehydration of each alcohol, follow the steps in Example 14.3.

- First find the carbon bonded to the OH group, and then identify all carbons with H's bonded to this carbon.
- Remove the elements of H and OH from two adjacent C's, and draw a double bond between these C's in the product.
- When two different alkenes are formed, the major product has more C's bonded to the C=C.

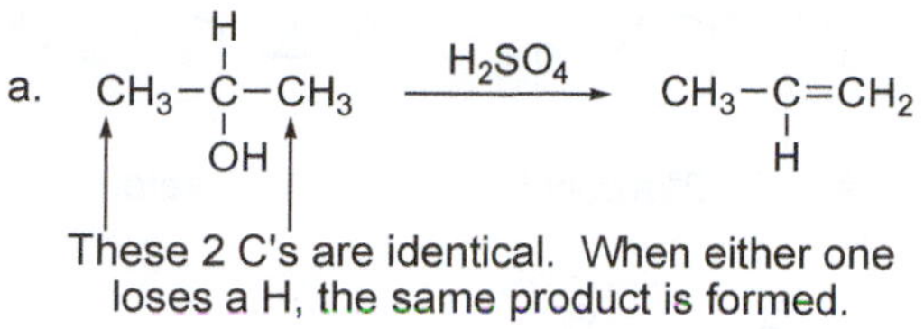

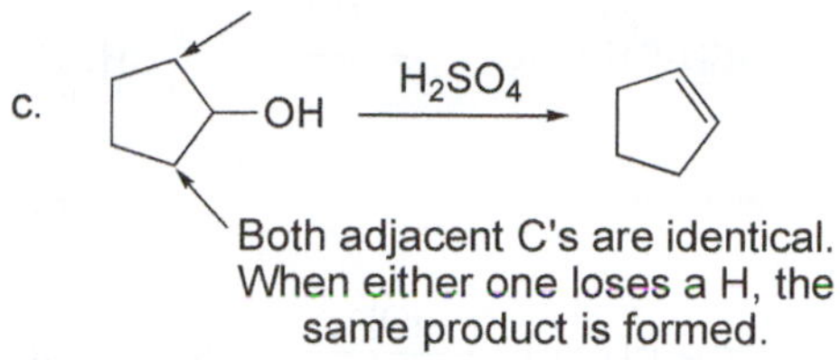

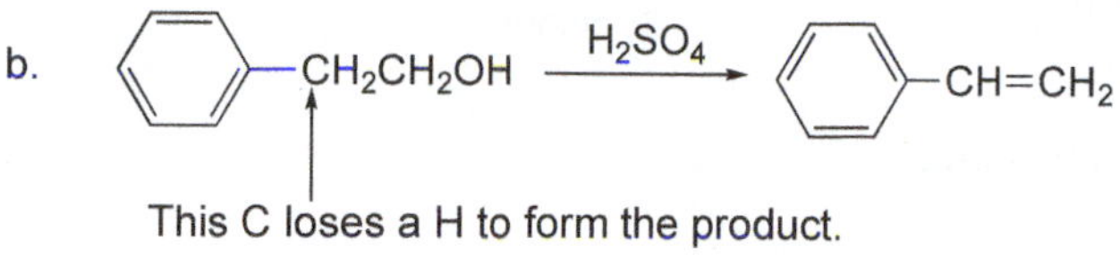

a.

These 2 C's are identical.  When either one
loses a H, the same product is formed.

c.

Both adjacent C's are identical.
When either one loses a H, the
same product is formed.

b.

This C loses a H to form the product.

**14.10**   The Zaitsev rule states that when two different alkenes are formed, the major product has more C's bonded to the C=C.

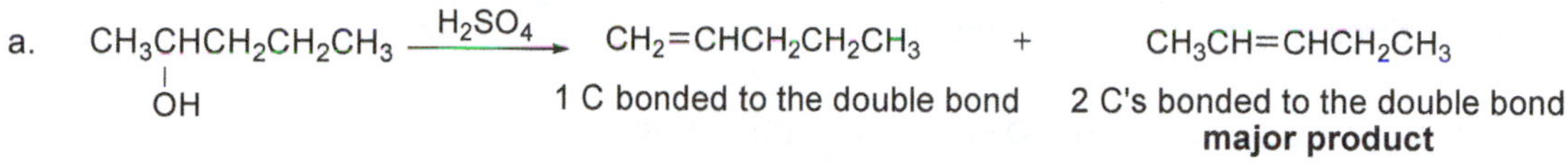

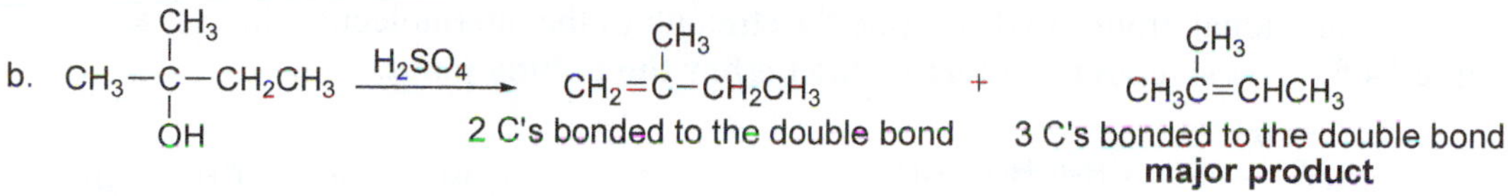

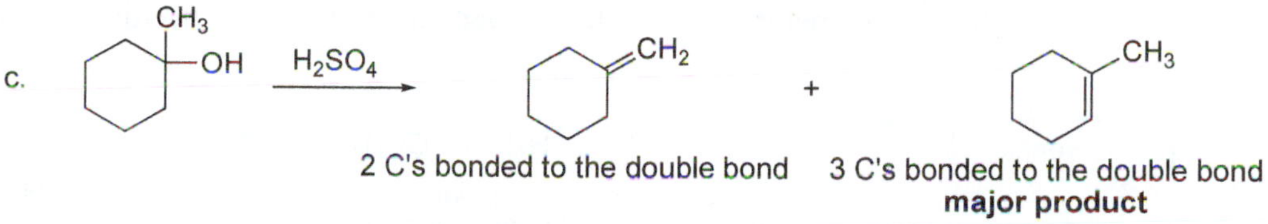

a.   $CH_3CHCH_2CH_2CH_3 \xrightarrow{H_2SO_4} CH_2=CHCH_2CH_2CH_3 \quad + \quad CH_3CH=CHCH_2CH_3$

  $\underset{\text{OH}}{|}$   

  1 C bonded to the double bond    2 C's bonded to the double bond
  **major product**

b.   $CH_3-\underset{\underset{\text{OH}}{|}}{\overset{\overset{\text{CH}_3}{|}}{C}}-CH_2CH_3 \xrightarrow{H_2SO_4} CH_2=\overset{\overset{\text{CH}_3}{|}}{C}-CH_2CH_3 \quad + \quad CH_3\overset{\overset{\text{CH}_3}{|}}{C}=CHCH_3$

  2 C's bonded to the double bond    3 C's bonded to the double bond
  **major product**

c.

  2 C's bonded to the double bond    3 C's bonded to the double bond
  **major product**

**14.11**   **B** has no H on the carbon adjacent to the carbon with the OH group, so $H_2O$ cannot be lost.

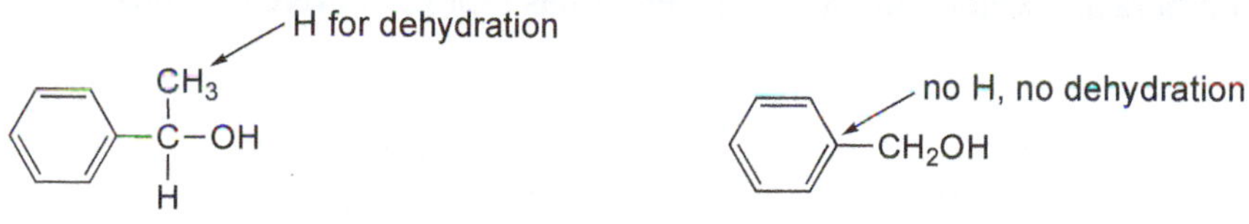

H for dehydration

no H, no dehydration

A                                    B

**14.12**   Draw the products when each alcohol is oxidized, as in Example 14.4.
  • $RCH_2OH$ (1° alcohols) are oxidized to $RCHO$, which are then oxidized to $RCOOH$.
  • $R_2CHOH$ (2° alcohols) are oxidized to $R_2CO$.
  • $R_3COH$ (3° alcohols) are not oxidized since they have no H atom on the C with the OH.

a. $CH_3CHCH_2CH_2CH_3$ $\xrightarrow{[O]}$ $CH_3CCH_2CH_2CH_3$
      |                                        ||
      OH                                       O

    2° alcohol                  ketone

c. (cyclopentane)—OH $\xrightarrow{[O]}$ (cyclopentanone)=O

    2° alcohol                 ketone

b. $(CH_3)_3CCH_2CH_2OH$ $\xrightarrow{[O]}$ $(CH_3)_3CCH_2\overset{\overset{\displaystyle O}{||}}{C}-OH$

    1° alcohol              carboxylic acid

d. (cyclohexane with CH$_3$ and OH) $\xrightarrow{[O]}$ No reaction

    3° alcohol

**14.13** Ethylene glycol is oxidized to a dicarboxylic acid.

    $HOCH_2CH_2OH$ $\xrightarrow{[O]}$ $HO-\overset{\overset{\displaystyle O}{||}}{C}-\overset{\overset{\displaystyle O}{||}}{C}-OH$

    ethylene glycol

**14.14** Draw the three constitutional isomers.

    $CH_3-O-CH_2CH_2CH_3$     $CH_3CH_2-O-CH_2CH_3$     $(CH_3)_2CHOCH_3$

**14.15** Look at the functional groups to determine the strength of the intermolecular forces, as in Example 14.5. **The stronger the forces, the higher the boiling point.**

a.   $CH_3(CH_2)_6CH_3$  or  $CH_3(CH_2)_5OCH_3$

      hydrocarbon           polar
                      **higher boiling point**

c.   $CH_3(CH_2)_6OH$   or   $CH_3(CH_2)_5OCH_3$

    polar, can hydrogen bond       polar
     **higher boiling point**     no hydrogen bonding

b.   (cyclohexane)  or  (tetrahydropyran, ring with O)

     hydrocarbon         polar
                **higher boiling point**

d.   $CH_3(CH_2)_5OCH_3$   or   $CH_3OCH_3$

       polar              polar
    larger molecule
   **higher boiling point**

**14.16** Low molecular weight ethers are water soluble. If the ether has more than five carbons, the ether is insoluble in water.

a.  $CH_3CH_2-O-CH_3$

    3 C ether
   **water soluble**

b. (cyclohexane)—O—(cyclohexane)

    12 C ether
   **water insoluble**

c. (decalin ring with OCH$_3$)

    11 C ether
   **water insoluble**

**14.17** Name each ether as in Example 14.6.

a.  $CH_3-O-\boxed{CH_2CH_2CH_2CH_3}$ $----\rightarrow$ $CH_3-O-\overset{1\ \ 2\ \ 3\ \ 4}{\boxed{CH_2CH_2CH_2CH_3}}$ $----\rightarrow$ **1-methoxybutane**
                                        (or butyl methyl ether)

    4 C's in the longer chain            ↑
        butane               methoxy group

b.

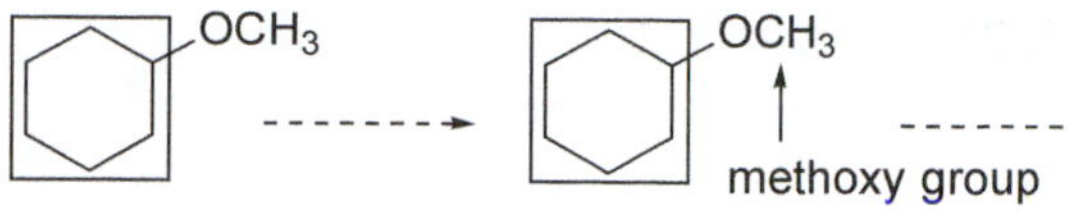

6 C's in the ring
cyclohexane

methoxy group → **methoxycyclohexane**

c. $CH_3CH_2CH_2-O-CH_2CH_2CH_3$ ------→ **dipropyl ether**

two propyl groups

**14.18**   Work backwards from the IUPAC name to the structure.

a. **dibutyl** ether ---------→ $CH_3CH_2CH_2CH_2-O-CH_2CH_2CH_2CH_3$

two butyl groups
on the ether

butyl group        butyl group

b. **ethyl propyl** ether ---------→ $CH_3CH_2CH_2-O-CH_2CH_3$

one ethyl, one propyl
group on the ether

propyl group    ethyl group

c. 1-methoxy**pentane** - - - - - - →

5  4  3  2  1
$CH_3CH_2CH_2CH_2CH_2$ $O-CH_3$

5 carbon chain

methoxy group

d. 3-ethoxy**hexane** - - - - - - →

$OCH_2CH_3$ ← ethoxy group at C3

$CH_3CH_2CHCH_2CH_2CH_3$

6 carbon chain

1  2  3        6

**14.19**   Draw the three-dimensional structure of halothane.

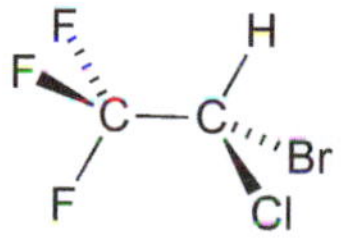

**14.20**   Classify each alkyl halide as 1°, 2°, or 3°.
   • A primary (1°) alkyl halide has a halogen on a carbon bonded to one carbon.
   • A secondary (2°) alkyl halide has a halogen on a carbon bonded to two carbons.
   • A tertiary (3°) alkyl halide has a halogen on a carbon bonded to three carbons.

a.   $CH_3CH_2CH_2CH_2CH_2-Br$

C bonded to 1 C
**1° alkyl halide**

b.

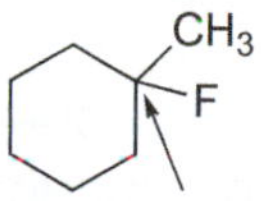

C bonded to 3 C's
**3° alkyl halide**

c. 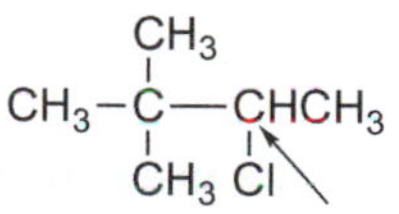   $CH_3-C-CHCH_3$ with $CH_3$ above, $CH_3$ and $Cl$ below

C bonded to 2 C's
**2° alkyl halide**

**14.21** Label the alkyl halide and each alcohol as 1°, 2°, or 3°.

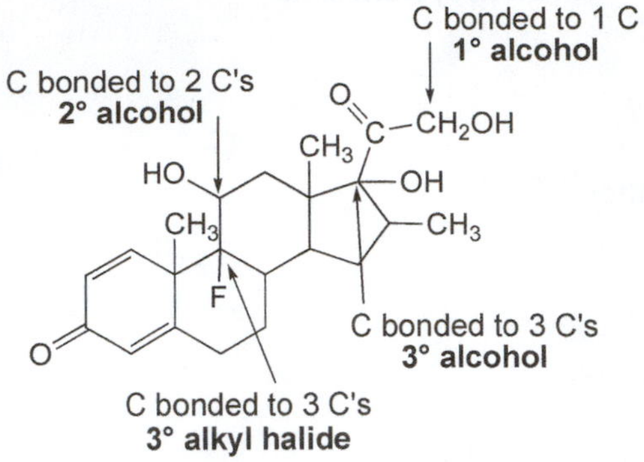

**14.22** The boiling point of alkyl halides increases with the size of the alkyl group as well as the size of the halogen.

a. $CH_3CH_2CH_2F$, $CH_3CH_2CH_2Cl$, $CH_3CH_2CH_2I$

Increasing boiling point
Increasing size of the halogen

b. $CH_3(CH_2)_4CH_3$, $CH_3(CH_2)_5Cl$, $CH_3(CH_2)_5Br$

Increasing boiling point

**14.23** Give the IUPAC name for each compound. Always start by finding the longest chain.

a. $CH_3CHCH_2CH_2CH_2CH_3$ $- - - - \rightarrow$ $CH_3CHCH_2CH_2CH_2CH_3$ $- - - - \rightarrow$ **2-bromohexane**

Br | Br

6 carbon chain | bromo at C2
**hexane**

b. $(CH_3)_2CHCHCH_2CH_3$ $- - \rightarrow$ $CH_3CHCHCH_2CH_3$ $- - \rightarrow$ $CH_3CHCHCH_2CH_3$ $- - \rightarrow$ **3-chloro-2-methylpentane**

Cl | Cl | Cl

$CH_3$ | $CH_3 \leftarrow$ methyl at C2

5 carbon chain | chloro at C3
**pentane**

c.

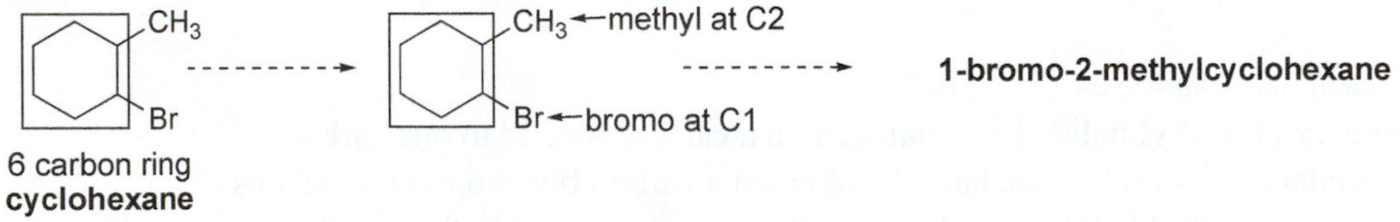

6 carbon ring
**cyclohexane**

$CH_3 \leftarrow$ methyl at C2

Br $\leftarrow$ bromo at C1

**1-bromo-2-methylcyclohexane**

**14.24** Work backwards from the name to the structure.

a. 3-chloro-2-methyl**hexane** $- - - - - - - \rightarrow$

6 carbon chain

Cl $\leftarrow$ chloro at C3

$CH_3CHCHCH_2CH_2CH_3$

$CH_3$

methyl at C2

b. 4-ethyl-5-iodo-2,2-dimethyl**octane**  - - - - - - - - ->

$$CH_3 \quad CH_2CH_3 \longleftarrow \text{ethyl at C4}$$
$$CH_3CCH_2CHCHCH_2CH_2CH_3$$
$$CH_3 \quad\quad I \longleftarrow \text{iodo at C5}$$

**8 carbon chain**

↑ two methyl groups at C2

c. 1,1,3-tribromo**cyclohexane**  - - - - - - - - ->

Br  Br ← two bromo groups at C1

Br ← one bromo group at C3

**6 carbon ring**

d. **propyl** chloride  - - - - - - - - ->

H
CH₃CH₂C—Cl ← one chloro group
H

**3 carbon chain**

**14.25**  **Chlorofluorocarbons (CFCs)** have the general molecular structure $CF_xCl_{4-x}$. **Hydrochlorofluorocarbons (HCFCs)** contain the elements of H, Cl, and F bonded to carbon, and **hydrofluorocarbons (HFCs)** contain the elements of H and F bonded to carbon.

a. $CF_3Cl$ is a CFC.      b. $CHFCl_2$ is an HCFC.      c. $CH_2F_2$ is an HFC.

**14.26**  Name the thiols using the steps in Example 14.7.

a.  $\boxed{CH_3CH_2CHCH_2CH_3}$  - - - - - ->  $\boxed{CH_3CH_2CHCH_2CH_3}$  - - - - - ->  **3-pentanethiol**
      SH                                1   2   SH
                                      SH at C3

**5 C's in the longest chain**
**pentane**

b.  
SH                     SH
                       SH at C1  - - - - - ->  **cyclohexanethiol**

**6 C's in the ring**
**cyclohexane**

c.  $(CH_3CH_2)_2CHCH_2CH_2CH_2CH_2SH$

CH₂CH₃
$\boxed{CH_3CH_2CHCH_2CH_2CH_2}SH$  - - - ->  $\boxed{CH_3CH_2CHCH_2CH_2CH_2}SH$  - - - ->  **5-ethyl-1-heptanethiol**

CH₂CH₃ ← ethyl at C5

5          2   1
SH at C1

**7 C's in the longest chain**
**heptane**

**14.27** Work backwards to draw the structure from the IUPAC name.

trans-2-**butene**-1-thiol ------>
4 C chain

**14.28** When thiols are oxidized, they form disulfides. Disulfides are converted back to thiols with a reducing agent.

a. $2 \ CH_3CH_2CH_2{-}SH \xrightarrow{[O]} CH_3CH_2CH_2{-}S{-}S{-}CH_2CH_2CH_3$

b. $CH_3CH_2CH_2CH_2{-}S{-}S{-}CH_2CH_3 \xrightarrow{[H]} CH_3CH_2CH_2CH_2SH \ + \ CH_3CH_2SH$

## Solutions to Odd-Numbered End-of-Chapter Problems

**14.29** To determine whether an alcohol is 1°, 2°, or 3°, locate the C with the OH group and count the number of C's bonded to it. A 1° alcohol has the OH group on a C bonded to one C, and so forth, as in Example 14.1.

a. $CH_3CH_2CH_2OH$

C bonded to 1 C
**1° alcohol**

b. $(CH_3CH_2)_3COH$

C bonded to 3 C's
**3° alcohol**

c.

C bonded to 2 C's
**2° alcohol**

d. $CH_3CH_2CHCHCH_3$ with OH and $CH_3$

C bonded to 2 C's
**2° alcohol**

**14.31** Classify each alkyl halide as 1°, 2°, or 3°.
- A primary (1°) alkyl halide has a halogen on a carbon bonded to one carbon.
- A secondary (2°) alkyl halide has a halogen on a carbon bonded to two carbons.
- A tertiary (3°) alkyl halide has a halogen on a carbon bonded to three carbons.

a. $CH_3(CH_2)_5CHCH_3$ with Cl

C bonded to 2 C's
**2° alkyl halide**

b. $CH_3CH_2CCH_2CH_3$ with $CH_3$ and Cl

C bonded to 3 C's
**3° alkyl halide**

c. I

C bonded to 2 C's
**2° alkyl halide**

d. $CH_3CH_2{-}C{-}CH_2Br$ with two $CH_3$

C bonded to 1 C
**1° alkyl halide**

**14.33** Draw a structure that fits each description.

a. $CH_3CH_2CHCH_2CH_2CH_3$ with OH

**2° alcohol**

b. $CH_3{-}O{-}CH_2CH_2CH_2CH_2CH_3$

ether with a methoxy group

c. $CH_3CH_2C{-}Br$ with two $CH_3$

**3° alkyl halide**

**14.35** Draw the structure of the six constitutional isomers of molecular formula C$_5$H$_{12}$O that contain an ether functional group.

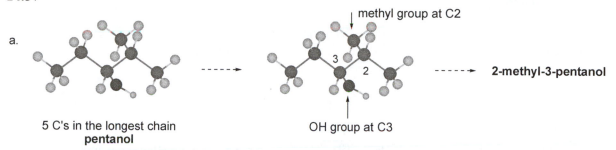

**14.37**

a.

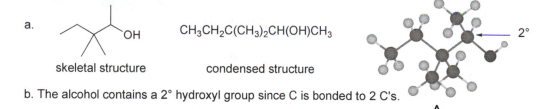

5 C's in the longest chain
**pentanol**

methyl group at C2

OH group at C3

**2-methyl-3-pentanol**

b.

chloro group at C1

5 C's in a ring
**cyclopentane**

methyl group at C2

**1-chloro-2-methylcyclopentane**

**14.39**

a.

skeletal structure

OH

CH$_3$CH$_2$C(CH$_3$)$_2$CH(OH)CH$_3$

condensed structure

2°

A

b. The alcohol contains a 2° hydroxyl group since C is bonded to 2 C's.

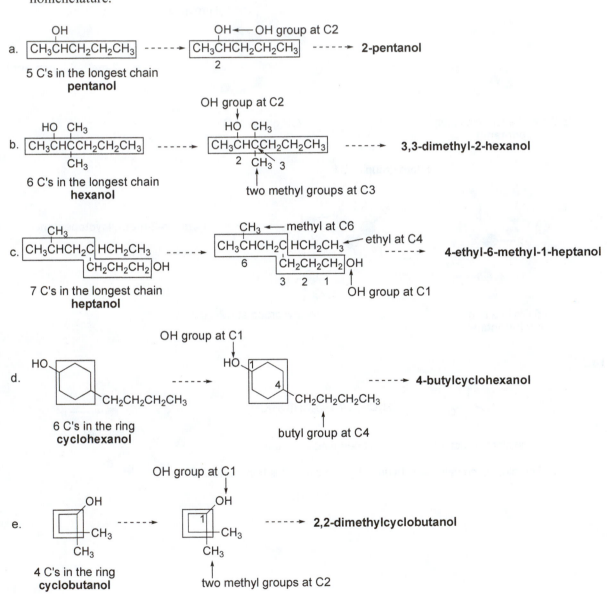

**14.41** To name alcohols using the IUPAC system, follow the steps in Example 14.2:

[1] Find the longest carbon chain that contains the carbon bonded to the OH group.

[2] Number the carbon chain to give the OH group the lower number, and apply all other rules of nomenclature.

f.

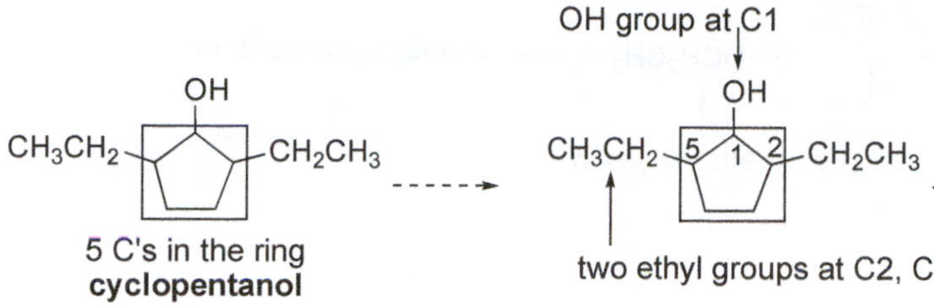

OH group at C1

CH₃CH₂⟨ring with OH⟩CH₂CH₃ — — — ➤ CH₃CH₂−5/1\2−CH₂CH₃ — — — ➤ **2,5-diethylcyclopentanol**

5 C's in the ring
**cyclopentanol**

two ethyl groups at C2, C5

**14.43** Work backwards to draw the structure corresponding to each name.

a. **3-hexanol** — — — — — ➤ 
OH
|
CH₃CH₂CHCH₂CH₂CH₃
3

6 C chain with OH at C3

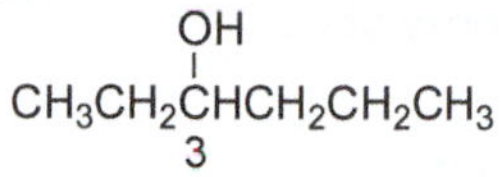

b. **propyl** alcohol — — — — — ➤ CH₂CH₂CH₂OH

3 C chain with OH at C1

c. 2-methyl**cyclopropanol** — — — — — ➤ OH

3 C ring with OH group at C1

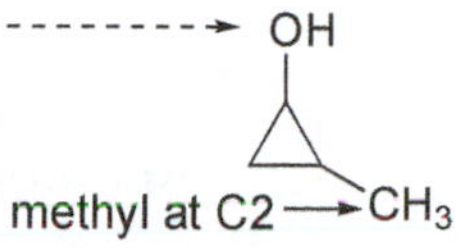

methyl at C2 ➤ CH₃

d. 1,2-**butanediol** — — — — — ➤ 
OH OH
| |
CH₂CHCH₂CH₃

4 C chain with OH groups at C1 and C2

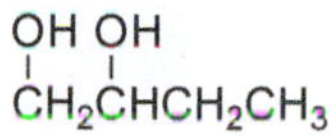

e. 4,4,5-trimethyl-3-**heptanol** — — — ➤ 
CH₃ CH₃ H    3
|   |   |  /
CH₃CH₂−C—C—C−CH₂CH₃
|   |   |
H   CH₃ OH

7 C chain with OH at C3

three methyl groups at C4 and C5

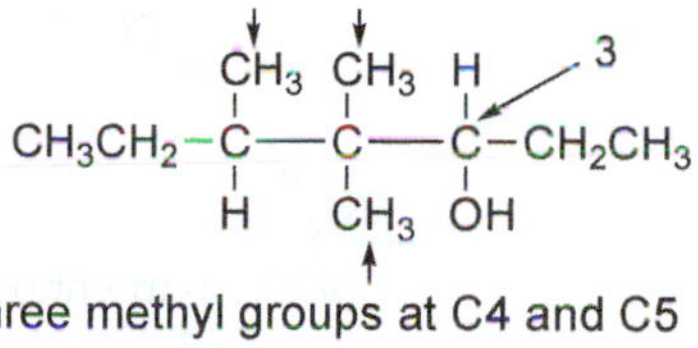

f. 3,5-dimethyl-1-**heptanol** — — — ➤ 
3      5
\  H   H  /
1  |   |
HOCH₂CH₂CCH₂CCH₂CH₃
|      |
CH₃    CH₃

7 C chain with OH at C1

two methyl groups at C3 and C5

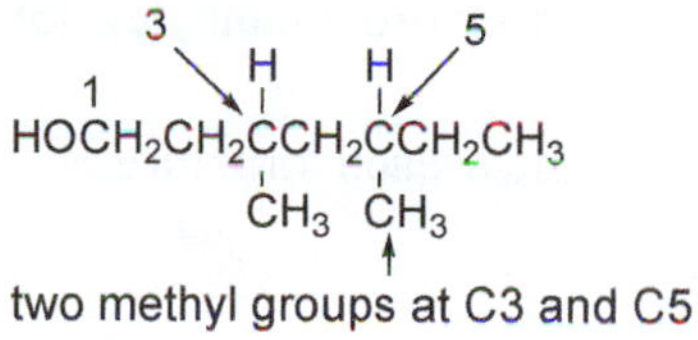

**14.45** Name each ether as in Example 14.6.

a. CH₃CH₂O|CH₂CH₂CH₂CH₃| — — — ➤ 
1  2  3  4
CH₃CH₂O|CH₂CH₂CH₂CH₃| — — — ➤ **1-ethoxybutane**
(or butyl ethyl ether)

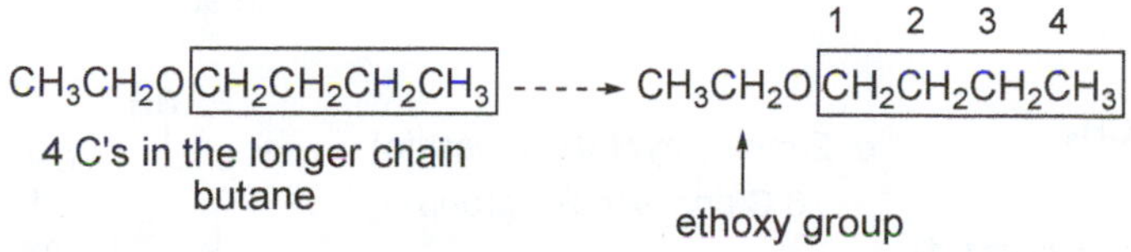

4 C's in the longer chain
butane

ethoxy group

b. 
OCH₂CH₃
|
|CH₃CH₂CH₂CHCH₃| — — — ➤ 
OCH₂CH₃ ⟵ ethoxy group at C2
|
|CH₃CH₂CH₂CHCH₃| — — — ➤ **2-ethoxypentane**
2  1

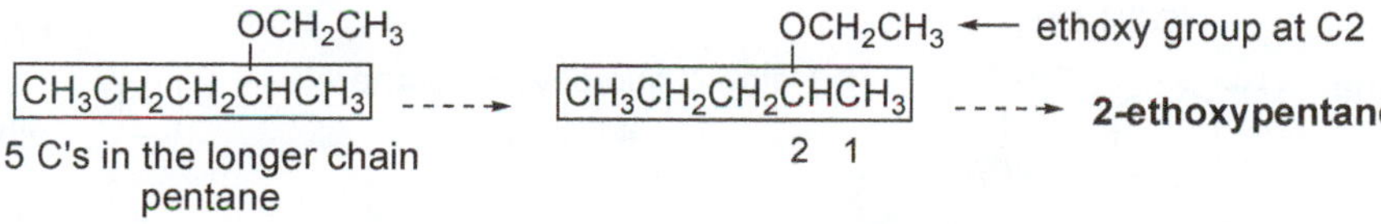

5 C's in the longer chain
pentane

c.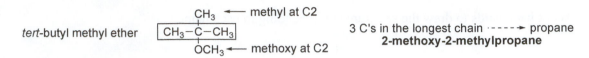

5 C's in the ring
cyclopentane

ethoxy group

ethoxycyclopentane

**14.47** Give the IUPAC name.

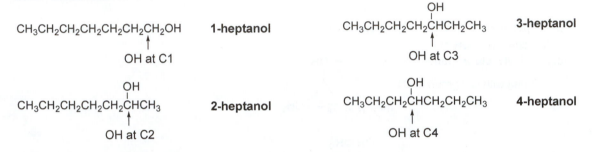

*tert*-butyl methyl ether

CH₃ ← methyl at C2

CH₃-C-CH₃

OCH₃ ← methoxy at C2

3 C's in the longest chain ----→ propane
**2-methoxy-2-methylpropane**

**14.49** Draw the structures and then name the isomers.

CH₃CH₂CH₂CH₂CH₂CH₂CH₂OH    **1-heptanol**

OH at C1

OH
|
CH₃CH₂CH₂CH₂CHCH₂CH₃    **3-heptanol**

OH at C3

OH
|
CH₃CH₂CH₂CH₂CH₂CHCH₃    **2-heptanol**

OH at C2

OH
|
CH₃CH₂CH₂CHCH₂CH₂CH₃    **4-heptanol**

OH at C4

**14.51** Name each compound using the IUPAC system.

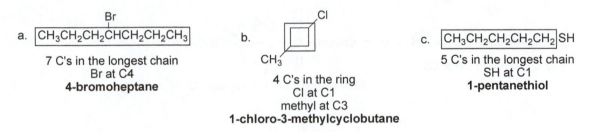

Br
|
a. CH₃CH₂CH₂CHCH₂CH₂CH₃

7 C's in the longest chain
Br at C4
**4-bromoheptane**

b.

Cl

CH₃

4 C's in the ring
Cl at C1
methyl at C3
**1-chloro-3-methylcyclobutane**

c. CH₃CH₂CH₂CH₂CH₂ SH

5 C's in the longest chain
SH at C1
**1-pentanethiol**

**14.53** Work backwards from the name to draw each structure.

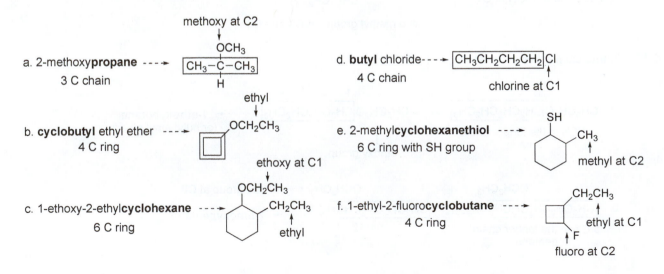

methoxy at C2

OCH₃
|
a. 2-methoxy**propane** ---→ CH₃-C-CH₃
3 C chain                    |
                            H

d. **butyl** chloride ---→ CH₃CH₂CH₂CH₂ Cl
4 C chain
chlorine at C1

ethyl
↓
OCH₂CH₃
b. **cyclobutyl** ethyl ether ---→
4 C ring

e. 2-methyl**cyclohexanethiol** ---→
6 C ring with SH group

SH

CH₃

methyl at C2

ethoxy at C1
↓
OCH₂CH₃
c. 1-ethoxy-2-ethyl**cyclohexane** ---→        CH₂CH₃
6 C ring                                        ↑
                                            ethyl

f. 1-ethyl-2-fluoro**cyclobutane** ---→
4 C ring

CH₂CH₃
↑
ethyl at C1
F
fluoro at C2

**14.55** Alcohols have higher boiling points than hydrocarbons of comparable size and shape. The boiling point of alkyl halides increases with the size of the alkyl group and the size of the halogen.

a. $CH_3CH_2CH_2I$ (larger halogen)  c. $CH_3CH_2CH_2OH$ (can hydrogen bond)
b. $HOCH_2CH_2OH$ (two hydroxyl groups)

**14.57** Ethanol is a polar molecule capable of hydrogen bonding with itself and water. Dimethyl ether is polar, but cannot hydrogen bond with itself. The stronger intermolecular forces make the boiling point of ethanol higher. Both ethanol and dimethyl ether have only two carbons and can hydrogen bond to $H_2O$, so both are water soluble.

**14.59** 1-Butanol is capable of hydrogen bonding with $H_2O$, so 1-butanol is water soluble. 1-Butene is not polar and cannot hydrogen bond with $H_2O$, so 1-butene is water insoluble.

**14.61** To draw the products of dehydration of each alcohol, follow the steps in Example 14.3.

a.
This C loses a H to form the product.

b.
This C loses a H to form the product.

c. 
$CH_3CH{=}CHCH_2CH_2CH_3$  +  $CH_2{=}CH(CH_2)_3CH_3$
2 C's bonded to the double bond    1 C bonded to the double bond
**major product**
Either C can lose a H.

d.
$CH_3CH{=}CCH_2CH_3$  +  $CH_3CH_2CCH_2CH_3$
3 C's bonded to the double bond    2 C's bonded to the double bond
**major product**
Either C can lose a H.

**14.63** Draw the products of dehydration, as in Example 14.3.

a.
$CH_3CH{=}CHCH_2CH_2CH_3$  +  $CH_3CH_2CH{=}CHCH_2CH_3$

b. The carbons of both double bonds are bonded to an equal number of C's; therefore, equal amounts of both isomers are formed.

**14.65** Work backwards to determine what alcohol can be used to form each product.

a. (ring)—CH=CH—(ring) ← (ring)—C(OH)(H)—C(H)(H)—(ring)

b. (cyclooctene) ← (cyclooctanol, —OH)

**14.67** Work backwards to determine what alcohols can be used to form propene. The OH group can be bonded to either the middle or the end carbon.

$CH_3CH=CH_2$ ← $CH_3CH_2CH_2OH$   or   $CH_3\overset{OH}{\underset{|}{C}}HCH_3$

**14.69** Draw the products when each alcohol is oxidized, as in Example 14.4.

- $RCH_2OH$ (1° alcohols) are oxidized to RCHO, which are then oxidized to RCOOH.
- $R_2CHOH$ (2° alcohols) are oxidized to $R_2CO$.
- $R_3COH$ (3° alcohols) are not oxidized since they have no H atom on the C with the OH.

a.  $CH_3(CH_2)_6CH_2OH$  $\xrightarrow{[O]}$  $CH_3(CH_2)_6\overset{O}{\overset{||}{C}}OH$
    1° alcohol                    carboxylic acid

c.  $CH_3CH_2\underset{\underset{CH_3}{|}}{C}HCH_2OH$  $\xrightarrow{[O]}$  $CH_3CH_2\underset{\underset{CH_3}{|}}{\overset{O}{\overset{||}{C}}}HCOH$
    1° alcohol                    carboxylic acid

b.  $(CH_3CH_2)_2CHOH$  $\xrightarrow{[O]}$  $CH_3CH_2\overset{O}{\overset{||}{C}}CH_2CH_3$
    2° alcohol                    ketone

d.  $CH_3CH_2\underset{\underset{CH_3}{|}}{\overset{CH_3}{\overset{|}{C}}}OH$  $\xrightarrow{[O]}$  No reaction
    3° alcohol

**14.71** Draw the product of oxidation.

(cortisol structure) — 3° OH not oxidized  $\xrightarrow{[O]}$  (oxidized product structure)

cortisol

**14.73** Work backwards to determine what alcohol can be used to prepare each carbonyl compound.

a. (cycloheptanone, =O) ← (cycloheptanol, —OH)
   ketone                    2° alcohol

b. $CH_3-\underset{\underset{CH_3}{|}}{\overset{CH_3}{\overset{|}{C}}}-\overset{O}{\overset{||}{C}}-CH_3$  ←  $CH_3-\underset{\underset{CH_3}{|}}{\overset{CH_3}{\overset{|}{C}}}-\underset{\underset{CH_3}{|}}{\overset{OH}{\overset{|}{C}}}H$
   ketone                    2° alcohol

c.

carboxylic acid         1° alcohol

**14.75** Draw the products of each reaction.

a. $CH_3CH_2CH_2\overset{\text{OH}}{\underset{\text{}}{CH}}CH_2CH_2CH_3$ $\xrightarrow{H_2SO_4}$ $CH_3CH_2CH_2CH{=}CHCH_2CH_3$

b. $CH_3CH_2CH_2\overset{\text{OH}}{\underset{\text{}}{CH}}CH_2CH_2CH_3$ $\xrightarrow{K_2Cr_2O_7}$ $CH_3CH_2CH_2\overset{\text{O}}{\overset{\|}{C}}CH_2CH_2CH_3$

**14.77** Draw the disulfides formed by thiol oxidation.

a. (cyclohexyl)—SH $\xrightarrow{[O]}$ (cyclohexyl)—S—S—(cyclohexyl)      b. $CH_3(CH_2)_4SH$ $\xrightarrow{[O]}$ $CH_3(CH_2)_4S{-}S(CH_2)_4CH_3$

**14.79** Draw the thiols formed by disulfide reduction.

a. $CH_3S{-}SCH_2CH_2CH_3$ $\xrightarrow{[H]}$ $CH_3SH$   +   $HSCH_2CH_2CH_3$

b. $CH_2{=}CHCH_2S{-}SCH_2CH_2CH_3$ $\xrightarrow{[H]}$ $CH_2{=}CHCH_2SH$   +   $HSCH_2CH_2CH_3$

**14.81** Write a balanced equation for the combustion of diethyl ether.

$C_4H_{10}O$  +  $6\,O_2$ $\longrightarrow$ $4\,CO_2$  +  $5\,H_2O$
diethyl ether

**14.83** a. CFCs are chlorofluorocarbons with a general formula of $CF_xCl_{4-x}$; HCFCs have fluorine, chlorine, and hydrogen bonded to carbon; and HFCs have only hydrogen and fluorine bonded to carbon.
b. CFCs destroy the ozone layer whereas HCFCs and HFCs decompose more readily before ascending to the ozone layer.

**14.85** PEG is capable of hydrogen bonding with water, so PEG is water soluble. PVC cannot hydrogen bond to water, so PVC is water insoluble, even though it has many polar bonds.

**14.87** The greater the blood alcohol level, the greater the color change from red-orange to green with the Breathalyzer.

$CH_3CH_2OH$  +  $K_2Cr_2O_7$ $\longrightarrow$ $CH_3\overset{\text{O}}{\overset{\|}{C}}OH$  +  $Cr^{3+}$
ethanol          (orange)                              (green)

**14.89** Draw the product of the oxidation of propylene glycol.

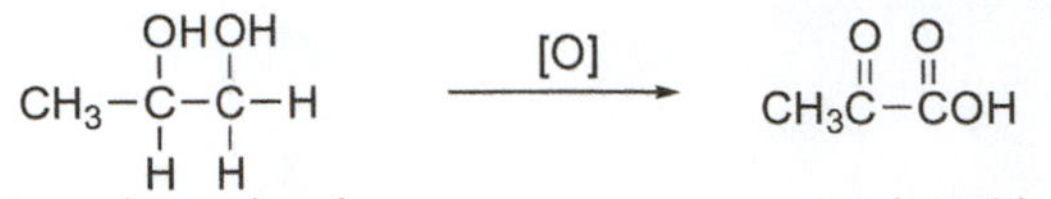

$$CH_3-\underset{\underset{H}{|}}{\overset{\overset{OH}{|}}{C}}-\underset{\underset{H}{|}}{\overset{\overset{OH}{|}}{C}}-H \quad \xrightarrow{[O]} \quad CH_3\overset{\overset{O}{||}}{C}-\overset{\overset{O}{||}}{C}OH$$

propylene glycol                         pyruvic acid

**14.91** Draw the oxidation products formed from ethanol. Antabuse blocks the conversion of acetaldehyde to acetic acid, and the accumulation of acetaldehyde makes people ill.

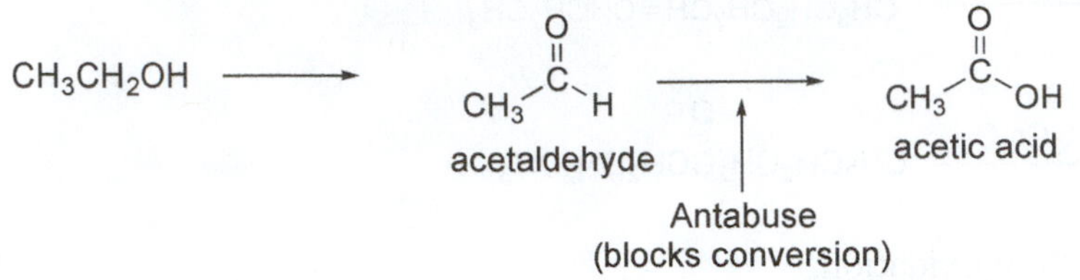

$$CH_3CH_2OH \longrightarrow CH_3\overset{\overset{O}{||}}{C}H \longrightarrow CH_3\overset{\overset{O}{||}}{C}OH$$

acetaldehyde                         acetic acid

Antabuse
(blocks conversion)

**14.93** Draw the structure of the alcohol that fits the description. Since no reaction occurs with an oxidizing agent, the compound must be a 3° alcohol. Since only one alkene is formed when treated with sulfuric acid, all carbons adjacent to the C–OH must be identical.

$$(CH_3)_3COH$$

3° alcohol
All 3 C's adjacent to the C–OH are identical.

**14.95** Ether molecules have no H's on O's capable of hydrogen bonding to other ether molecules, but the hydrogens in water can hydrogen bond to the oxygen of the ether.

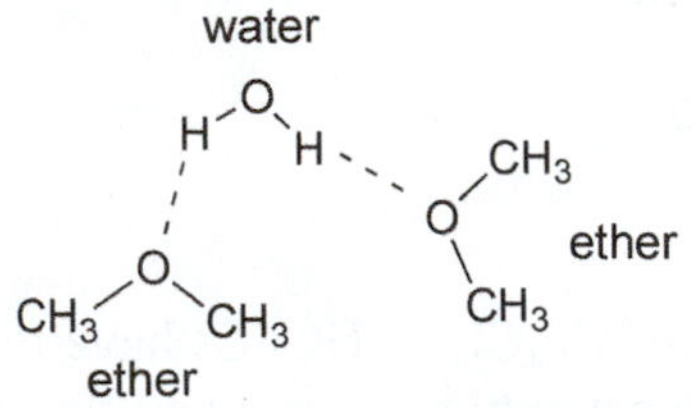

water

ether

ether

**14.97** Answer each question about the alcohol.

a. and b. 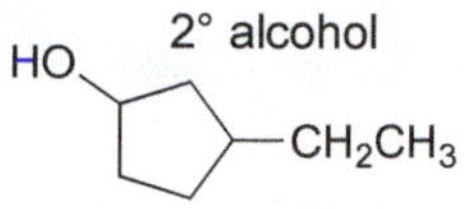   HO    2° alcohol

—CH₂CH₃

5 C's in the ring = cyclopentanol
ethyl group at C3
**3-ethylcyclopentanol**

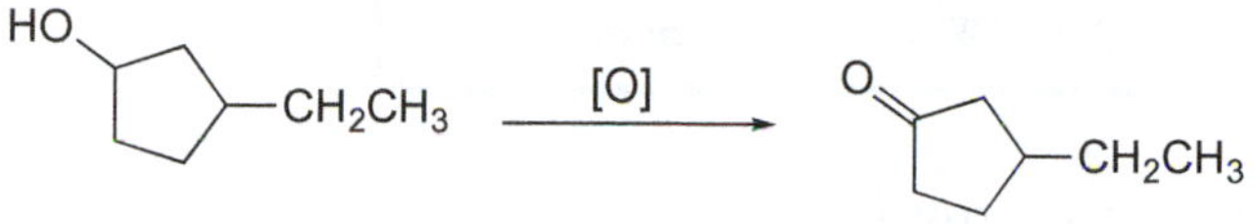

e. 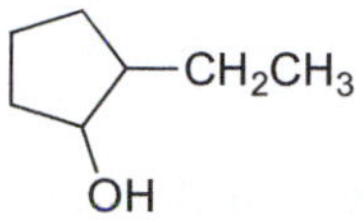   [cyclopentane ring]—CH₂CH₃   constitutional isomer
     OH

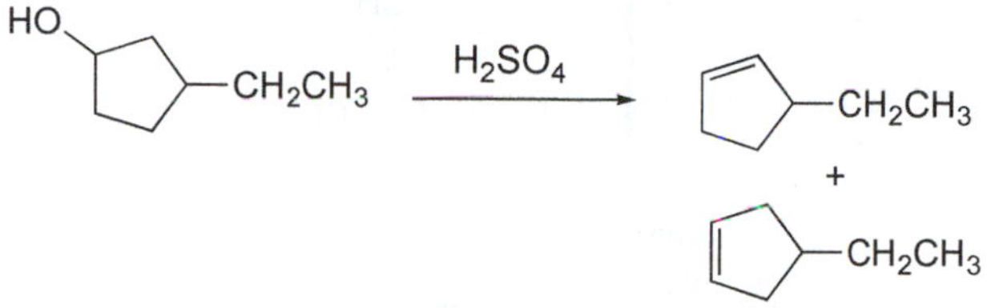

f. 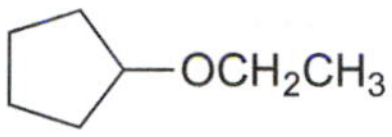   [cyclopentane ring]—OCH₂CH₃   constitutional isomer

c.   HO [cyclopentane]—CH₂CH₃   $\xrightarrow{H_2SO_4}$   [cyclopentene]—CH₂CH₃

+

[cyclopentene]—CH₂CH₃

d.   HO [cyclopentane]—CH₂CH₃   $\xrightarrow{[O]}$   O=[cyclopentanone]—CH₂CH₃

**14.99**

a. and b.   **A**   CH₃CH₂ $\overset{\overset{\displaystyle CH_3}{|}}{\underset{\underset{\displaystyle H}{|}}{C}}$ CH₂OH   $\xrightarrow{H_2SO_4}$   CH₃CH₂ $\overset{\overset{\displaystyle CH_3}{|}}{C}$ =CH₂

**2-methyl-1-butene**

**B**   CH₃CH₂ $\overset{\overset{\displaystyle CH_3}{|}}{\underset{\underset{\displaystyle OH}{|}}{C}}$ CH₃   $\xrightarrow{H_2SO_4}$   CH₃CH₂ $\overset{\overset{\displaystyle CH_3}{|}}{C}$ =CH₂   +   CH₃CH= $\overset{\overset{\displaystyle CH_3}{|}}{C}$ CH₃

2-methyl-1-butene          **major product**
minor product

# Chapter 15 The Three-Dimensional Shape of Molecules

## Chapter Review

**[1] When is a molecule chiral or achiral? (15.2)**

- A chiral molecule is not superimposable on its mirror image.

$CH_3CHBrCl$

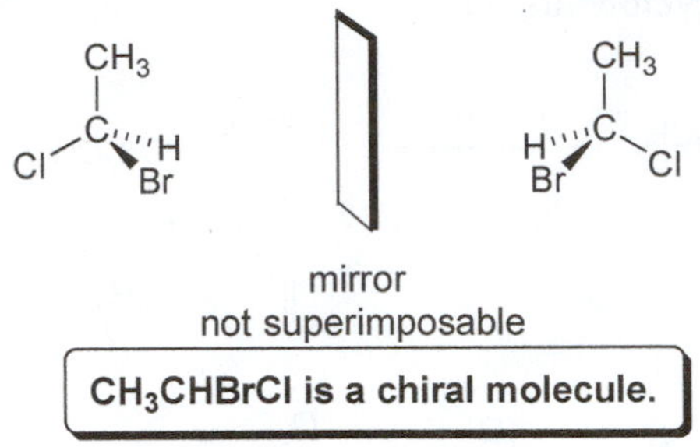

mirror
not superimposable

**$CH_3CHBrCl$ is a chiral molecule.**

- An achiral molecule is superimposable on its mirror image.

$CH_2BrCl$

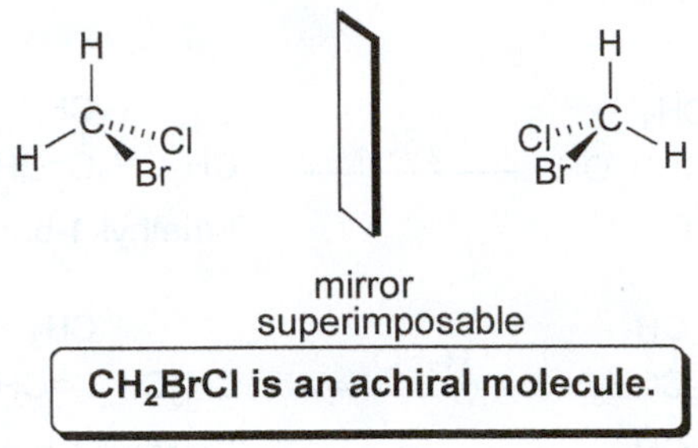

mirror
superimposable

**$CH_2BrCl$ is an achiral molecule.**

- To see whether a molecule is chiral, draw it and its mirror image and see if all the atoms and bonds align.

**[2] What is a chirality center? (15.2, 15.3)**

- A chirality center is a carbon with four different groups around it.

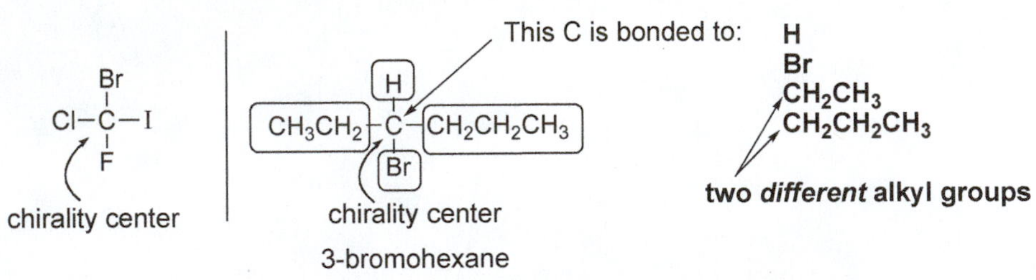

This C is bonded to: H
Br
$CH_2CH_3$
$CH_2CH_2CH_3$

**two *different* alkyl groups**

chirality center

chirality center

3-bromohexane

- A molecule with one chirality center is chiral.

**[3] What are enantiomers? (15.2, 15.3)**

- Enantiomers are mirror images that are not superimposable on each other. To draw two enantiomers, draw one molecule in three dimensions around the chirality center. Then draw the mirror image so that the substituents are a reflection of the groups in the first molecule.

Draw the molecule...then the mirror image.

$$CH_3$$

mirror

not superimposable

**enantiomers**

## [4] Why do some chiral drugs have different properties from their mirror image isomers? (15.5)

- When a chiral drug must interact with a chiral receptor, only one enantiomer fits the receptor properly and evokes a specific response. Ibuprofen, naproxen, and L-dopa are examples of chiral drugs in which the two enantiomers have very different properties.

## [5] What is a Fischer projection? (15.6)

- A Fischer projection is a specific way of depicting a chirality center. The chirality center is located at the intersection of a cross. The horizontal lines represent bonds that come out of the plane on wedges, and the vertical lines represent bonds that go back on dashed lines.

Draw a tetrahedron as:                    Abbreviate it as a cross formula:

Horizontal bonds come forward, on wedges.

Vertical bonds go back, on dashes.

Fischer projection formula

## [6] How do the physical properties of two enantiomers compare? (15.7)

- Two enantiomers have identical physical properties except for their interaction with plane-polarized light.
- The specific rotation of two enantiomers is the same numerical value, but opposite in sign.

## [7] What is the difference between an enantiomer and a diastereomer? (15.8)

- Enantiomers are stereoisomers that are nonsuperimposable mirror images of each other.
- Diastereomers are stereoisomers that are *not* mirror images of each other.

## [8] How is the shape of a molecule related to its odor? (15.9)

- It is thought that the odor of a molecule is determined more by its shape than the presence of a particular functional group, so that compounds of similar shape have similar odors. For an odor to be perceived, a molecule must bind to an olfactory receptor, resulting in a nerve impulse that travels to the brain. Enantiomers may have different odors because they bind with chiral receptors and each enantiomer fits the chiral receptors in a different way.

## Problem Solving

## [1] Chirality Centers (15.3)

**Example 15.1** Locate the chirality centers in each molecule.

a.

$$O=\overset{OH}{\underset{C}{|}}$$
$$CH_3-\overset{|}{\underset{|}{C}}-H$$
$$CH_3-\overset{|}{\underset{|}{C}}-H$$
$$H-\overset{|}{\underset{Cl}{C}}-H$$

b.

(cyclopentyl)$-\overset{H}{\underset{Cl}{C}}-\overset{Cl}{\underset{CH_3}{C}}-H$

### Analysis

In compounds with many C's, look at each C individually and eliminate those C's that can't be chirality centers. Thus, omit all $CH_2$ and $CH_3$ groups and all multiply bonded C's. Check all remaining C's to see if they are bonded to four different groups.

### Solution

a.

$$O=\overset{OH}{\underset{C}{|}}$$
$$CH_3-\overset{|}{\underset{|}{C}}-H \quad \text{chirality center}$$
$$CH_3-\overset{|}{\underset{|}{C}}-H \quad \text{chirality center}$$
$$H-\overset{|}{\underset{Cl}{C}}-H \quad \begin{array}{l}\text{This C is bonded to two H's.}\\ \text{not a chirality center}\end{array}$$

b.

(cyclopentyl)$-\overset{H}{\underset{Cl}{C}}-\overset{Cl}{\underset{CH_3}{C}}-H$

chirality center
chirality center

---

## [2] Fischer Projections (15.6)

**Example 15.2** Draw both enantiomers of the following compound using Fischer projection formulas.

$$HOCH_2\overset{|}{\underset{Br}{C}}HCl$$

### Analysis

- Draw the tetrahedron of one enantiomer with the horizontal bonds on wedges and the vertical bonds on dashed lines. Arrange the four groups on the chirality center—H, Br, Cl, and $CH_2OH$—arbitrarily in the first enantiomer.
- Draw the second enantiomer by arranging the substituents in the mirror image so they are a reflection of the groups in the first molecule.
- Replace the chirality center with a cross to draw the Fischer projections.

## Solution

HOCH$_2$CHCl — chirality center
Br

Draw the compound.

Cl
|
H►C◄Br
|
CH$_2$OH

• Horizontal bonds come forward.
• Vertical bonds go back.

Cl
|
H———Br
|
CH$_2$OH

Fischer projection of one enantiomer

Then, draw the mirror image.

Cl
|
Br►C◄H
|
CH$_2$OH

Cl
|
Br———H
|
CH$_2$OH

Fischer projection of the second enantiomer

## [3] Compounds With Two or More Chirality Centers (15.8)

**Example 15.3** Considering the four stereoisomers of 3,4-heptanediol (**A–D**):

      a. Which compound is an enantiomer of **A?**

      b. Which compound is an enantiomer of **B?**

      c. What two compounds are diastereomers of **C?**

Four stereoisomers of 3,4-heptanediol:

| CH$_2$CH$_3$ | CH$_2$CH$_3$ | CH$_2$CH$_3$ | CH$_2$CH$_3$ |
| H►C◄OH | H►C◄OH | HO►C◄H | HO►C◄H |
| H►C◄OH | HO►C◄H | HO►C◄H | H►C◄OH |
| CH$_2$ | CH$_2$ | CH$_2$ | CH$_2$ |
| CH$_2$ | CH$_2$ | CH$_2$ | CH$_2$ |
| CH$_3$ | CH$_3$ | CH$_3$ | CH$_3$ |
| **A** | **B** | **C** | **D** |

## Analysis

Keep in mind the definitions:

- **Enantiomers are stereoisomers that are nonsuperimposable mirror images of each other.** Look for two compounds that when placed side-by-side have the groups on both chirality centers drawn as a *reflection* of each other. A given compound has only one possible enantiomer.

- **Diastereomers are stereoisomers that are *not* mirror images of each other.**

## Solution

a. **A** and **C** are enantiomers because they are nonsuperimposable mirror images of each other.

b. **B** and **D** are enantiomers for the same reason.

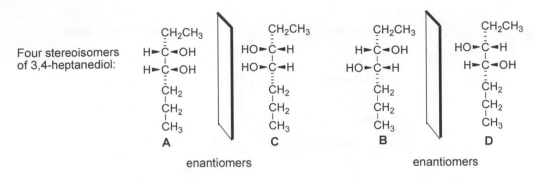

Four stereoisomers of 3,4-heptanediol:

A        C        B        D

enantiomers                    enantiomers

c. Any compound that is a stereoisomer of **C** but not its mirror image is a diastereomer. Thus, **B** and **D** are diastereomers of **C**.

## Self-Test

**[1] Fill in the blank with one of the terms listed below.**

Achiral (15.2)              Diastereomers (15.8)          Stereochemistry (15.1)
Chiral (15.2)               Enantiomers (15.2)            Stereoisomers (15.1)
Chirality center (15.2)     Racemic mixture (15.5)

1. An equal mixture of two enantiomers is called a _____.
2. A molecule (or object) that is *not* superimposable on its mirror image is said to be _____.
3. _____ are stereoisomers but they are *not* mirror images of each other.
4. _____ differ *only* in the three-dimensional arrangement of atoms.
5. A molecule (or object) that *is* superimposable on its mirror image is said to be _____.
6. _____ is the three-dimensional structure of molecules.
7. _____ are mirror images that are not superimposable.
8. A carbon atom surrounded by four different groups is a _____.

**[2] Label the molecules as chiral or achiral.**

9.  $I-\underset{\underset{F}{|}}{\overset{\overset{Br}{|}}{C}}-CH_3$

10. $Cl-\underset{\underset{CH_3}{|}}{\overset{\overset{OH}{|}}{C}}-CH_2CH_3$  $H-\overset{|}{\underset{|}{C}}-H$

11. $Br-\underset{\underset{CH_3}{|}}{\overset{\overset{CH_3}{|}}{C}}-\underset{\underset{H}{|}}{\overset{\overset{Cl}{|}}{C}}-\underset{\underset{CH_3}{|}}{\overset{\overset{CH_3}{|}}{C}}-CH_3$

12. $Br-\underset{\underset{CH_3}{|}}{\overset{\overset{OH}{|}}{C}}-\underset{\underset{H}{|}}{\overset{\overset{Cl}{|}}{C}}-H$

**[3] How are the compounds related? Are they enantiomers or diastereomers?**

A        B        C        D

13. What is the relationship between **A** and **B**?
14. What is the relationship between **A** and **C**?
15. What is the relationship between **B** and **C**?
16. What is the relationship between **C** and **D**?
17. What is the relationship between **A** and **D**?

**[4] How are the compounds in each pair related? Are they identical molecules or enantiomers?**

18.

19.

20.

21.

**[5] Label each object as chiral or achiral.**

22. wine glass  23. pencil  24. iPod  25. face

## Answers to Self-Test

| | | | | |
|---|---|---|---|---|
| 1. racemic mixture | 6. Stereochemistry | 11. chiral | 16. diastereomers | 21. enantiomers |
| 2. chiral | 7. Enantiomers | 12. chiral | 17. enantiomers | 22. achiral |
| 3. Diastereomers | 8. chirality center | 13. diastereomers | 18. enantiomers | 23. achiral |
| 4. Stereoisomers | 9. chiral | 14. diastereomers | 19. enantiomers | 24. chiral |
| 5. achiral | 10. achiral | 15. enantiomers | 20. identical | 25. chiral |

## Solutions to In-Chapter Problems

**15.1** *Constitutional isomers* differ in the way the atoms are connected to one another. *Stereoisomers* differ *only* in the three-dimensional arrangement of atoms.

c. $CH_3CH=CHCH_2CH_3$ and $CH_3CH_2CH=CHCH_3$

identical

d. 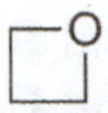 and

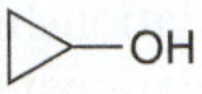

different connection of atoms
constitutional isomers

**15.2** Draw the isomers of *trans*-2-hexene.

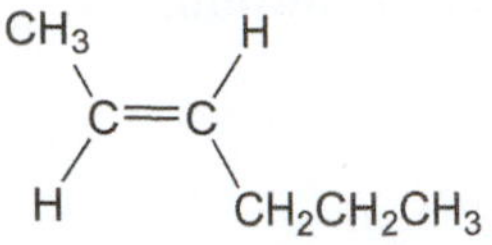

*trans*-2-hexene

a.

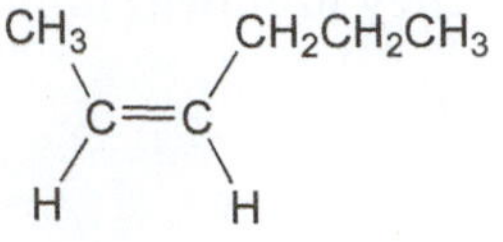

*cis*-2-hexene
stereoisomer

b. $CH_2=CHCH_2CH_2CH_2CH_3$

different connection of atoms
constitutional isomer

c.

different connection of atoms
constitutional isomer

**15.3** A molecule (or object) that is *not* superimposable on its mirror image is said to be *chiral*.
A molecule (or object) that *is* superimposable on its mirror image is said to be *achiral*.

a. nail—achiral
b. screw—chiral

c. glove—chiral
d. pencil—achiral

**15.4** A molecule that is chiral is not superimposable on its mirror image. A chirality center is a carbon with four different groups bonded to it.

**15.5** To locate the chirality centers, look at each C individually and eliminate those C's that can't be chirality centers. Thus, omit all $CH_2$ and $CH_3$ groups and all multiply bonded C's. Check all remaining C's to see if they are bonded to four different groups, as in Example 15.1.

a.

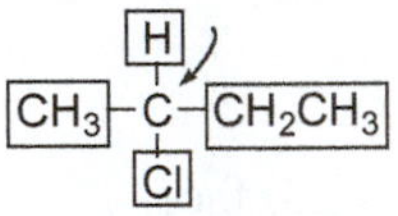

b. $(CH_3)_3CH$

c.

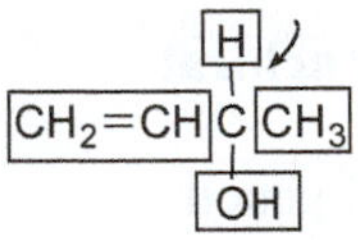

d.

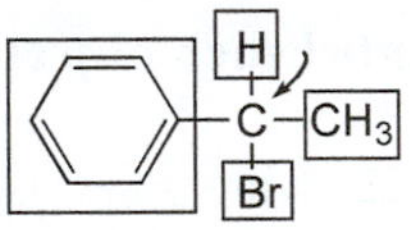

**15.6** Label each chirality center, as in Example 15.1.

a.

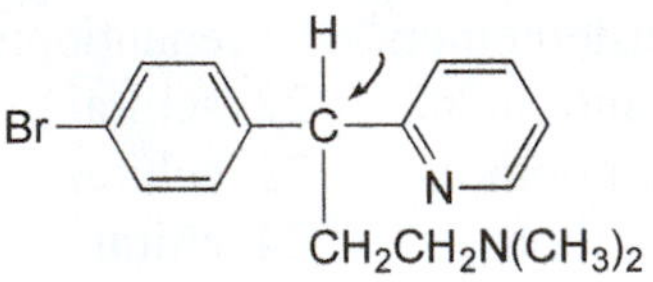

b.

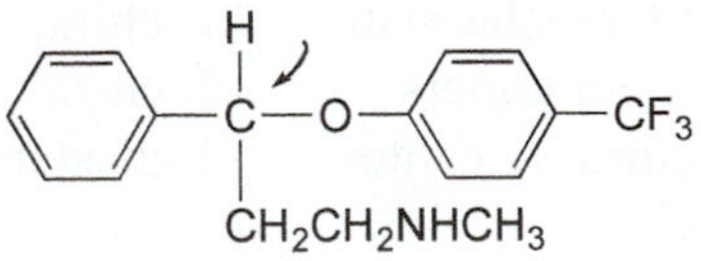

**15.7** Label each chirality center, as in Example 15.1.

a.

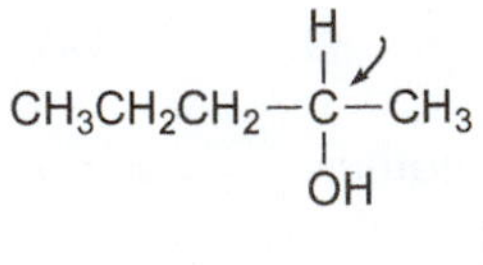

b.

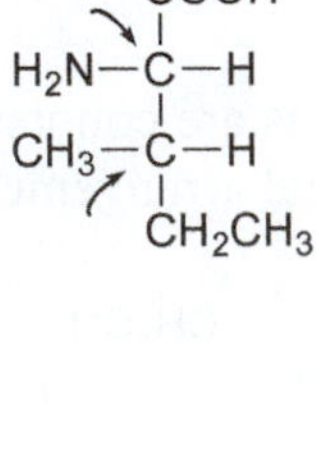

c.

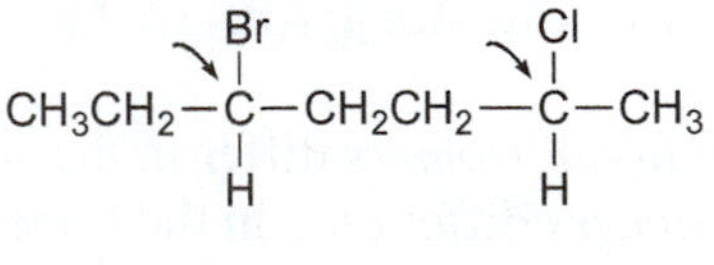

**15.8** Label the two chirality centers in vitamin K.

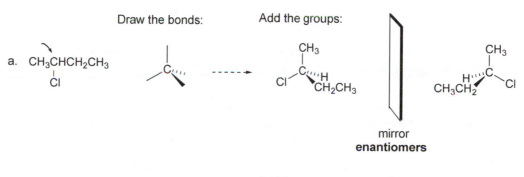

Structure showing naphthoquinone ring with:
$CH_2CH=CCH_2CH_2CH_2CHCH_2CH_2CH_2CHCH_2CH_2CH_2CH(CH_3)_2$ with $CH_3$, $CH_3$, $CH_3$ substituents below.

**15.9** To draw both enantiomers:

[1] Draw one enantiomer by arbitrarily placing the four different groups on the bonds to the chirality center.

[2] Draw a mirror plane and arrange the substituents in the mirror image so that they are a reflection of the groups in the first molecule.

Draw the bonds:        Add the groups:

a.  $CH_3CHCH_2CH_3$
         |
         Cl

mirror
**enantiomers**

Draw the bonds:        Add the groups:

b. $CH_3CH_2CHCH_2OH$
              |
              OH

mirror
**enantiomers**

**15.10** Locate the chirality centers in each compound.

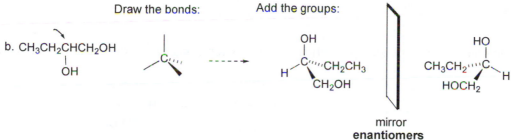

a.  (cyclohexene ring)—OH

b.  (benzene ring with $CH_2CH_3$)

none
All C's are part of multiple
bonds or are bonded to
two or three H's.

c.  (cyclobutane ring with Cl, Cl)

**15.11** Label the two chirality centers.

$CH_3NH$—(structure with tetrahydronaphthalene and dichlorophenyl, Cl, Cl)

**15.12** Answer each question about propranolol.

a.

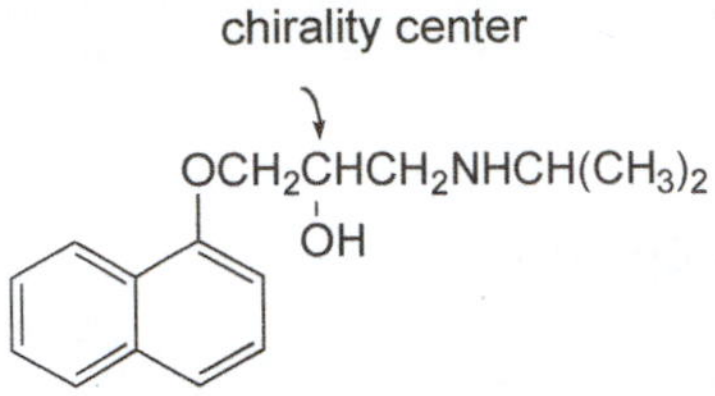

chirality center

OCH$_2$CHCH$_2$NHCH(CH$_3$)$_2$
OH

b.

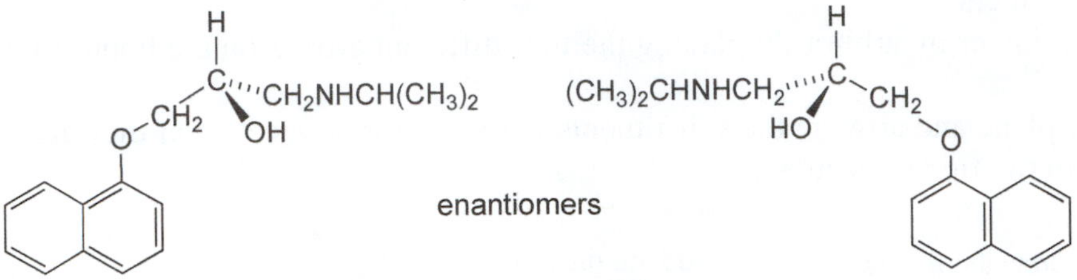

H
C''''CH$_2$NHCH(CH$_3$)$_2$
CH$_2$  OH
O

(CH$_3$)$_2$CHNHCH$_2$''''C
H
CH$_2$
HO
O

enantiomers

c.

receptor

H
C''''CH$_2$NHCH(CH$_3$)$_2$
CH$_2$  OH
O

This enantiomer fits the receptor.

receptor

(CH$_3$)$_2$CHNHCH$_2$''''C
H
CH$_2$
HO
O

This enantiomer does not fit the receptor.

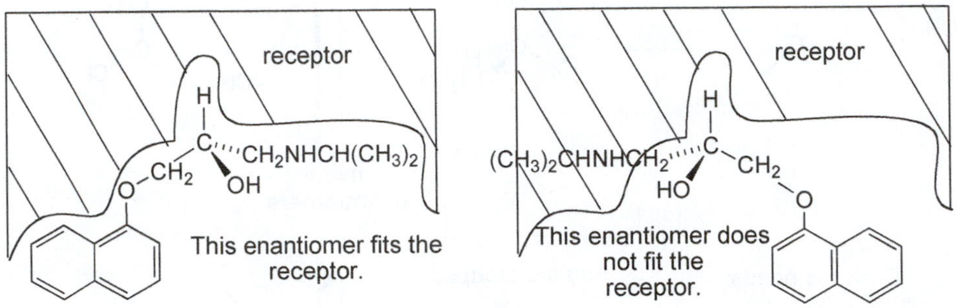

**15.13** Convert each molecule to a Fisher projection formula by replacing the chirality center with a cross.

a.

CH$_3$
H—C—Br   ----->   H—|—Br
Cl                      Cl

b.

COOH
H—C—OH   ----->   H—|—OH
CH$_2$Cl                  CH$_2$Cl

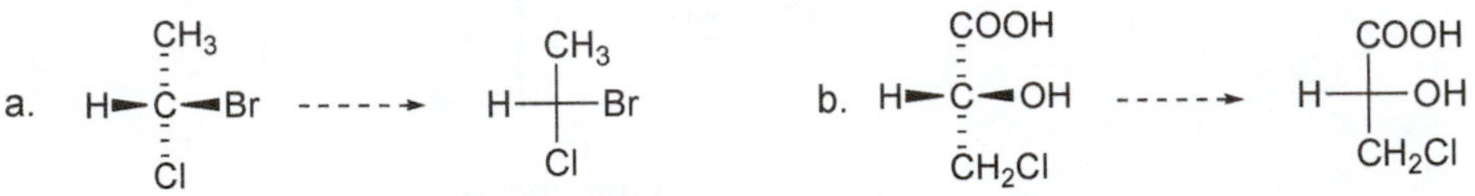

**15.14** Work backwards to draw a structure with wedges (horizontal bonds) and dashes (vertical bonds).

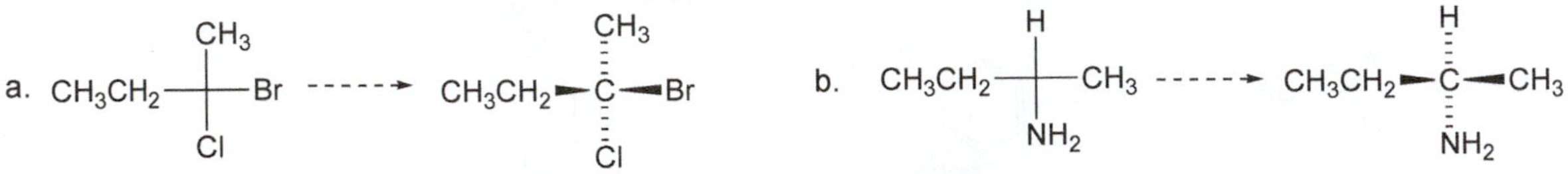

a.

CH$_3$
CH$_3$CH$_2$—|—Br   ----->   CH$_3$CH$_2$—C—Br
Cl                           Cl

b.

H
CH$_3$CH$_2$—|—CH$_3$   ----->   CH$_3$CH$_2$—C—CH$_3$
NH$_2$                         NH$_2$

**15.15** Convert each molecule to a Fisher projection formula as in Example 15.2.
- Draw the tetrahedron of one enantiomer with the horizontal bonds on wedges and the vertical bonds on dashed lines. Arrange the four groups on the chirality center arbitrarily in the first enantiomer.
- Draw the second enantiomer by arranging the substituents in the mirror image so they are a reflection of the groups in the first molecule.
- Replace the chirality center with a cross to draw the Fischer projections.

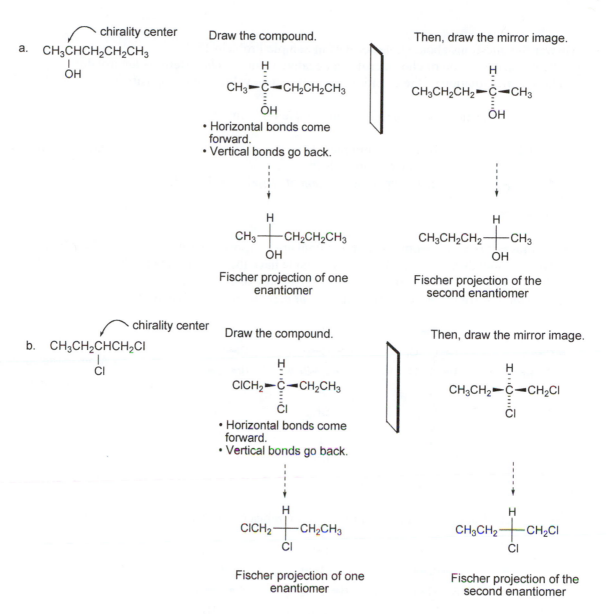

**15.16** A chiral compound is optically active and an achiral compound is optically inactive.

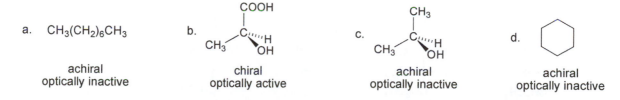

a. $CH_3(CH_2)_6CH_3$

achiral
optically inactive

b.

chiral
optically active

c.

achiral
optically inactive

d.

achiral
optically inactive

**15.17** Because two enantiomers rotate the plane of polarized light to an equal extent but in opposite directions, the rotation of one enantiomer is cancelled by the rotation of the other enantiomer and the sample is optically inactive.

**15.18** Answer the questions about cholesterol as in Sample Problem 15.3.
a. The specific rotation of cholesterol is a negative value, so cholesterol is levorotatory.
b. The specific rotation of the enantiomer has the same value, but is opposite in sign—that is, +32.

**15.19** Answer each question about alanine, as in Sample Problem 15.3.

a.

COO⁻
H⁣⁣⁣C
H₃N⁺  CH₃

(R)-alanine

b. Two enantiomers have the same melting point, so the melting point
of (R)-alanine is 297 °C.
c. The specific rotation of (R)-alanine is –8.5.

**15.20** **Enantiomers are stereoisomers that are nonsuperimposable mirror images of each other.** Look
for two compounds that when placed side-by-side have the groups on both chirality centers drawn as
a *reflection* of each other. A given compound has only one possible enantiomer.
**Diasteromers are stereoisomers that are *not* mirror images of each other.**

$$CH_3 \qquad CH_3 \qquad CH_3 \qquad CH_3$$
$$H–C–Cl \quad H–C–Cl \quad Cl–C–H \quad Cl–C–H$$
$$H–C–Br \quad Br–C–H \quad H–C–Br \quad Br–C–H$$
$$CH_2 \qquad CH_2 \qquad CH_2 \qquad CH_2$$
$$CH_3 \qquad CH_3 \qquad CH_3 \qquad CH_3$$
$$\textbf{W} \qquad\quad \textbf{X} \qquad\quad \textbf{Y} \qquad\quad \textbf{Z}$$

c. **Z** and **W** are diastereomers of **Y**.

d. **X** and **Y** are diastereomers of **Z**.

b. **X** and **Y** are enantiomers.

a. **W** and **Z** are enantiomers.

**15.21** Convert each structure to a Fischer projection by replacing each chirality center with a cross.

$$CH_3 \qquad CH_3 \qquad CH_3 \qquad CH_3$$
$$H–C–Cl \quad H–C–Cl \quad Cl–C–H \quad Cl–C–H$$
$$H–C–Br \quad Br–C–H \quad H–C–Br \quad Br–C–H$$
$$CH_2 \qquad CH_2 \qquad CH_2 \qquad CH_2$$
$$CH_3 \qquad CH_3 \qquad CH_3 \qquad CH_3$$

$$CH_3 \qquad\quad CH_3 \qquad\quad CH_3 \qquad\quad CH_3$$
$$H—\!\!—Cl \quad H—\!\!—Cl \quad Cl—\!\!—H \quad Cl—\!\!—H$$
$$H—\!\!—Br \quad Br—\!\!—H \quad H—\!\!—Br \quad Br—\!\!—H$$
$$CH_2CH_3 \quad CH_2CH_3 \quad CH_2CH_3 \quad CH_2CH_3$$

$$\textbf{W} \qquad\qquad \textbf{X} \qquad\qquad \textbf{Y} \qquad\qquad \textbf{Z}$$

**15.22**   Identify the chirality center and draw the enantiomers.

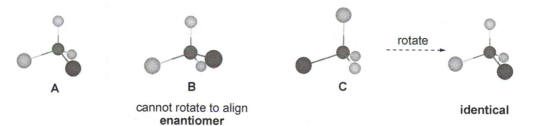

## Solutions to Odd-Numbered End-of-Chapter Problems

**15.23**   Use the definitions in Answer 15.3 to label each object as chiral or achiral.

a. chalk—achiral
b. shoe—chiral

c. baseball glove—chiral
d. soccer ball—achiral

**15.25**   All objects and molecules have a mirror image. Butane does not have an enantiomer since the molecule and its mirror image *are* superimposable, so they are identical.

**15.27**   Draw the mirror image for each compound.

a.

molecule        mirror image
   nonsuperimposable
        chiral

b.

molecule      mirror image
    superimposable
        achiral

**15.29**   **A** and **B** are mirror images of each other—enantiomers.  **A** and **C** are superimposable and identical.

A              B              C              rotate

cannot rotate to align
     **enantiomer**

**identical**

**15.31**   To locate the chirality centers, look at each C individually and eliminate those C's that can't be chirality centers. Thus, omit all $CH_2$ and $CH_3$ groups and all multiply bonded C's. Check all remaining C's to see if they are bonded to four different groups, as in Example 15.1.

a.   $CH_3CH_2CHCl$
                      |
                     $Cl$

two Cl's bonded to end C
  no chirality center

b.   $CH_3CHCHCl_2$
              |
             $Cl$

four different
groups bonded to C

c.   $CH_3CHCH_3$
              |
             $Cl$

two $CH_3$'s bonded
  to middle C
no chirality center

d.   $CH_3CCH_2CHCH_3$
         |         |
        $CH_3$   $Cl$, $H$

four different
groups bonded to C

**15.33** Label each chirality center.

a. ⬡—OH    b. ⬡—OH    c. ⬠—CH₃    d. ⬠—CH₃

    none                                                 none

**15.35** Draw a compound to fit each description.

a.
$$CH_3CH_2-\underset{\underset{H}{|}}{\overset{\overset{CH_3}{|}}{C}}-CH_2CH_2CH_3$$

      C₇H₁₆
one chirality center

b.
$$CH_3CH_2-\underset{\underset{H}{|}}{\overset{\overset{OH}{|}}{C}}-CH_2CH_2CH_3$$

      C₆H₁₄O
one chirality center

**15.37** A chirality center must have a carbon bonded to four groups and a carbonyl carbon has only three groups. Therefore, a carbonyl carbon can never be a chirality center.

**15.39** Locate the chirality centers in each compound.

a. (structure: pyridine ring attached to N-methylpyrrolidine, with chirality center indicated)

b. 
$$HO_2CCH_2\underset{|}{\overset{\overset{NH_2}{|}}{CH}}C\underset{\underset{O}{\|}}{N}H\overset{\overset{CO_2CH_3}{|}}{CH}CH_2-⬡$$

**15.41** Locate the chirality center and draw the enantiomers.

Draw the bonds:      Add the groups:

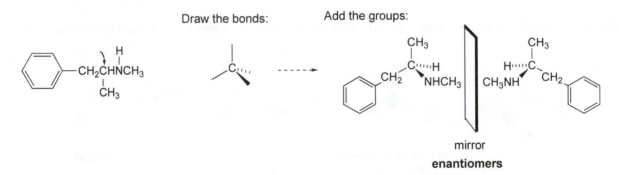

                                                      mirror

                                                 **enantiomers**

**15.43** Enantiomers are stereoisomers that are nonsuperimposable mirror images of each other.

a.
$$H-\underset{\underset{OH}{|}}{\overset{\overset{CH_3}{|}}{C}}\blacktriangleleft CH_3 \quad \text{and} \quad CH_3\blacktriangleright\underset{\underset{OH}{|}}{\overset{\overset{CH_3}{|}}{C}}-H$$

The central C is bonded to two CH₃ groups.
There is no chirality center.
**identical molecules**

b.
$$CH_3\underset{\underset{H}{}}{\overset{\overset{NH_2}{|}}{C}}^{''''}CO_2H \quad \text{and} \quad HO_2C^{''''}\underset{\underset{H}{}}{\overset{\overset{NH_2}{|}}{C}}CH_3$$

The central C is bonded to four different groups.
The compounds are nonsuperimposable
mirror images.
**enantiomers**

c.

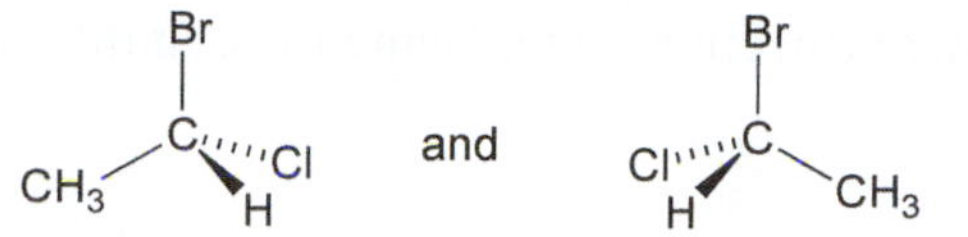

The central C is bonded to four different groups.
The compounds are nonsuperimposable
mirror images.
**enantiomers**

**15.45** Enantiomers are stereoisomers that are nonsuperimposable mirror images of each other. Diastereomers are stereoisomers that are *not* mirror images of each other.

a. $CH_3CH_2CH_2OCH_3$ and $CH_3CH_2OCH_2CH_3$
different connectivity
**constitutional isomers**

c. 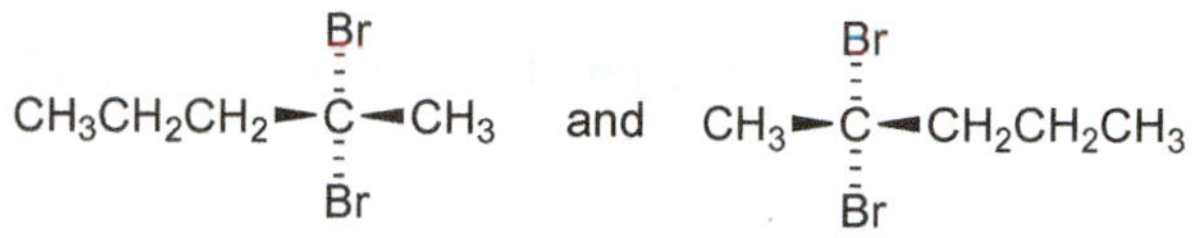 $CH_3CH_2CH_2$—C—$CH_3$ (Br, Br) and $CH_3$—C—$CH_2CH_2CH_3$ (Br, Br)
There is no chirality center.
**identical molecules**

b. 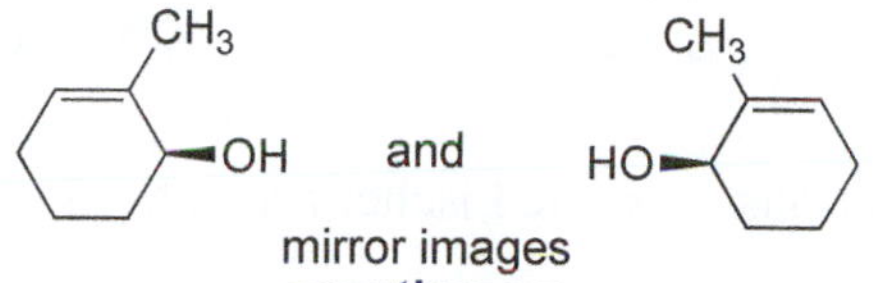
mirror images
**enantiomers**

d.

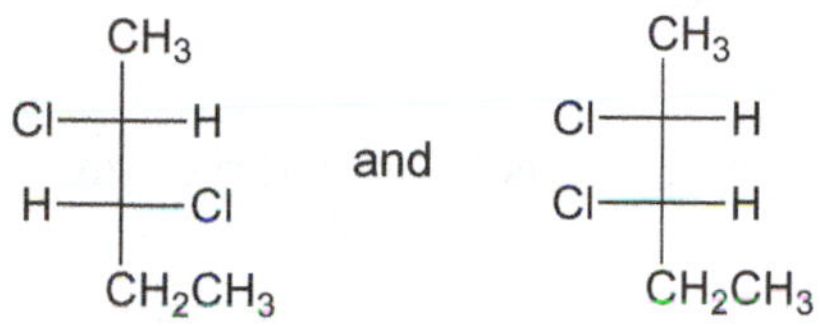

| | $CH_3$ | | | $CH_3$ | |
|---|---|---|---|---|---|
| Cl— | | —H | Cl— | | —H |
| H— | | —Cl | Cl— | | —H |
| | $CH_2CH_3$ | | | $CH_2CH_3$ | |

and

two chirality centers
not mirror images
**diastereomers**

**15.47** Answer each question with a compound of molecular formula $C_5H_{10}O_2$.

a. 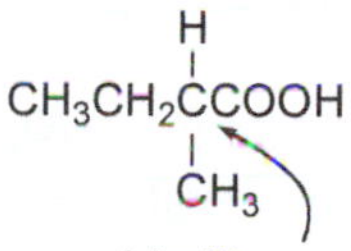  $CH_3CH_2\overset{\displaystyle H}{\underset{\displaystyle CH_3}{C}}COOH$

chirality center

b.  $CH_3CH_2CH_2CH_2COOH$
no chirality center
same molecular formula
**constitutional isomer**

c. 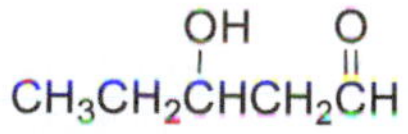  $CH_3CH_2\overset{\displaystyle OH}{C}HCH_2\overset{\displaystyle O}{C}H$
no COOH
same molecular formula
**constitutional isomer**

**15.49** Draw the constitutional isomers and then label the chirality centers.

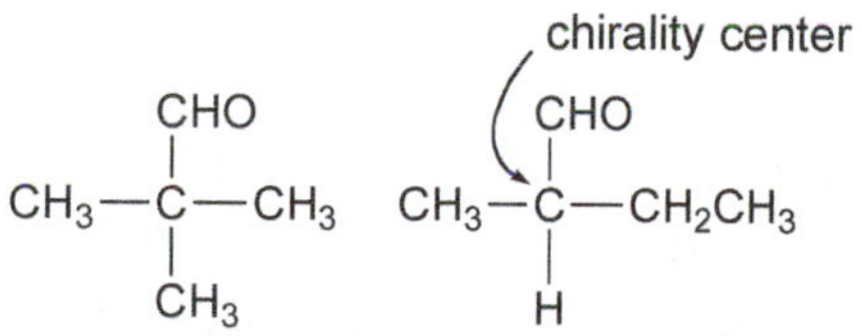

chirality center

$CH_3$—C(CHO)(CH_3)—$CH_3$      $CH_3$—C(CHO)(H)—$CH_2CH_3$      $CH_3CH_2CH_2CH_2CHO$      $CH_3CHCH_2CHO$ ($CH_3$)

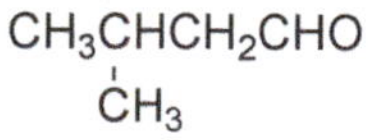

**15.51** Convert the given ball-and-stick model to a Fischer projection by replacing the chirality center with a cross.

**15.53** Convert each molecule to a Fisher projection formula by replacing the chirality center with a cross.

a. H—C—OH ------> H——OH with CHO on top and CH₃ on bottom

b. top OCH₃, CH₃—C—H, CH₃—C—OH, CH₂CH₃ ------> OCH₃, CH₃——H, CH₃——OH, CH₂CH₃

**15.55** Work backwards to draw three-dimensional representations from the Fischer projections.

a. CH₃ on top, H——F, CH(CH₃)₂ on bottom ------> H—C—F with CH₃ on top and CH(CH₃)₂ on bottom

b. COOH on top, H——NH₂, CH₃——H, CH₂CH₃ ------> COOH, H—C—NH₂, CH₃—C—H, CH₂CH₃

**15.57** Enantiomers are stereoisomers that are nonsuperimposable mirror images of each other.

a. CH₃ on top, Cl——H, CH₃ on bottom    and    CH₃ on top, H——Cl, CH₃ on bottom

The central C is bonded to two CH₃ groups.
There is no chirality center.
**identical molecules**

c. CO₂H on top, H——NH₂, CH₂OH on bottom    and    CO₂H on top, H₂N——H, CH₂OH on bottom

mirror images
**enantiomers**

b. COOH on top, CH₃——OH, H on bottom    and    COOH on top, HO——CH₃, H on bottom

mirror images
**enantiomers**

d. C(CH₃)₃ on top, (CH₃)₃C——OH, H on bottom    and    C(CH₃)₃ on top, HO——C(CH₃)₃, H on bottom

The central C is bonded to two C(CH₃)₃ groups.
There is no chirality center.
**identical molecules**

**15.59**

a. An optically active compound rotates the plane of polarized light. An optically inactive compound does not rotate the plane of polarized light.

b. One possibility: CH₃CH₂CH₂CH₂OH is achiral and optically inactive.

c. One possibility:

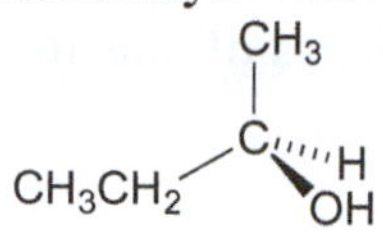

$$CH_3$$

$$CH_3CH_2—C_{,,,,}^H$$
$$OH$$

chiral and optically active

**15.61** A chiral compound is optically active and an achiral compound is optically inactive.

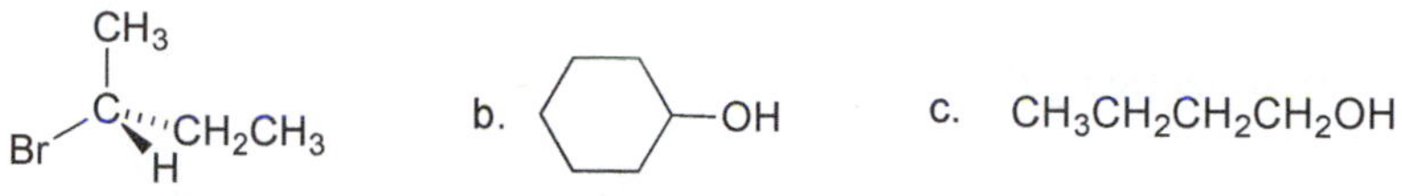

a.
$$CH_3$$
$$Br—C_{,,,}^{CH_2CH_3}$$
$$H$$

chiral
optically active

b. ⬡—OH

achiral
optically inactive

c. $CH_3CH_2CH_2CH_2OH$

achiral
optically inactive

**15.63**

a. Lactose is dextrorotatory.
b. The specific rotation of two enantiomers is the same numerical value, but opposite in sign. Therefore, the specific rotation of the enantiomer of lactose is –52.
c. Two enantiomers have identical physical properties except for their interaction with plane-polarized light, and therefore, 22 g of the enantiomer of lactose dissolves in 100 mL of water.

**15.65** Enantiomers are stereoisomers that are nonsuperimposable mirror images of one another, whereas diastereomers are stereoisomers that are *not* mirror images of one another.

**15.67** **Enantiomers are stereoisomers that are nonsuperimposable mirror images of each other.** Look for two compounds that when placed side-by-side have the groups on both chirality centers drawn as a *reflection* of each other. A given compound has only one possible enantiomer. **Diastereomers are stereoisomers that are *not* mirror images of each other.**

a.
CHO
H—C—OH
H—C—OH
CH₂OH
A

CHO
HO—C—H
H—C—OH
CH₂OH
B

CHO
H—C—OH
HO—C—H
CH₂OH
C

CHO
HO—C—H
HO—C—H
CH₂OH
D

B and C, enantiomers; A and D, enantiomers

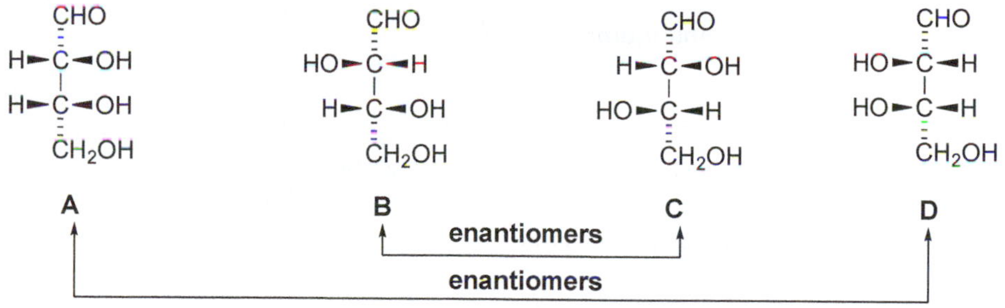

The figure above shows the enantiomers. All other relationships are diastereomers.

[1] **A** and **B**, diastereomers; [2] **A** and **C**, diastereomers; [3] **A** and **D**, enantiomers; [4] **B** and **C**, enantiomers; [5] **B** and **D**, diastereomers; [6] **C** and **D**, diastereomers.

b.
CH₂OH
C=O
H—C—OH
CH₂OH

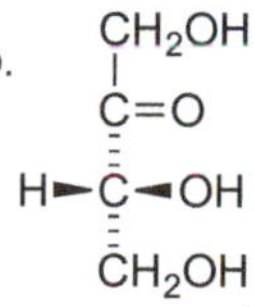

**15.69** Enantiomers are stereoisomers that are mirror images. *cis*-2-Butene and *trans*-2-butene are diastereomers because they are stereoisomers that are not mirror images of one another.

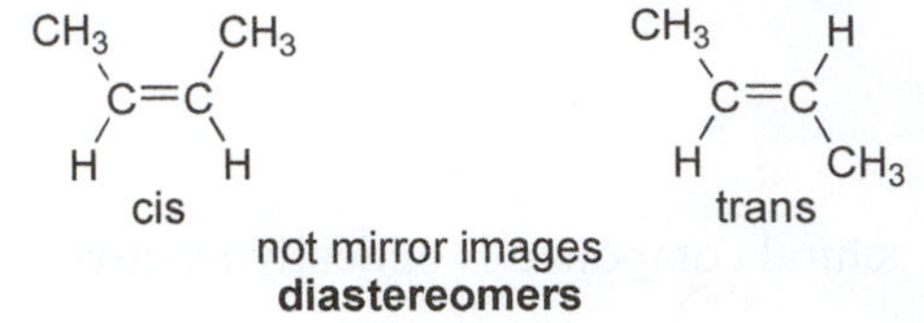

not mirror images
**diastereomers**

**15.71** Answer each question about lactic acid.

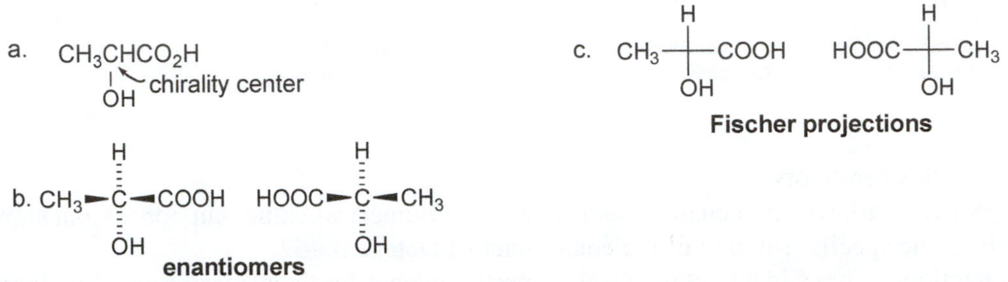

a. $CH_3CHCO_2H$
   chirality center
   OH

**enantiomers**

c. 
**Fischer projections**

**15.73** Answer each question about hydroxydihydrocitronellal.

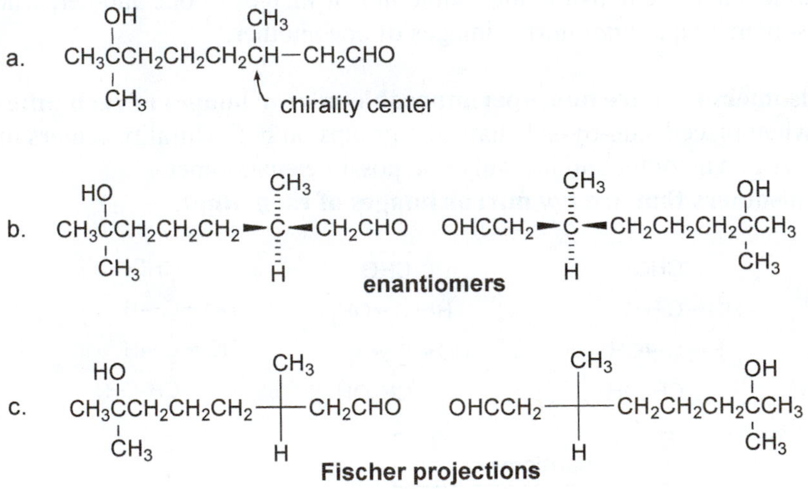

a. 
   chirality center

b. 
**enantiomers**

c. 
**Fischer projections**

**15.75** Answer each question about Darvon.

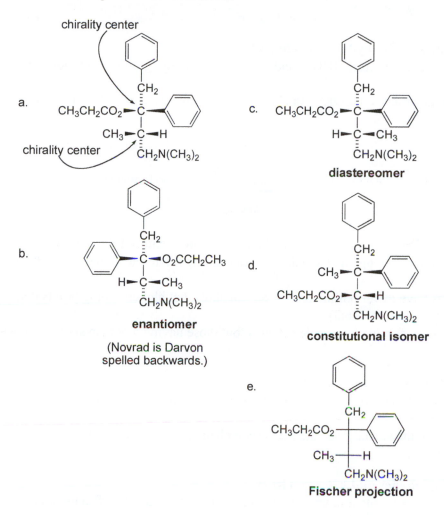

**15.77** Each chirality center in sucrose is labeled with a star.

★ = chirality center

# Chapter 16 Aldehydes and Ketones

## Chapter Review

**[1] What are the characteristics of aldehydes and ketones?**
- An aldehyde has the general structure RCHO, and contains a carbonyl group (C=O) bonded to at least one hydrogen atom. (16.1)
- A ketone has the general structure RCOR', and contains a carbonyl group (C=O) bonded to two carbon atoms. (16.1)

- The carbonyl carbon is trigonal planar, with bond angles of approximately $120°$. (16.1)
- The carbonyl group is polar, giving an aldehyde or ketone stronger intermolecular forces than hydrocarbons. (16.3)
- Aldehydes and ketones have lower boiling points than alcohols, but higher boiling points than hydrocarbons of comparable size. (16.3)
- Aldehydes and ketones are soluble in organic solvents, but those having less than six C's are water soluble, too. (16.3)

**[2] How are aldehydes and ketones named? (16.2)**
- Aldehydes are identified by the suffix *-al*, and the carbon chain is numbered to put the carbonyl group at C1.
- Ketones are identified by the suffix *-one*, and the carbon chain is numbered to give the carbonyl group the lower number.

**[3] Give examples of useful aldehydes and ketones. (16.4)**
- Formaldehyde ($CH_2$=O) is an irritant in smoggy air, a preservative, and a disinfectant, and it is a starting material for synthesizing polymers.
- Acetone [$(CH_3)_2$C=O] is an industrial solvent and polymer starting material.
- Several aldehydes—cinnamaldehyde, vanillin, geranial, and citronellal—have characteristic odors and occur naturally in fruits and plants.
- Dihydroxyacetone [$(HOCH_2)_2$C=O] is the active ingredient in artificial tanning agents, and many other ketones are useful sunscreens.
- Amygdalin, a naturally occurring carbonyl derivative, forms toxic HCN on hydrolysis.

**[4] What products are formed when aldehydes are oxidized? (16.5)**
- Aldehydes are oxidized to carboxylic acids (RCOOH) with $K_2Cr_2O_7$ or Tollens reagent.

$$RCHO \xrightarrow[\text{Tollens reagent}]{K_2Cr_2O_7 \text{ or }} RCOOH$$
carboxylic acid

- Ketones are not oxidized since they contain no H atom on the carbonyl carbon.

**[5] What products are formed when aldehydes and ketones are reduced? (16.6)**
- Aldehydes are reduced to $1°$ alcohols with $H_2$ and a Pd catalyst.

$$RCHO \xrightarrow[\text{Pd}]{\text{H}_2} RCH_2OH$$

1° alcohol

- Ketones are reduced to 2° alcohols with $H_2$ and a Pd catalyst.

$$R_2CO \xrightarrow[\text{Pd}]{\text{H}_2} R_2CHOH$$

2° alcohol

- Biological reduction occurs with the coenzyme NADH.

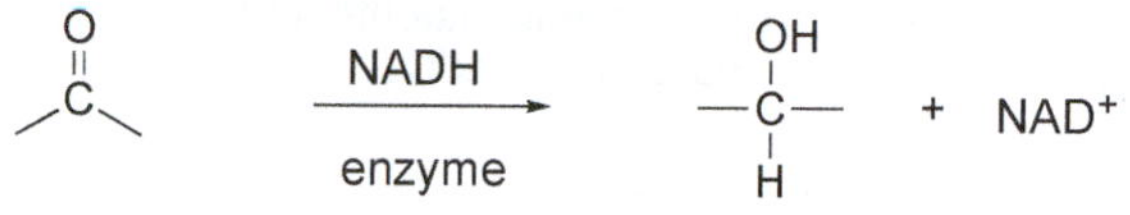

## [6] What reactions occur during vision? (16.7)
- When light hits the retina, the crowded C=C in 11-*cis*-retinal is isomerized to a more stable trans double bond, and a nerve impulse is generated. All-*trans*-retinal is converted back to the cis isomer by a three-step sequence—reduction, isomerization, and oxidation.

## [7] What are hemiacetals and acetals, and how are they prepared? (16.8)
- Hemiacetals contain an OH group and an OR group bonded to the same carbon.
- Acetals contain two OR groups bonded to the same carbon.
- Treatment of an aldehyde or ketone with an alcohol (ROH) first forms an unstable hemiacetal, which reacts with more alcohol to form an acetal.

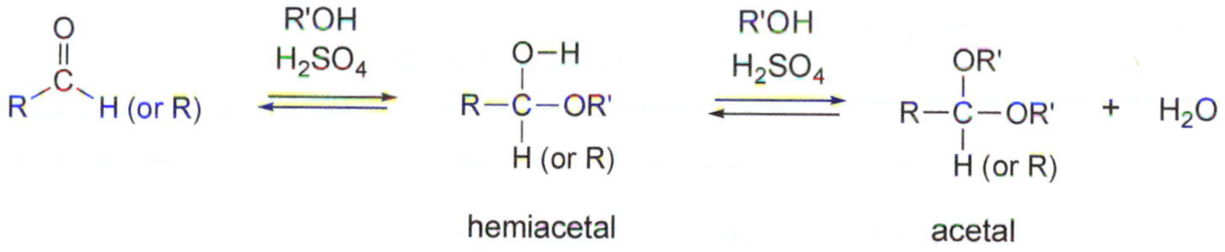

- Cyclic hemiacetals are stable compounds present in carbohydrates and some drugs.

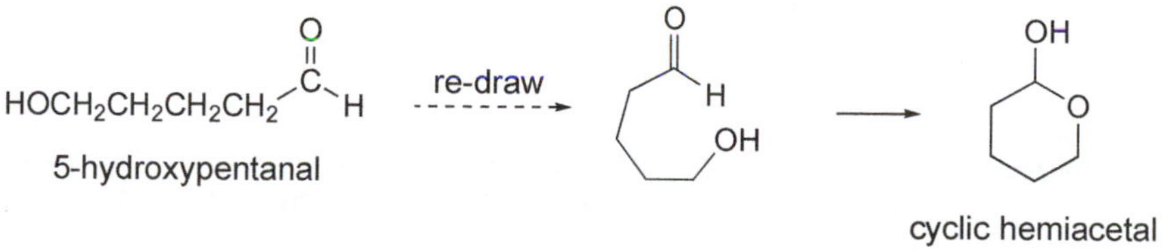

- Acetals are converted back to aldehydes and ketones by hydrolysis with water and acid.

## Problem Solving

## [1] Nomenclature (16.2)

**Example 16.1** Give the IUPAC name for each aldehyde.

a. 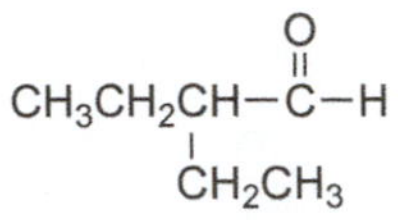   CH₃CH₂CH−C−H
          ‖
          O
          |
          CH₂CH₃

b. 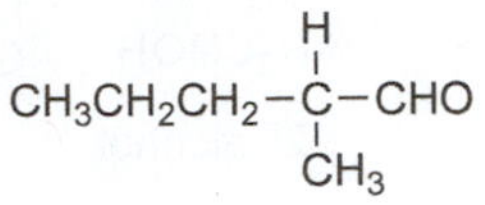   CH₃CH₂CH₂−C−CHO
              |
              H (top)
              CH₃

### Analysis and Solution

a.   **[1]** Find and name the longest chain containing the CHO:

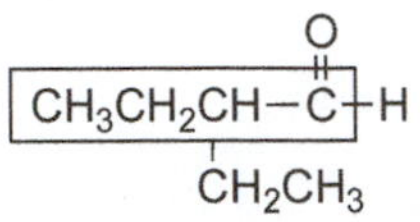

CH₃CH₂CH−C−H
       ‖
       O
    |
    CH₂CH₃

butane  - - - - → butanal
(4 C's)

**[2]** Number and name substituents, making sure the CHO group is at C1:

3   2   O
CH₃CH₂CH−C−H
        |       1
        CH₂CH₃

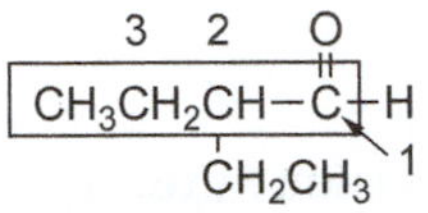

**Answer: 2-ethylbutanal**

b.   **[1]** Find and name the longest chain containing the CHO:

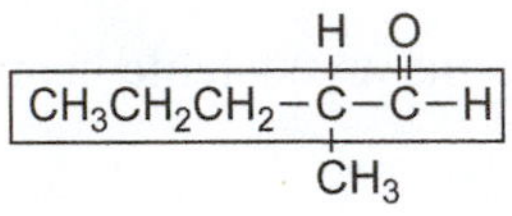

H   O
|   ‖
CH₃CH₂CH₂−C−C−H
        |
        CH₃

pentane  - - - - → pentanal
(5 C's)

**[2]** Number and name substituents, making sure the CHO group is at C1:

2  H   O   1
CH₃CH₂CH₂−C−C−H
         |
         CH₃

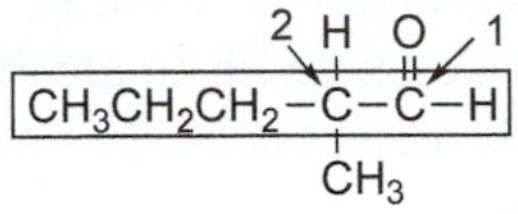

**Answer: 2-methylpentanal**

**Example 16.2** Give the IUPAC name for each ketone.

a. 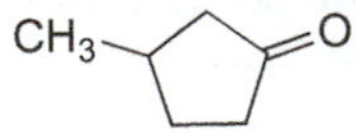   CH₃⟨ring⟩=O

b. 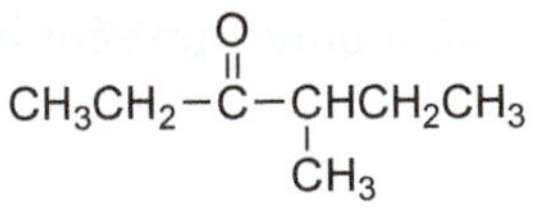   CH₃CH₂−C−CHCH₂CH₃
            ‖        |
            O        CH₃

### Analysis and Solution

a.   **[1]** Name the ring:

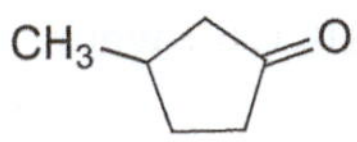

CH₃⟨ring⟩=O

cyclopentane - - - - → cyclopentanone
(5 C's)

**[2]** Number and name the substituents, making sure the carbonyl carbon is at C1:

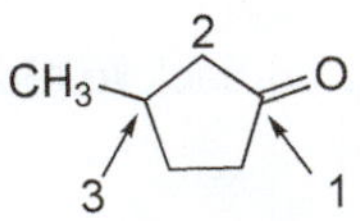

CH₃⟨ring with 2, 3, 1 labels⟩=O

**Answer: 3-methylcyclopentanone**

b. **[1] Find and name the longest chain containing the carbonyl group:**

$$CH_3CH_2-\overset{\overset{\displaystyle O}{\|}}{C}-\underset{\underset{\displaystyle CH_3}{|}}{CH}CH_2CH_3$$

hexane ----→ hexan*one*
(6 C's)

**[2] Number and name the substituents, making sure the carbonyl carbon has the lower number:**

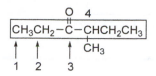

**Answer: 4-methyl-3-hexanone**

---

## [2] Reactions of Aldehydes and Ketones (16.5)

**Example 16.3** What product is formed when each carbonyl compound is treated with $K_2Cr_2O_7$?

a. $CH_3-\overset{\overset{\displaystyle O}{\|}}{C}-\underset{\underset{\displaystyle CH_3}{|}}{CH}CH_2CH_3$

b.

### Analysis
Compounds that contain a C–H and C–O bond on the *same* carbon are oxidized with $K_2Cr_2O_7$. Thus:
- Aldehydes (RCHO) are oxidized to $RCO_2H$.
- Ketones ($R_2CO$) are *not* oxidized with $K_2Cr_2O_7$.

### Solution
The ketone in part (a) is inert to oxidation, but the aldehyde in part (b) is oxidized with $K_2Cr_2O_7$ to a carboxylic acid.

a. $CH_3-\overset{\overset{\displaystyle O}{\|}}{C}-\underset{\underset{\displaystyle CH_3}{|}}{CH}CH_2CH_3$ $\xrightarrow{K_2Cr_2O_7}$ No reaction

This C is bonded only to other C's.

b. $\xrightarrow{K_2Cr_2O_7}$

Replace 1 C–H bond by 1 C–O bond.

---

**Example 16.4** What product is formed when each compound is treated with Tollens reagent ($Ag_2O$, $NH_4OH$)?

a.

b. $CH_3-\overset{\overset{\displaystyle O}{\|}}{C}-CH_2CH_3$

### Analysis
**Only aldehydes (RCHO) react with Tollens reagent.** Ketones and alcohols are inert to oxidation.

### Solution
The aldehyde in part (a) is oxidized to $RCO_2H$, but the ketone in part (b) does not react with Tollens reagent.

a. [structure: cyclohexyl-CHO] $\xrightarrow[\text{NH}_4\text{OH}]{\text{Ag}_2\text{O}}$ [structure: cyclohexyl-COOH]

b. [structure: $CH_3-C(=O)-CH_2CH_3$] $\xrightarrow[\text{NH}_4\text{OH}]{\text{Ag}_2\text{O}}$ No reaction

| Replace 1 C–H bond by 1 C–O bond. |

| This C=O is bonded only to other C's. |

## [3] Reduction of Aldehydes and Ketones (16.6)

**Example 16.5** What alcohol is formed when each aldehyde or ketone is treated with $H_2$ in the presence of a Pd catalyst?

a. $CH_3CH_2CH_2-C\overset{O}{\underset{H}{\Vert}}$

b. [cyclopentane ring]=O

### Analysis
To draw the products of reduction:
- Locate the C=O and mentally break one bond in the double bond.
- Mentally break the H–H bond of the reagent.
- Add one H atom to each atom of the C=O, forming new C–H and O–H single bonds.

### Solution
The aldehyde (RCHO) in part (a) forms a 1° alcohol (RCH$_2$OH) and the ketone in part (b) forms a 2° alcohol (R$_2$CHOH).

Break one bond.

a. $CH_3CH_2CH_2-C\overset{O}{\underset{H}{\Vert}}$ + $\underset{H}{\overset{H}{|}}$ $\xrightarrow{\text{Pd}}$ $CH_3CH_2CH_2-\overset{O-H}{\underset{H}{\overset{|}{C}}}-H$ = $CH_3CH_2CH_2CH_2OH$

Break the single bond.

1° alcohol

Break one bond.

b. [cyclopentane]=O

H—H

Break the single bond.

$\xrightarrow{\text{Pd}}$ [cyclopentane]$-\overset{O-H}{\underset{H}{C}}$ = [cyclopentane]—OH

2° alcohol

## [4] Acetals (16.8)

**Example 16.6** Draw the hemiacetal and acetal formed when the following ketone is treated with methanol (CH$_3$OH) in the presence of H$_2$SO$_4$.

$CH_3CH_2-\overset{O}{\overset{\Vert}{C}}-CH_3$ + CH$_3$OH $\xrightarrow[\phantom{xxx}]{\text{H}_2\text{SO}_4}$

methanol
(two equivalents)

**Analysis**

To form a hemiacetal and acetal from a carbonyl compound:

- Locate the C=O in the starting material.
- Break one C–O bond and add one equivalent of $CH_3OH$ across the double bond, placing the $OCH_3$ group on the carbonyl carbon. This forms the hemiacetal.
- Replace the OH group of the hemiacetal by $OCH_3$ to form the acetal.

**Solution**

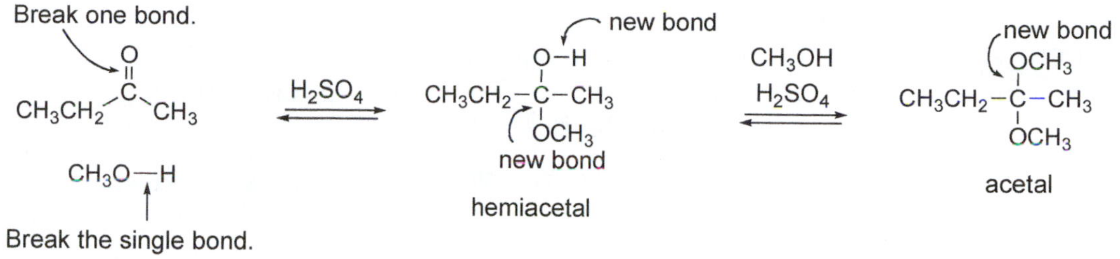

---

**Example 16.7** Identify each compound as an ether, hemiacetal, or acetal.

a. $CH_3-O-CH_2CH_2CH_3$

b.

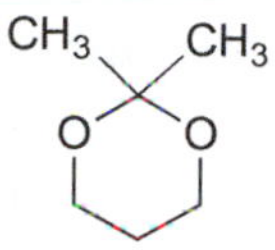

**Analysis**

Recall the definitions to identify the functional groups:

- An ether has the general structure ROR.
- A hemiacetal has one C bonded to OH and OR.
- An acetal has one C bonded to two OR groups.

**Solution**

a. $CH_3-O-CH_2CH_2CH_3$

2 C's on 1 O atom

ether

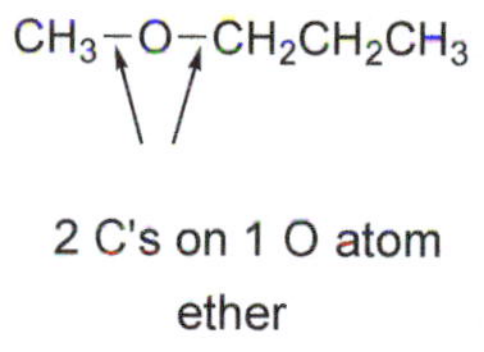

b.

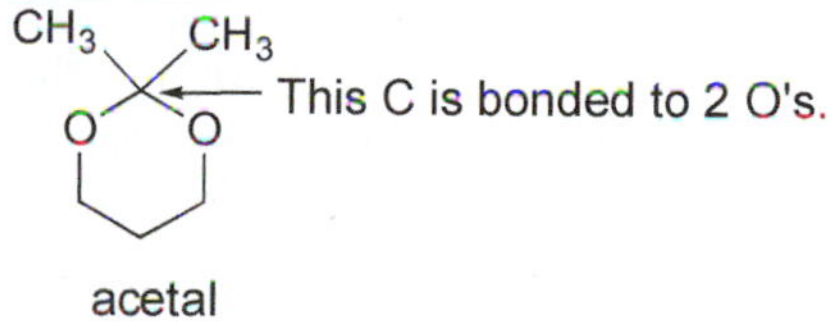

This C is bonded to 2 O's.

acetal

Note that the acetal in part (b) is part of a ring. It contains one ring carbon bonded to two oxygens, making it an acetal.

---

**Example 16.8** What hydrolysis products are formed when the given acetal is treated with $H_2SO_4$ in $H_2O$?

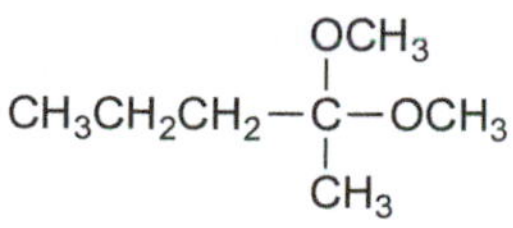

## Analysis

To draw the products of hydrolysis:

- Locate the two C–OR bonds on the same carbon.
- Replace the two C–O single bonds with a carbonyl group (C=O).
- Each OR group then becomes a molecule of alcohol (ROH) product.

## Solution

---

## Self-Test

**[1] Fill in the blank with one of the terms listed below.**

Acetals (16.8)          Hydrolysis (16.8)          Reduction (16.6)
Aldehyde (16.1)         Ketone (16.1)              Tollens reagent (16.5)
Hemiacetal (16.8)       NADH (16.6)

1. When bonds are cleaved by reaction with water, the reaction is a _____ reaction.
2. One reagent used to oxidize aldehydes selectively is the _____.
3. _____ are compounds that contain two OR groups (alkoxy groups) bonded to the same carbon.
4. An _____ has at least one H atom bonded to a carbonyl group.
5. A _____ contains an OH group (hydroxyl) and an OR group (alkoxy) bonded to the same carbon.
6. A _____ has two alkyl groups bonded to the carbonyl group.
7. _____ results in a decrease in the number of C–O bonds or an increase in the number of C–H bonds.
8. Biological systems use _____ as a reducing agent.

**[2] Label each compound as a hemiacetal, acetal, ether, aldehyde, or ketone.**

9.   $CH_3CH_2CH_2-\overset{\overset{OH}{|}}{\underset{\underset{CH_3}{|}}{C}}-OCH_3$

10.

11.  $CH_3CH_2-\overset{\overset{OCH_3}{|}}{\underset{\underset{CH_3}{|}}{C}}-OCH_3$

12.

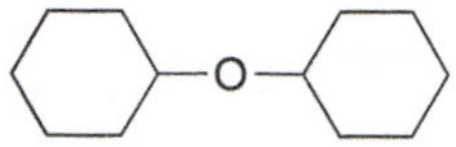

13. 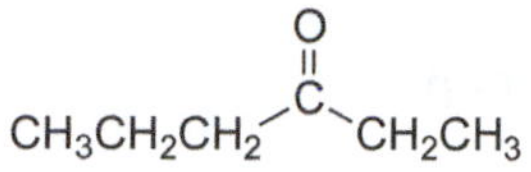 CH$_3$CH$_2$CH$_2$—C(=O)—CH$_2$CH$_3$

## [3] Match the products with the correct reaction.

14. CH$_3$CH$_2$CH$_2$CHO $\xrightarrow{K_2Cr_2O_7}$

a. CH$_3$CH$_2$CH$_2$CH$_2$OH

15. 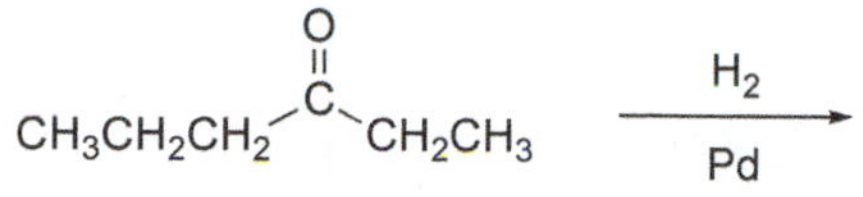 CH$_3$CH$_2$CH$_2$—C(=O)—CH$_2$CH$_3$ $\xrightarrow[Pd]{H_2}$

b. CH$_3$CH$_2$CH$_2$COOH

16. 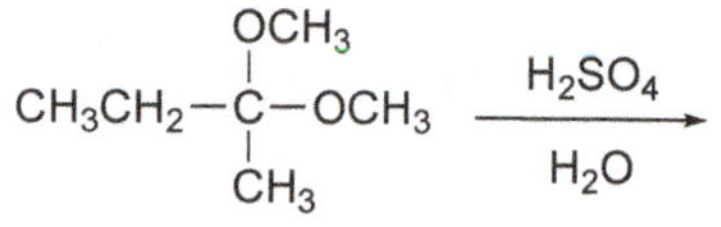 CH$_3$CH$_2$—C(OCH$_3$)(OCH$_3$)(CH$_3$) $\xrightarrow[H_2O]{H_2SO_4}$

c. CH$_3$CH$_2$CH$_2$—C(OCH$_3$)(OCH$_3$)—CH$_2$CH$_3$

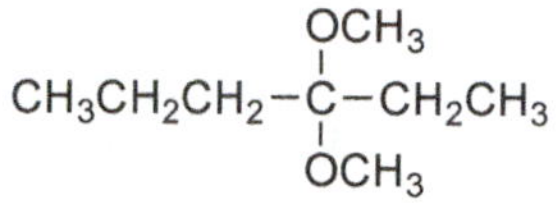

17. 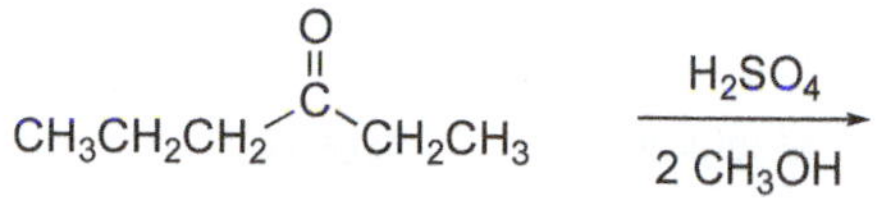 CH$_3$CH$_2$CH$_2$—C(=O)—CH$_2$CH$_3$ $\xrightarrow[2\ CH_3OH]{H_2SO_4}$

d. 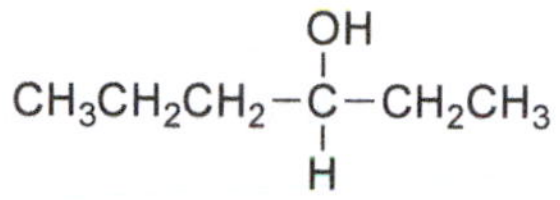 CH$_3$CH$_2$CH$_2$—C(OH)(H)—CH$_2$CH$_3$

18. 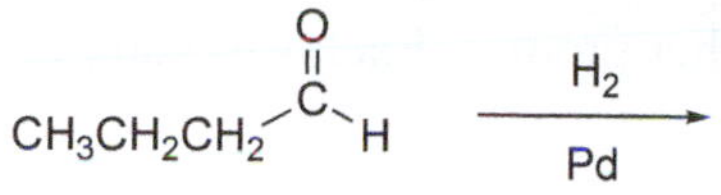 CH$_3$CH$_2$CH$_2$—C(=O)—H $\xrightarrow[Pd]{H_2}$

e. 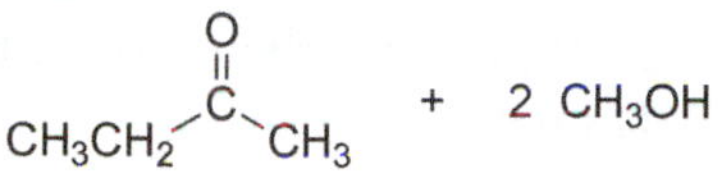 CH$_3$CH$_2$—C(=O)—CH$_3$ + 2 CH$_3$OH

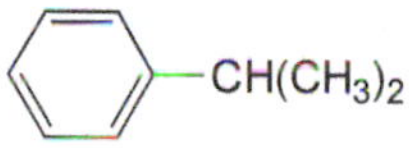

## [4] Which compound in each pair has the higher boiling point?

19. CH$_3$(CH$_2$)$_6$CHO   or   CH$_3$(CH$_2$)$_7$OH

      **A**               **B**

20. 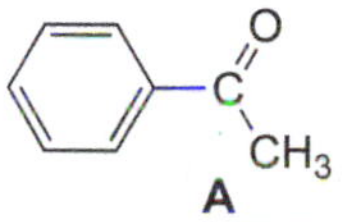 (phenyl)—C(=O)—CH$_3$   or   (phenyl)—CH(CH$_3$)$_2$

      **A**               **B**

21. CH$_3$(CH$_2$)$_6$CHO   or   CH$_3$(CH$_2$)$_2$CHO

      **A**               **B**

## [5] Label each compound as water soluble or water insoluble.

22. CH$_3$(CH$_2$)$_7$CHO

23. 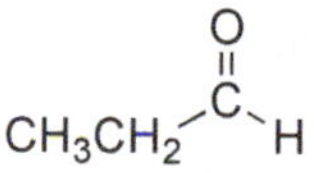 CH$_3$CH$_2$—C(=O)—H

24. CH$_3$(CH$_2$)$_2$CHO

25. 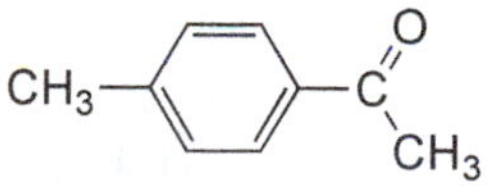 CH$_3$—(phenyl)—C(=O)—CH$_3$

## Answers to Self-Test

| | | | | |
|---|---|---|---|---|
| 1. hydrolysis | 6. ketone | 11. acetal | 16. e | 21. A |
| 2. Tollens reagent | 7. Reduction | 12. ether | 17. c | 22. insoluble |
| 3. Acetals | 8. NADH | 13. ketone | 18. a | 23. soluble |
| 4. aldehyde | 9. hemiacetal | 14. b | 19. B | 24. soluble |
| 5. hemiacetal | 10. aldehyde | 15. d | 20. A | 25. insoluble |

## Solutions to In-Chapter Problems

**16.1**   An **aldehyde** has at least one H atom bonded to the carbonyl group.
A **ketone** has two alkyl groups bonded to the carbonyl group.

a.   $CH_3CH_2$—C(=O)—H      aldehyde          c.   $(CH_3)_3C$—C(=O)—$CH_3$      ketone

b.   $CH_3CH_2$—C(=O)—$CH_3$      ketone          d.   $(CH_3CH_2)_2CH$—C(=O)—H      aldehyde

**16.2**   Draw the constitutional isomers of molecular formula $C_4H_8O$ and then label each compound using the definitions from Answer 16.1.

$CH_3CH_2$—C(=O)—$CH_3$   ketone          $CH_3CH_2CH_2$—C(=O)—H   aldehyde          $(CH_3)_2CH$—C(=O)—H   aldehyde

**16.3**   Trigonal planar carbons are carbons bonded to three other groups.  Each trigonal planar carbon is labeled with an arrow.

**16.4**   To name an aldehyde using the IUPAC system, use the steps in Example 16.1:
[1] Find the longest chain containing the CHO group, and change the *-e* ending of the parent alkane to the suffix *-al.*
[2] Number the chain or ring to put the CHO group at C1, but omit this number from the name.
Apply all of the other usual rules of nomenclature.

a.   $(CH_3)_2CHCH_2CH_2CH_2CHO$

$CH_3$
$CH_3CHCH_2CH_2CH_2CH$(=O)     - - - - - - - - →

5-methyl
$CH_3$
$CH_3CHCH_2CH_2CH_2CH$(=O)     - - - - - - - - → **Answer: 5-methylhexanal**

hexane   - - - - →   hexan*al*
(6 C's)

b.  $(CH_3)_3CC(CH_3)_2CH_2CHO$

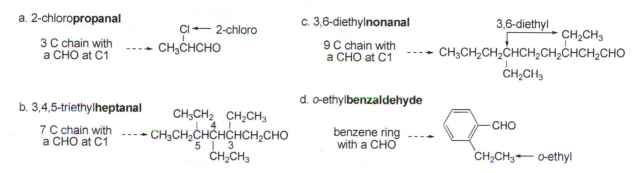

pentane ----> pentanal
(5 C's)

**Answer:**
**3,3,4,4-tetramethylpentanal**

c.

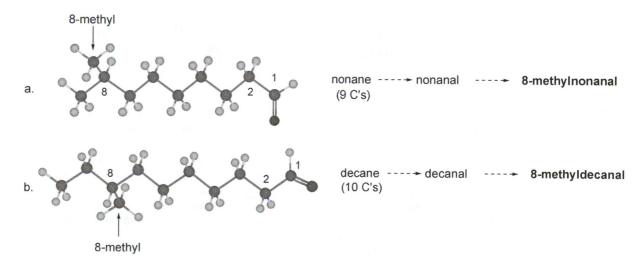

octane ----> octanal
(8 C's)

**Answer:**
**2,5,6-trimethyloctanal**

**16.5**   Work backwards from the name to draw each structure.

a. 2-chloro**propanal**

3 C chain with a CHO at C1 ----> CH₃CHCHO, Cl ← 2-chloro

c. 3,6-diethyl**nonanal**

9 C chain with a CHO at C1 ---- CH₃CH₂CH₂CHCH₂CH₂CHCH₂CHO
3,6-diethyl, CH₂CH₃, CH₂CH₃

b. 3,4,5-triethyl**heptanal**

7 C chain with a CHO at C1 ----> CH₃CH₂CHCHCHCH₂CHO
CH₃CH₂, CH₂CH₃, CH₂CH₃

d. *o*-ethyl**benzaldehyde**

benzene ring with a CHO ----> —CHO, CH₂CH₃ ← *o*-ethyl

**16.6**   To name an aldehyde using the IUPAC system, use the steps in Example 16.1.

a.

8-methyl

nonane ----> nonanal ----> **8-methylnonanal**
(9 C's)

b.

8-methyl

decane ----> decanal ----> **8-methyldecanal**
(10 C's)

**16.7**  To name a ketone using IUPAC rules, use the steps in Example 16.2:

[1] Find the longest chain containing the carbonyl group, and change the *-e* ending of the parent alkane to the suffix *-one.*

[2] Number the carbon chain to give the carbonyl carbon the lower number. Apply all of the other usual rules of nomenclature.

a. $CH_3CH_2\overset{O}{\overset{\|}{C}}CHCH_2CH_2CH_3$  - - - - ➤  $\overset{O\ 4}{CH_3CH_2\overset{\|}{C}CHCH_2CH_2CH_3}$  - - - - ➤  **Answer: 4-methyl-3-heptanone**

$\qquad\qquad\quad CH_3$ $\qquad\qquad\qquad$ 1  2  3 $CH_3$

$\qquad$ heptane - - - - ➤ heptan*one* $\qquad\qquad\qquad\qquad$ ↑
$\qquad\quad$ (7 C's) $\qquad\qquad\qquad\qquad\qquad\qquad\quad$ 4-methyl

b.

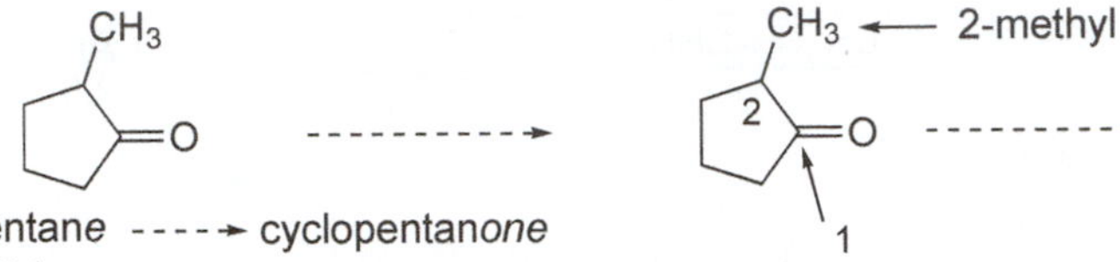

$CH_3$

$\qquad\qquad$ cyclopentane - - - - ➤ cyclopentan*one* $\qquad$ CH₃ ⟵ 2-methyl

$\qquad\qquad$ (5 C's) $\qquad\qquad\qquad\qquad\qquad\qquad$ 2 ⟶ O  - - - - - - - ➤  **Answer: 2-methylcyclopentanone**

$\qquad\qquad\qquad\qquad\qquad\qquad\qquad\qquad\qquad\qquad\qquad$ 1

c. $CH_3C\overset{CH_3\ O}{\overset{\ \ \ \|}{—}}CCH_2CH_2CH_3$  - - - - ➤ $\overset{2\ \ CH_3\ O}{CH_3C—CCH_2CH_2CH_3}$  - - - - ➤  **Answer: 2,2-dimethyl-3-heptanone**

$\qquad\qquad CH_3$ $\qquad\qquad\qquad\qquad$ 1  $CH_3$  3

$\qquad$ heptane - - - - ➤ heptan*one* $\qquad\qquad\qquad\qquad$ ↑
$\qquad\quad$ (7 C's) $\qquad\qquad\qquad\qquad\qquad\qquad$ 2,2-dimethyl

**16.8**  Work backwards from the name to draw each structure.

a. butyl ethyl ketone $\qquad$ $CH_3CH_2CH_2CH_2\overset{O}{\overset{\|}{C}}CH_2CH_3$ $\qquad$ c. *p*-ethyl**acetophenone**

$\qquad\qquad\qquad\qquad\qquad\qquad\qquad$ ↑ $\qquad\qquad\qquad$ ↑ $\qquad\qquad\qquad\qquad$ *p*-ethyl ⟶ CH₃CH₂—[ring]—$\overset{O}{\overset{\|}{C}}$—CH₃

$\qquad\qquad\qquad\qquad\qquad\qquad$ butyl $\qquad\quad$ ethyl $\qquad\qquad\qquad\qquad\qquad\qquad\qquad\qquad\qquad\qquad$ acetophenone

b. 2-methyl-**3-pentanone**

$\qquad$ 5 C chain with $\qquad$ - - - ➤ $CH_3CH_2\overset{O}{\overset{\|}{C}}CHCH_3$ $\qquad$ d. 2-propyl**cyclobutanone**

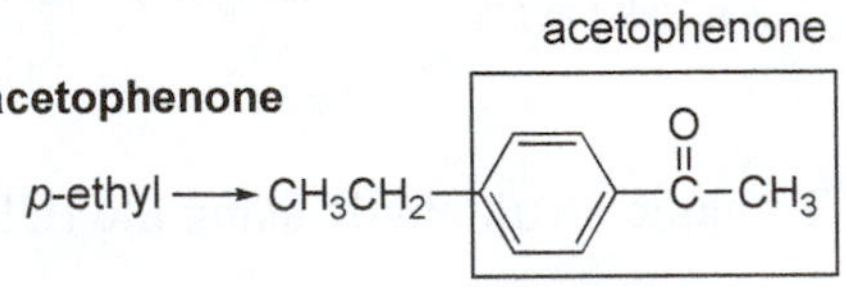

$\qquad$ C=O at C3 $\qquad\qquad\qquad\qquad\qquad\qquad$ $CH_3$ ⟵ 2-methyl $\qquad$ 4 C ring with $\qquad$ - - - ➤
$\qquad\qquad\qquad\qquad\qquad\qquad\qquad\qquad\qquad\qquad\qquad\qquad\qquad$ C=O at C1 $\qquad\qquad\qquad\qquad$ $CH_2CH_2CH_3$ ⟵ 2-propyl

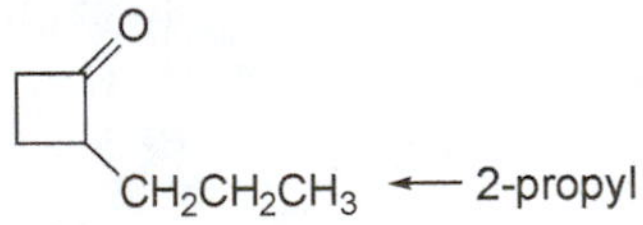

**16.9**  Aldehydes and ketones have *higher* boiling points than hydrocarbons of comparable size. Aldehydes and ketones have *lower* boiling points than alcohols of comparable size.

a. 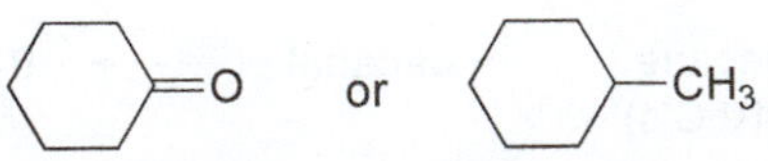  [ring]=O $\quad$ or $\quad$ [ring]—CH₃ $\qquad\qquad$ c.  [ring]=O $\quad$ or $\quad$ [ring]—OH

$\qquad$ ketone $\qquad\qquad$ hydrocarbon $\qquad\qquad\qquad\qquad$ ketone $\qquad\qquad\qquad$ alcohol
$\quad$ **higher boiling point** $\qquad\qquad\qquad\qquad\qquad\qquad\qquad\qquad\qquad$ **higher boiling point**

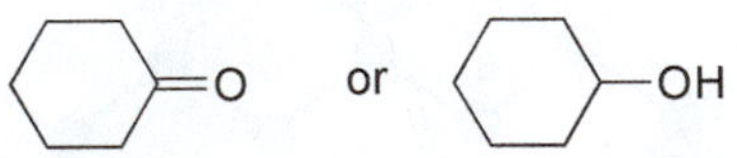

b.  $(CH_3CH_2)_2CO$ $\quad$ or $\quad$ $(CH_3CH_2)_2C=CH_2$ $\qquad$ d.  $CH_3(CH_2)_6CH_3$ $\quad$ or $\quad$ $CH_3(CH_2)_5CHO$

$\qquad$ ketone $\qquad\qquad$ hydrocarbon $\qquad\qquad\qquad\qquad$ hydrocarbon $\qquad\qquad$ aldehyde
$\quad$ **higher boiling point** $\qquad\qquad\qquad\qquad\qquad\qquad\qquad\qquad\qquad$ **higher boiling point**

**16.10** Acetone will be soluble in water and organic solvents since it is a low molecular weight ketone (less than six carbons). Progesterone will be soluble only in organic solvents since it has many carbons and only two polar functional groups.

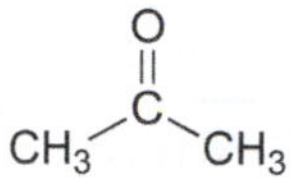

small ketone
acetone

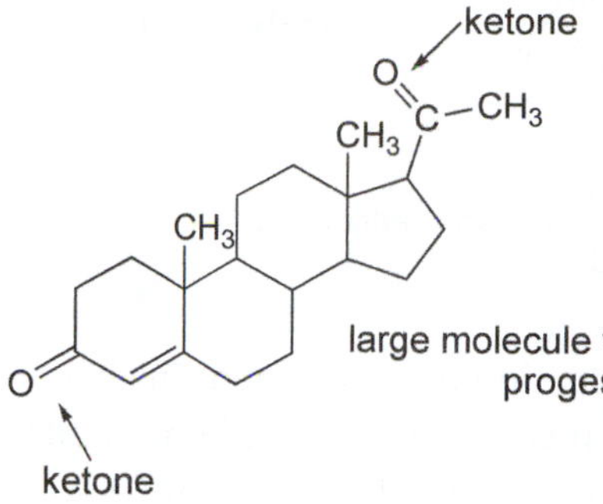

large molecule with two ketones
progesterone

**16.11** Hexane is soluble in acetone because both compounds are organic and "like dissolves like." Water is soluble in acetone because acetone has a short hydrocarbon chain and is capable of hydrogen bonding with water.

**16.12** Compare the functional groups in each sunscreen. Dioxybenzone will most likely be washed off in water because it contains two hydroxyl groups and is the most water soluble.

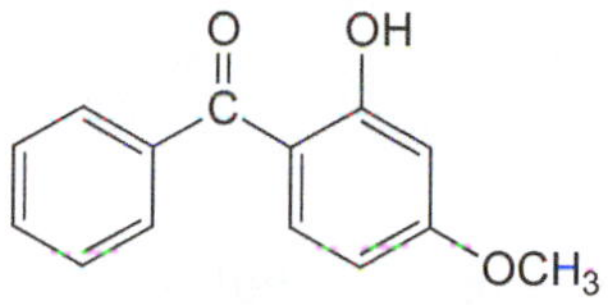

oxybenzone
one hydroxyl group
one ketone
one ether

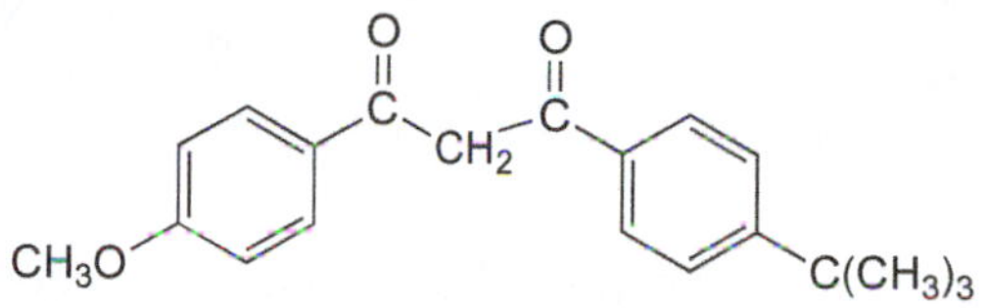

avobenzone
two ketones
one ether

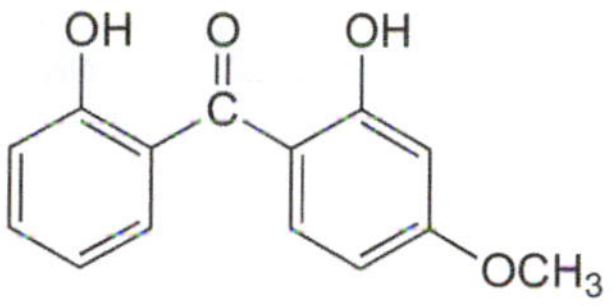

dioxybenzone
two hydroxyl groups
one ketone
one ether
**most water soluble**

**16.13** Draw the product of each reaction using the guidelines in Example 16.3. Compounds that contain a C–H and C–O bond on the *same* carbon are oxidized with $K_2Cr_2O_7$.
   - Aldehydes (RCHO) are oxidized to $RCO_2H$.
   - Ketones ($R_2CO$) are *not* oxidized with $K_2Cr_2O_7$.

a. 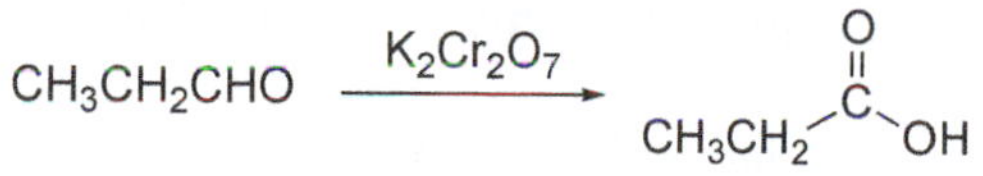 $CH_3CH_2CHO$ $\xrightarrow{K_2Cr_2O_7}$

b. $(CH_3CH_2)_2C=O$ $\xrightarrow{K_2Cr_2O_7}$ No reaction

c. 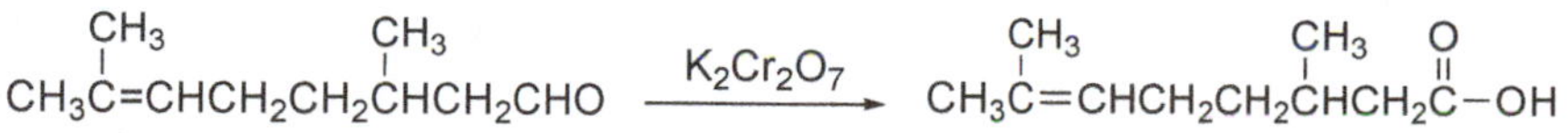 $\xrightarrow{K_2Cr_2O_7}$

**16.14** Draw the product of each reaction using Example 16.4 as a guide. **Only aldehydes (RCHO) react with Tollens reagent, and they are oxidized to RCO$_2$H.** Ketones and alcohols are inert to oxidation.

a. $CH_3(CH_2)_6CHO \xrightarrow[NH_4OH]{Ag_2O} CH_3(CH_2)_6\overset{\overset{O}{\|}}{C}-OH$

c. (cyclopentyl)—CHO $\xrightarrow[NH_4OH]{Ag_2O}$ (cyclopentyl)—$\overset{\overset{O}{\|}}{C}$—OH

b. (cyclopentanone) =O $\xrightarrow[NH_4OH]{Ag_2O}$ No reaction

d. (cyclohexyl)—OH $\xrightarrow[NH_4OH]{Ag_2O}$ No reaction

**16.15** Draw the products of reduction using the steps in Example 16.5.
- Locate the C=O and mentally break one bond in the double bond.
- Mentally break the H–H bond of the reagent.
- Add one H atom to each atom of the C=O, forming new C–H and O–H single bonds.

a. $CH_3CH_2CH_2\overset{\overset{O}{\|}}{C}_H \xrightarrow[Pd]{H_2} CH_3CH_2CH_2\overset{\overset{OH}{|}}{C}H_2$

c. $CH_3\overset{\overset{O}{\|}}{C}_{CH_2CH_3} \xrightarrow[Pd]{H_2} CH_3\overset{\overset{OH}{|}}{C}HCH_2CH_3$

b. (cyclopentanone with CH$_3$) $\xrightarrow[Pd]{H_2}$ (cyclopentanol with CH$_3$)

d. (phenyl)—CHO $\xrightarrow[Pd]{H_2}$ (phenyl)—CH$_2$OH

**16.16** Work backwards to determine what carbonyl compound is needed to prepare alcohol **A**.

$(CH_3)_2CHCH_2$—(phenyl)—$\overset{\overset{O}{\|}}{C}CH_3$ $\xrightarrow[Pd]{H_2}$ $(CH_3)_2CHCH_2$—(phenyl)—$\overset{\overset{OH}{|}}{C}HCH_3$

**A**

**16.17** Recall that stereoisomers differ only in the three-dimensional arrangement of atoms in space, but all connectivity is identical. Constitutional isomers have the same molecular formula, but atoms are connected differently.

a. All-*trans*-retinal and 11-*cis*-retinal are stereoisomers, and differ only in the arrangement of groups around one double bond.
b. All-*trans*-retinal and vitamin A are not isomers. They have different molecular formulas.
c. Vitamin A and 11-*cis*-retinol are stereoisomers, and differ only in the arrangement of groups around one double bond.

**16.18** To form a hemiacetal and acetal from a carbonyl compound, use the steps in Example 16.6.

- Locate the C=O in the starting material.
- Break one C–O bond and add one equivalent of ROH across the double bond, placing the OR group on the carbonyl carbon. This forms the hemiacetal.
- Replace the OH group of the hemiacetal by OR to form the acetal.

a.
$$CH_3-\underset{\overset{\|}{O}}{C}-H \;+\; CH_3OH \;\underset{}{\overset{H_2SO_4}{\rightleftharpoons}}\; CH_3\underset{\underset{H}{|}}{\overset{OCH_3}{|}}COH \;\underset{}{\overset{CH_3OH \;\; H_2SO_4}{\rightleftharpoons}}\; CH_3\underset{\underset{H}{|}}{\overset{OCH_3}{|}}COCH_3$$

                                                                     hemiacetal                                        acetal

b.
$$(CH_3CH_2)_2C=O \;+\; CH_3OH \;\overset{H_2SO_4}{\rightleftharpoons}\; (CH_3CH_2)_2\overset{OCH_3}{\underset{}{C}}OH \;\overset{CH_3OH \;\; H_2SO_4}{\rightleftharpoons}\; (CH_3CH_2)_2\overset{OCH_3}{\underset{}{C}}OCH_3$$

                                                   hemiacetal                                  acetal

c.
(benzaldehyde, C$_6$H$_5$CHO) $+$ CH$_3$CH$_2$OH $\overset{H_2SO_4}{\rightleftharpoons}$ C$_6$H$_5$CHOH with OCH$_2$CH$_3$ (hemiacetal) $\overset{CH_3CH_2OH \;\; H_2SO_4}{\rightleftharpoons}$ C$_6$H$_5$CHOCH$_2$CH$_3$ with OCH$_2$CH$_3$ (acetal)

**16.19** Recall the definitions from Example 16.7 to identify the functional groups:

- An ether has the general structure ROR.
- A hemiacetal has one C bonded to OH and OR.
- An acetal has one C bonded to two OR groups.

a. (cyclohexane–OCH$_3$)    **ether**

b. (cyclohexane with two OCH$_3$)    **acetal**

c. $CH_3CH_3CH_2CH_2-\underset{\underset{H}{|}}{\overset{\overset{OH}{|}}{C}}-OCH_3$    **hemiacetal**

d. (bicyclic with O–C(CH$_3$)$_2$–O)    **acetal**

**16.20** Label the acetal or hemiacetal in each compound using the definitions in Example 16.7.

a. (sugar ring structure) — **hemiacetal**

HOCH$_2$, HO, HO, OH, NH$_2$

b. (sugar ring structure with aromatic OCH$_2$OH) — **acetal**

HOCH$_2$, HO, OH, OH, CH$_2$OH

**16.21** Draw the products of each reaction using the steps in Example 16.6.

a. (tetrahydropyran–OH) $+$ CH$_3$CH$_2$OH $\overset{H_2SO_4}{\longrightarrow}$ (tetrahydropyran–OCH$_2$CH$_3$)

Replace OH by OCH$_2$CH$_3$

b.

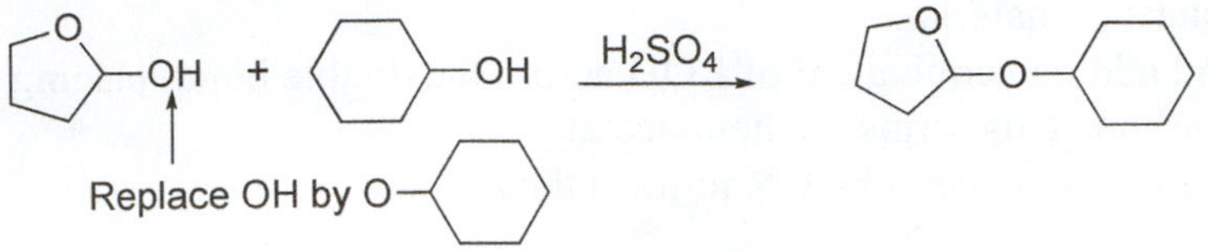

Replace OH by O—⬡

**16.22** To draw the products of hydrolysis, use the steps in Example 16.8.
- Locate the two C–OR bonds on the same carbon.
- Replace the two C–O single bonds with a carbonyl group (C=O).
- Each OR group then becomes a molecule of alcohol (ROH) product.

a.

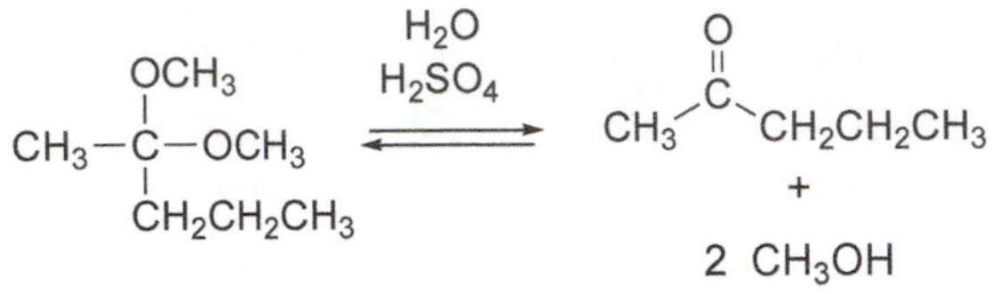

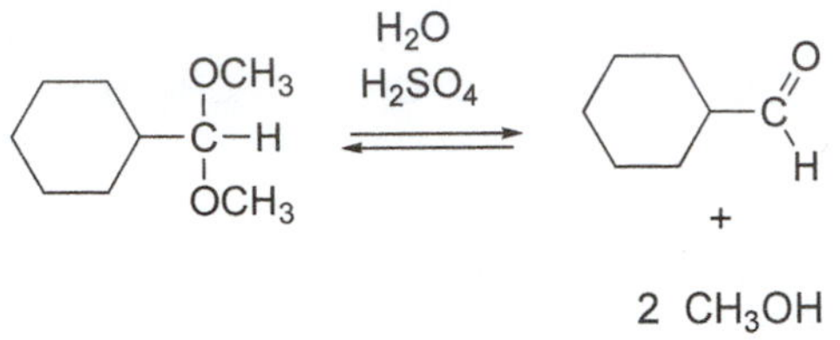

b.

c.

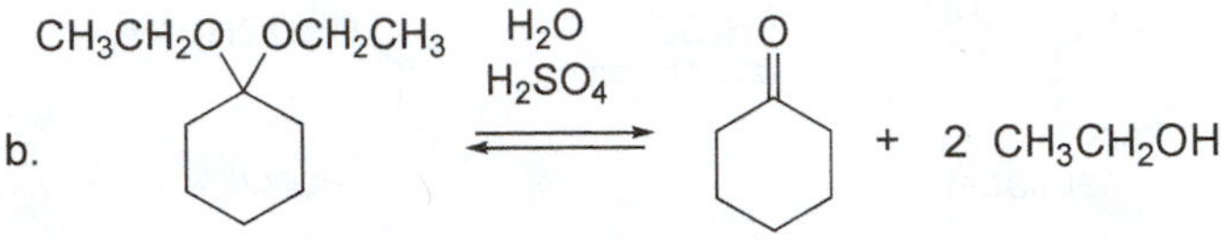

## Solutions to Odd-Numbered End-of-Chapter Problems

**16.23** Draw a structure to fit each description.

a. 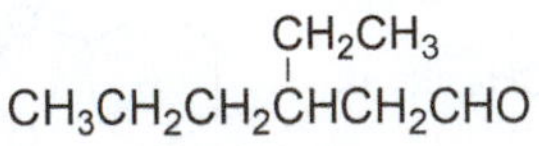 $CH_3CH_2CH_2CHCH_2CHO$ with $CH_2CH_3$ substituent
aldehyde
$C_8H_{16}O$

b. 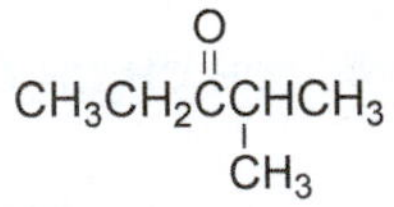 $CH_3CH_2CCHCH_3$ with $CH_3$ substituent
ketone
$C_6H_{12}O$

c. 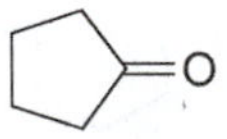 cyclopentanone
ketone
$C_5H_8O$

d. 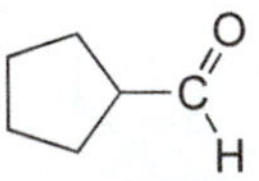 aldehyde
$C_6H_{10}O$

**16.25** Compare C=O and C=C bonds.
a. Both are trigonal planar.
b. A C=O is polar and a C=C is *not* polar.
c. Both functional groups undergo addition reactions.

**16.27** An aldehyde cannot have the molecular formula $C_5H_{12}O$. $C_5H_{12}$ has too many H's. Since an aldehyde has a double bond, the number of C's and H's resembles an alkene, not an alkane. An aldehyde with 5 C's would have the molecular formula $C_5H_{10}O$.

**16.29**   To name the aldehyde and ketone, use the IUPAC rules in Examples 16.1 and 16.2.

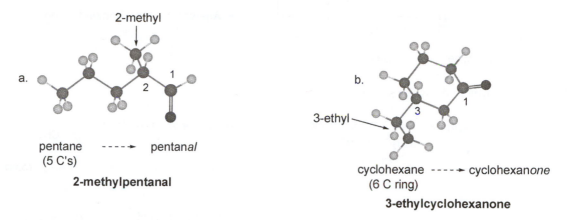

a.

2-methyl

2

1

pentane ----→ pentanal
(5 C's)

**2-methylpentanal**

b.

3-ethyl

3

1

cyclohexane ----→ cyclohexanone
(6 C ring)

**3-ethylcyclohexanone**

**16.31**   To name an aldehyde using the IUPAC system, use the steps in Example 16.1:
[1] Find the longest chain containing the CHO group, and change the *-e* ending of the parent alkane to the suffix *-al.*
[2] Number the chain or ring to put the CHO group at C1, but omit this number from the name. Apply all other usual rules of nomenclature.

a.   $CH_3CH_2CH_2CHCH_2CHO$ --------→ $CH_3CH_2CH_2CHCH_2CHO$ --------→ **Answer: 3-methylhexanal**
          |                                                          |
          $CH_3$                                                    $CH_3$ ← 3-methyl

hexane ----→ hexanal
(6 C's)

b.            $CH_3$                                    $CH_3$   3   1
          |                                          |
$CH_3CH_2CHCH_2CHCH_2CHO$ ----→ $CH_3CH_2CHCH_2CHCH_2CHO$ ----→ **Answer: 3,5-dimethylheptanal**
                      |                          5       |
                      $CH_3$                            $CH_3$ ← 3,5-dimethyl

heptane ----→ heptanal
(7 C's)

c.      $O{=}C{-}H$                              $O{=}C{-}H$
              |                                        |
        $H{-}C{-}H$                              $H{-}C{-}H$
              |                                        |
$CH_3CH_2CH_2{-}C{-}CH_2CH_2CH_3$ ----→ $CH_3CH_2CH_2{-}C{-}CH_2CH_2CH_3$ ----→ **Answer: 3-propylhexanal**
              |                                    ↑   |
              H                                3-propyl  H   3

hexane ----→ hexanal
(6 C's)

d.      $CH_2CH_3$        $CH_3$              $CH_2CH_3$        $CH_3$ ← 2,2-dimethyl
              |              |                      |              |
$CH_3CH_2CCH_2CH_2CH_2{-}C{-}CHO$ ----→ $CH_3CH_2CCH_2CH_2CH_2{-}C{-}CHO$ ----→ **Answer:**
              |              |                      $CH_2CH_3$     $CH_3$        **6,6-diethyl-2,2-dimethyloctanal**
        $CH_2CH_3$        $CH_3$                  6              2       1

octane ----→ octanal
(8 C's)

e.   Cl—⟨benzene⟩—CHO  ----→  Cl—⟨benzene⟩—CHO  ----→  **Answer: *p*-chlorobenzaldehyde**

        benzaldehyde              *p*-chloro

**16.33**   Work backwards to draw the structure.

a. 3,3-dichloro**pentanal**       Cl ← 3,3-dichloro          c. *o*-bromo**benzaldehyde**   ⟨benzene⟩—CHO
                                  CH₃CH₂CCH₂CHO
        5 C chain                 Cl ←                       benzene ring with CHO       Br ← *o*-bromo

b. 3,4-dimethyl**hexanal**       CH₃ ← 3,4-dimethyl         d. 4-hydroxy**heptanal**      OH ← 4-hydroxy
                                 CH₃CH₂CHCHCH₂CHO                                         CH₃CH₂CH₂CHCH₂CH₂CHO
        6 C chain                CH₃ ←                              7 C chain

**16.35**   To name a ketone using IUPAC rules, use the steps in Example 16.2:
   [1]  Find the longest chain containing the carbonyl group, and change the *-e* ending of the parent alkane to the suffix *-one.*
   [2]  Number the carbon chain to give the carbonyl carbon the lower number.  Apply all of the other usual rules of nomenclature.

a.
$$CH_3CHCH_2-\overset{\overset{O}{\|}}{C}-CH_3$$
        CH₃
   ----→
$$\underset{4}{CH_3CHCH_2}-\overset{\overset{O}{\|}}{\underset{2}{C}}-\overset{1}{CH_3}$$
        CH₃
   ----→  **Answer: 4-methyl-2-pentanone**

   pentan*e* ----→ pentan*one*
        (5 C's)
        ↑
   4-methyl

b.
   CH₃ ⟨cyclohexanone ring⟩ CH₃
   ---------→
   CH₃ ⟨6 1 2⟩ CH₃
   --------→  **Answer: 2,6-dimethylcyclohexanone**

   cyclohexan*e* ----→ cyclohexan*one*
        (6 C's)                    2,6-dimethyl

c.
   ⟨benzene⟩—C(=O)—CH₃
              CH₂CH₂CH₂CH₃
   ----------→
   ⟨benzene⟩—C(=O)—CH₃
              CH₂CH₂CH₂CH₃
   ----------→  **Answer: *o*-butylacetophenone**

   benzene ring with CH₃C=O
        acetophenone                        *o*-butyl

d.

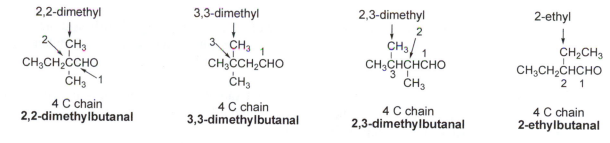

O
‖
CH₃CH—C—CHCH₂CH₃      ----→      CH₃CH  C  CHCH₂CH₃      ----→      **Answer: 2,4-dimethyl-3-hexanone**
     |              |                          1  2  3  4
    CH₃            CH₃                        CH₃     CH₃

hexane ----→ hexan*one*                                   ↑       ↑
(6 C's)                                                   2,4-dimethyl

e.

Cl                                 3-chloro
 |                                    ↗
[cyclopentane ring with Cl and =O]   Cl  3  2
                                         [ring]  1
                    ----→                        =O      ----→      **Answer: 3-chlorocyclopentanone**

cyclopentane ----→ cyclopentan*one*
(5 C's)

**16.37**    Work backwards from the name to draw each structure.

a. 3,3-dimethyl-2-**hexanone**

↓

6 C chain

       3  CH₃    O
          |      ‖
CH₃CH₂CH₂C——C←— 2
          |      |
         CH₃    CH₃
                 1
          ↑
       3,3-dimethyl

c. *m*-ethyl**acetophenone**

↓

benzene ring with a
CH₃C=O

[benzene ring with]      O
                         ‖
                         C
                          \
                           CH₃

*m*-ethyl→ CH₂CH₃

b. methyl propyl **ketone**

↓

two alkyl groups with a
C=O in the middle

        O
        ‖
CH₃——C——CH₂CH₂CH₃
 ↑                ↑
                propyl
methyl

d. 2,4,5-triethyl**cyclohexanone**

↓

6 C ring

                       O
                       ‖
              [ring]  1  2  CH₂CH₃
             5      4              ↑
CH₃CH₂——            
                CH₃CH₂——  CH₂CH₃
2,4,5-triethyl →

**16.39**    Draw the four aldehydes and then name them using the steps in Example 16.1.

2,2-dimethyl
↓
2    CH₃
  \  |
CH₃CH₂CCHO
     |    \
    CH₃    1

4 C chain
**2,2-dimethylbutanal**

3,3-dimethyl
↓
3   CH₃  1
  \  |
CH₃CCH₂CHO
     |
    CH₃

4 C chain
**3,3-dimethylbutanal**

2,3-dimethyl
↓        2
CH₃  1
  \  |
CH₃CHCHCHO
   3    |
       CH₃

4 C chain
**2,3-dimethylbutanal**

2-ethyl
↓
CH₂CH₃
|
CH₃CH₂CHCHO
            2   1

4 C chain
**2-ethylbutanal**

**16.41**    Draw the structure and correct each name.

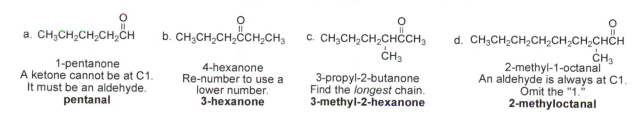

        O
        ‖
a.  CH₃CH₂CH₂CH₂CH

1-pentanone
A ketone cannot be at C1.
It must be an aldehyde.
**pentanal**

        O
        ‖
b.  CH₃CH₂CH₂CCH₂CH₃

4-hexanone
Re-number to use a
lower number.
**3-hexanone**

        O
        ‖
c.  CH₃CH₂CH₂CHCCH₃
               |
              CH₃

3-propyl-2-butanone
Find the *longest* chain.
**3-methyl-2-hexanone**

                O
                ‖
d.  CH₃CH₂CH₂CH₂CH₂CH₂CHCH
                         |
                        CH₃

2-methyl-1-octanal
An aldehyde is always at C1.
Omit the "1."
**2-methyloctanal**

**16.43** Draw benzaldehyde and then the hydrogen bond.

**16.45** Aldehydes and ketones have *higher* boiling points than hydrocarbons of comparable size. Aldehydes and ketones have *lower* boiling points than alcohols of comparable size.

a.　$(CH_3)_3CCH_2CH_2CH_3$　or　$(CH_3)_3CCH_2CHO$　　　b.

|  hydrocarbon | aldehyde<br>**higher boiling point** | ketone | alcohol<br>**higher boiling point** |

**16.47** Aldehydes and ketones have *higher* melting points than hydrocarbons of comparable size. Aldehydes and ketones have *lower* melting points than alcohols of comparable size.

Increasing melting point

**16.49** Low molecular weight aldehydes and ketones (less than six carbons) are water soluble.

|  a. 7 C aldehyde<br>**insoluble** | b. 4 C ketone<br>**soluble** | c. hydrocarbon<br>**insoluble** |

**16.51** 2,3-Butanedione has two carbonyl groups capable of hydrogen bonding whereas acetone has one carbonyl group. This makes 2,3-butanedione more water soluble than acetone. 2,3-Butanedione would also be soluble in an organic solvent like diethyl ether by the "like dissolves like" rule.

**16.53** Draw the product of each reaction using the steps in Example 16.3. Compounds that contain a C–H and C–O bond on the *same* carbon are oxidized with $K_2Cr_2O_7$.

- Aldehydes (RCHO) are oxidized to $RCO_2H$.
- Ketones ($R_2CO$) are *not* oxidized with $K_2Cr_2O_7$.
- 1° Alcohols ($RCH_2OH$) are oxidized to $RCO_2H$ (Section 14.5B).

a. $CH_3(CH_2)_4CHO \xrightarrow{\;K_2Cr_2O_7\;} CH_3(CH_2)_4COOH$　　　b.

c. $\xrightarrow{K_2Cr_2O_7}$ No reaction    d. $CH_3(CH_2)_4CH_2OH \xrightarrow{K_2Cr_2O_7} CH_3(CH_2)_4COOH$

**16.55** Draw the product of each reaction using Example 16.4 as a guide. **Only aldehydes (RCHO) react with Tollens reagent, and they are oxidized to RCO$_2$H.** Ketones and alcohols are inert to oxidation.

a. $CH_3(CH_2)_4CHO \xrightarrow[NH_4OH]{Ag_2O} CH_3(CH_2)_4COOH$

c. $\xrightarrow[NH_4OH]{Ag_2O}$ No reaction

b. $\xrightarrow[NH_4OH]{Ag_2O}$

d. $CH_3(CH_2)_4CH_2OH \xrightarrow[NH_4OH]{Ag_2O}$ No reaction

**16.57** Answer each question about erythrulose.

a, b.

c. $\xrightarrow{\text{Tollens reagent}}$ No reaction

d. $\xrightarrow{K_2Cr_2O_7}$

**16.59** Work backwards to determine what aldehyde can be used to prepare each carboxylic acid.

a. $CH_3CH_2\overset{\overset{\displaystyle CH_3}{|}}{C}HCH_2CO_2H \longleftarrow CH_3CH_2\overset{\overset{\displaystyle CH_3}{|}}{C}HCH_2CHO$

c. $CH_3CH_2\overset{\underset{\displaystyle CO_2H}{|}}{C}HCH_2CH_3 \longleftarrow CH_3CH_2\overset{\underset{\displaystyle CHO}{|}}{C}HCH_2CH_3$

b. $CH_3\text{—} \bigcirc \text{—}CO_2H \longleftarrow CH_3\text{—}\bigcirc\text{—}CHO$

**16.61** Draw the products of reduction using the steps in Example 16.5.
• Locate the C=O and mentally break one bond in the double bond.
• Mentally break the H–H bond of the reagent.
• Add one H atom to each atom of the C=O, forming new C–H and O–H single bonds.

a. $CH_3CH_2\text{—}\bigcirc\text{—}CHO \xrightarrow[Pd]{H_2} CH_3CH_2\text{—}\bigcirc\text{—}CH_2OH$

b. $\xrightarrow[Pd]{H_2}$

**16.63**

a, b: $CH_3CH_2\text{—}\overset{\overset{\displaystyle CH_3}{|}}{C}H\text{—}(CH_2)_4CHO$ <br>chirality center

c. $CH_3CH_2\text{—}\overset{\overset{\displaystyle CH_3}{|}}{C}H\text{—}(CH_2)_4CHO \xrightarrow[Pd]{H_2} CH_3CH_2\text{—}\overset{\overset{\displaystyle CH_3}{|}}{C}H\text{—}(CH_2)_4CH_2OH$

**16.65**   Work backwards to determine what carbonyl compound is needed to make each alcohol.

a.   $CH_3CH_2CH_2CH_2CH_2OH$ ⟵ $CH_3CH_2CH_2CH_2$ $\overset{\overset{\displaystyle O}{\|}}{C}$ $H$          b.

**16.67**   1-Methylcyclohexanol is a 3° alcohol and cannot be produced from the reduction of a carbonyl compound because only 1° or 2° alcohols can be formed in these reactions.

**16.69**   Recall the definitions from Example 16.7 to draw a compound of molecular formula $C_5H_{12}O_2$ that fits each description:
- An ether has the general structure ROR.
- A hemiacetal has one C bonded to OH and OR.
- An acetal has one C bonded to two OR groups.

a.   $CH_3CH_2{-}O{-}\overset{\overset{\displaystyle H}{|}}{\underset{\underset{\displaystyle H}{|}}{C}}{-}O{-}CH_2CH_3$
**acetal**

c.   $CH_3CH_2{-}O{-}\overset{\overset{\displaystyle H}{|}}{\underset{\underset{\displaystyle H}{|}}{C}}\overset{\overset{\displaystyle H}{|}}{\underset{\underset{\displaystyle H}{|}}{C}}{-}O{-}CH_3$
**two ethers**

b.   $CH_3{-}\overset{\overset{\displaystyle H}{|}}{\underset{\underset{\displaystyle OH}{|}}{C}}{-}O{-}CH_2CH_2CH_3$
**hemiacetal**

d.   $CH_3{-}\overset{\overset{\displaystyle H}{|}}{\underset{\underset{\displaystyle H}{|}}{C}}{-}O{-}\overset{\overset{\displaystyle H}{|}}{\underset{\underset{\displaystyle H}{|}}{C}}\overset{\overset{\displaystyle OH}{|}}{\underset{\underset{\displaystyle H}{|}}{C}}{-}CH_3$  ⟵ **alcohol**
**ether**

**16.71**   Label the functional groups using the definitions from Example 16.7.

a.   $CH_3{-}\overset{\overset{\displaystyle OCH_3}{|}}{\underset{\underset{\displaystyle OCH_3}{|}}{C}}{-}H$   **acetal**

b.   **hemiacetal**   $CH_3{-}\overset{\overset{\displaystyle OCH_2CH_3}{|}}{\underset{\underset{\displaystyle OH}{|}}{C}}{-}H$

c.   **ether** $HOCH_2\overset{\overset{\displaystyle OCH_3}{|}}{C}HCH_2CH_3$   **alcohol**

d.   **ether**

**16.73**   To form a hemiacetal and acetal from a carbonyl compound, use the steps in Example 16.6.
- Locate the C=O in the starting material.
- Break one C–O bond and add one equivalent of $CH_3OH$ across the double bond, placing the $OCH_3$ group on the carbonyl carbon. This forms the hemiacetal.
- Replace the OH group of the hemiacetal by $OCH_3$ to form the acetal.

a.   $\xrightarrow[\text{}]{\begin{array}{c}\text{2 CH}_3\text{OH}\\ \text{H}_2\text{SO}_4\end{array}}$

c.   $CH_3{-}\overset{\overset{\displaystyle O}{\|}}{C}{-}CH_2CH_2CH_3$ $\xrightarrow[\text{}]{\begin{array}{c}\text{2 CH}_3\text{OH}\\ \text{H}_2\text{SO}_4\end{array}}$ $CH_3{-}\overset{\overset{\displaystyle CH_3O}{|}}{\underset{\underset{\displaystyle CH_2CH_2CH_3}{|}}{C}}{-}OCH_3$

b.   $CH_2{=}O$ $\xrightarrow[\text{}]{\begin{array}{c}\text{2 CH}_3\text{OH}\\ \text{H}_2\text{SO}_4\end{array}}$ $CH_2(OCH_3)_2$

d.   ${-}CH_2CHO$ $\xrightarrow[\text{}]{\begin{array}{c}\text{2 CH}_3\text{OH}\\ \text{H}_2\text{SO}_4\end{array}}$ ${-}CH_2\overset{\overset{\displaystyle OCH_3}{|}}{\underset{\underset{\displaystyle OCH_3}{|}}{C}}H$

**16.75** Draw the products of each reaction.

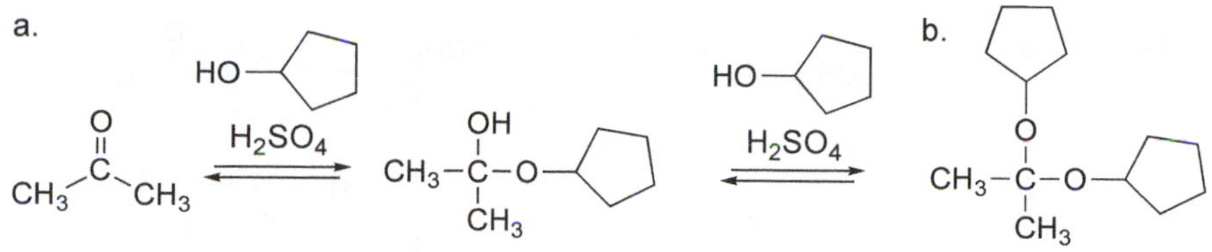

a.

b.

**16.77** Answer each question.

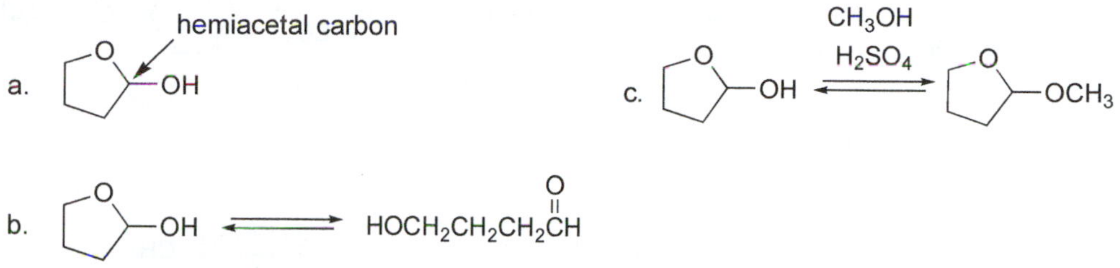

a.

hemiacetal carbon

b.

c.

**16.79** Draw the product of cyclization.

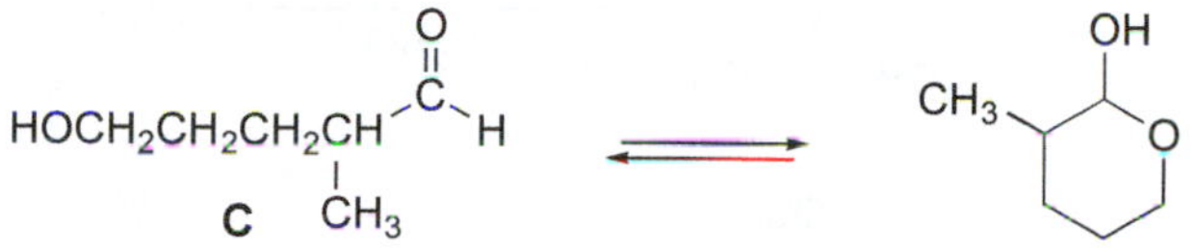

**16.81** To draw the products of hydrolysis, use the steps in Example 16.8.
- Locate the two C–OR bonds on the same carbon.
- Replace the two C–O single bonds with a carbonyl group (C=O).
- Each OR group then becomes a molecule of alcohol (ROH) product.

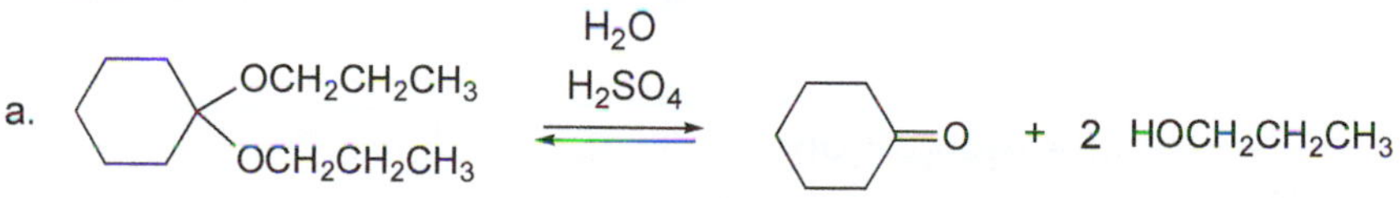

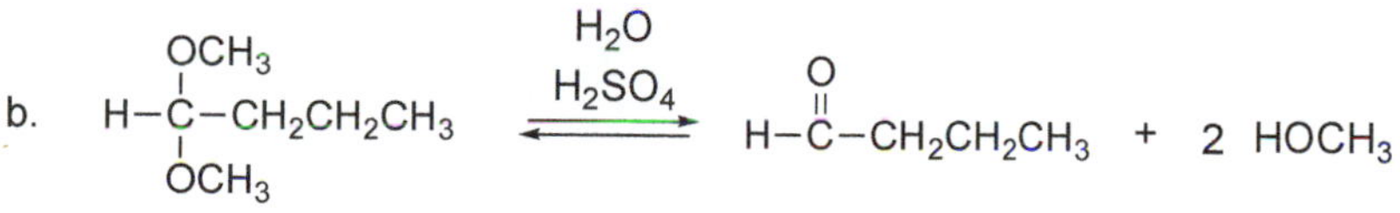

a.

b.

**16.83** Answer each question about compound **A**.

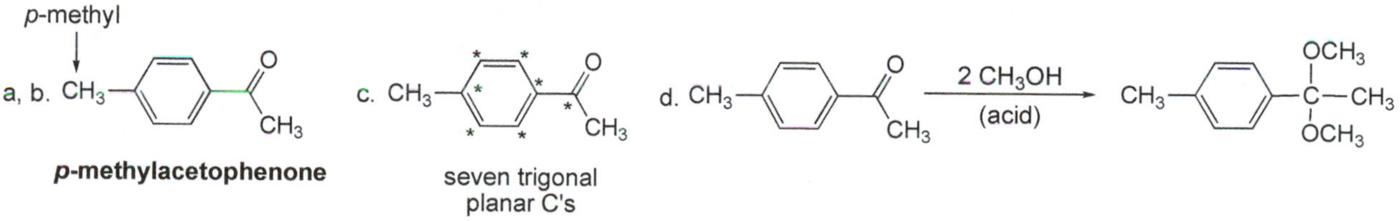

*p*-methyl

a, b. 

**p-methylacetophenone**

c.

seven trigonal planar C's

d.

**16.85** Draw the products of each reaction.

a. (benzaldehyde) $\xrightarrow[\text{Pd}]{H_2}$ (benzyl alcohol) —CH₂OH   d. (benzaldehyde) $\xrightarrow[\text{H}_2\text{SO}_4]{2\ CH_3OH}$ product with OCH₃, C—H, OCH₃

b. (benzaldehyde) $\xrightarrow{K_2Cr_2O_7}$ —COH   e. (benzaldehyde) $\xrightarrow[\text{H}_2\text{SO}_4]{2\ CH_3CH_2OH}$ product with OCH₂CH₃, C—H, OCH₂CH₃

c. (benzaldehyde) $\xrightarrow[\text{NH}_4\text{OH}]{Ag_2O}$ —COH   f. (acetal with OCH₂CH₃, C—H, OCH₂CH₃) $\xrightarrow[\text{H}_2\text{SO}_4]{H_2O}$ benzaldehyde + 2 CH₃CH₂OH

**16.87** Draw the products of each reaction.

a. $CH_3\overset{O}{\underset{}{C}}(CH_2)_4CH_3$ $\xrightarrow[\text{Pd}]{H_2}$ $CH_3-\overset{OH}{\underset{H}{C}}-(CH_2)_4CH_3$   d. $CH_3\overset{O}{\underset{}{C}}(CH_2)_4CH_3$ $\xrightarrow[\text{H}_2\text{SO}_4]{2\ CH_3OH}$ $CH_3-\overset{OCH_3}{\underset{OCH_3}{C}}-(CH_2)_4CH_3$

b. $CH_3\overset{O}{\underset{}{C}}(CH_2)_4CH_3$ $\xrightarrow{K_2Cr_2O_7}$ No reaction   e. $CH_3\overset{O}{\underset{}{C}}(CH_2)_4CH_3$ $\xrightarrow[\text{H}_2\text{SO}_4]{2\ CH_3CH_2OH}$ $CH_3-\overset{OCH_2CH_3}{\underset{OCH_2CH_3}{C}}-(CH_2)_4CH_3$

c. $CH_3\overset{O}{\underset{}{C}}(CH_2)_4CH_3$ $\xrightarrow[\text{NH}_4\text{OH}]{Ag_2O}$ No reaction   f. $CH_3-\overset{OCH_2CH_3}{\underset{OCH_2CH_3}{C}}-(CH_2)_4CH_3$ $\xrightarrow[\text{H}_2\text{SO}_4]{H_2O}$ $CH_3\overset{O}{\underset{}{C}}(CH_2)_4CH_3$ + 2 CH₃CH₂OH

**16.89** Draw the three constitutional isomers that can be converted to 1-pentanol. The starting material needs a C=O at C1 and a C=C.

$CH_2=CHCH_2CH_2CHO$
or
$CH_3CH=CHCH_2CHO$  $\xrightarrow[\text{Pd}]{H_2}$  $CH_3CH_2CH_2CH_2CH_2OH$
or
$CH_3CH_2CH=CHCHO$

**16.91** Draw the products of each reaction.

a. (steroid with ketone) $\xrightarrow[\text{Pd}]{H_2}$ (steroid with alcohol)

b.

c.

d.

**16.93** Draw the product of oxidation.

**16.95** Label each hemiacetal or alcohol.

**16.97** The main reaction that occurs in the rod cells in the retina is conversion of 11-*cis*-retinal to its trans isomer. The cis double bond in 11-*cis*-retinal produces crowding, making the molecule unstable. Light energy converts this to the more stable trans isomer, and with this conversion an electrical impulse is generated in the optic nerve.

**16.99** Identify the alcohol, acetal, hemiacetal, ether, and carboxylic acid functional groups.

# Chapter 17 Carboxylic Acids, Esters, and Amides

## Chapter Review

**[1] What are the characteristics of carboxylic acids, esters, and amides?**
- Carboxylic acids have the general structure RCOOH; esters have the general structure RCOOR'; amides have the general structure RCONR'$_2$, where R' = H or alkyl. (17.1)

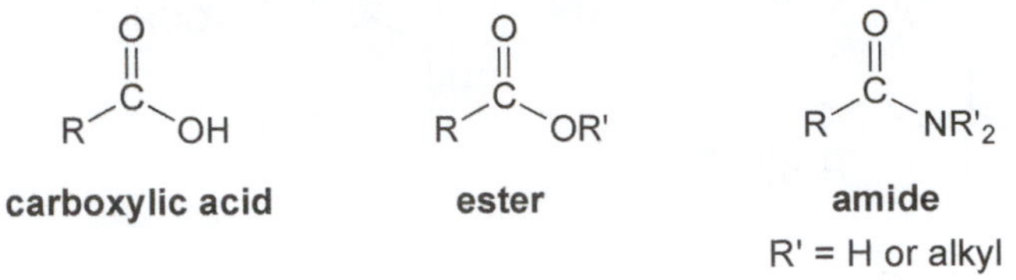

carboxylic acid  ester  amide

R' = H or alkyl

- The carbonyl carbon is trigonal planar and bond angles are 120°. (17.1)

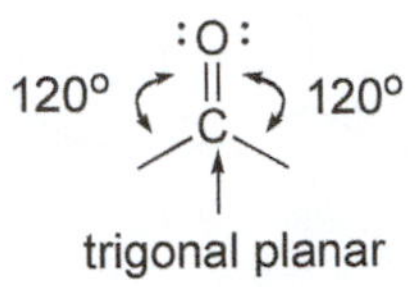

120°  120°

trigonal planar

- All acyl compounds have a polar C=O. RCO$_2$H, RCONH$_2$, and RCONHR' are capable of intermolecular hydrogen bonding. (17.3)

**[2] How are carboxylic acids, esters, and amides named? (17.2)**
- Carboxylic acids are identified by the suffix *-oic acid*.
- Esters are identified by the suffix *-ate*.
- Amides are identified by the suffix *-amide*.

**[3] Give examples of useful carboxylic acids. (17.4)**
- α-Hydroxy carboxylic acids are used in skin care products.

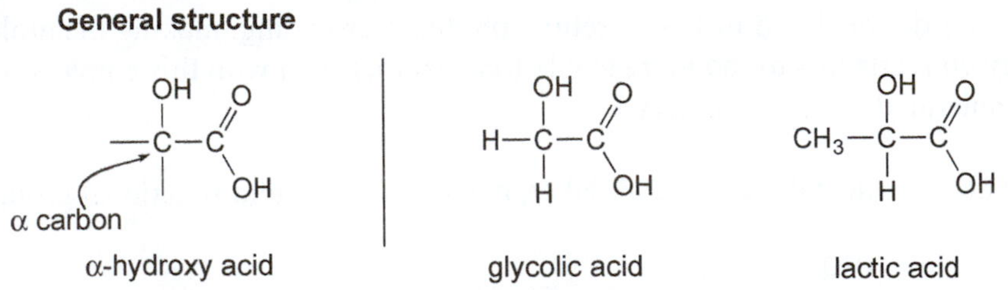

**General structure**

α-hydroxy acid  glycolic acid  lactic acid

- Aspirin, ibuprofen, and naproxen are pain relievers and anti-inflammatory agents.
- Aspirin is an anti-inflammatory agent because it blocks the synthesis of prostaglandins from arachidonic acid.

**[4] Give examples of useful esters and amides. (17.5)**
- Some esters—ethyl butanoate, pentyl butanoate, and methyl salicylate—have characteristic odors and flavors.
- Benzocaine is the active ingredient in over-the-counter oral topical anesthetics.
- Acetaminophen is the active ingredient in Tylenol.

**[5] What products are formed when carboxylic acids are treated with base? (17.6)**

- Carboxylic acids react with bases to form carboxylate anions (RCOO⁻).

$$R-C(=O)-O-H \quad + \quad Na^+\ OH^- \quad \longrightarrow \quad R-C(=O)-O^-\ Na^+ \quad + \quad H_2O$$

- Carboxylate anions are water soluble and commonly used as preservatives.

**[6] How does soap clean away dirt? (17.6C)**

- Soaps are salts of carboxylic acids that have many carbon atoms in a long hydrocarbon chain. A soap molecule has an ionic head and a nonpolar hydrocarbon tail.

**Structure of a soap molecule**

$$Na^+\ {}^-O-C(=O)-CH_2CH_2CH_2CH_2CH_2CH_2CH_2CH_2CH_2CH_2CH_2CH_2CH_2CH_2CH_2CH_2CH_3$$

ionic end · · · · · · · · · · · · · · · long, hydrocarbon chain

**polar head** · · · · · · · · · · · · · **nonpolar tail**

- Soap forms micelles in water with the polar heads on the surface and the hydrocarbon tails in the interior. Grease and dirt dissolve in the nonpolar tails, making it possible to wash them away with water.

**[7] Discuss the acid–base chemistry of aspirin. (17.7)**

- Aspirin remains in its neutral form in the stomach, and in this form it can cross a cell membrane and serve as an anti-inflammatory agent.
- A proton is removed from aspirin in the basic environment of the intestines to form an ionic carboxylate anion. This form is not absorbed.

**[8] How are carboxylic acids converted to esters and amides? (17.8)**

- Carboxylic acids are converted to esters by reaction with alcohols (R'OH) and acid ($H_2SO_4$).

$$R-C(=O)-OH \quad + \quad R'OH \quad \underset{}{\overset{H_2SO_4}{\rightleftharpoons}} \quad R-C(=O)-OR' \quad + \quad H_2O$$
$$\text{ester}$$

- Carboxylic acids are converted to amides by heating with ammonia ($NH_3$) or amines (R'NH₂ or R'₂NH).

$$R-C(=O)-OH \quad + \quad R'_2NH \quad \overset{\Delta}{\longrightarrow} \quad R-C(=O)-NR'_2 \quad + \quad H_2O$$

R' = H or alkyl · · · · · · · · · · · · · · · · · amide

**[9] What hydrolysis products are formed from esters and amides? (17.9)**

- Esters are hydrolyzed to carboxylic acids (RCOOH) in the presence of an acid catalyst ($H_2SO_4$). Esters are converted to carboxylate anions (RCOO⁻) with aqueous base (NaOH in $H_2O$).

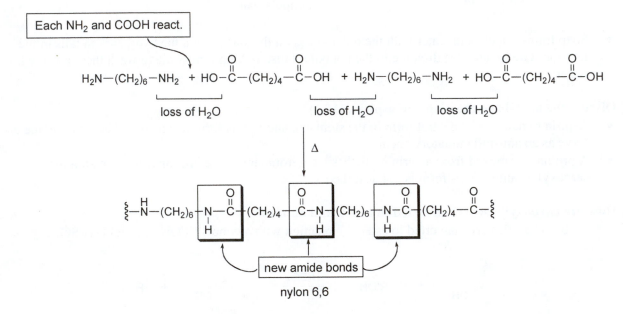

- Amides are hydrolyzed to carboxylic acids (RCOOH) in the presence of an acid catalyst (HCl). Amides are converted to carboxylate anions (RCOO⁻) with aqueous base (NaOH in $H_2O$).

**[10] What are polyamides and polyesters and how are they formed? (17.10)**
  - Polyamides like nylon are polymers that contain many amide bonds. They are formed when a dicarboxylic acid is heated with a diamine.

Each NH$_2$ and COOH react.

$H_2N-(CH_2)_6-NH_2$ + $HO-\overset{O}{\underset{}{C}}-(CH_2)_4-\overset{O}{\underset{}{C}}-OH$ + $H_2N-(CH_2)_6-NH_2$ + $HO-\overset{O}{\underset{}{C}}-(CH_2)_4-\overset{O}{\underset{}{C}}-OH$

loss of H$_2$O    loss of H$_2$O    loss of H$_2$O

Δ

new amide bonds

nylon 6,6

- Polyesters like PET are polymers that contain many ester bonds. They are formed when a dicarboxylic acid is treated with a diol in the presence of acid ($H_2SO_4$).

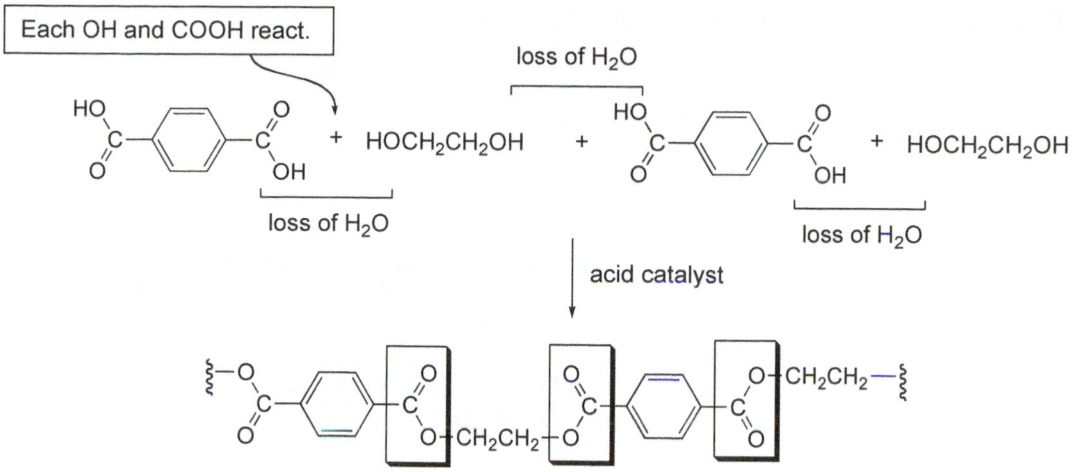
Each OH and COOH react.
loss of H2O
HO
C
O
O
C
OH
+ HOCH2CH2OH
+
HO
C
O
O
C
OH
+ HOCH2CH2OH
loss of H2O
loss of H2O
acid catalyst
O
C
O
O
C
O
CH2CH2
O
C
O
O
C
O
CH2CH2

polyethylene terephthalate
PET

**[11] How does penicillin act as an antibiotic? (17.11)**

- The β-lactam of penicillin is more reactive than a regular amide and it reacts with an enzyme needed to synthesize the cell wall of a bacterium. Without a cell wall, the bacterium dies.

## Problem Solving

### [1] Nomenclature (17.2)

**Example 17.1** Give the IUPAC name of the following carboxylic acid.

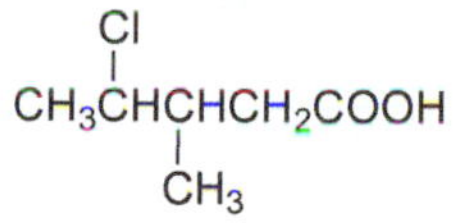
Cl
CH3CHCHCH2COOH
CH3

**Analysis and Solution**

[1] Find and name the longest chain containing COOH:

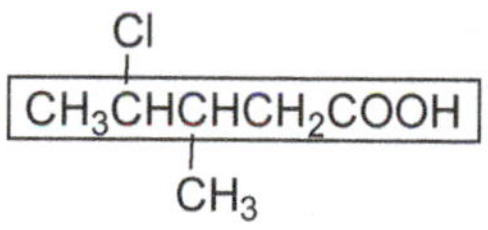
Cl
CH3CHCHCH2COOH
CH3

pentane ---→ pentan*oic acid*
(5 C's)

The COOH contributes one C to the longest chain.

[2] Number and name the substituents, making sure the COOH group is at C1:

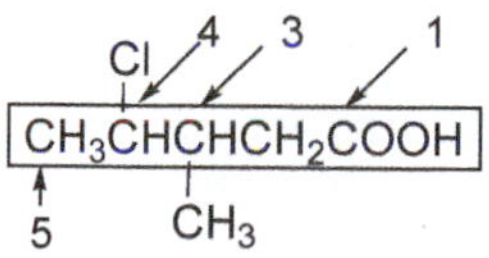
Cl
4
3
1
CH3CHCHCH2COOH
5
CH3

one methyl substituent on C3
one chloro substituent on C4

**Answer: 4-chloro-3-methylpentanoic acid**

**Example 17.2** Give the IUPAC name of the following ester.

$$\overset{O}{\underset{CH_3CH_2O}{\overset{\|}{C}}}-CH_2CH_2CH_2CH_3$$

**Analysis and Solution**

[1] Name the alkyl group on the O atom:

$$\overset{O}{\underset{CH_3CH_2O}{\overset{\|}{C}}}-CH_2CH_2CH_2CH_3$$

ethyl group

The word *ethyl* becomes the first part of the name.

[2] To name the acyl group, find and name the longest chain containing the carbonyl group, placing the C=O at C1:

$$\overset{O}{\underset{CH_3CH_2O}{\overset{\|}{C}}}\boxed{-CH_2CH_2CH_2CH_3}$$

pentano*ic acid* ---→ pentano*ate*
(5 C's)

**Answer: ethyl pentanoate**

---

## [2] The Acidity of Carboxylic Acids (17.6)

**Example 17.3** What products are formed when butanoic acid ($CH_3CH_2CH_2COOH$) reacts with potassium hydroxide (KOH)?

**Analysis**
In any acid–base reaction with a carboxylic acid:
- Remove a proton from the carboxyl group (COOH) and form the carboxylate anion ($RCOO^-$).
- Add a proton to the base. If the base has a ($-1$) charge to begin with, it becomes a neutral product when a proton ($H^+$) is added to it.
- Balance the charge of the carboxylate anion by drawing it as a salt with a metal cation.

**Solution**

$$\underset{\text{butanoic acid}}{\overset{O}{\underset{CH_3CH_2CH_2}{\overset{\|}{C}}}\diagup_{O-H}} \quad + \quad \underset{\text{base}}{K^+\ OH^-} \longrightarrow \underset{\substack{\text{The carboxylate anion is formed} \\ \text{as a potassium salt.}}}{\overset{O}{\underset{CH_3CH_2CH_2}{\overset{\|}{C}}}\diagup_{O^-\ K^+}} \quad + \quad H-O-H$$

This proton is transferred from the acid to the base.

Thus, $CH_3CH_2CH_2COOH$ loses a proton to form $CH_3CH_2CH_2COO^-$, which is present in solution as its potassium salt, $CH_3CH_2CH_2COO^-\ K^+$. Hydroxide ($OH^-$) gains a proton to form $H_2O$.

---

**Example 17.4** Give an acceptable name for each salt.

a.
$$\underset{\underset{CH_3}{|}}{\overset{O}{\underset{CH_3CH}{\overset{\|}{C}}}}\diagup_{O^-\ K^+}$$

b.
$$\overset{O}{\underset{CH_3CH_2CH_2}{\overset{\|}{C}}}\diagup_{O^-\ Na^+}$$

**Analysis**

Name the carboxylate salt by putting three parts together:

- the name of the metal cation
- the parent name that indicates the number of carbons in the parent chain
- the suffix, *-ate*

**Solution**

a.

potassium cation

$CH_3CH$ (with $CH_3$ substituent) — $C(=O)$ — $O^-$ $K^+$

2-methyl substituent
parent + suffix
propano- -ate

- The first part of the name is the metal cation, potassium.
- The parent name is derived from the IUPAC name, propanoic acid. Change the *-ic acid* ending to *-ate*; propanoic acid → propanoate.
- **Answer: potassium 2-methylpropanoate**

b.

sodium cation

$CH_3CH_2CH_2$ — $C(=O)$ — $O^-$ $Na^+$

parent + suffix
butano- -ate

- The first part of the name is the metal cation, sodium.
- The parent name is derived from the IUPAC name, butanoic acid. Change the *-ic acid* ending to *-ate*; butanoic acid → butanoate.
- **Answer: sodium butanoate**

---

## [3] The Conversion of Carboxylic Acids to Esters and Amides (17.8)

**Example 17.5** What ester is formed when butanoic acid ($CH_3CH_2CH_2COOH$) is treated with methanol ($CH_3OH$) in the presence of $H_2SO_4$?

**Analysis**

To draw the products of any acyl substitution, arrange the carboxyl group of the carboxylic acid next to the functional group with which it reacts—the OH group of the alcohol in this case. Then replace the OH group of the carboxylic acid by the OR' group of the alcohol, forming a new C–O bond at the carbonyl carbon.

Remove OH and H to form $H_2O$.

carboxylic acid + H–OR' ⟶ ester + H–OH     (new C–O bond)

**Solution**

Replace the OH group of butanoic acid by the $OCH_3$ group of methanol to form the ester.

$CH_3CH_2CH_2$—C(=O)—OH + H–OCH$_3$ $\xrightarrow{H_2SO_4}$ $CH_3CH_2CH_2$—C(=O)—OCH$_3$ + H–OH     (new C–O bond)

$OCH_3$ replaces OH.

**Example 17.6** What amide is formed when butanoic acid ($CH_3CH_2CH_2COOH$) is heated with propylamine ($CH_3CH_2CH_2NH_2$)?

**Analysis**

To draw the products of amide formation, arrange the carboxyl group of the carboxylic acid next to the H–N bond of the amine. Then replace the OH group of the carboxylic acid by the $NHCH_2CH_2CH_3$ group of the amine, forming a new C–N bond at the carbonyl carbon.

**Solution**

The reaction of RCOOH with an amine that has one alkyl group on the N atom ($R'NH_2$) forms a $2°$ amide ($RCONHR'$).

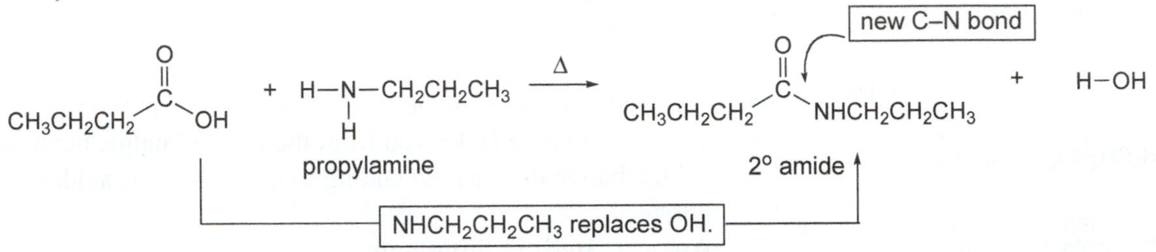

## [4] Hydrolysis of Esters and Amides (17.9)

**Example 17.7** What products are formed when ethyl acetate ($CH_3CO_2CH_2CH_3$) is hydrolyzed with water in the presence of $H_2SO_4$?

**Analysis**

To draw the products of hydrolysis in acid, replace the OR' group of the ester by an OH group from water, forming a new C–O bond at the carbonyl carbon. A molecule of alcohol (R'OH) is also formed from the alkoxy group (OR') of the ester.

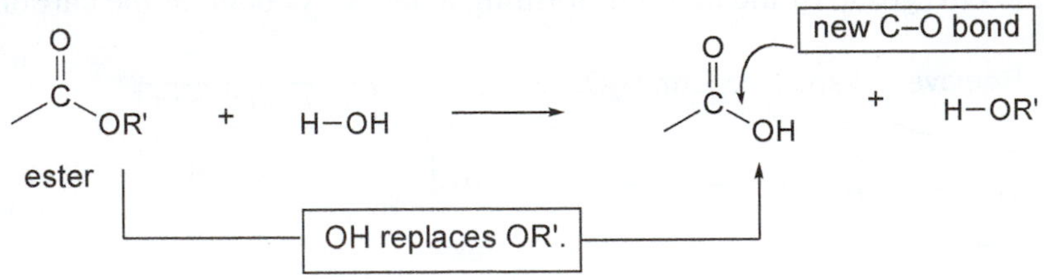

**Solution**

Replace the $OCH_2CH_3$ group of ethyl acetate by the OH group of water to form acetic acid ($CH_3CO_2H$) and ethanol ($CH_3CH_2OH$).

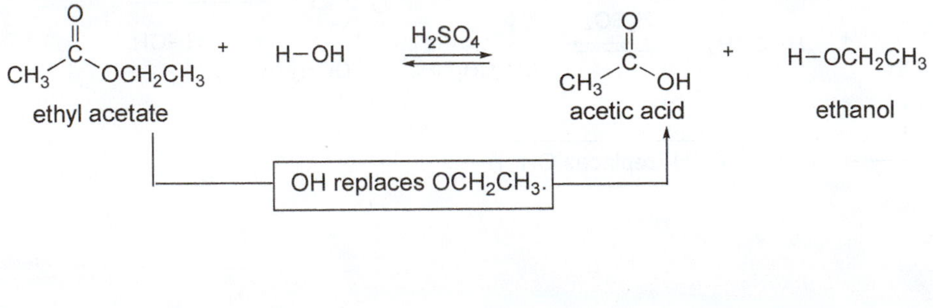

**Example 17.8** What products are formed when $N$-ethylacetamide ($CH_3CONHCH_2CH_3$) is hydrolyzed with water in the presence of NaOH?

**Analysis**

To draw the products of amide hydrolysis in base, replace the NHR' group of the amide by an oxygen anion ($O^-$), forming a new C–O bond at the carbonyl carbon. A molecule of amine ($R'NH_2$) is also formed from the nitrogen group (NHR') of the amide.

**Solution**

Replace the $NHCH_2CH_3$ group of $N$-ethylacetamide by a negatively charged oxygen atom ($O^-$) to form sodium acetate ($CH_3CO_2^-\ Na^+$) and ethylamine ($CH_3CH_2NH_2$).

## Self-Test

**[1] Fill in the blank with one of the terms listed below.**

Amides (17.1)
Carboxylic acids (17.10)
Condensation polymer (17.10)
Esters (17.1)
Hydrolysis (17.9)

α-Hydroxy acids (17.4)
Lactam (17.1)
Lactone (17.1)
Lipids (17.9)

Micelles (17.6)
Saponification (17.9)
Soaps (17.6)
Triacylglycerols (17.9)

1. A cyclic ester is called a _____.
2. When bonds are cleaved on reaction with water, this reaction is called _____.
3. _____ are carbonyl compounds that contain a nitrogen atom bonded to the carbonyl carbon.
4. _____ are spherical droplets having the ionic heads on the surface and the nonpolar tails packed together in the interior.
5. _____ contain a hydroxyl group on the α (alpha) carbon to the carboxyl group.
6. _____ contain three ester groups, each having a long carbon chain bonded to the carbonyl group.
7. _____ are organic compounds containing a carboxyl group.
8. _____ are salts of carboxylic acids that have many carbon atoms in a long hydrocarbon chain.
9. A cyclic amide is called a _____.
10. Basic hydrolysis of an ester is called _____.

11. _____ are carbonyl compounds that contain an alkoxy group (OR') bonded to the carbonyl carbon.
12. _____ are water-insoluble organic compounds found in biological systems.
13. A _____ is a polymer formed when monomers containing two functional groups come together with loss of a small molecule such as water.

## [2] Label each compound as an amide, ester, ether, or carboxylic acid.

14. $CH_3CH_2-O-CH_2CH_2CH_3$

16.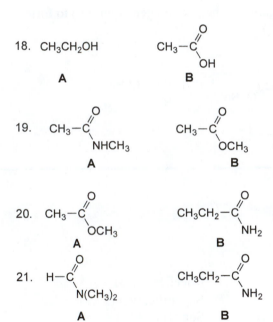

15.

17.

## [3] Which compound in each pair has the higher boiling point?

18. $CH_3CH_2OH$          $CH_3-C$ (O, OH)

   A          B

19. $CH_3-C$ (O, $NHCH_3$)          $CH_3-C$ (O, $OCH_3$)

   A          B

20. $CH_3-C$ (O, $OCH_3$)          $CH_3CH_2-C$ (O, $NH_2$)

   A          B

21. $H-C$ (O, $N(CH_3)_2$)          $CH_3CH_2-C$ (O, $NH_2$)

   A          B

**[4] Match the reaction with the products.**

22.

23.

24.

25.

## Answers to Self-Test

| | | | | |
|---|---|---|---|---|
| 1. lactone | 6. Triacylglycerols | 11. Esters | 16. ester | 21. B |
| 2. hydrolysis | 7. Carboxylic acids | 12. Lipids | 17. carboxylic acid | 22. b |
| 3. Amides | 8. Soaps | 13. condensation polymer | 18. B | 23. d |
| 4. Micelles | 9. lactam | 14. ether | 19. A | 24. c |
| 5. $\alpha$-Hydroxy acids | 10. saponification | 15. amide | 20. B | 25. a |

## Solutions to In-Chapter Problems

**17.1 Carboxylic acids** are organic compounds containing a carboxyl group (COOH).
**Esters** are carbonyl compounds that contain an alkoxy group (OR') bonded to the carbonyl carbon.
**Amides** are carbonyl compounds that contain a nitrogen atom bonded to the carbonyl carbon.

**17.2** A **primary ($1^0$) amide** contains one C–N bond. A $1^0$ amide has the structure $RCONH_2$.
A **secondary ($2^0$) amide** contains two C–N bonds. A $2^0$ amide has the structure RCONHR'.
A **tertiary ($3^0$) amide** contains three C–N bonds. A $3^0$ amide has the structure RCONR'$_2$.

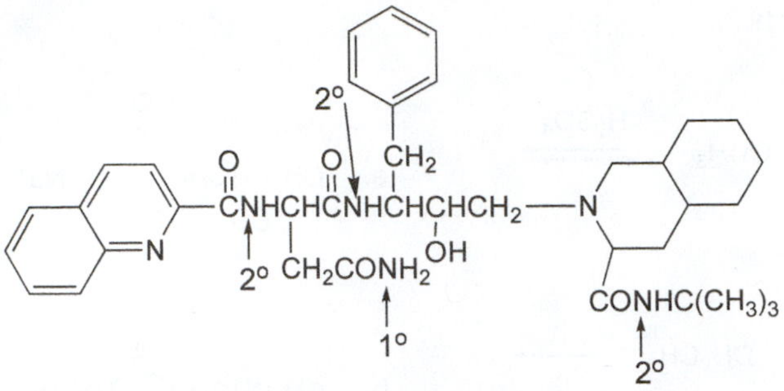

**17.3** Label the functional groups and the trigonal planar carbons in lisinopril.

a, b: The nine trigonal planar carbons are labeled (*).

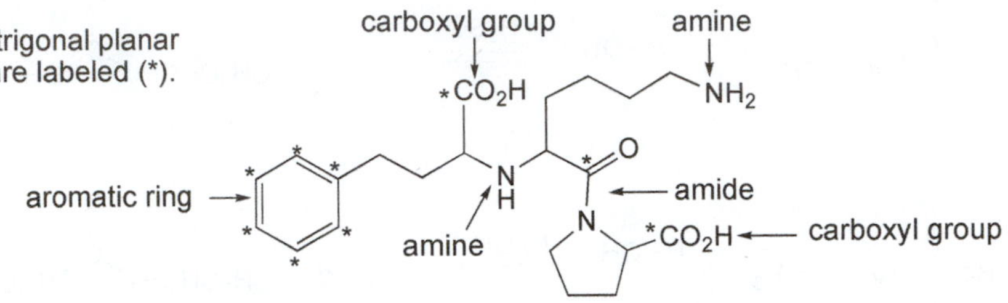

**17.4** Name the carboxylic acids as in Example 17.1.

a.

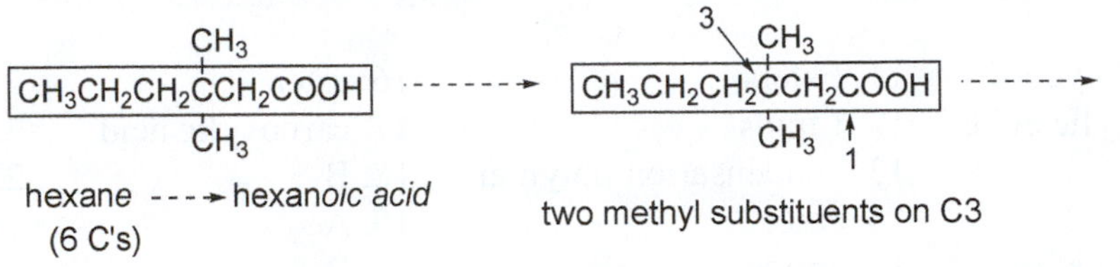

$$CH_3CH_2CH_2CCH_2COOH$$
with $CH_3$ above and $CH_3$ below

hexane ---→ hexanoic acid
(6 C's)

$$\overset{3}{C}CH_3CH_2CH_2CCH_2COOH$$
with $CH_3$ above, $CH_3$ below, 1

two methyl substituents on C3

**Answer: 3,3-dimethylhexanoic acid**

b.

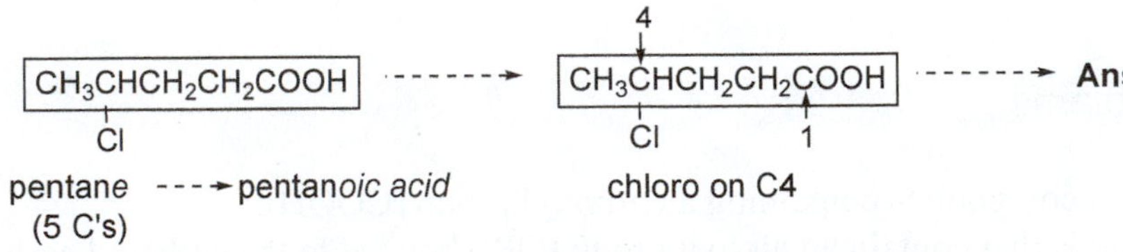

$$CH_3CHCH_2CH_2COOH$$
with $Cl$ below

pentane ---→ pentanoic acid
(5 C's)

$$\overset{4}{CH_3}CHCH_2CH_2COOH$$
with $Cl$ below, 1

chloro on C4

**Answer: 4-chloropentanoic acid**

c.

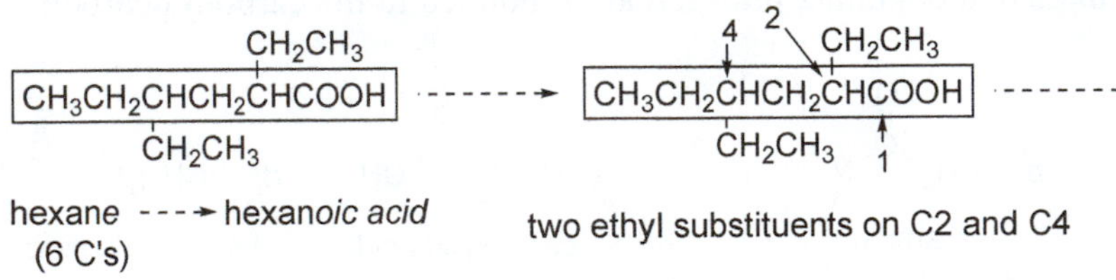

$$CH_3CH_2CHCH_2CHCOOH$$
with $CH_2CH_3$ above and $CH_2CH_3$ below

hexane ---→ hexanoic acid
(6 C's)

$$\overset{4}{CH_3}CH_2\overset{2}{CH}CH_2CHCOOH$$
with $CH_2CH_3$ above, $CH_2CH_3$ below, 1

two ethyl substituents on C2 and C4

**Answer: 2,4-diethylhexanoic acid**

**17.5** Work backwards from the IUPAC name to draw the structure of each carboxylic acid.

a. 2-bromo**butanoic acid**

4 C chain with
COOH at C1   ---->

$$CH_3CH_2CH \overset{\overset{\displaystyle O}{\|}}{-}C-OH$$
$$|$$
$$Br$$

2-bromo

c. 2-ethyl-5,5-dimethyl**octanoic acid**

8 C chain with
COOH at C1   ---->

$$CH_3 \qquad\qquad O$$
$$| \qquad\qquad \|$$
$$CH_3CH_2CH_2CCH_2CH_2CH-C-OH$$
$$| \qquad\qquad |$$
$$CH_3 \qquad\quad CH_2CH_3$$

5,5-dimethyl     2-ethyl

b. 2,3-dimethyl**pentanoic acid**

5 C chain with
COOH at C1   ---->

$$CH_3 \quad O$$
$$| \quad \|$$
$$CH_3CH_2CHCH-C-OH$$
$$|$$
$$CH_3$$

2,3-dimethyl

d. 3,4,5,6-tetraethyl**decanoic acid**

10 C chain with
COOH at C1   ---->

$$CH_3CH_2 \quad CH_2CH_3 \qquad O$$
$$| \qquad\quad | \qquad\qquad \|$$
$$CH_3CH_2CH_2CH_2CHCHCHCHCH_2-C-OH$$
$$| \qquad\quad |$$
$$CH_3CH_2 \quad CH_2CH_3$$

3,4,5,6-tetraethyl

**17.6** Use the steps in Example 17.2 to give the IUPAC name for each ester.

a.

$$CH_3CH_2CH_2CH_2CH_2-\overset{\overset{\displaystyle O}{\|}}{C}-OCH_3$$

methyl group

The word *methyl* becomes the first
part of the name.

---->

$$\boxed{CH_3CH_2CH_2CH_2CH_2-\overset{\overset{\displaystyle O}{\|}}{C}}\!-OCH_3$$

hexan*oic acid*  ----> hexan*oate*
(6 C's)

----> **Answer: methyl hexanoate**

b.

$$\text{(benzene ring)}-\overset{\overset{\displaystyle O}{\|}}{C}-OCH_2CH_3$$

ethyl group

---->

$$\boxed{\text{(benzene ring)}-\overset{\overset{\displaystyle O}{\|}}{C}}\!-OCH_2CH_3$$

benz*oic acid* ----> benz*oate*

----> **Answer: ethyl benzoate**

c.

$$CH_3CH_2CH_2CH_2\overset{\overset{\displaystyle O}{\|}}{C}OCH_2CH_2CH_3$$

propyl group

---->

$$\boxed{CH_3CH_2CH_2CH_2\overset{\overset{\displaystyle O}{\|}}{C}}\!OCH_2CH_2CH_3$$

pentan*oic acid* ----> pentan*oate*
(5 C's)

----> **Answer: propyl pentanoate**

**17.7** Work backwards to draw the structure from the IUPAC name.

a. propyl propanoate

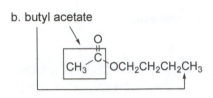

c. ethyl hexanoate

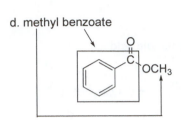

b. butyl acetate

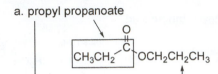

d. methyl benzoate

**17.8** To give each amide an IUPAC name, use these steps:

Step [1]  Name the alkyl group (or groups) bonded to the N atom of the amide.  Use the prefix "*N-*" preceding the name of each alkyl group.

Step [2]  Name the acyl group (RCO–) with the suffix -*amide*.

a.

CH₃CH₂CH₂CH₂ —C(=O)— NH₂    **Answer: pentanamide**

derived from
pentan*oic acid* - - - - - → pentan*amide*

c.

H —C(=O)— N(CH₂CH₂CH₃)₂    **Answer: *N,N*-dipropylformamide**

two propyl groups

derived from
form*ic acid*  - - - - - → form*amide*

b.

—C(=O)— NHCH₃    **Answer: *N*-methylbenzamide**

methyl group

derived from
benz*oic acid*  - - - - - → benz*amide*

**17.9** Work backwards from the name to draw each amide.

a. propanamide

CH₃CH₂ —C(=O)— NH₂

b. *N*-ethylhexanamide

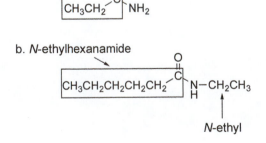

*N*-ethyl

c. *N,N*-dimethylacetamide

CH₃ —C(=O)— N(CH₃)₂

*N,N*-dimethyl

d. *N*-butyl-*N*-methylbutanamide

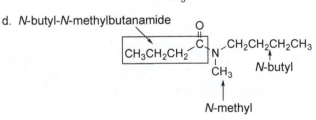

*N*-butyl

*N*-methyl

**17.10** Use the following rules to determine which compound has a higher boiling point:
- Carboxylic acids have stronger intermolecular forces than esters, giving them higher boiling points and melting points when comparing compounds of comparable size.
- Carboxylic acids have higher boiling points and melting points than alcohols of comparable size.
- Primary (1°) and 2° amides have higher boiling points and melting points than esters and 3° amides of comparable size.

a.   CH$_3$COOH   or   CH$_3$CH$_2$CHO
     carboxylic acid        aldehyde
     **higher boiling point**

c.   [benzene ring]—COOH   or   [benzene ring]—CHCH$_3$ with OH
     carboxylic acid        alcohol
     **higher boiling point**

b.   CH$_3$CH$_2$CH$_2$CONH$_2$   or   CH$_3$CH$_2$CO$_2$CH$_3$
       1° amide                          ester
     **higher boiling point**

**17.11** Use the rules in Answer 17.10 to rank the compounds in order of increasing boiling point.

[cyclohexane]—CH$_2$CH$_2$CH$_2$CH$_3$      [cyclohexane]—COOCH$_3$      [cyclohexane]—CH$_2$COOH

hydrocarbon                      ester                 carboxylic acid
**lowest boiling point**                              **highest boiling point**
least water soluble                                    most water soluble

**17.12** α-Hydroxy acids contain a hydroxyl group on the α (alpha) carbon to the carboxyl group.

a.   [structure: HO–C(=O)–CHCH–C(=O)–OH with HO OH]
     both hydroxyl groups
     bonded to α carbons

b.   CH$_3$CHCH$_2$–C(=O)OH with OH
     no OH group on the α carbon

c.   [benzene ring with OH and COOH]
     no OH group on the α carbon

**17.13** PGF$_{2\alpha}$ is more water soluble than arachidonic acid since PGF$_{2\alpha}$ has a COOH and three OH groups, whereas arachidonic acid has no hydroxyl groups.

**17.14** Work backwards from the name to draw each ester.

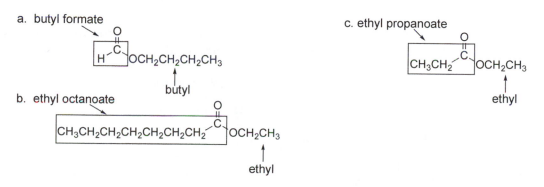

a. butyl formate

   H—C(=O)—OCH$_2$CH$_2$CH$_2$CH$_3$
                        butyl

b. ethyl octanoate

   CH$_3$CH$_2$CH$_2$CH$_2$CH$_2$CH$_2$CH$_2$—C(=O)—OCH$_2$CH$_3$
                                      ethyl

c. ethyl propanoate

   CH$_3$CH$_2$—C(=O)—OCH$_2$CH$_3$
                        ethyl

**17.15** Draw the products of each acid–base reaction as in Example 17.3.

a. [cyclohexyl—C(=O)—OH] + NaOH ⟶ [cyclohexyl—C(=O)—O⁻ Na⁺] + $H_2O$

b. $CH_3CH_2CH_2$—C(=O)—OH + $Na_2CO_3$ ⟶ $CH_3CH_2CH_2$—C(=O)—O⁻ Na⁺ + $Na^+ HCO_3^-$

**17.16** Draw the products of each reaction.

a. [phenyl—C(=O)—OH] + NaOH ⟶ [phenyl—C(=O)—O⁻ Na⁺] + $H_2O$

b. [phenyl—C(=O)—OH] + $Na_2CO_3$ ⟶ [phenyl—C(=O)—O⁻ Na⁺] + $Na^+ HCO_3^-$

c. [phenyl—C(=O)—OH] + $NaHCO_3$ ⟶ [phenyl—C(=O)—O⁻ Na⁺] + $H_2CO_3$

**17.17** Name each salt of a carboxylic acid using the steps in Example 17.4.

a. $CH_3CH_2CH_2CO_2^-$ Na⁺ ← sodium cation

  parent + suffix
  butano-   -ate

  **sodium butanoate**

b. [phenyl]—COO⁻ Li⁺ ← lithium cation

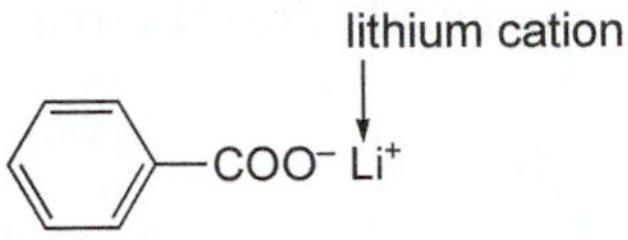

  parent + suffix
  benzo-   -ate

  **lithium benzoate**

**17.18** Work backwards to draw the structure from the name.

**sodium propanoate**

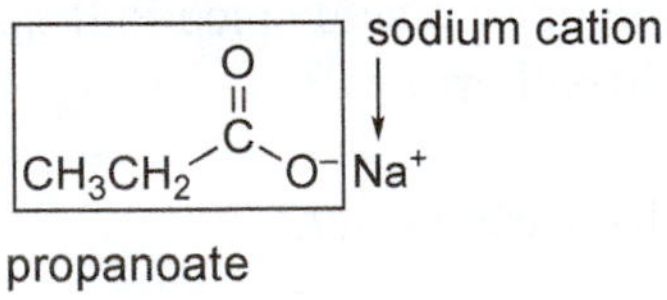

$CH_3CH_2$—C(=O)—O⁻ Na⁺ ← sodium cation

propanoate

**17.19** Soaps are salts of carboxylic acids that have many carbon atoms in a long hydrocarbon chain. A soap has two parts: a long hydrocarbon chain and an ionic end (polar head).

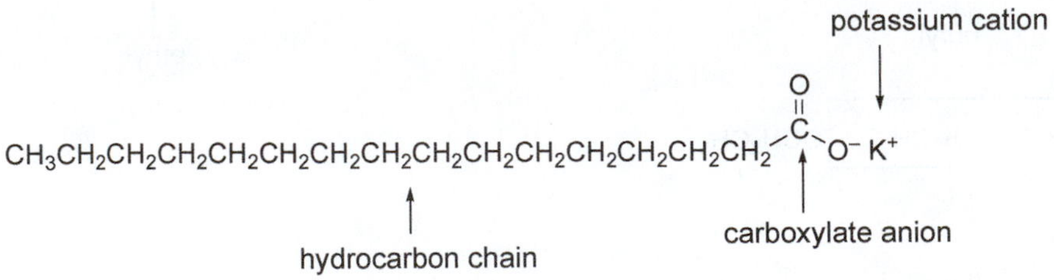

potassium cation

$CH_3CH_2CH_2CH_2CH_2CH_2CH_2CH_2CH_2CH_2CH_2CH_2CH_2CH_2$—C(=O)—O⁻ K⁺

↑ hydrocarbon chain

↑ carboxylate anion

**17.20** Draw the acid–base reaction.

**17.21** Answer each question about ibuprofen.

a. $(CH_3)_2CHCH_2$—⬡—$CHCOOH$ + NaOH ⟶ $(CH_3)_2CHCH_2$—⬡—$CHCOO^- Na^+$ + $H_2O$

b. The neutral form of ibuprofen is present in the stomach since the stomach is acidic.
c. The ionized form of ibuprofen is present in the intestines since the intestines are basic.

**17.22** Draw the products of each reaction as in Example 17.5.

**17.23** Identify **A** in the reaction.

**17.24** Work backwards to determine what carboxylic acid and alcohol are needed to prepare benzocaine.

$H_2N$—⬡—$CO_2H$ + $CH_3CH_2OH$ ⟶ $H_2N$—⬡—$CO_2CH_2CH_3$

benzocaine

**17.25** Draw the products of each reaction as in Example 17.6.

b. $CH_3CH_2CH_2CH_2$—$\overset{O}{\overset{||}{C}}$—OH $\xrightarrow[\Delta]{(CH_3)_2NH}$ $CH_3CH_2CH_2CH_2$—$\overset{O}{\overset{||}{C}}$—$\underset{CH_3}{\overset{}{N}}$—$CH_3$

c. $CH_3CH_2CH_2CH_2$—$\overset{O}{\overset{||}{C}}$—OH $\xrightarrow[\Delta]{}$ ⬡—$NH_2$ $CH_3CH_2CH_2CH_2$—$\overset{O}{\overset{||}{C}}$—$\underset{H}{\overset{}{N}}$—⬡

d. $CH_3CH_2CH_2CH_2$—$\overset{O}{\overset{||}{C}}$—OH $\xrightarrow[\Delta]{CH_3NH_2}$ $CH_3CH_2CH_2CH_2$—$\overset{O}{\overset{||}{C}}$—$\underset{H}{\overset{}{N}}$—$CH_3$

**17.26** Work backwards to determine what carboxylic acid and amine are needed to make acetaminophen.

HO—⬡—$NH_2$ + $CH_3$—$\overset{O}{\overset{||}{C}}$—OH $\longrightarrow$ HO—⬡—$\underset{\overset{|}{\overset{C-CH_3}{\overset{||}{O}}}}{\overset{H}{N}}$

This bond must be formed.

acetaminophen

**17.27** Draw the products of each reaction as in Example 17.7.

a. $CH_3(CH_2)_8$—$\overset{O}{\overset{||}{C}}$—$OCH_3$ $\xrightarrow[H_2SO_4]{H_2O}$ $CH_3(CH_2)_8$—$\overset{O}{\overset{||}{C}}$—OH + $CH_3OH$

b. $CH_3\underset{CH_3}{\overset{}{C}}HCH_2$—$\overset{O}{\overset{||}{C}}$—$OCH_2CH_3$ $\xrightarrow[H_2SO_4]{H_2O}$ $CH_3\underset{CH_3}{\overset{}{C}}HCH_2$—$\overset{O}{\overset{||}{C}}$—OH + $CH_3CH_2OH$

c. ⬡—$CO_2CH_2CH_2CH_3$ $\xrightarrow[H_2SO_4]{H_2O}$ ⬡—$\overset{O}{\overset{||}{C}}$OH + $CH_3CH_2CH_2OH$

**17.28** Basic hydrolysis of esters forms carboxylate anions and alcohols.

a. $CH_3(CH_2)_8$—$\overset{O}{\overset{||}{C}}$—$OCH_3$ $\xrightarrow[NaOH]{H_2O}$ $CH_3(CH_2)_8$—$\overset{O}{\overset{||}{C}}$—$O^-$ $Na^+$ + $CH_3OH$

b. $CH_3\underset{CH_3}{\overset{}{C}}HCH_2$—$\overset{O}{\overset{||}{C}}$—$OCH_2CH_3$ $\xrightarrow[NaOH]{H_2O}$ $CH_3\underset{CH_3}{\overset{}{C}}HCH_2$—$\overset{O}{\overset{||}{C}}$—$O^-$ $Na^+$ + $CH_3CH_2OH$

c.

**17.29** Draw the products of hydrolysis of aspirin.

aspirin

**17.30** Draw the products formed when each amide is treated with $H_2O$ and $H_2SO_4$.

a.

b.

c.

**17.31** Draw the products formed when each amide is treated with $H_2O$ and NaOH as in Example 17.8.

a.

b.

c.

**17.32** The triacylglycerol is hydrolyzed to glycerol and three carboxylic acids.

**17.33** Work backwards to determine what two monomers are needed to prepare nylon 6,10.

**17.34** Draw the structure of Kodel. Since many bonds in the polymer backbone are part of a ring, the polymer is less flexible. This results in a stiffer fabric.

**17.35** Polyesters can be converted back to their monomers by acid hydrolysis. The strong C–C bonds in polymers like polyethylene are not easily broken.

**17.36** The three chirality centers in penicillin G are indicated.

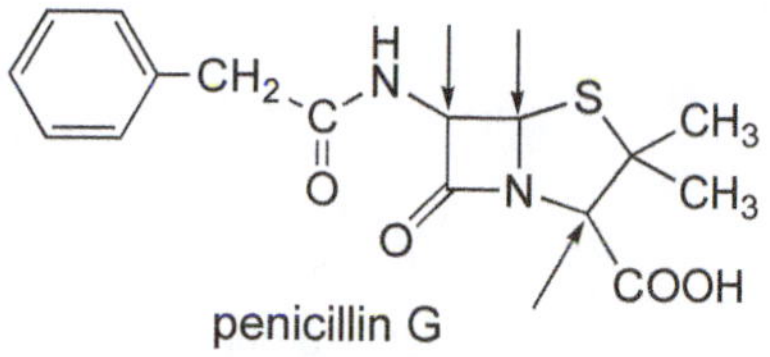

penicillin G

three chirality centers

## Solutions to Odd-Numbered End-of-Chapter Problems

**17.37** Draw a structure that fits each description.

a.

$CH_3CCH_2CH_2CH_2$—$\overset{O}{\overset{\|}{C}}$—OH

with CH$_3$ groups

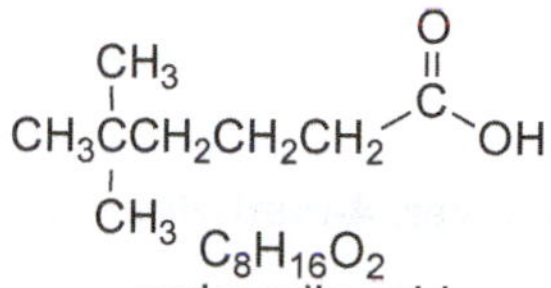

$C_8H_{16}O_2$
carboxylic acid

c.

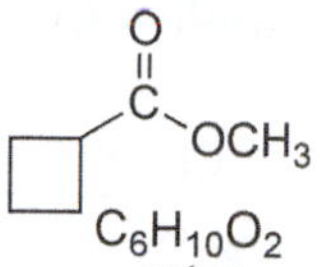

$C_6H_{10}O_2$
ester

b. 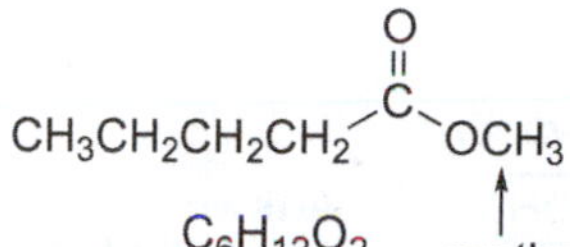 $CH_3CH_2CH_2CH_2$—$\overset{O}{\overset{\|}{C}}$—$OCH_3$

$C_6H_{12}O_2$
ester          methoxy group

d.

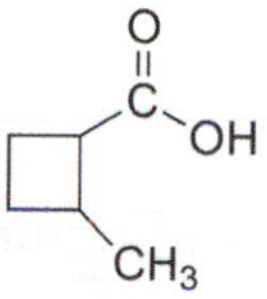

$CH_3$

$C_6H_{10}O_2$
carboxylic acid

**17.39** Draw a structure with molecular formula $C_5H_{11}NO$ that fits each description.

a. 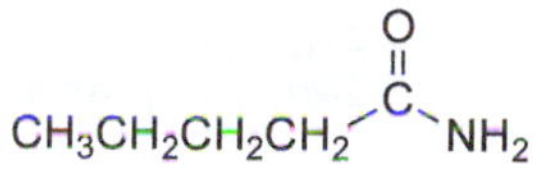 $CH_3CH_2CH_2CH_2$—$\overset{O}{\overset{\|}{C}}$—$NH_2$

1° amide

b. 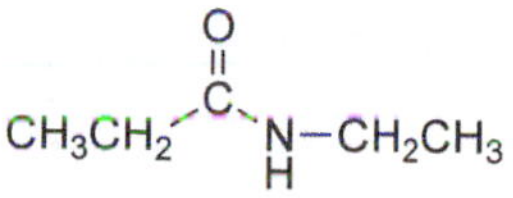 $CH_3CH_2$—$\overset{O}{\overset{\|}{C}}$—$\underset{H}{N}$—$CH_2CH_3$

2° amide

c. 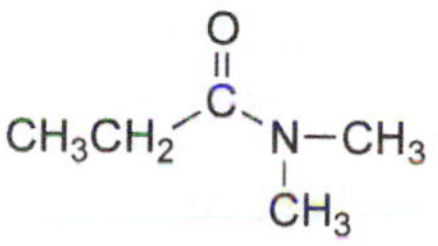 $CH_3CH_2$—$\overset{O}{\overset{\|}{C}}$—$N$—$CH_3$

$CH_3$

3° amide

**17.41** Draw the four esters with molecular formula $C_4H_8O_2$.

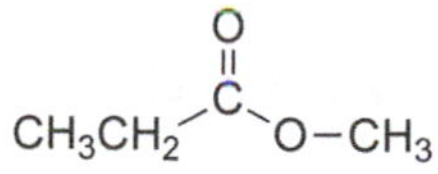

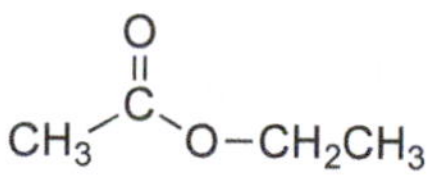

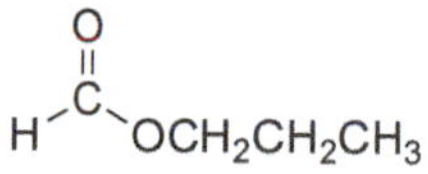

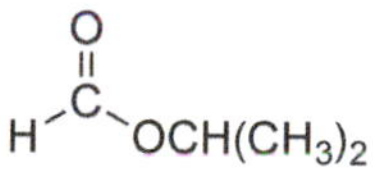

$CH_3CH_2$—$\overset{O}{\overset{\|}{C}}$—$O$—$CH_3$          $CH_3$—$\overset{O}{\overset{\|}{C}}$—$O$—$CH_2CH_3$          $H$—$\overset{O}{\overset{\|}{C}}$—$OCH_2CH_2CH_3$          $H$—$\overset{O}{\overset{\|}{C}}$—$OCH(CH_3)_2$

**17.43** A lactone is a cyclic ester and a lactam is a cyclic amide.

**17.45** Name each compound using the rules in Examples 17.1 and 17.2.

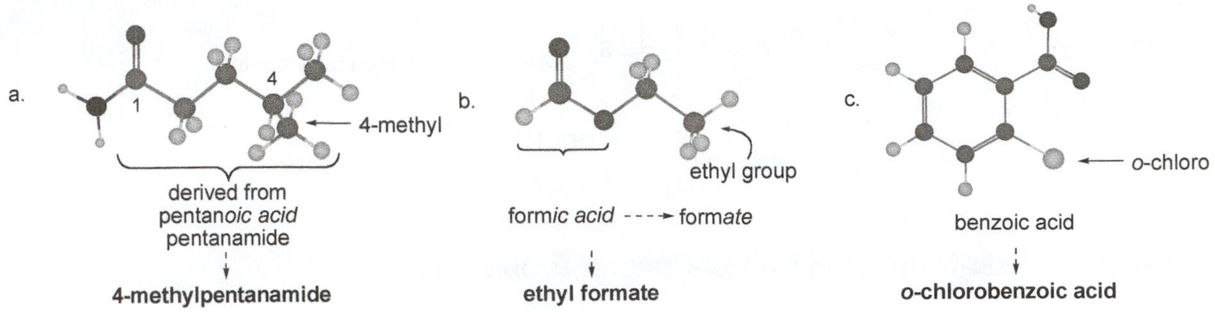

a.

4-methyl

derived from
pentan*oic acid*
pentanamide

**4-methylpentanamide**

b.

ethyl group

form*ic acid* ----→ form*ate*

**ethyl formate**

c.

*o*-chloro

benzoic acid

**o-chlorobenzoic acid**

**17.47** Name the carboxylic acids as in Example 17.1.

a.

$CH_3$

$CH_3CHCH_2CH_2COOH$ ------→

$CH_3$ ←— 4-methyl

$CH_3CHCH_2CH_2COOH$ ------→ **Answer: 4-methylpentanoic acid**

4          1

pentane ----→ pentan*oic acid*
(5 C's)

b.

$CH_2CH_3$

$CH_3CH_2CH_2CHCHCH_2CH_2COOH$ ---→

$CH_2CH_3$

octane ----→ octan*oic acid*
(8 C's)

5  $CH_2CH_3$ ←— 4,5-diethyl

$CH_3CH_2CH_2CHCHCH_2CH_2COOH$ ---→ **Answer:
4,5-diethyloctanoic acid**

$CH_2CH_3$       1

4

c.

$CH_3CH_2CH_2CH_2CH_2C HCH_2CH_3$ ---→

$CO_2H$

heptane ----→ heptan*oic acid*
(7 C's)

2    2-ethyl

$CH_3CH_2CH_2CH_2CH_2C HCH_2CH_3$ ---→ **Answer:
2-ethylheptanoic acid**

$CO_2H$

1

**17.49** Use the steps in Example 17.2 to give the IUPAC name for each ester.

a. 

O
||

$CH_3CH_2CH_2COCH_2CH_2CH_3$ -----→

propyl group

O
||

$CH_3CH_2CH_2COCH_2CH_2CH_3$ -----→ **Answer: propyl butanoate**

butan*oic acid* ---→butano*ate*
(4 C's)

b.

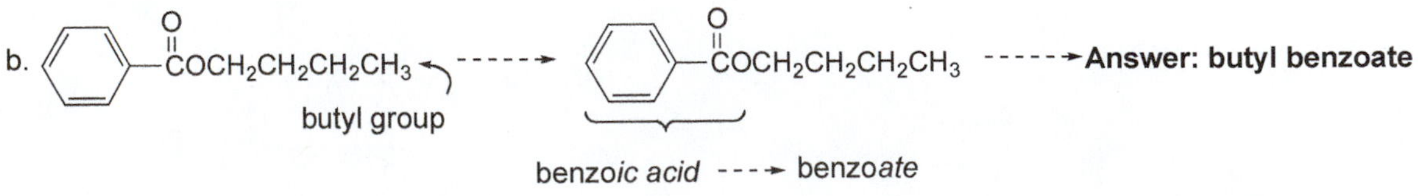

O
||

$COCH_2CH_2CH_2CH_3$ -----→

butyl group

O
||

$COCH_2CH_2CH_2CH_3$ -----→**Answer: butyl benzoate**

benz*oic acid* ----→ benzo*ate*

**17.51** To give each amide an IUPAC name, use the following steps:
Step [1] Name the alkyl group (or groups) bonded to the N atom of the amide. Use the prefix "*N-*" preceding the name of each alkyl group.
Step [2] Name the acyl group (RCO–) with the suffix -*amide*.

a. $CH_3CH_2CH_2CH_2CH_2CONH_2$

derived from
hexan*oic acid* - - - - - ➤ hexan*amide*
**Answer: hexanamide**

b.

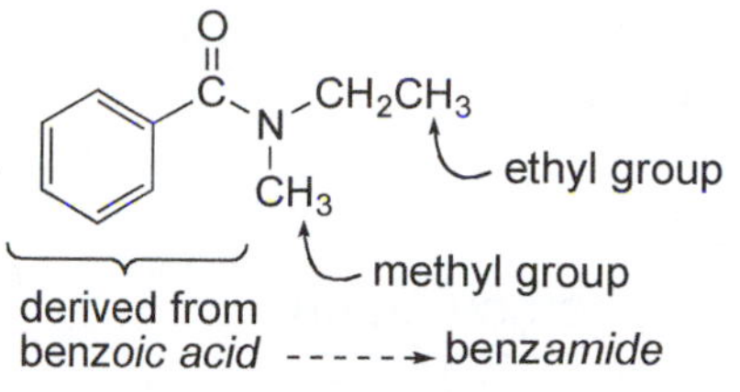

ethyl group
methyl group
derived from
benz*oic acid* - - - - - ➤ benz*amide*
**Answer: *N*-ethyl-*N*-methylbenzamide**

**17.53** Give the IUPAC name for each compound.

a. $HCONH_2$

derived from
form*ic acid* - - - - - ➤ form*amide*
**Answer: formamide**

b.

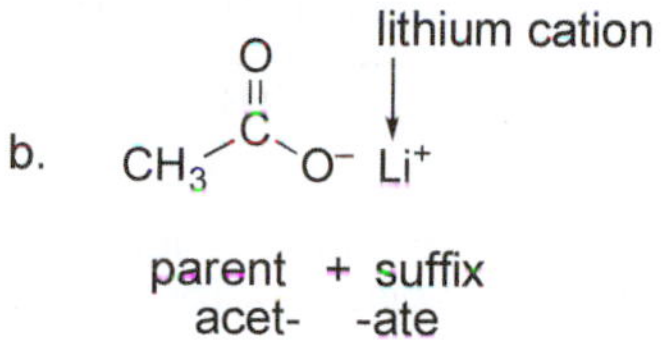

lithium cation

parent + suffix
acet-    -ate

**Answer: lithium acetate**

c. $CH_3CH_2CHCOOH$    4 C's = butanoic acid
2-hydroxy ──➤ OH
**Answer: 2-hydroxybutanoic acid**

d. $CH_3CH_2COCH_2CH_2CH_3$

propano*ate*    propyl group
**Answer: propyl propanoate**

**17.55** Work backwards to draw each structure from the given name.

a. 2-hydroxy**heptanoic acid**

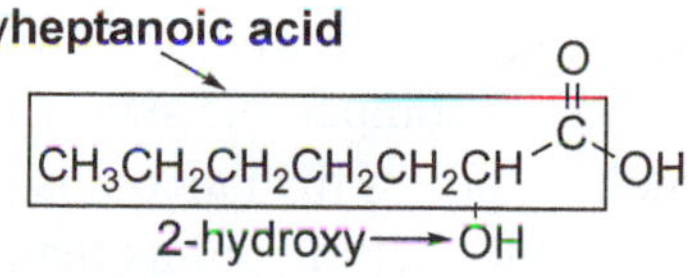

$CH_3CH_2CH_2CH_2CH_2CH$ — C — OH
2-hydroxy ──➤ OH

b. 4-chloro**nonanoic acid**

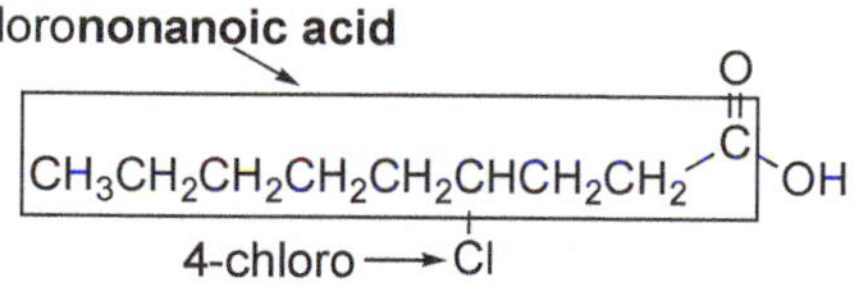

$CH_3CH_2CH_2CH_2CH_2CHCH_2CH_2$ — C — OH
4-chloro ──➤ Cl

c. 3,4-dibromo**benzoic acid**

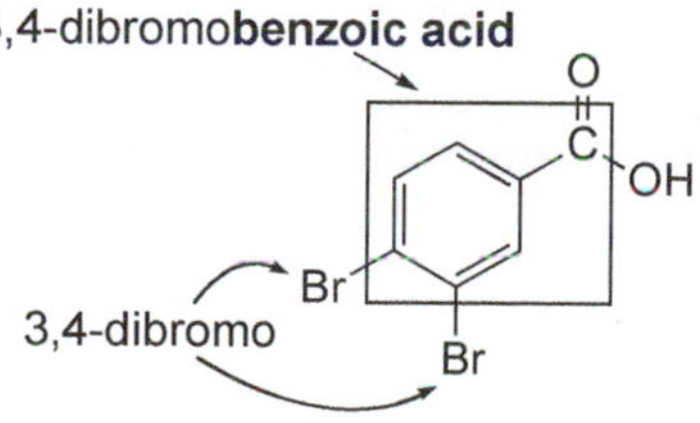

3,4-dibromo
Br
Br

d. lithium **propanoate**

lithium
$CH_3CH_2$ — C — O⁻ Li⁺
propanoate

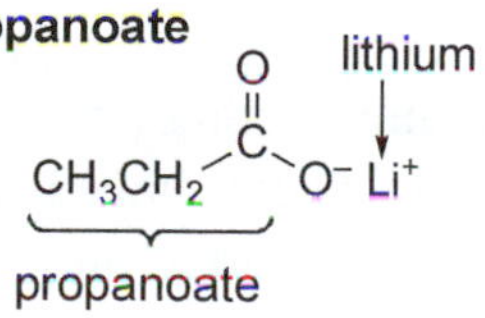

e. 2,2-dibromo**butanoic acid**

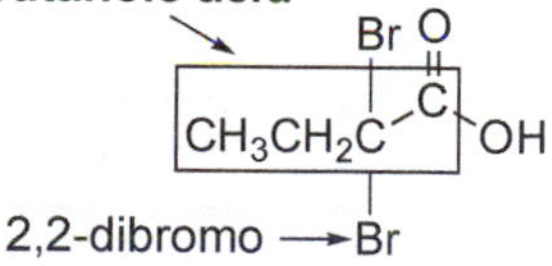

Br O
$CH_3CH_2C$ — C — OH
2,2-dibromo ──➤ Br

f. ethyl 2-methyl**propanoate**

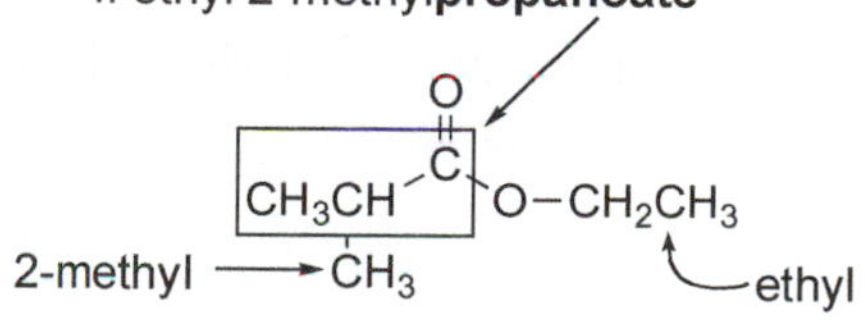

$CH_3CH$ — C — O–CH$_2$CH$_3$
2-methyl ──➤ CH$_3$
ethyl

**17.57** Work backwards to draw each structure from the given name.

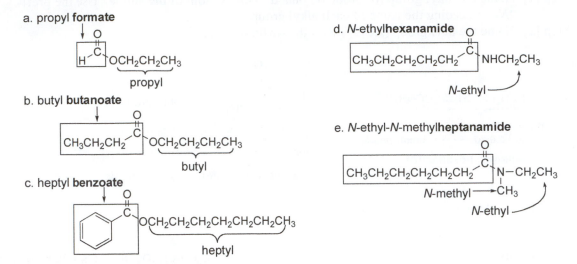

a. propyl **formate**

propyl

b. butyl **butanoate**

butyl

c. heptyl **benzoate**

heptyl

d. *N*-ethyl**hexanamide**

*N*-ethyl

e. *N*-ethyl-*N*-methyl**heptanamide**

*N*-methyl — CH₃

*N*-ethyl

**17.59** Draw the four carboxylic acids of molecular formula $C_5H_{10}O_2$ and give the IUPAC name of each one.

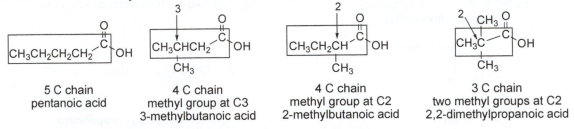

5 C chain
pentanoic acid

4 C chain
methyl group at C3
3-methylbutanoic acid

4 C chain
methyl group at C2
2-methylbutanoic acid

3 C chain
two methyl groups at C2
2,2-dimethylpropanoic acid

**17.61** (a) $HCO_2CH_3$ can hydrogen bond to water since it has oxygen atoms. (b) $CH_3CH_2COOH$ can hydrogen bond to itself and to water since it has an OH group.

**17.63** Use the following rules to rank the compounds in order of increasing boiling point:
- Carboxylic acids have stronger intermolecular forces than esters, giving them higher boiling points and melting points when comparing compounds of comparable size.
- Carboxylic acids have higher boiling points and melting points than alcohols of comparable size.
- Primary (1°) and 2° amides have higher boiling points and melting points than esters and 3° amides of comparable size.

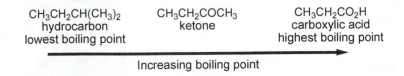

CH₃CH₂CH(CH₃)₂
hydrocarbon
lowest boiling point

CH₃CH₂COCH₃
ketone

CH₃CH₂CO₂H
carboxylic acid
highest boiling point

Increasing boiling point

**17.65** $CH_3CH_2CONH_2$ can intermolecularly hydrogen bond, so it has stronger intermolecular forces than $CH_3CO_2CH_3$, which cannot.

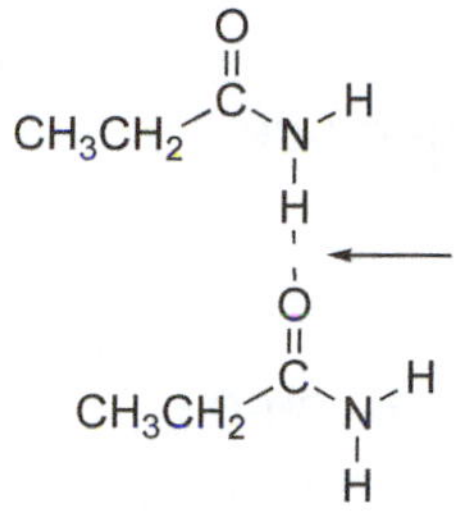

CH₃CH₂—C(=O)—N(H)—H

hydrogen bond

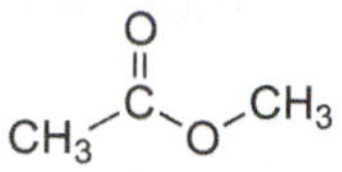

CH₃—C(=O)—O—CH₃

no H bonded to the O atom

**17.67** Draw the products of each acid–base reaction.

a. $CH_3(CH_2)_3$—C(=O)—OH + KOH ⟶ $CH_3(CH_2)_3$—C(=O)—O⁻ K⁺ + $H_2O$

b. $(CH_3)_2CHCH_2CH_2COOH$ + $Na_2CO_3$ ⟶ $(CH_3)_2CHCH_2CH_2\overset{O}{\overset{\|}{C}}O^- Na^+$ + $NaHCO_3$

**17.69** Draw the products of each reaction as in Example 17.5.

a. $CH_3CH_2CH_2$—C(=O)—OH + $CH_3OH$ $\xrightarrow{H_2SO_4}$ $CH_3CH_2CH_2$—C(=O)—$OCH_3$

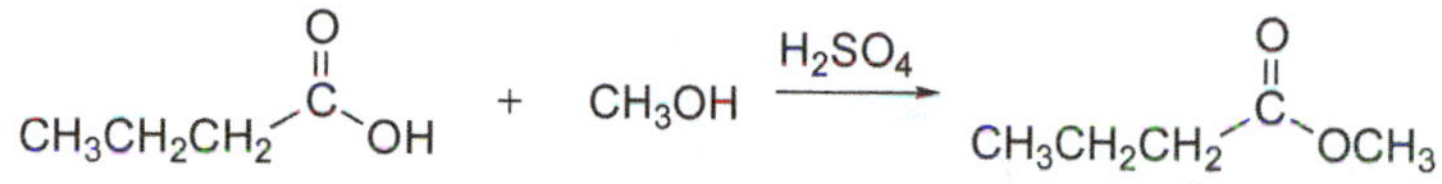

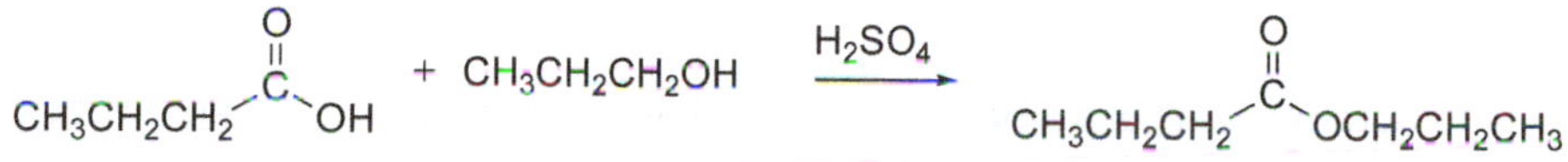

b. $CH_3CH_2CH_2$—C(=O)—OH + $CH_3CH_2CH_2OH$ $\xrightarrow{H_2SO_4}$ $CH_3CH_2CH_2$—C(=O)—$OCH_2CH_2CH_3$

c. $CH_3CH_2CH_2$—C(=O)—OH + ⬡—OH $\xrightarrow{H_2SO_4}$ $CH_3CH_2CH_2$—C(=O)—O—⬡

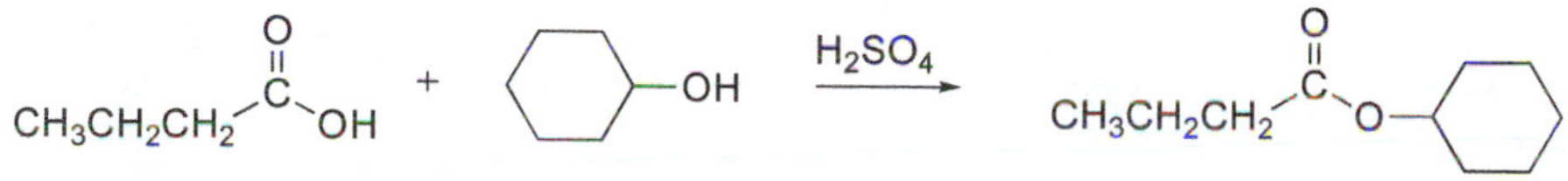

d. $CH_3CH_2CH_2$—C(=O)—OH + ⬡—$CH_2CH_2OH$ $\xrightarrow{H_2SO_4}$ $CH_3CH_2CH_2$—C(=O)—O—$CH_2CH_2$—⬡

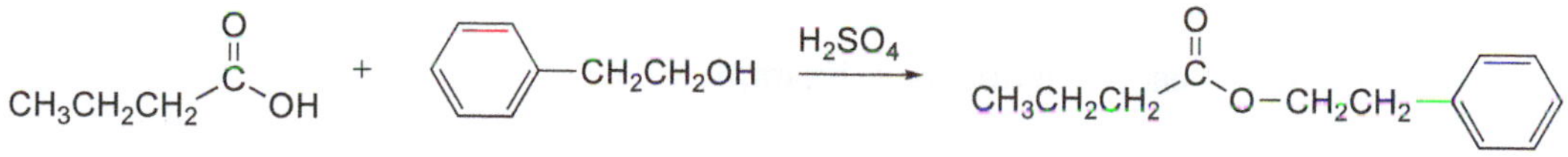

**17.71** Work backwards to determine what carboxylic acid and alcohol are needed to synthesize methylparaben.

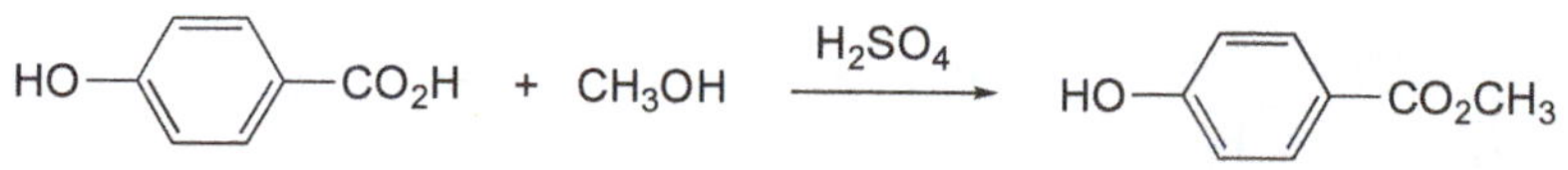

HO—⬡—$CO_2H$ + $CH_3OH$ $\xrightarrow{H_2SO_4}$ HO—⬡—$CO_2CH_3$

methylparaben

**17.73** Draw the products of each reaction as in Example 17.6.

a. $CH_3CH_2CH_2$—C(=O)—OH + $NH_3$ ⟶ $CH_3CH_2CH_2$—C(=O)—$NH_2$

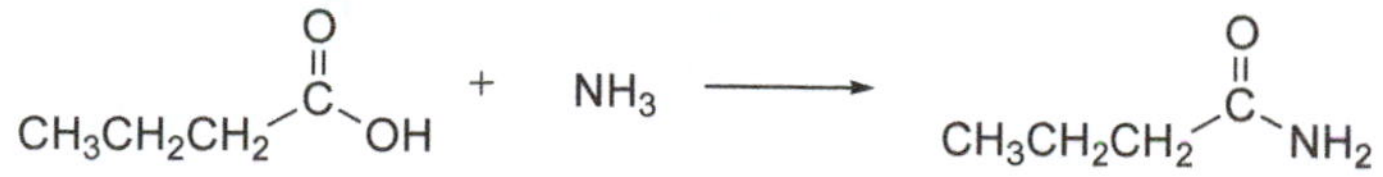

b.

$$CH_3CH_2CH_2\overset{\overset{O}{\|}}{C}OH \ + \ CH_3CH_2NH_2 \ \longrightarrow \ CH_3CH_2CH_2\overset{\overset{O}{\|}}{C}\underset{H}{N}-CH_2CH_3$$

c.

$$CH_3CH_2CH_2\overset{\overset{O}{\|}}{C}OH \ + \ (CH_3CH_2)_2NH \ \longrightarrow \ CH_3CH_2CH_2\overset{\overset{O}{\|}}{C}\underset{CH_2CH_3}{N}-CH_2CH_3$$

d.

$$CH_3CH_2CH_2\overset{\overset{O}{\|}}{C}OH \ + \ CH_3NHCH_2CH_3 \ \longrightarrow \ CH_3CH_2CH_2\overset{\overset{O}{\|}}{C}\underset{CH_2CH_3}{N}-CH_3$$

**17.75** Work backwards to determine what carboxylic acid and alcohol are needed to synthesize each ester.

a.

$$\text{(cyclohexane)}\overset{\overset{O}{\|}}{C}OH \ + \ CH_3OH \ \longrightarrow \ \text{(cyclohexane)}-CO_2CH_3$$

b.

$$HO\overset{\overset{O}{\|}}{C}CH_3 \ + \ \text{(cyclohexane)}-OH \ \longrightarrow \ \text{(cyclohexane)}-O-\overset{\overset{O}{\|}}{C}CH_3$$

**17.77** Work backwards to determine what carboxylic acid and amine are needed to synthesize phenacetin.

$$CH_3CH_2O-\text{(benzene)}-NH_2 \ + \ HO\overset{\overset{O}{\|}}{C}CH_3 \ \longrightarrow \ CH_3CH_2O-\text{(benzene)}-\underset{H}{N}-\overset{\overset{O}{\|}}{C}CH_3$$

phenacetin

**17.79** Draw the products of each reaction as in Example 17.7.

a.

$$CH_3CH_2CH_2\overset{\overset{O}{\|}}{C}OCH(CH_3)_2 \ \xrightarrow[\text{H}_2\text{SO}_4]{\text{H}_2\text{O}} \ CH_3CH_2CH_2\overset{\overset{O}{\|}}{C}OH \ + \ HOCH(CH_3)_2$$

b.

$$\xrightarrow[\text{H}_2\text{SO}_4]{\text{H}_2\text{O}}$$

$$+ \ HO\overset{\overset{O}{\|}}{C}CH_3$$

c.

$$\text{(cyclohexane)}-CH_2CH_2-O-\overset{\overset{O}{\|}}{C}H \ \xrightarrow[\text{H}_2\text{SO}_4]{\text{H}_2\text{O}} \ \text{(cyclohexane)}-CH_2CH_2-OH \ + \ HO-\overset{\overset{O}{\|}}{C}H$$

**17.81**  Draw the products when each amide is treated with $H_2O$ and HCl.

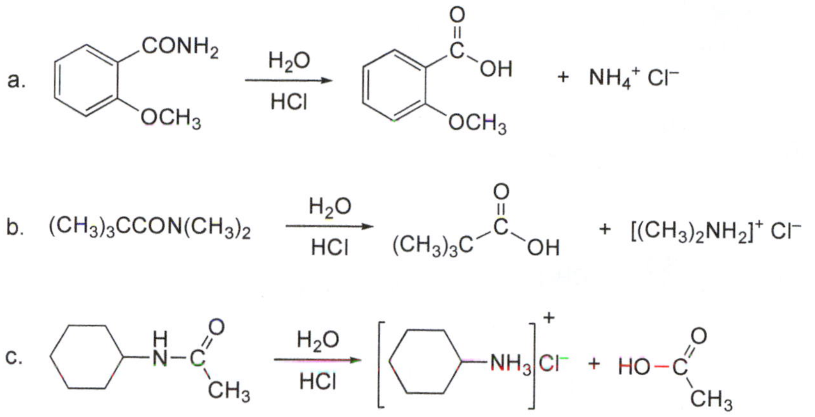

a.

b.  $(CH_3)_3CCON(CH_3)_2$ $\xrightarrow[\text{HCl}]{H_2O}$ $(CH_3)_3C\overset{\overset{O}{\|}}{C}\!-\!OH$ + $[(CH_3)_2NH_2]^+\,Cl^-$

c.

**17.83**  Draw the products of ester hydrolysis.

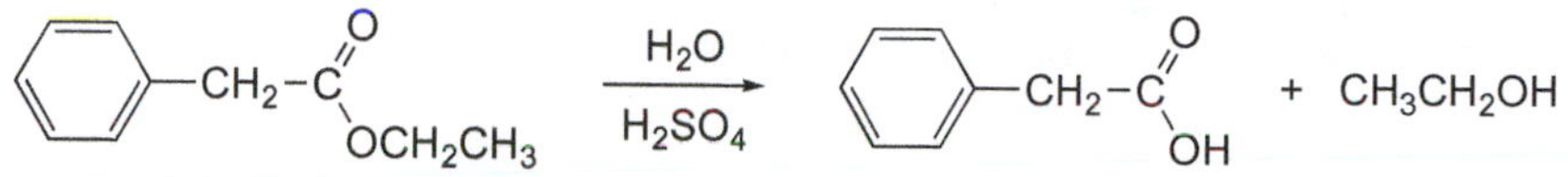

**17.85**  Work backwards to determine what two monomers are needed to prepare the polyamide.

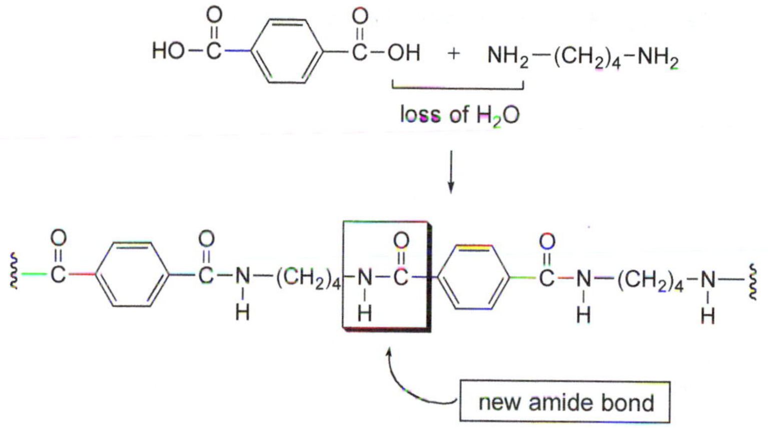

loss of $H_2O$

new amide bond

**17.87**  Draw the structure of the polyester PTT.

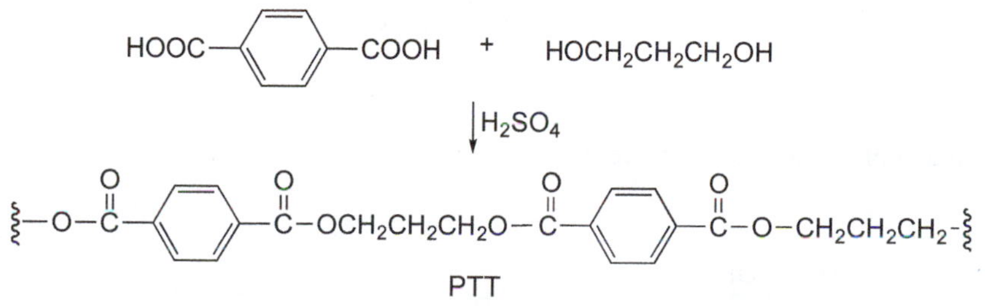

PTT

**17.89**  Saponification is the hydrolysis of an ester with strong base, which forms a metal salt of a carboxylate anion, $RCOO^- M^+$. Esterification forms a new ester, $RCOOR'$, from a carboxylic acid and an alcohol.

**17.91**  Draw the products of each reaction.

a. $CH_3-\overset{O}{\overset{\|}{C}}-OH$  +  $CH_3OH$  $\underset{}{\overset{H_2SO_4}{\rightleftharpoons}}$  $CH_3-\overset{O}{\overset{\|}{C}}-OCH_3$  +  $H_2O$

b. $(CH_3)_2CHO-\overset{O}{\overset{\|}{C}}-CH_3$  +  $H_2O$  $\underset{}{\overset{H_2SO_4}{\rightleftharpoons}}$  $(CH_3)_2CHOH$  +  $HO-\overset{O}{\overset{\|}{C}}-CH_3$

c. $(CH_3)_2CHCH_2O-\overset{O}{\overset{\|}{C}}-CH(CH_3)_2$  +  $H_2O$  $\overset{NaOH}{\longrightarrow}$  $Na^+ {}^-O-\overset{O}{\overset{\|}{C}}-CH(CH_3)_2$  +  $(CH_3)_2CHCH_2OH$

d. $CH_3(CH_2)_4-\overset{O}{\overset{\|}{C}}-OH$  +  NaOH  $\longrightarrow$  $CH_3(CH_2)_4-\overset{O}{\overset{\|}{C}}-O^- Na^+$  +  $H_2O$

**17.93**  Answer each question.

a. $CH_3CHCH_2CH_2CH_2-\overset{O}{\overset{\|}{C}}-OH$  $\dashrightarrow$  **5-methylhexanoic acid**

  5-methyl $\longrightarrow CH_3$
  6 C's
  hexanoic acid

b. $CH_3CH_2\overset{CH_3}{\overset{|}{C}H}CH_2CH_2-\overset{O}{\overset{\|}{C}}-OH$
  isomer

c. $CH_3CH_2CH_2CH_2CH_2-\overset{O}{\overset{\|}{C}}-OCH_3$
  isomer

d. $CH_3\overset{}{C}HCH_2CH_2CH_2-\overset{O}{\overset{\|}{C}}-OH$  $\overset{NaOH}{\longrightarrow}$  $CH_3\overset{}{C}HCH_2CH_2CH_2-\overset{O}{\overset{\|}{C}}-O^- Na^+$  +  $H_2O$
  $\overset{|}{C}H_3$  $\overset{|}{C}H_3$

e. **A** is insoluble in $H_2O$, but soluble in an organic solvent.

f. $CH_3\overset{}{C}HCH_2CH_2CH_2-\overset{O}{\overset{\|}{C}}-OH$  $\underset{H_2SO_4}{\overset{CH_3CH_2OH}{\longrightarrow}}$  $CH_3\overset{}{C}HCH_2CH_2CH_2-\overset{O}{\overset{\|}{C}}-OCH_2CH_3$  +  $H_2O$
  $\overset{|}{C}H_3$  $\overset{|}{C}H_3$

g.

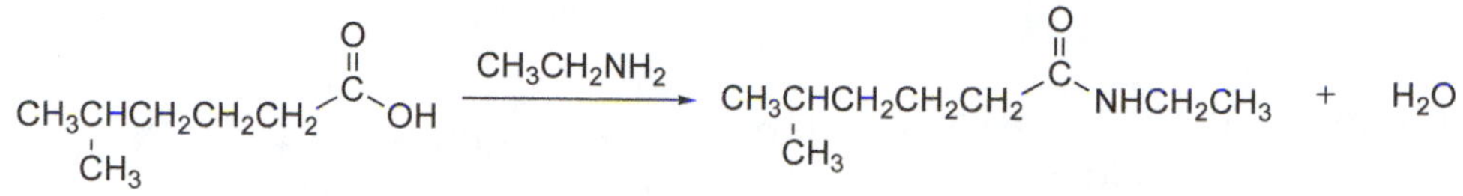

$$CH_3CHCH_2CH_2CH_2\overset{\overset{O}{\|}}{C}OH \xrightarrow{CH_3CH_2NH_2} CH_3CHCH_2CH_2CH_2\overset{\overset{O}{\|}}{C}NHCH_2CH_3 + H_2O$$

with $CH_3$ groups on the chains

**17.95** A soap contains both a long chain hydrocarbon and a carboxylic acid salt.

sodium cation

a. $CH_3CO_2^- Na^+$

short chain
carboxylic acid

b. $CH_3(CH_2)_{14}CO_2^- Na^+$

long chain carboxylate anion
This is a soap because it contains both
a long chain and a carboxylic acid salt.

c. $CH_3(CH_2)_{12}COOH$

no salt

**17.97** Answer each question about naproxen.

a.

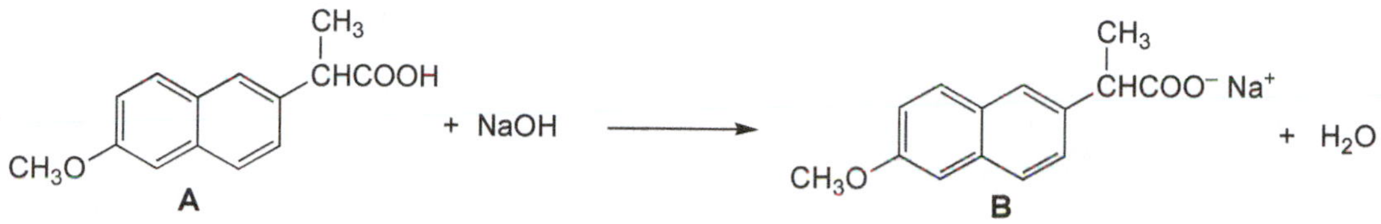

A + NaOH ⟶ B + $H_2O$

(naproxen structures A and B with $CH_3$, CHCOOH / CHCOO⁻ Na⁺, and $CH_3O$ groups)

b. In the stomach naproxen exists as the neutral carboxylic acid (**A**).
c. In the intestines naproxen exists as the ionized carboxylate anion (**B**).

**17.99** Aspirin acts as an anti-inflammatory agent by inhibiting the production of prostaglandins, which are part of the inflammatory cascade of reactions.

**17.101** Soap is able to dissolve nonpolar hydrocarbons in water because the nonpolar end of a soap molecule binds to dirt and hydrocarbons while the polar end then hydrogen bonds with water, thereby rendering the dirt soluble in water.

**Structure of a soap molecule**

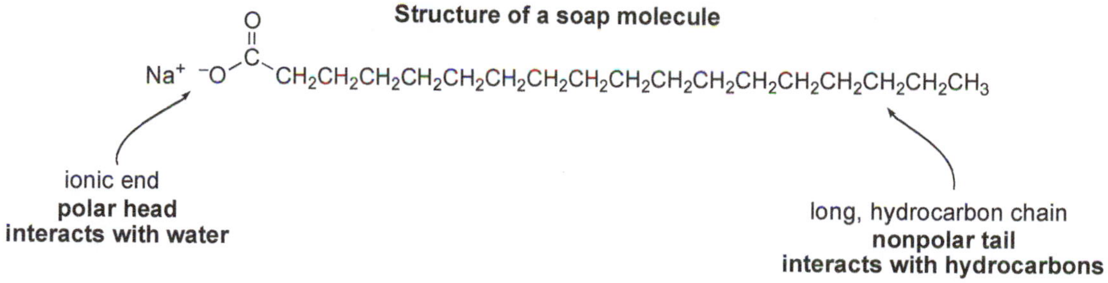

$$Na^+ \ ^-O\overset{\overset{O}{\|}}{C}CH_2CH_2CH_2CH_2CH_2CH_2CH_2CH_2CH_2CH_2CH_2CH_2CH_2CH_2CH_2CH_2CH_3$$

ionic end
**polar head
interacts with water**

long, hydrocarbon chain
**nonpolar tail
interacts with hydrocarbons**

**17.103** Draw the products of the hydrolysis of aspartame by breaking the amide and ester bonds.

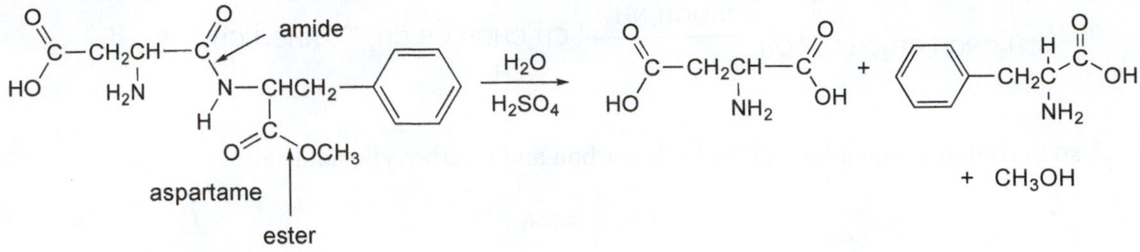

aspartame

amide

ester

$$\xrightarrow[\text{H}_2\text{SO}_4]{\text{H}_2\text{O}}$$

+ CH$_3$OH

**17.105** Draw the products of an intramolecular reaction.

a.

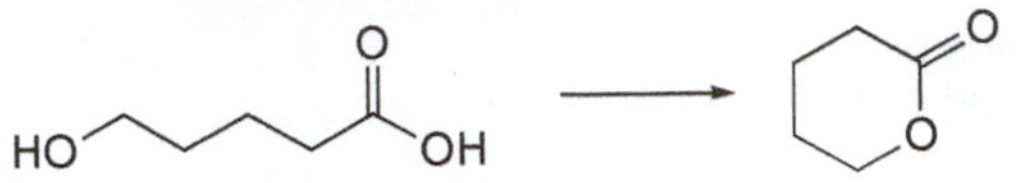

b.

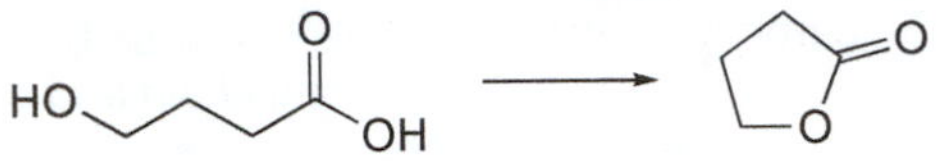

# Chapter 18 Amines and Neurotransmitters

## Chapter Review

### [1] What are the characteristics of amines? (18.1, 18.3)

- Amines are organic nitrogen compounds, formed by replacing one or more hydrogen atoms of $NH_3$ by alkyl groups.
- Primary ($1°$) amines have one C–N bond; $2°$ amines have two C–N bonds; $3°$ amines have three C–N bonds.

R—N̈—H      R—N̈—H      R—N̈—R
|H      |R      |R

1° amine      2° amine      3° amine

- An amine has a lone pair of electrons on the N atom, and is trigonal pyramidal in shape.
- Amines contain polar C–N and N–H bonds. Primary ($1°$) and $2°$ amines can hydrogen bond. Intermolecular hydrogen bonds between N and H are weaker than those between O and H.
- Tertiary ($3°$) amines have lower boiling points than $1°$ and $2°$ amines of comparable size, since $3°$ amines cannot hydrogen bond.

### [2] How are amines named? (18.2)

- Primary ($1°$) amines are identified by the suffix *-amine*.

$CH_3NH_2$          $CH_3CH_2CH_2CH_2NH_2$
                               4   3   2   1

Common name:    methylamine      Common name:      butylamine
Systematic name:   methanamine     Systematic name:   1-butanamine

- Secondary ($2°$) and $3°$ amines with identical alkyl groups are named by adding the prefix *di-* or *tri-* to the name of the $1°$ amine.
- Secondary ($2°$) and $3°$ amines with different alkyl groups are named as *N*-substituted $1°$ amines.

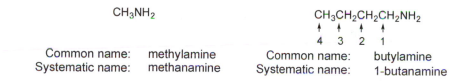

$CH_2CH_3$             H                      $CH_3CHCH_3$
|                |                            |
$CH_3CH_2$—N—$CH_2CH_3$     $CH_3CH_2CH_2$—N—$CH_2CH_2CH_3$     N—H
                                                   |
                                                   $CH_3$

triethylamine             dipropylamine             *N*-methyl-2-propanamine

### [3] What are alkaloids? Give examples of common alkaloids. (18.4, 18.5)

- Alkaloids are naturally occurring amines isolated from plant sources.
- Examples of alkaloids include: caffeine (from coffee and tea), nicotine (from tobacco), morphine and codeine (from the opium poppy), quinine (from the Cinchona tree), and atropine (from the deadly nightshade plant).

caffeine                    nicotine

## [4] What products are formed when an amine is treated with acid? (18.6)

- Amines act as proton acceptors in water and acid. For example, the reaction of $RNH_2$ with HCl forms the water-soluble ammonium salt $RNH_3^+ Cl^-$.

## [5] What are the characteristics of ammonium salts and how are they named? (18.6, 18.7)

- An ammonium salt consists of a positively charged ammonium ion and an anion.
- An ammonium salt is named by changing the suffix -amine of the parent amine to the suffix -ammonium, followed by the name of the anion.

chloride

**ethylammonium chloride**

derived from

ethylamine

- Ammonium salts are water-soluble solids.
- Water-insoluble amine drugs are sold as their ammonium salts to increase their solubility in the aqueous environment of the blood.

octylamine                                    octylammonium chloride

**water-insoluble amine** - - - - - - - - - - - - - - - - - - - → **water-soluble salt**

| The solubility properties change. |

## [6] What are neurotransmitters and how do they differ from hormones? (18.8, 18.9)

- A neurotransmitter is a chemical messenger that transmits a nerve impulse from a neuron to another cell.
- A hormone is a compound produced by an endocrine gland that travels through the bloodstream to a target tissue or organ.

## [7] What roles do dopamine and serotonin play in the body? (18.8)

- Dopamine affects movement, emotions, and pleasure. Too little dopamine causes Parkinson's disease. Too much dopamine causes schizophrenia. Dopamine plays a role in addiction.

- Serotonin is important in mood, sleep, perception, and temperature regulation. A deficiency of serotonin causes depression. SSRIs are antidepressants that effectively increase the concentration of serotonin.

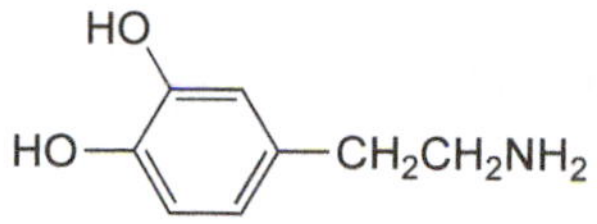

dopamine 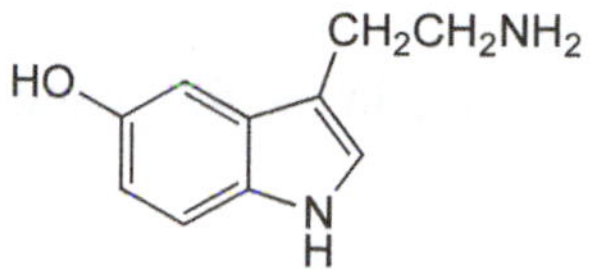      serotonin

## [8] Give examples of important derivatives of 2-phenylethylamine. (18.9)

- Derivatives of 2-phenylethylamine contain a benzene ring bonded to a two-carbon chain that is bonded to a nitrogen atom.

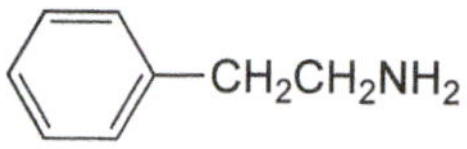

2-phenylethylamine 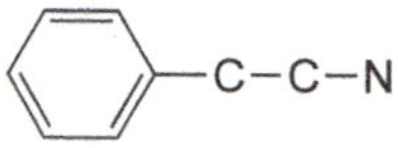      common structural feature

- Examples of 2-phenylethylamine derivatives include epinephrine, norepinephrine, amphetamine, and methamphetamine. Albuterol and salmeterol are also derivatives of 2-phenylethylamine that are used to treat asthma.

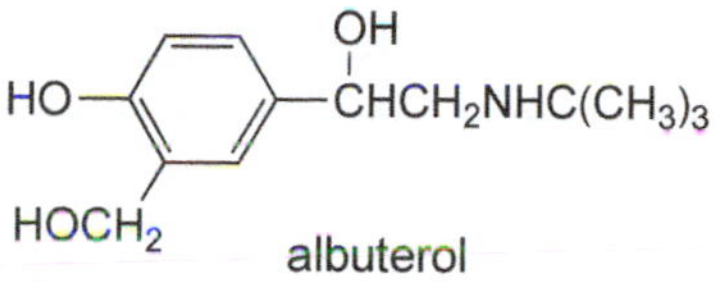

albuterol
(Trade names: Ventolin, Proventil)

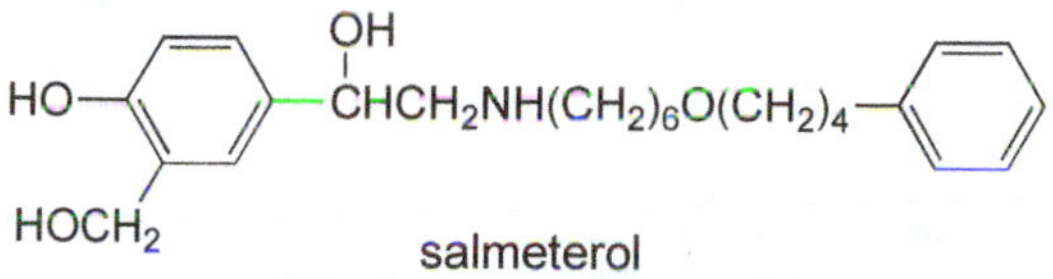

salmeterol
(Trade name: Serevent)

## [9] What is histamine, and how do antihistamines and anti-ulcer drugs work? (18.10)

- Histamine is an amine with a wide range of physiological effects. Histamine dilates capillaries, is responsible for the runny nose and watery eyes of allergies, and stimulates the secretion of stomach acid.

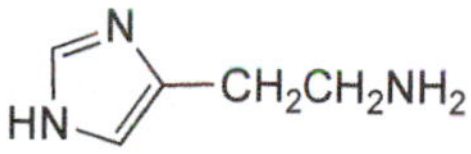

histamine

- Antihistamines bind to the H1 histamine receptor and inhibit vasodilation, so they are used to treat the symptoms of colds and allergies.
- Anti-ulcer drugs bind to the H2 histamine receptor and reduce the production of stomach acid.

## Problem Solving

## [1] Structure and Bonding (18.1)

**Example 18.1** Classify each amine in the following compounds as $1°$, $2°$, or $3°$.

$$\begin{array}{c} CH_3 \\ | \end{array}$$

a. $CH_3\overset{\underset{|}{CH_2CH_3}}{\overset{|}{C}}NHCH_3$

b. $CH_3CH_2\overset{\underset{|}{CH_3}}{C}HCH_2NH_2$

**Analysis**
To determine whether an amine is $1°$, $2°$, or $3°$, count the number of carbons bonded to the nitrogen atom. A $1°$ amine has one C–N bond, and so forth.

**Solution**
Draw out the structure or add H's to the skeletal structure to clearly see how many C–N bonds the amine contains.

a. $CH_3\overset{\underset{|}{CH_2CH_3}}{\overset{|}{C}}NHCH_3$

This N is bonded to 2 C's, making it a **2° amine**.

b. $CH_3CH_2\overset{\underset{|}{CH_3}}{C}HCH_2NH_2$

This N is bonded to only 1 C, making it a **1° amine**.

---

## [2] Nomenclature (18.2)

**Example 18.2** Give a systematic name for each amine.

a. (cyclohexyl)–N(CH_2CH_2CH_3)(CH_2CH_3)

b. $CH_3CH_2\overset{\underset{|}{Cl}}{\underset{\underset{|}{CH_2CH_3}}{C}}CH_2CH_2CH_2NH_2$

**Analysis and Solution**
a. For a 3° amine, one alkyl group on N is the principal R group and the others are substituents.

[1] Name the ring bonded to the N:

(cyclohexyl ring)–N(CH_2CH_2CH_3)(CH_2CH_3)

6 C's in the ring

↓

cyclohexanamine

[2] Name the substituents:

(cyclohexyl)–N(CH_2CH_2CH_3)(CH_2CH_3)   a propyl and ethyl group on N

• 2 N's are needed, one for each alkyl group.
• Alphabetize the *e* of ethyl before the *p* of **propyl**.

**Answer: *N*-ethyl-*N*-propylcyclohexanamine**

b. [1] For a 1° amine, find and name the longest chain containing the amine nitrogen:

$$CH_3CH_2\underset{\overset{|}{CH_2CH_3}}{\overset{\overset{Cl}{|}}{C}}CH_2CH_2CH_2NH_2$$

hexane ----→ hexan*amine*
(6 C's)

[2] Number the carbon skeleton:

4-chloro
4-ethyl
1

You must use a number to show the location of the $NH_2$ group.

**Answer: 4-chloro-4-ethyl-1-hexanamine**

---

## [3] Physical Properties (18.3)

**Example 18.3** Which compound in each pair has the higher boiling point?
a. $CH_3CH_2N(CH_3)_2$ or $CH_3CH_2CH_2CH_2NH_2$
b. $CH_3OCH_2CH_2CH_3$ or $CH_3CH_2CH_2CH_2NH_2$

**Analysis**

Keep in mind the general rule: For compounds of comparable size, **the stronger the intermolecular forces, the higher the boiling point.** Compounds that can hydrogen bond have higher boiling points than compounds that are polar but cannot hydrogen bond. Polar compounds have higher boiling points than nonpolar compounds.

**Solution**

a. The 1° amine ($CH_3CH_2CH_2CH_2NH_2$) has N–H bonds, so intermolecular hydrogen bonding is possible. The 3° amine [$CH_3CH_2N(CH_3)_2$] has only C–H bonds, so there is no possibility of intermolecular hydrogen bonding. $CH_3CH_2CH_2CH_2NH_2$ has a higher boiling point because it has stronger intermolecular forces.

$$CH_3-\underset{\overset{|}{CH_3}}{N}-CH_2CH_3$$

a 3° amine with only C–H bonds

$CH_3CH_2CH_2CH_2NH_2$

a 1° amine with N–H bonds
intermolecular hydrogen bonding
**higher boiling point**

b. The 1° amine ($CH_3CH_2CH_2CH_2NH_2$) has N–H bonds, so intermolecular hydrogen bonding is possible. The ether ($CH_3OCH_2CH_2CH_3$) has only C–H bonds, so there is no possibility of intermolecular hydrogen bonding. $CH_3CH_2CH_2CH_2NH_2$ has a higher boiling point because it has stronger intermolecular forces.

$CH_3-O-CH_2CH_2CH_3$

an ether with only C–H bonds

$CH_3CH_2CH_2CH_2NH_2$

a 1° amine with two N–H bonds
intermolecular hydrogen bonding
**higher boiling point**

## [4] Amines as Bases (18.6)

**Example 18.4** What products are formed when $CH_3CH_2NHCH_3$ reacts with HCl?

**Analysis**
In any acid–base reaction with an amine:
- Locate the N atom of the amine and add a proton to it. Since the amine nitrogen is neutral to begin with, adding a proton gives it a (+1) charge.
- Remove a proton from the acid (HCl) and form its conjugate base (Cl⁻).

**Solution**
Transfer a proton from the acid to the base. Use the lone pair on the N atom to form the new bond to the proton of the acid.

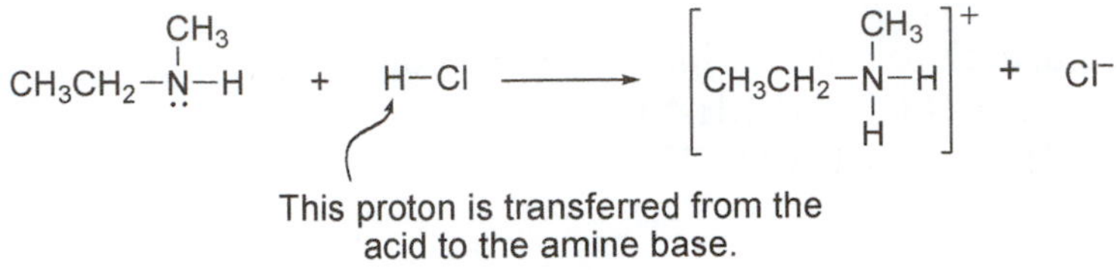

This proton is transferred from the acid to the amine base.

Thus, HCl loses a proton to form Cl⁻, and the N atom of the amine gains a proton to form an ammonium cation.

---

**Example 18.5** Name each ammonium salt.

a. $\left[(CH_3CH_2)_3NH\right]^+$ $CH_3COO^-$     b.    $(CH_3NH_3)^+$ $Br^-$

**Analysis**
To name an ammonium salt, draw out the four groups bonded to the N atom. Remove one hydrogen from the N atom to draw the structure of the parent amine. Then put two parts of the name together.
- Name the ammonium ion by changing the suffix *-amine* of the parent amine to the suffix *-ammonium*.
- Add the name of the anion.

**Solution**

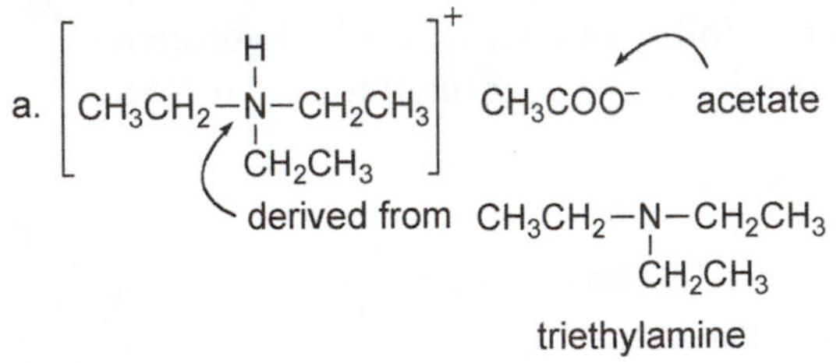

a.

- Change the name triethyl*amine* to triethyl*ammonium*.
- Add the name of the anion, acetate.
- Answer: **triethylammonium acetate.**

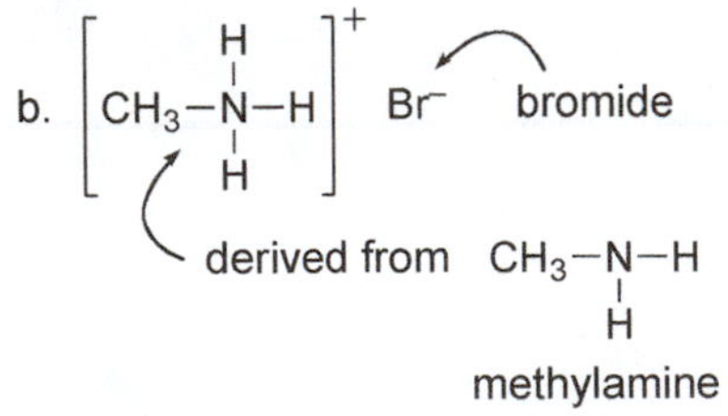

b.

- Change the name methyl*amine* to methyl*ammonium*.
- Add the name of the anion, bromide.
- Answer: **methylammonium bromide.**

## Self-Test

**[1] Fill in the blank with one of the terms listed below.**

Amines (18.1)
Ammonium salts (18.6)
Antihistamines (18.10)
Anti-ulcer drugs (18.10)
Axon (18.8)

Heterocycle (18.1)
Hormone (18.9)
Neuron (18.8)
Neurotransmitter (18.8)
Postsynaptic neuron (18.8)

Presynaptic neuron (18.8)
Quaternary ammonium salts (18.1)
SSRIs (18.8)
Synapse (18.8)

1. _____ are water-soluble solids.
2. _____ act by inhibiting the reuptake of serotonin by the presynaptic neuron, effectively increasing the concentration of serotonin.
3. A _____ is a compound produced by an endocrine gland, which then travels through the bloodstream to a target tissue or organ.
4. _____ are organic nitrogen compounds that are considered to be derivatives of $NH_3$.
5. A _____ consists of many short filaments called dendrites connected to a cell body.
6. A _____ is a cell that releases a neurotransmitter.
7. _____ are ammonium salts with nitrogen atoms bonded to *four* alkyl groups.
8. A _____ is a chemical messenger that transmits nerve impulses from one neuron to another.
9. A _____ is a cell that contains the receptors that bind a neurotransmitter.
10. Dendrites are separated from each other by a small gap called a _____.
11. A _____ is a ring that contains a heteroatom such as N, O, or S.
12. _____ bind to the H2 histamine receptor, reducing acid secretion in the stomach.
13. _____ bind to the H1 histamine receptor and inhibit vasodilation. They are used to treat the symptoms of the common cold and environmental allergies.
14. A long stem called an _____ protrudes from the cell body of a neuron.

**[2] Label each compound as a 1°, 2°, or 3° amine.**

15. $CH_3NH_2$

16. $CH_3-N-CH_2CH_3$ (with cyclopentane ring on N)

17. $H-N-H$ with $CH_2CH_3$

18. $H-N-CH_2CH_3$ with $CH_2CH_3$

**[3] Pick the compound in each pair with the higher boiling point.**

19. $CH_3CH_2NH_2$   or   $CH_3CH_2OH$
    **A**              **B**

21. $CH_3OCH_3$   or   $H-N-CH_3$ with $CH_3$
    **A**              **B**

20. $CH_3-\overset{H}{\underset{}{C}}-CH_2CH_3$ (with cyclopentane ring)   or   $CH_3-N-CH_2CH_3$ (with cyclopentane ring)
    **A**                                    **B**

## [4] Match the reactants and products.

22. 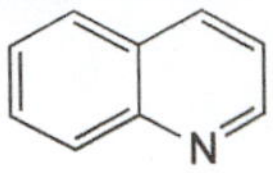  (quinoline) + HCl ⟶

a. $\left[ \text{(phenyl)}-CH_2NH_2CH_3 \right]^+$ + $Cl^-$

23. $CH_3CH_2CHCH_2CH_3$ + $H_2O$ ⟶
       |
       $NHCH_3$

b. 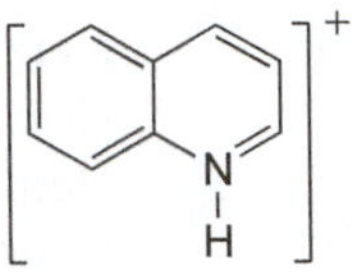 $\left[ \text{(quinolinium)} \right]^+$ + $Cl^-$

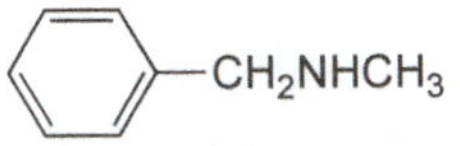

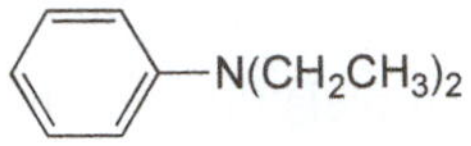

24. (phenyl)$-CH_2NHCH_3$ + HCl ⟶

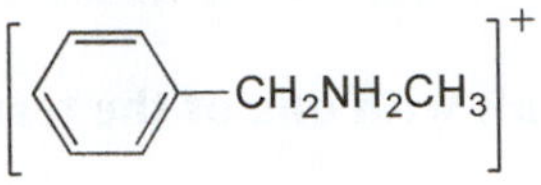

c. $\left[ CH_3CH_2CHCH_2CH_3 \atop \quad\; NH_2CH_3 \right]^+$ + $OH^-$

25. (phenyl)$-N(CH_2CH_3)_2$ + HCl ⟶

d. 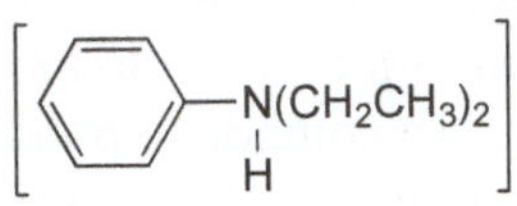 $\left[ \text{(phenyl)}-N(CH_2CH_3)_2 \atop \qquad\qquad\quad H \right]^+$ + $Cl^-$

## Answers to Self-Test

| | | | | |
|---|---|---|---|---|
| 1. Ammonium salts | 6. presynaptic neuron | 11. heterocycle | 16. 3° | 21. B |
| 2. SSRIs | 7. Quaternary ammonium salts | 12. Anti-ulcer drugs | 17. 1° | 22. b |
| 3. hormone | 8. neurotransmitter | 13. Antihistamines | 18. 2° | 23. c |
| 4. Amines | 9. postsynaptic neuron | 14. axon | 19. B | 24. a |
| 5. neuron | 10. synapse | 15. 1° | 20. B | 25. d |

## Solutions to In-Chapter Problems

**18.1** To determine whether an amine is 1°, 2°, or 3°, count the number of carbons bonded to the nitrogen atom as in Example 18.1. A 1° amine has one C–N bond, and so forth.

a.   1°   2°   2°   1°
   $H_2N(CH_2)_3NH(CH_2)_4NH(CH_2)_3NH_2$

b. 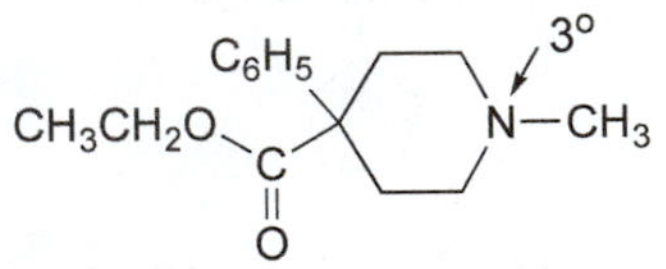   $CH_3CH_2O$ — (with $C_6H_5$ group, 3° amine) $N—CH_3$

**18.2** Label the amine and hydroxyl group in scopolamine, a drug used to treat motion sickness, as 1°, 2°, or 3° as in Example 18.1.

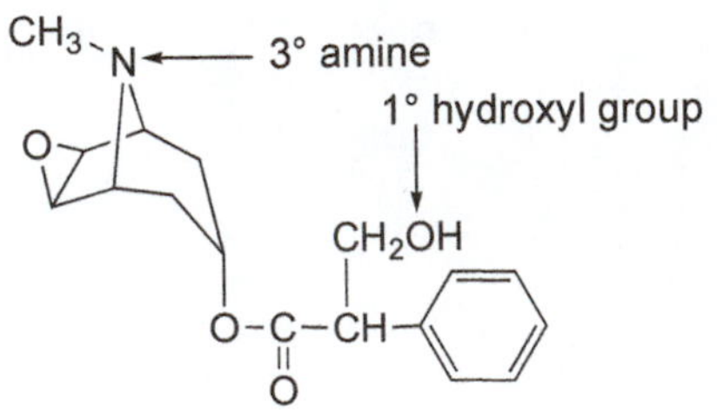

$CH_3$–N ← 3° amine

1° hydroxyl group

$CH_2OH$

$O-C-CH$—(phenyl)
   ‖
   $O$

scopolamine

**18.3**  Methamphetamine is a 2° amine.  Give the molecular shape around each atom by counting groups.

a. 2° amine

b.

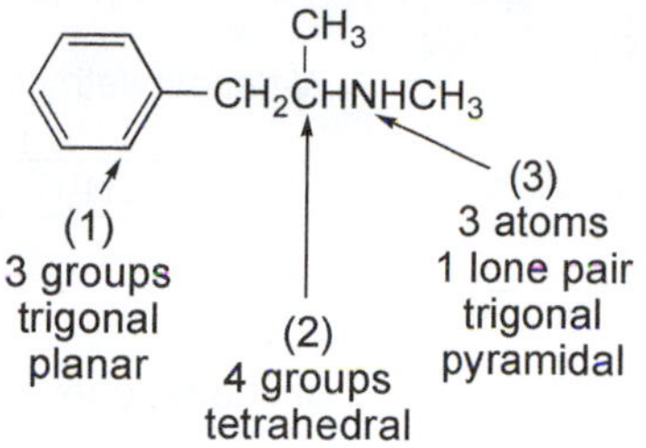

(1)
3 groups
trigonal
planar

(2)
4 groups
tetrahedral

(3)
3 atoms
1 lone pair
trigonal
pyramidal

**18.4**  Name each amine as in Example 18.2.

a.

$CH_3CH_2CHCH_3$
  |
  $NH_2$

------→

2  1

$CH_3CH_2CHCH_3$
  |
  $NH_2$

------→  **Answer: 2-butanamine**

butane ---→ butanamine
(4 C's)

b.

$CH_3CH_2CH_2$ $NHCH_3$ -------→

1

$CH_3CH_2CH_2$ $NHCH_3$ -------→  **Answer: N-methyl-1-propanamine**

propane ---→ propanamine
3 C's

N-methyl

c.

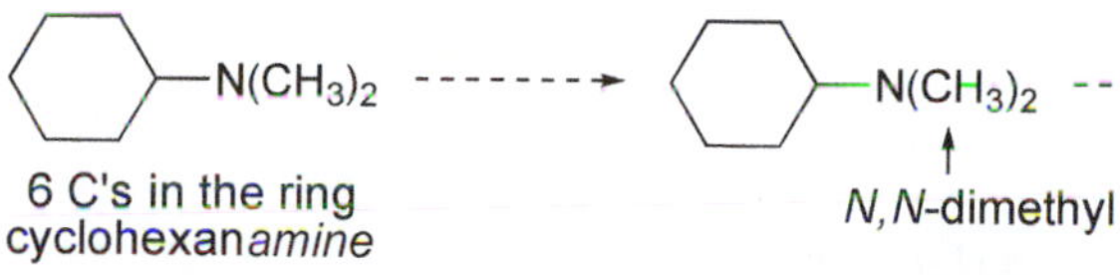

⬡—$N(CH_3)_2$ -------→  ⬡—$N(CH_3)_2$ -------→  **Answer: N,N-dimethylcyclohexanamine**

6 C's in the ring
cyclohexanamine

N,N-dimethyl

d.  $(CH_3CH_2CH_2CH_2)_2NH$ -------→  **Answer: dibutylamine**

2 butyl groups
(4 C's)

**18.5**  Work backwards to draw the structure corresponding to each name.

a. N-methylaniline

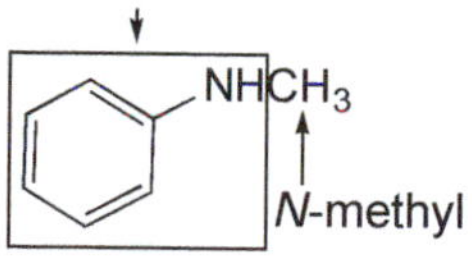

N-methyl

b. m-ethylaniline

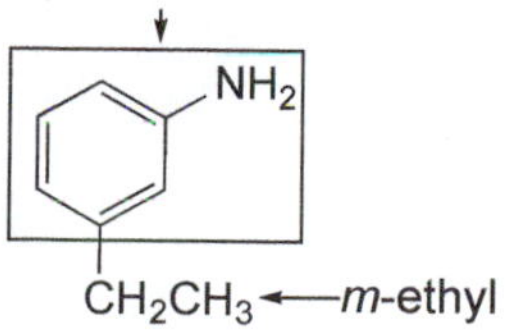

$CH_2CH_3$ ←—m-ethyl

c. 3,5-diethylaniline

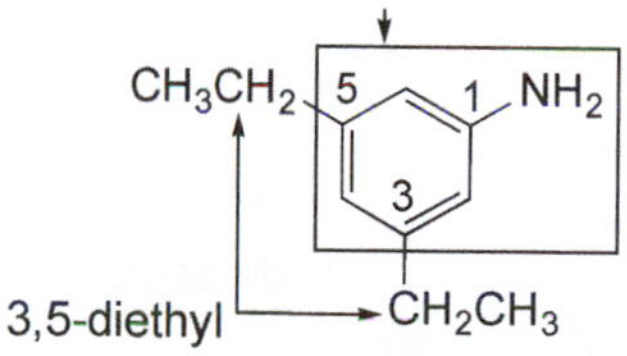

3,5-diethyl ———→ $CH_2CH_3$

d. N,N-diethylaniline

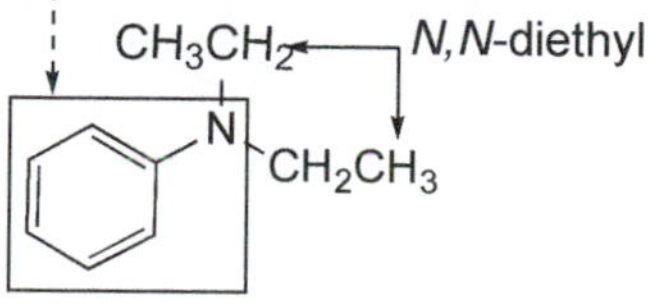

**18.6**  Work backwards to draw the structure corresponding to each name.

a.  3-hexanamine  $NH_2$

$CH_3CH_2CHCH_2CH_2CH_3$

3

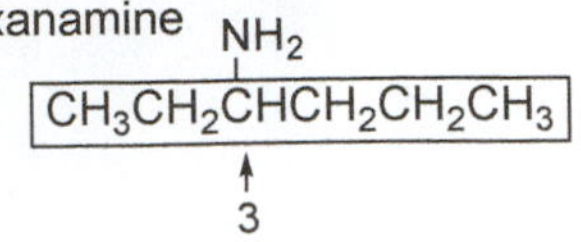

b.  *N*-methylpentylamine

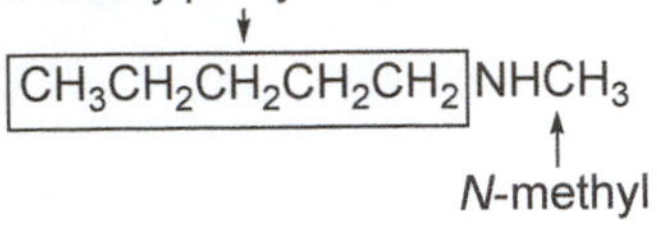

*N*-methyl

$CH_3CH_2CH_2CH_2CH_2$ $NHCH_3$

*N*-methyl

c.  *p*-nitroaniline

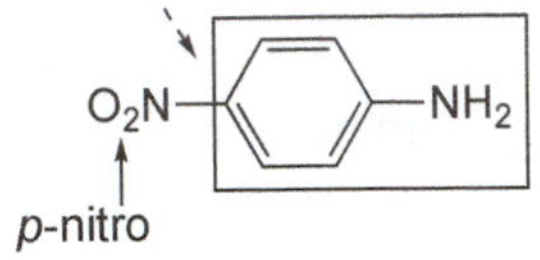

$O_2N-$ ⬡ $-NH_2$

*p*-nitro

d.  *N*-methylpiperidine

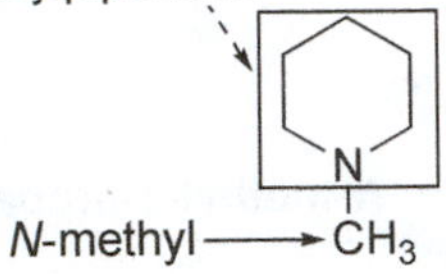

*N*-methyl ⟶ $CH_3$

e.  *N,N*-dimethylethylamine

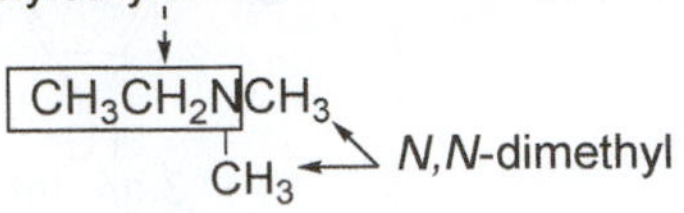

$CH_3CH_2NCH_3$

$CH_3$ ⟵ *N,N*-dimethyl

f.  2-aminocyclohexanone

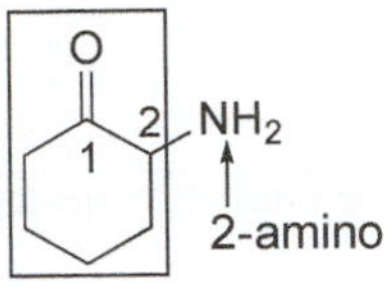

O

2 $NH_2$

1

2-amino

g.  1-propylcyclohexanamine

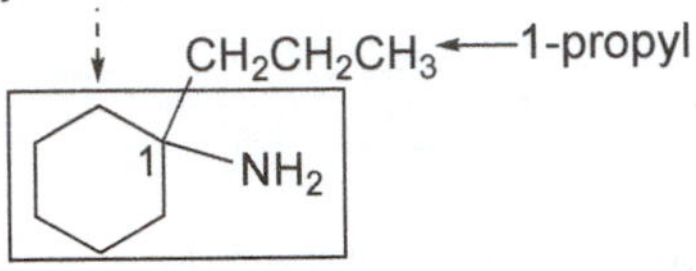

$CH_2CH_2CH_3$ ⟵ 1-propyl

1  $NH_2$

h.  *N*-propylaniline

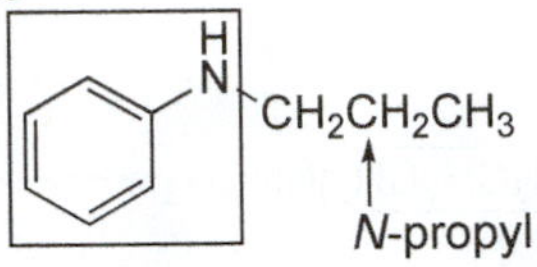

H
N $CH_2CH_2CH_3$

*N*-propyl

**18.7**  Determine the compound in each pair with a higher boiling point, using the rules in Example 18.3.

a.

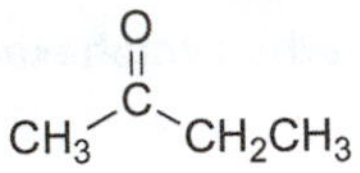

O
‖
$CH_3{-}C{-}CH_2CH_3$     or     $(CH_3)_2CHCH_2NH_2$

a ketone with only C–H bonds

a 1° amine with N–H bonds
intermolecular hydrogen bonding
**higher boiling point**

b.     $(CH_3)_2CHCH_2NH_2$     or     $(CH_3)_2CHCH_2OH$

a 1° amine with N–H bonds
intermolecular hydrogen bonding

a 1° alcohol with an O–H bond
intermolecular hydrogen bonding
stronger intermolecular forces
**higher boiling point**

c. 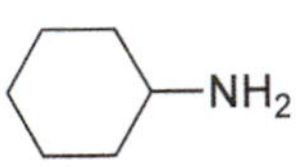     ⬡$-NH_2$     or     ⬡$-CH_3$

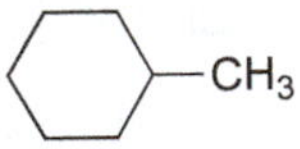

a 1° amine with N–H bonds
intermolecular hydrogen bonding
**higher boiling point**

hydrocarbon

**18.8**  Caffeine is soluble in the organic solvent $CH_2Cl_2$ because caffeine is organic and "like dissolves like."

**18.9**  Nicotine has two heterocycles: pyridine and pyrrolidine.

**18.10** Draw both enantiomers of nicotine.

**18.11** Morphine contains a 3° amine.

**18.12** Identify the functional groups in heroin.

**18.13** Quinine has 11 trigonal planar carbon atoms [part (a), labeled with *] and nine tetrahedral carbon atoms [part (b), all other C's].

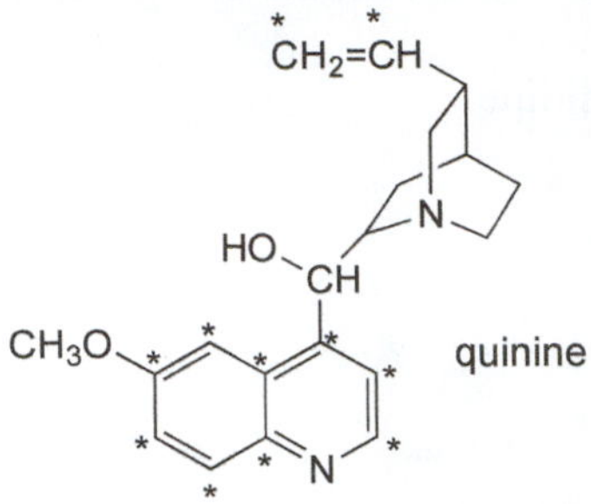

quinine

**18.14** Draw the products when the amines are treated with HCl as in Example 18.4.

a. $CH_3CH_2NH_2$ $\xrightarrow{\text{HCl}}$ $(CH_3CH_3NH_3)^+ + Cl^-$

c. $(CH_3CH_2)_3N$ $\xrightarrow{\text{HCl}}$ $\left[(CH_3CH_2)_3NH\right]^+ + Cl^-$

b. $(CH_3CH_2)_2NH$ $\xrightarrow{\text{HCl}}$ $\left[(CH_3CH_2)_2NH_2\right]^+ + Cl^-$

**18.15** Draw the products of each reaction.

a. $CH_3CH_2CH_2CH_2-NH_2$ + HCl $\longrightarrow$ $(CH_3CH_2CH_2CH_2-NH_3)^+$ + $Cl^-$

b. $(CH_3)_2NH$ + $C_6H_5COOH$ $\longrightarrow$ $\left[(CH_3)_2NH_2\right]^+$ + $C_6H_5COO^-$

c. 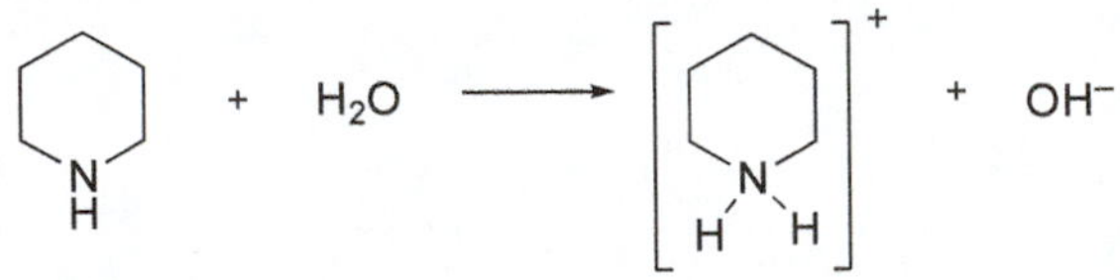 + $H_2O$ $\longrightarrow$ + $OH^-$

**18.16** Draw the products of the reaction.

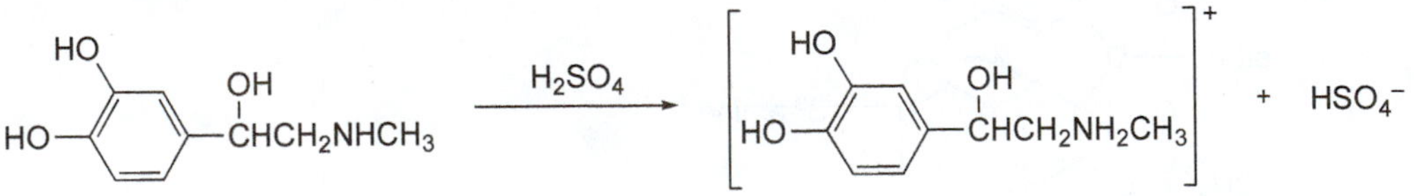

**18.17** Name each ammonium salt as in Example 18.5.

a. $(CH_3NH_3)^+$ $Cl^-$
chloride
derived from methylamine
**methylammonium chloride**

b. $\left[(CH_3CH_2CH_2)_2NH_2\right]^+$ $Br^-$
bromide
derived from dipropylamine
**dipropylammonium bromide**

c. $\left[(CH_3)_2NHCH_2CH_3\right]^+$ $CH_3COO^-$
acetate
derived from ethyldimethylamine
**ethyldimethylammonium acetate**

**18.18** Ammonium salts are water-soluble solids. A water-insoluble amine can be converted to a water-soluble ammonium salt by treatment with acid.

a.   $(CH_3CH_2)_3N$

3° amine with 6 C's
**water insoluble**

b.   $[(CH_3CH_2)_3NH]^+ \ Br^-$

ammonium salt
**water soluble**

c.   $CH_3CH_2NH_2$

1° amine with 2 C's
**water soluble**

d. 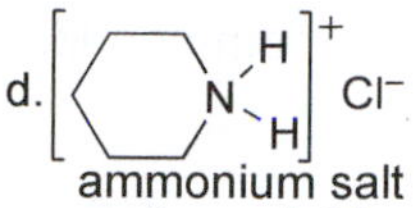   ammonium salt
**water soluble**

**18.19** Draw the products of each reaction.

a.   $[(CH_3CH_2)_3NH]^+ \ Br^- \xrightarrow{\text{NaOH}} (CH_3CH_2)_3N \ + \ H_2O \ + \ NaBr$

c.   $\xrightarrow{\text{NaOH}}$ NH

$+ \ H_2O \ + \ NaCl$

b.   $(CH_3CH_2NH_3)^+ \ HSO_4^- \xrightarrow{\text{NaOH}} CH_3CH_2NH_2 \ + \ H_2O \ + \ NaHSO_4$

**18.20** Draw the structure of each amine from which the ammonium salt is derived.

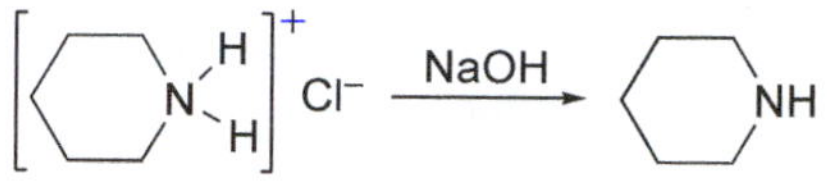

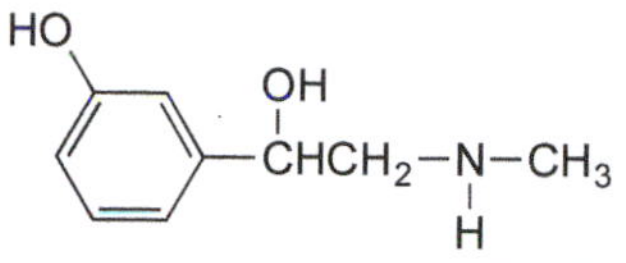

phenylephrine

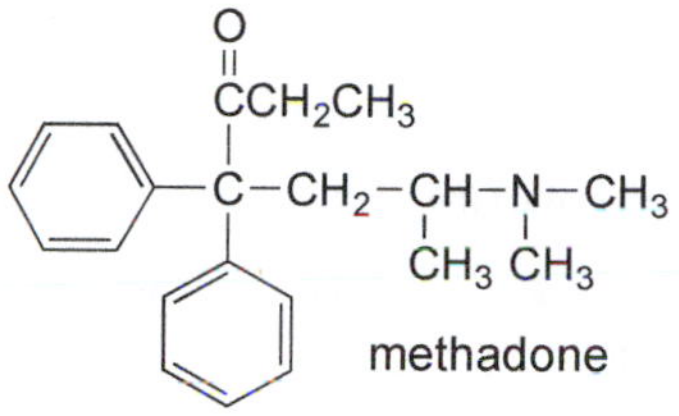

methadone

**18.21** A quaternary ammonium salt like Bitrex has four R groups bonded to N, so there is no proton available that can be removed to form an amine.

**18.22** The chirality center in each compound is labeled.

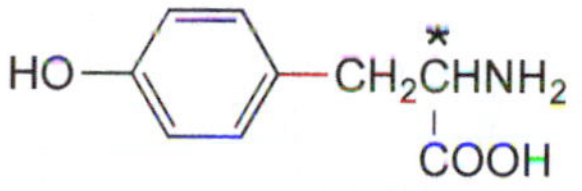

tyrosine

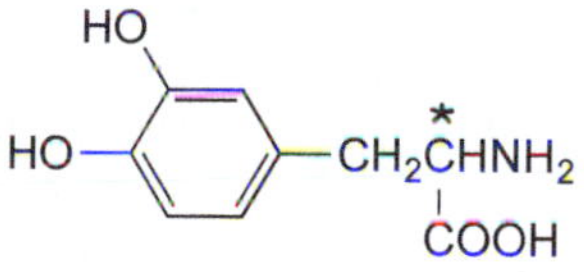

L-dopa

\* = chirality center

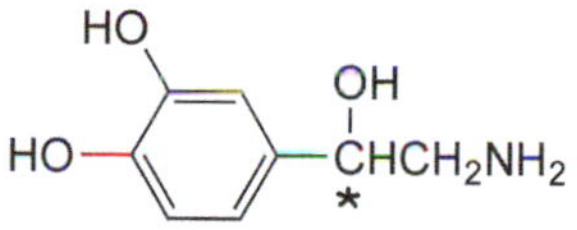

norepinephrine

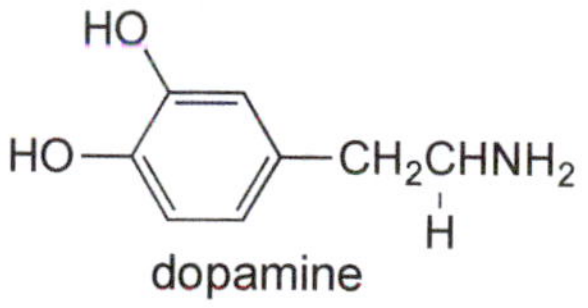

dopamine

achiral

**18.23** COOH must be removed and OH must be added.

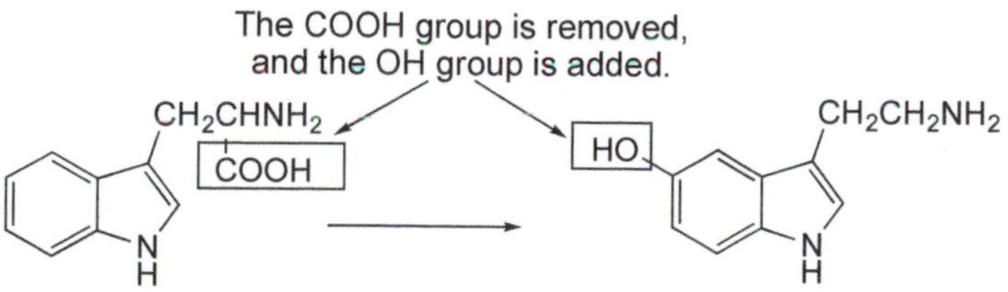

The COOH group is removed,
and the OH group is added.

tryptophan

serotonin

**18.24** Serotonin, bufotenin, and psilocin have the same ring system with a two-carbon chain containing a N atom bonded to it. They all contain an OH group bonded to the six-membered ring.

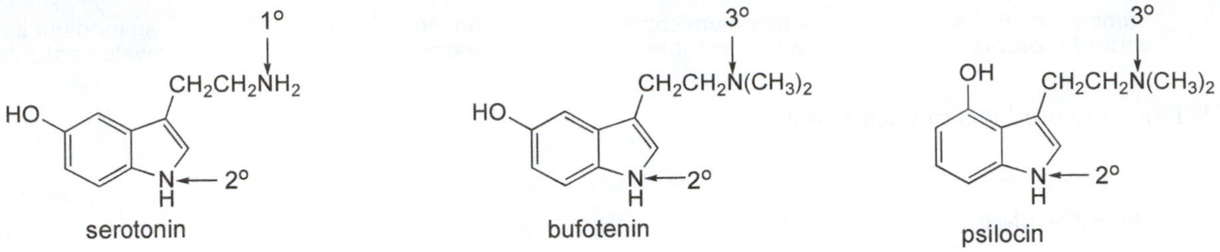

1°
$CH_2CH_2NH_2$
HO
N — 2°
H
serotonin

3°
$CH_2CH_2N(CH_3)_2$
HO
N — 2°
H
bufotenin

3°
OH   $CH_2CH_2N(CH_3)_2$
N — 2°
H
psilocin

**18.25** To convert phenylephrine to methamphetamine requires the addition of a methyl group and removal of two OH groups.

**18.26** The atoms of 2-phenylethylamine are in bold in each compound.

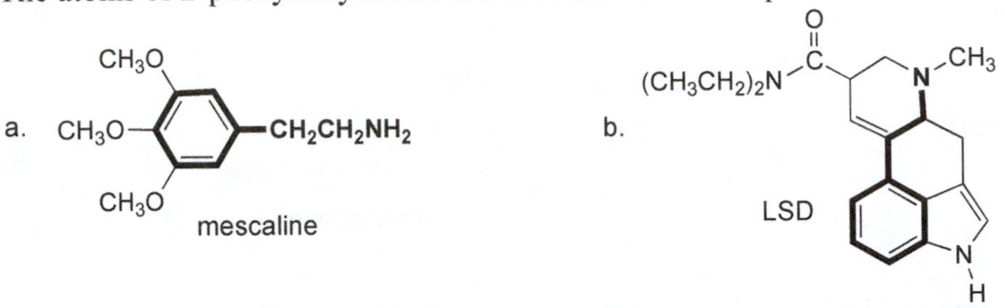

a.   $CH_3O$
$CH_3O$ — —$CH_2CH_2NH_2$
$CH_3O$
mescaline

b.
$(CH_3CH_2)_2N$—C— ... —$CH_3$
O
LSD
N
H

**18.27** Classify each alcohol and amine as 1°, 2°, or 3°. OH groups not labeled are phenols.

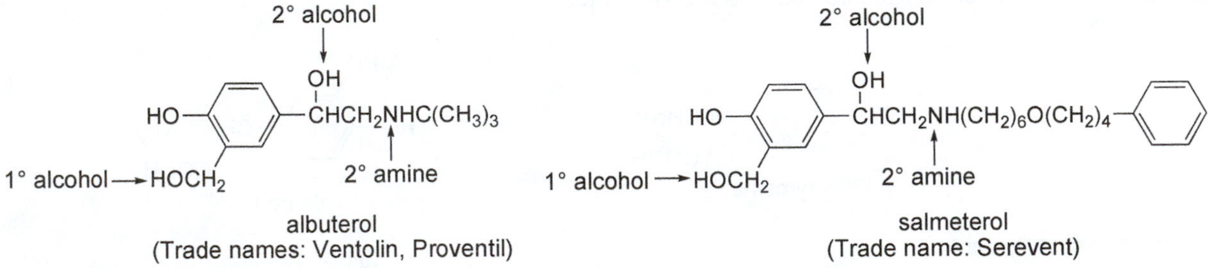

2° alcohol
OH
HO— —$CHCH_2NHC(CH_3)_3$
1° alcohol → $HOCH_2$        2° amine
albuterol
(Trade names: Ventolin, Proventil)

2° alcohol
OH
HO— —$CHCH_2NH(CH_2)_6O(CH_2)_4$—
1° alcohol → $HOCH_2$        2° amine
salmeterol
(Trade name: Serevent)

**18.28** Draw the complete structure for cimetidine.

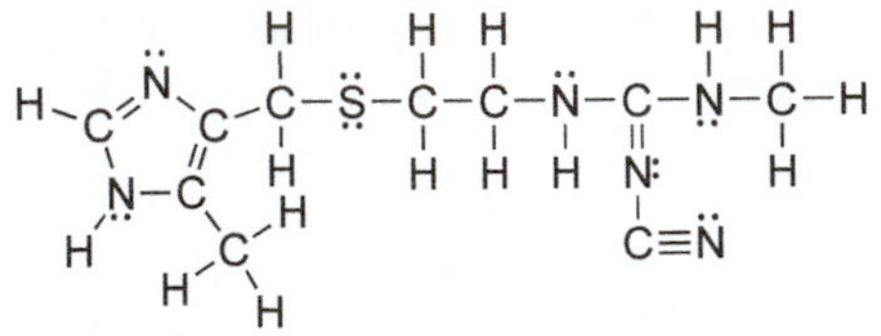

H    H    H H            H H
H—C⸗N   C—C—S̈—C—C—N̈—C—N—C—H
N—C    H    H H H  N:    H
H    C    H                  C≡N̈
H   H

## Solutions to Odd-Numbered End-of-Chapter Problems

**18.29** To determine whether an amine or amide is 1°, 2°, or 3°, count the number of carbons bonded to the nitrogen atom as in Example 18.1.

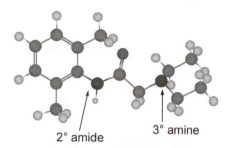

2° amide      3° amine

**18.31** To determine whether an amine is 1°, 2°, or 3°, count the number of carbons bonded to the nitrogen atom as in Example 18.1.

         2°                       3°

a.    $CH_3CHCH_2CH_2NHCH_3$       b.    (ring)$N-CH_3$
             $CH_3$

**18.33** Draw a structure to fit each description.

a.   $CH_3CH_2CH_2CH_2CH_2NH_2$    b.   $CH_3CH_2CH_2NCH_2CH_2CH_3$    c.   (ring)$N-CH_3$    d.   $\left[ CH_3CH_2NCH_2CH_3 \right]^+$

                                                  H                                         $CH_2CH_3$

        1° amine                 2° amine                 3° amine           quaternary

        $C_5H_{13}N$                 $C_6H_{15}N$                 $C_6H_{13}N$         ammonium ion

                                                                               $C_8H_{20}N^+$

**18.35** Name each amine as in Example 18.2.

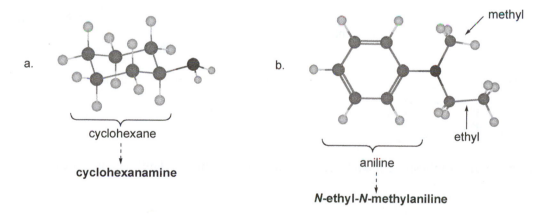

a.                                         b.                            methyl

       cyclohexane                                 ethyl

       **cyclohexanamine**                          aniline

                                     **N-ethyl-N-methylaniline**

**18.37**  Name each amine as in Example 18.2.

a.  CH$_3$CH$_2$—N(H)—CH$_2$CH$_3$  ----→ **Answer: diethylamine**

   2 ethyl groups

b.  |CH$_3$CH$_2$CHCH$_2$CH$_2$CH$_2$CH$_2$CH$_3$| (NH$_2$) --→ |CH$_3$CH$_2$CHCH$_2$CH$_2$CH$_2$CH$_2$CH$_3$| (NH$_2$) --→ **Answer: 3-octanamine**

   octane ---→ octan*amine*          1  2  3
   (8 C's)

c.  |CH$_3$CH$_2$CH$_2$CHCH$_2$CH$_3$| (NHCH$_3$) --→ |CH$_3$CH$_2$CH$_2$CHCH$_2$CH$_3$| (NHCH$_3$ ← *N*-methyl) --→ **Answer: *N*-methyl-3-hexanamine**

   hexane ---→ hexan*amine*          3  2  1
   (6 C's)

d.  |CH$_3$CHCH$_2$CH$_2$CH$_2$CH$_2$CH$_3$| (NHCH$_2$CH$_3$) --→ 1  2  |CH$_3$CHCH$_2$CH$_2$CH$_2$CH$_2$CH$_3$| (NHCH$_2$CH$_3$ ←—*N*-ethyl) --→ **Answer: *N*-ethyl-2-heptanamine**

   heptane ---→ heptan*amine*
   (7 C's)

**18.39**  Work backwards to draw the structure corresponding to each name.

a. **1-decan**amine

|CH$_3$CH$_2$CH$_2$CH$_2$CH$_2$CH$_2$CH$_2$CH$_2$CH$_2$CH$_2$NH$_2$|

b. **tricyclohexyl**amine

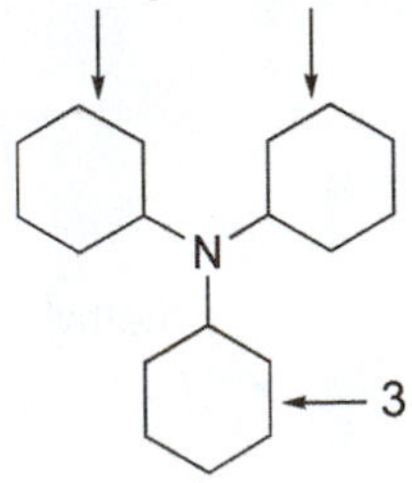

3 cyclohexyl groups

c. *p*-bromo**aniline**

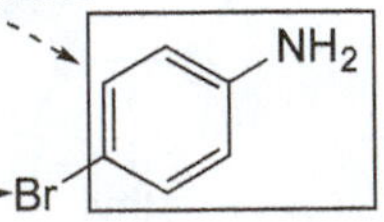

*p*-bromo —→ Br——NH$_2$

d. 3-amino**butan**oic acid

          3      1
   |CH$_3$CHCH$_2$COOH|

   3-amino —→NH$_2$

e. *N,N*-dipropyl-2-**octan**amine

   |CH$_3$CH$_2$CH$_2$CH$_2$CH$_2$CH$_2$CHCH$_3$|
                          N(CH$_2$CH$_2$CH$_3$)$_2$

                          *N,N*-dipropyl

f. *N*-ethyl**hexyl**amine

   CH$_3$CH$_2$NH|CH$_2$CH$_2$CH$_2$CH$_2$CH$_2$CH$_3$|

   *N*-ethyl

**18.41**  Draw the structures to illustrate the differences between *N,N*-dimethylaniline and 2,4-diethylaniline.

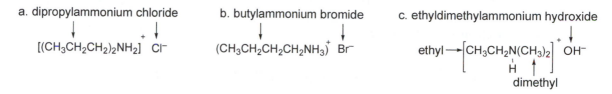

3° amine — N(CH$_3$)$_2$

1° amine — NH$_2$

CH$_3$CH$_2$ ... CH$_2$CH$_3$

*N,N*-dimethylaniline        2,4-diethylaniline

**18.43**  Work backwards from the name to draw each ammonium salt.

a. dipropylammonium chloride

[(CH$_3$CH$_2$CH$_2$)$_2$NH$_2$]$^+$  Cl$^-$

b. butylammonium bromide

(CH$_3$CH$_2$CH$_2$CH$_2$NH$_3$)$^+$  Br$^-$

c. ethyldimethylammonium hydroxide

ethyl → [CH$_3$CH$_2$N(CH$_3$)$_2$]$^+$  OH$^-$
                    |
                    H ↑
                dimethyl

**18.45**  Draw and name the four amines of molecular formula C$_3$H$_9$N.

CH$_3$—N—CH$_3$
       |
       CH$_3$

CH$_3$CH$_2$CH$_2$NH$_2$

CH$_3$CH$_2$NHCH$_3$

CH$_3$CHNH$_2$
       |
       CH$_3$

trimethylamine        1-propanamine        *N*-methylethanamine        2-propanamine

**18.47**  Pyridine is capable of hydrogen bonding with water, so it is more water soluble than benzene. Pyridine has a higher boiling point than benzene because it contains polar C–N bonds, while benzene is a nonpolar hydrocarbon.

**18.49**  Determine the compound in each pair with a higher boiling point, using the rules in Example 18.3.

a.   CH$_3$CH$_2$CH$_2$NH$_2$   or   CH$_3$(CH$_2$)$_7$NH$_2$
     3 C amine                  8 C amine
                                **higher
                                boiling point**

b.   CH$_3$(CH$_2$)$_6$OH   or   CH$_3$(CH$_2$)$_6$NH$_2$
     alcohol                amine
     **higher
     boiling point**

c.   ⬡—CH$_2$N(CH$_3$)$_2$   or   ⬡—CH$_2$CH$_2$CH$_2$NH$_2$
     3° amine                    1° amine
     no N–H bonds                N–H bonds
                                 **higher
                                 boiling point**

**18.51**  The hydrogen bond is drawn as a dashed line.

CH$_2$CH$_3$

H—N

CH$_3$CH$_2$    CH$_2$CH$_3$

:N—H

CH$_3$CH$_2$

**18.53**  Primary amines can hydrogen bond to each other, whereas 3° amines cannot. Therefore, 1° amines have higher boiling points than 3° amines of similar size. Since all amines contain nitrogen atoms, any amine can hydrogen bond to water. Therefore, 1° and 3° amines have similar solubility properties.

**18.55** Draw the reaction of each amine with water by transferring a proton from $H_2O$ to the amine.

a. $CH_3CH_2NH_2 + H_2O \longrightarrow (CH_3CH_2NH_3)^+ + OH^-$

b. $(CH_3CH_2)_2NH + H_2O \longrightarrow \left[(CH_3CH_2)_2NH_2\right]^+ + OH^-$

c. $(CH_3CH_2)_3N + H_2O \longrightarrow \left[(CH_3CH_2)_3NH\right]^+ + OH^-$

**18.57** Draw the products of each reaction.

a. $CH_3CH_2CH_2N(CH_3)_2 + HCl \longrightarrow \left[CH_3CH_2CH_2NH(CH_3)_2\right]^+ + Cl^-$

b.

c.

d.

**18.59** Draw the products of each reaction.

a.

b.

**18.61** The heterocycle in both coniine and morphine is piperidine (labeled in bold).

coniine

morphine

**18.63** Caffeine is a mild stimulant, imparting a feeling of alertness after consumption. It also increases heart rate, dilates airways, and stimulates the secretion of stomach acid. These effects are observed because caffeine increases glucose production, making an individual feel energetic.

**18.65** An alkaloid solution is slightly basic since its amine pulls off a proton from water, forming OH⁻ and an ammonium ion.

**18.67** Dopamine affects brain processes that control movement, emotions, and pleasure. Normal dopamine levels give an individual a pleasurable, satisfied feeling. Increased levels result in an intense "high." Drugs such as heroin, cocaine, and alcohol increase dopamine levels. When there is too little dopamine in the brain, an individual loses control of fine motor skills and Parkinson's disease results.

**18.69** Serotonin plays an important role in mood, sleep, perception, and temperature regulation. We get sleepy after eating a turkey dinner on Thanksgiving because the unusually high level of tryptophan in turkey is converted to serotonin. A deficiency of serotonin causes depression.

**18.71** Dopamine [part (a)] and norepinephrine [part (b)] are derived from tyrosine, serotonin [part (c)] is derived from tryptophan, and histamine [part (d)] is derived from histidine.

**18.73** The atoms of 2-phenylethylamine are labeled in bold in the compounds below.

b. not a derivative of 2-phenylethylamine

**18.75** The atoms of 2-phenylethylamine are labeled in bold.

**18.77** Chlorpheniramine is an example of an antihistamine. Antihistamines bind to the H1 histamine receptor, but they evoke a different response than histamine. An antihistamine like chlorpheniramine or diphenhydramine, for example, inhibits vasodilation, so it is used to treat the symptoms of the common cold and environmental allergies.

**18.79** Answer each question about benzphetamine.

a. and b.

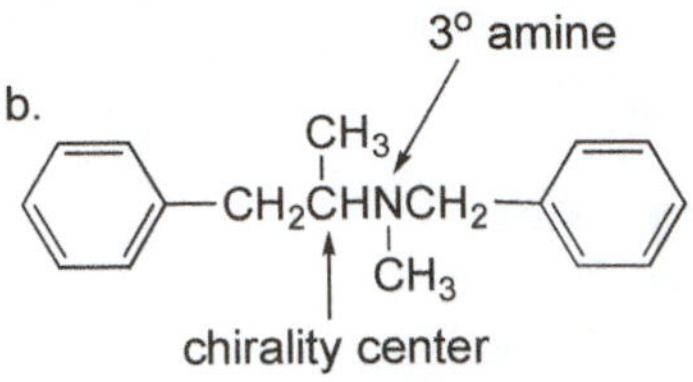

3° amine

$CH_3$
$-CH_2CHNCH_2-$
$CH_3$

chirality center

c.

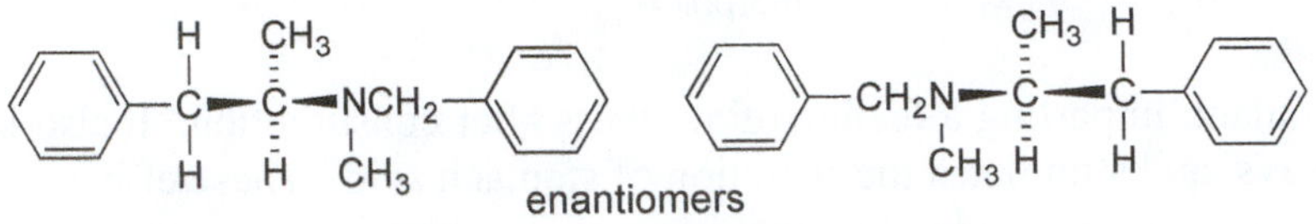

enantiomers

d.

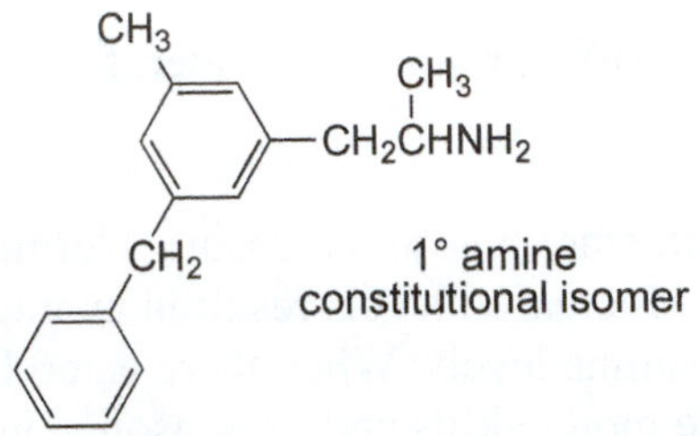

1° amine
constitutional isomer

e.

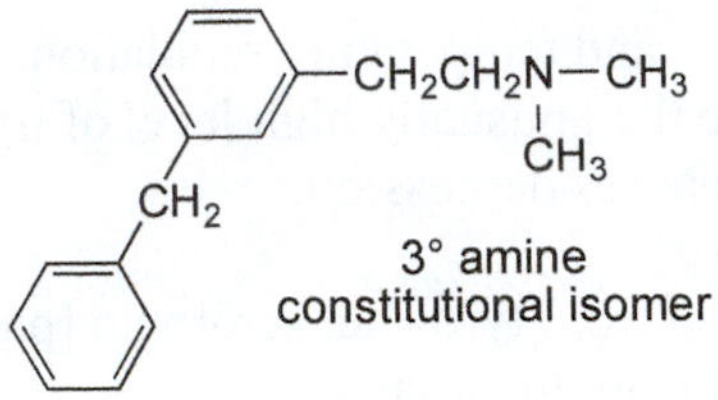

$-CH_2CH_2N-CH_3$
$CH_3$

3° amine
constitutional isomer

f.

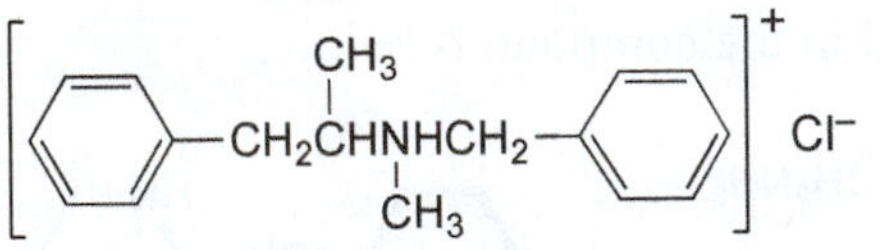

$[ -CH_2CHNHCH_2- ]^+$ $Cl^-$   benzphetamine hydrochloride
$CH_3$
$CH_3$

g.

$-CH_2CHNCH_2-$ $+ CH_3COOH \longrightarrow [ -CH_2CHNHCH_2- ]^+ + CH_3COO^-$
$CH_3$                                    $CH_3$
$CH_3$                                    $CH_3$

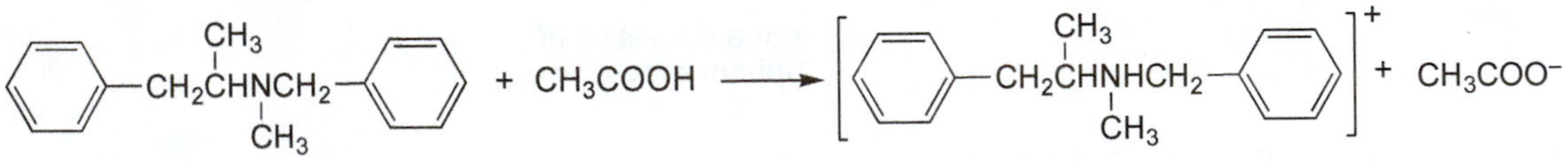

**18.81** Answer the questions about Ritalin.

a., b., and c.

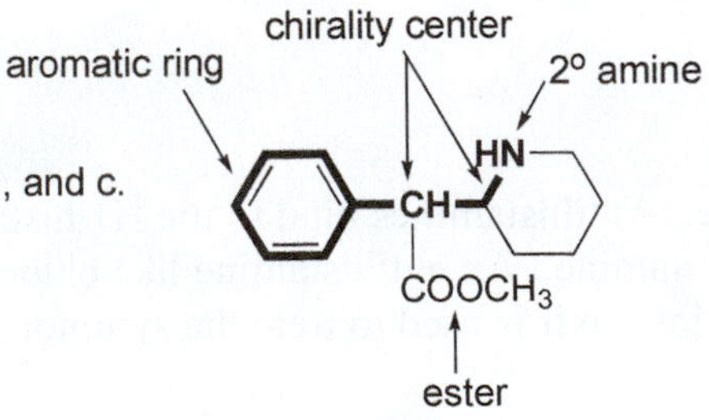

aromatic ring    chirality center    2° amine

HN
$CH$
$COOCH_3$
ester

e.

$[ \begin{array}{c} H \\ H-N \\ CH \\ COOCH_3 \end{array} ]^+$ $Cl^-$

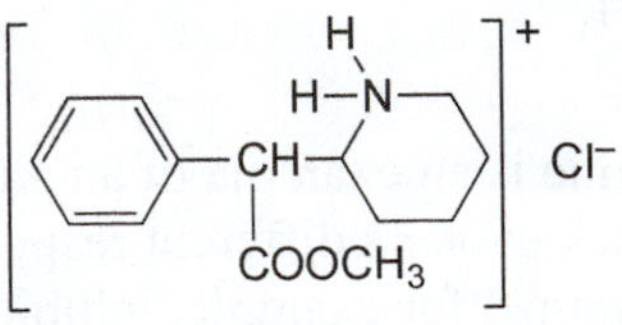

d. 2-phenylethylamine in bold

**18.83** Work backwards to draw the structure of the amine and carboxylic acid that are used to form Chlortrimeton, an ammonium salt.

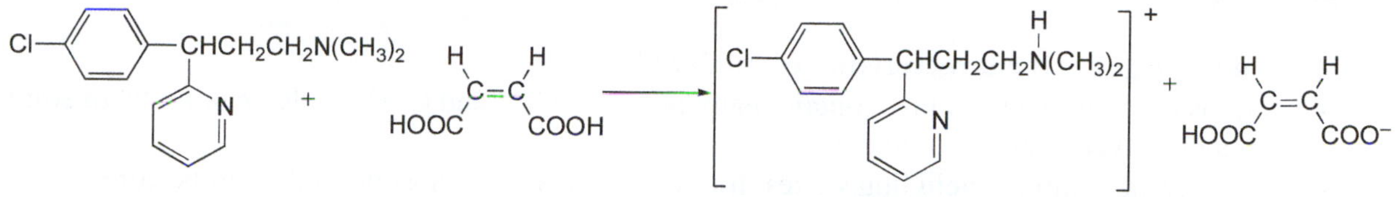

chlorpheniramine maleate
Chlortrimeton

**18.85** A vasodilator dilates blood vessels and a bronchodilator dilates airways in the lungs. Histamine is a vasodilator and albuterol is a bronchodilator.

**18.87** Albuterol will exist in the ionic form in the stomach (lower pH) and in the neutral form in the intestines.

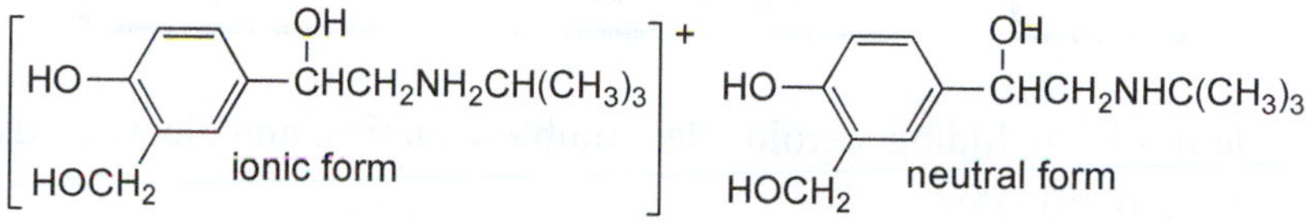

albuterol

**18.89** Heroin has two esters, which can be made from the two OH groups in morphine. Add acetic acid ($CH_3COOH$) and $H_2SO_4$ to make the two esters in heroin.

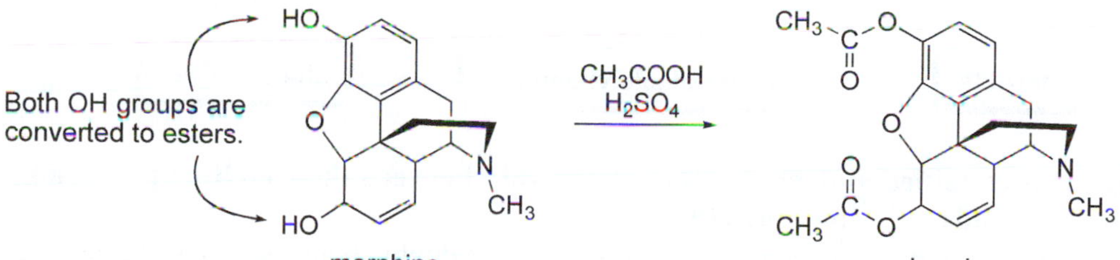

Both OH groups are converted to esters.

morphine

$CH_3COOH$
$H_2SO_4$

heroin

# Chapter 19 Lipids

## Chapter Review

**[1] What are the general characteristics of lipids? (19.1)**
- Lipids are biomolecules that contain many nonpolar C–C and C–H bonds, making them soluble in organic solvents and insoluble in water.
- Hydrolyzable lipids, including waxes, triacylglycerols, and phospholipids, can be converted to smaller molecules on reaction with water.

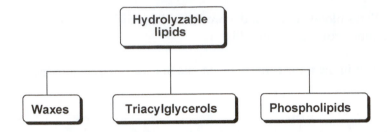

- Nonhydrolyzable lipids, including steroids, fat-soluble vitamins, and eicosanoids, cannot be cleaved into smaller units by hydrolysis.

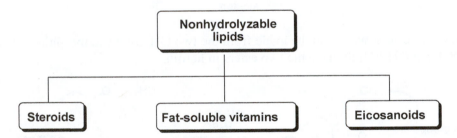

**[2] How are fatty acids classified and what is the relationship between their melting points and the number of double bonds they contain? (19.2)**
- Fatty acids are saturated if they contain no carbon–carbon double bonds and unsaturated if they contain one or more double bonds. Unsaturated fatty acids generally contain cis double bonds.
- As the number of double bonds in the fatty acid increases, its melting point decreases.

**[3] What are waxes? (19.3)**
- A wax is an ester (RCOOR') formed from a fatty acid (RCOOH) and a high molecular weight alcohol (R'OH). Since waxes contain many nonpolar C–C and C–H bonds, they are hydrophobic.
- Waxes (RCOOR') are hydrolyzed to fatty acids (RCOOH) and alcohols (R'OH).

**[4] What are triacylglycerols, and how do the triacylglycerols in a fat and oil differ? (19.4)**

- Triacylglycerols, or triglycerides, are triesters formed from glycerol and three molecules of fatty acids. A monounsaturated triacylglycerol contains one carbon–carbon double bond, whereas polyunsaturated triacylglycerols have more than one carbon–carbon double bond.

glycerol + fatty acids → triacylglycerol

R groups have 11–19 C's.

- Fats are triacylglycerols derived from fatty acids having few double bonds, making them solids at room temperature. Fats are generally obtained from animal sources.
- Oils are triacylglycerols derived from fatty acids having a larger number of double bonds, making them liquids at room temperature. Oils are generally obtained from plant sources.

**[5] What hydrolysis products are formed from a triacylglycerol? (19.5)**

- Triacylglycerols are hydrolyzed in acid or with enzymes (in biological systems) to form glycerol and three molecules of fatty acids.

triacylglycerol + 3 $H_2O$ → (H₂SO₄ or lipase) → glycerol + three fatty acids

- Base hydrolysis of a triacylglycerol forms glycerol and sodium salts of fatty acids—soaps.

triacylglycerol + 3 NaOH → ($H_2O$) → glycerol + soaps

**[6] What are the major types of phospholipids? (19.6)**

- All phospholipids contain a phosphorus atom, and have a polar (ionic) head and two nonpolar tails. Phosphoacylglycerols are derived from glycerol, two molecules of fatty acids, phosphate, and an alcohol (either ethanolamine or choline). Sphingomyelins are derived from sphingosine, a fatty acid (that forms an amide), phosphate, and an alcohol (either ethanolamine or choline).

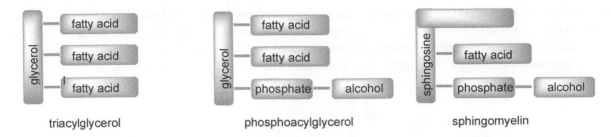

| triacylglycerol | phosphoacylglycerol | sphingomyelin |

**[7] Describe the structure of the cell membrane. How do molecules and ions cross the cell membrane? (19.7)**

- The main component of the cell membrane is phospholipids, arranged in a lipid bilayer with the ionic heads oriented towards the outside of the bilayer, and the nonpolar tails on the interior.
- Small molecules like $O_2$ and $CO_2$ diffuse through the membrane from the side of higher concentration to the side of lower concentration. Larger polar molecules and some ions travel through channels created by integral membrane proteins (facilitated diffusion). Some cations ($Na^+$, $K^+$, and $Ca^{2+}$) must travel against the concentration gradient, a process called active transport, which requires energy input.

**[8] What are the main structural features of steroids? What is the relationship between the steroid cholesterol and cardiovascular disease? (19.8)**

- Steroids are tetracyclic lipids that contain three six-membered rings and one five-membered ring.

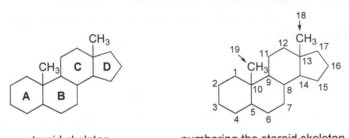

| steroid skeleton | numbering the steroid skeleton |

- Because cholesterol is insoluble in the aqueous medium of the blood, it is transported through the bloodstream in water-soluble particles called lipoproteins. Low-density lipoprotein particles (LDLs) transport cholesterol from the liver to the tissues. If the blood cholesterol level is high, it forms plaque on the walls of arteries, increasing the risk of heart attack and stroke. High-density lipoprotein particles (HDLs) transport cholesterol from the tissues to the liver, where it is metabolized or eliminated.

**[9] What is a hormone? Give examples of steroid hormones. (19.9)**

- A hormone is a molecule that is synthesized in one part of an organism, and elicits a response at a different site. Steroid hormones include estrogens and progestins (female sex hormones), androgens (male sex hormones), and adrenal cortical steroids such as cortisone, which are synthesized in the adrenal gland.

**[10] Which vitamins are fat soluble? (19.10)**

- Fat-soluble vitamins are lipids required in small quantities for normal cell function, and which cannot be synthesized in the body. Vitamins A, D, E, and K are fat soluble.

**[11] What are the general characteristics of prostaglandins and leukotrienes? (19.11)**

- Prostaglandins are a group of carboxylic acids that contain a five-membered ring and are derived from arachidonic acid. Prostaglandins cause inflammation, decrease gastric secretions, inhibit blood platelet aggregation, stimulate uterine contractions, and relax the smooth muscle of the uterus.
- Leukotrienes, acyclic molecules derived from arachidonic acid, contribute to the asthmatic response by constricting smooth muscles in the lungs.

$CH_3(CH_2)_4(CH=CHCH_2)_4(CH_2)_2COOH$

arachidonic acid
(four cis double bonds)

a fatty acid

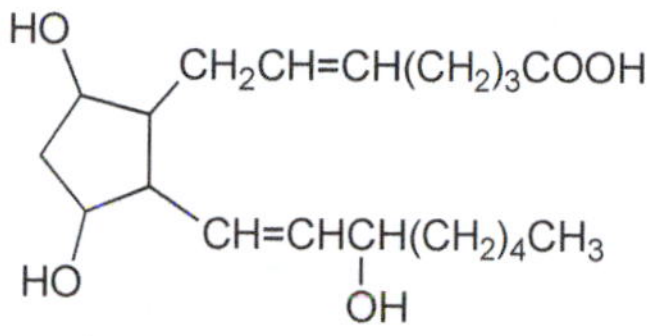

PGF$_{2\alpha}$
a prostaglandin

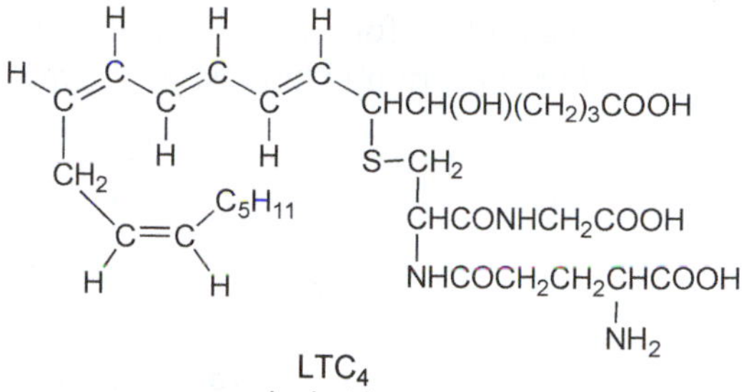

LTC$_4$
a leukotriene

## Problem Solving

## [1] Introduction to Lipids (19.1)

**Example 19.1** Which compounds are likely to be lipids?

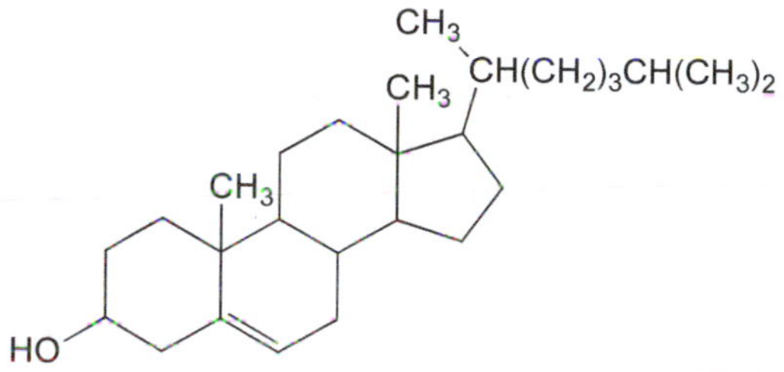

**Analysis**

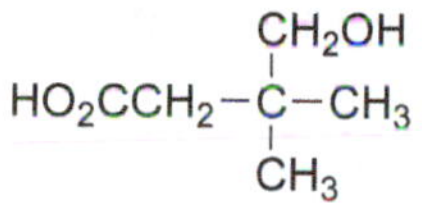

Lipids contain many nonpolar C–C and C–H bonds and few polar bonds.

**Solution**

a. Compound (a) is unlikely to be a lipid since it contains a polar OH group and a carboxyl group (COOH), and has only six carbon atoms.
b. Compound (b) is likely to be a lipid since it contains many nonpolar C–C and C–H bonds, and only one OH group.

## [2] Fatty Acids (19.2)

**Example 19.2** Draw the skeletal structure for palmitoleic acid (Table 19.2), and label the hydrophobic and hydrophilic portions.

**Analysis**
- Skeletal structures have a carbon at the intersection of two lines and at the end of every line. The double bond must have the cis arrangement in an unsaturated fatty acid.
- The nonpolar C–C and C–H bonds comprise the hydrophobic portion of a molecule and the polar bonds comprise the hydrophilic portion.

**Solution**
The skeletal structure for palmitoleic acid is drawn below. The hydrophilic portion is the COOH group and the hydrophobic portion is the rest of the molecule.

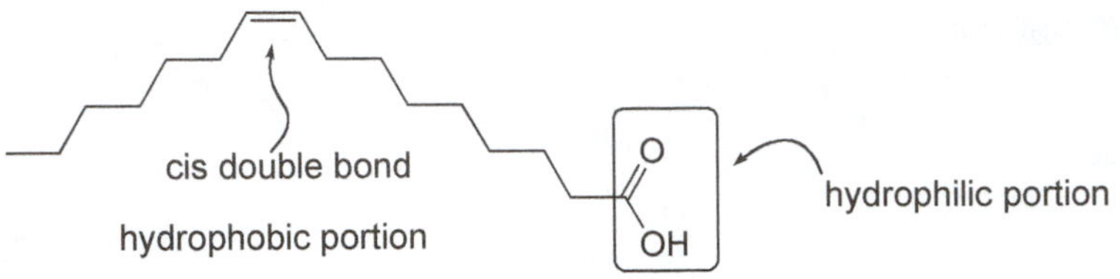

cis double bond

hydrophobic portion

hydrophilic portion

---

## [3] Waxes (19.3)

**Example 19.3** Draw the structure of a wax formed from an 18-carbon straight-chain alcohol and the fatty acid $CH_3(CH_2)_{10}COOH$.

**Analysis**
To draw the wax, arrange the carboxyl group of the fatty acid (RCOOH) next to the OH group of the alcohol (R'OH) with which it reacts. Then, replace the OH group of the fatty acid with the OR' group of the alcohol, forming an ester RCOOR' with a new C–O bond at the carbonyl carbon.

**Solution**
Draw the structures of the fatty acid and the alcohol, and replace the OH group of the 12-carbon acid with the $O(CH_2)_{17}CH_3$ group of the alcohol.

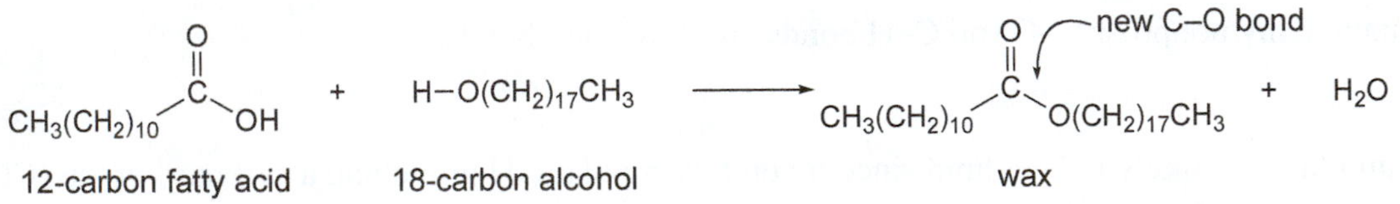

12-carbon fatty acid     18-carbon alcohol     wax

---

## [4] Triacylglycerols–Fats and Oils (19.4)

**Example 19.4** Draw the structure of a triacylglycerol formed from glycerol, one molecule of palmitic acid, and two molecules of linoleic acid. Bond the palmitic acid to the 2° OH group (OH on the middle carbon atom) of glycerol.

**Analysis**
To draw the triacylglycerol, arrange each OH group of glycerol next to the carboxyl group of a fatty acid. Then join each O atom of glycerol to a carbonyl carbon of a fatty acid, thus forming three new C–O bonds. Structures of fatty acids are located in Table 19.2.

**Solution**

Form three new ester bonds (RCOOR') from the OH groups of glycerol and the three fatty acids (RCOOH).

$$CH_2-OH$$
$$CH-OH \quad + \quad HO-\overset{O}{\underset{\parallel}{C}}-(CH_2)_7CH=CHCH_2CH=CH(CH_2)_4CH_3$$

palmitic acid

glycerol

2° OH

linoleic acid

$$CH_2-O-\overset{O}{\underset{\parallel}{C}}-(CH_2)_7CH=CHCH_2CH=CH(CH_2)_4CH_3$$
$$CH-O-\overset{O}{\underset{\parallel}{C}}-(CH_2)_{14}CH_3 \quad + \quad 3\ H_2O$$
$$CH_2-O-\overset{O}{\underset{\parallel}{C}}-(CH_2)_7CH=CHCH_2CH=CH(CH_2)_4CH_3$$

triacylglycerol

---

## [5] Hydrolysis of Triacylglycerols (19.5)

**Example 19.5** Draw the products formed when the given triacylglycerol is hydrolyzed with water in the presence of sulfuric acid.

$$CH_2-O-\overset{O}{\underset{\parallel}{C}}-(CH_2)_{14}CH_3$$
$$CH-O-\overset{O}{\underset{\parallel}{C}}-(CH_2)_{10}CH_3$$
$$CH_2-O-\overset{O}{\underset{\parallel}{C}}-(CH_2)_7CH=CH(CH_2)_7CH_3$$

**Analysis**

To draw the products of ester hydrolysis, cleave the three C–O single bonds at the carbonyl carbons to form glycerol and three fatty acids (RCOOH).

**Solution**

Bonds broken during hydrolysis

$$CH_2-O-\overset{O}{\underset{\parallel}{C}}-(CH_2)_{14}CH_3$$
$$CH-O-\overset{O}{\underset{\parallel}{C}}-(CH_2)_{10}CH_3$$
$$CH_2-O-\overset{O}{\underset{\parallel}{C}}-(CH_2)_7CH=CH(CH_2)_7CH_3$$

$$\xrightarrow[H_2SO_4]{3\ H-OH}$$

$$CH_2-OH$$
$$CH-OH \quad +$$
$$CH_2-OH$$

glycerol

$$HO-\overset{O}{\underset{\parallel}{C}}-(CH_2)_{14}CH_3$$

palmitic acid

$$HO-\overset{O}{\underset{\parallel}{C}}-(CH_2)_{10}CH_3$$

lauric acid

$$HO-\overset{O}{\underset{\parallel}{C}}-(CH_2)_7CH=CH(CH_2)_7CH_3$$

oleic acid

---

## [6] Phospholipids (19.6)

**Example 19.6** Draw the structure of a cephalin formed from two molecules of myristic acid.

**Analysis**

Substitute the 14-carbon saturated fatty acid myristic acid for the R and R' groups in the general structure of a cephalin molecule. In a cephalin, $-CH_2CH_2NH_3^+$ forms part of the phosphodiester.

## Solution

CH$_2$–O–C–R
CH–O–C–R'
CH$_2$–O–P–O–CH$_2$CH$_2$NH$_3$

from myristic acid

general structure

CH$_2$–O–C–(CH$_2$)$_{12}$CH$_3$
CH–O–C–(CH$_2$)$_{12}$CH$_3$
CH$_2$–O–P–O–CH$_2$CH$_2$NH$_3$

**Answer**

---

## Self-Test

**[1] Fill in the blank with one of the terms listed below.**

Cell membrane (19.7)     Lipids (19.1)     Soaps (19.5)
Fats (19.4)     Oils (19.4)     Triacylglycerols (19.4)
Fat-soluble vitamins (19.10)     Phospholipids (19.6)     Unsaturated fatty acids (19.2)
Fatty acids (19.2)     Saturated fatty acids (19.2)     Waxes (19.3)
Hormone (19.9)

1. _____ are triesters formed from glycerol and three molecules of fatty acids.
2. _____ are carboxylic acids that have no double bonds in their long hydrocarbon chains.
3. _____ are liquids at room temperature and derived from fatty acids having one or more carbon–carbon double bonds.
4. _____ are biomolecules that are soluble in organic solvents but insoluble in water.
5. _____ are esters formed from a fatty acid and a high molecular weight alcohol.
6. _____ are metal salts of fatty acids.
7. _____ are carboxylic acids with long carbon chains of 12–20 carbon atoms.
8. The vitamins A, D, E, and K are a group of lipids called _____.
9. _____ are solids at room temperature and are derived from fatty acids having few carbon–carbon double bonds.
10. A _____ is a molecule that is synthesized in one part of an organism, which then elicits a response at a different site.
11. _____ are carboxylic acids that have one or more carbon–carbon double bonds in their long hydrocarbon chains.
12. The cytoplasm is the aqueous medium inside the cell, separated from water outside the cell by the _____.
13. _____ are lipids that contain a phosphorus atom.

**[2] Which molecule in each pair has a higher melting point?**

14.     CH$_3$(CH$_2$)$_{14}$COOH        CH$_3$(CH$_2$)$_4$CH=CHCH$_2$CH=CH(CH$_2$)$_5$COOH
         **A**                     **B**

15.     CH$_3$(CH$_2$)$_2$CH=CH(CH$_2$)$_8$COOH        CH$_3$(CH$_2$)$_2$CH=CHCH$_2$CH=CH(CH$_2$)$_3$COOH
         **A**                     **B**

16.     $CH_3CH=CHCH_2CH=CH(CH_2)_7COOH$          $CH_3(CH_2)_6CH=CH(CH_2)_4COOH$

                            **A**                                                **B**

**[3] Choose the method of transport across a cell membrane for each species.**

   a. simple diffusion             b. facilitated transport          c. active transport

17. $HCO_3^-$                  18. $CO_2$                  19. $Na^+$                  20. glucose

**[4] Match the vitamin with the description.**

   a. vitamin A             b. vitamin D             c. vitamin E             d. vitamin K

21.   This vitamin is synthesized in the body from cholesterol.
22.   This vitamin is converted to 11-*cis*-retinal, the light-sensitive compound responsible for vision in all vertebrates.
23.   A deficiency of this vitamin leads to excessive and sometimes fatal bleeding because of inadequate blood clotting.

**[5] Match the hormone with a description of its actions.**

   a. estradiol and estrone          b. progesterone          c. testosterone and androsterone

24. This hormone is often called the "pregnancy hormone" because it prepares the uterus for the implantation of a fertilized egg.
25. These hormones control the development of secondary sex characteristics in males.

## Answers to Self-Test

| | | | | |
|---|---|---|---|---|
| 1. Triacylglycerols | 6. Soaps | 11. Unsaturated fatty acids | 16. **B** | 21. b |
| 2. Saturated fatty acids | 7. Fatty acids | 12. cell membrane | 17. b | 22. a |
| 3. Oils | 8. fat-soluble vitamins | 13. Phospholipids | 18. a | 23. d |
| 4. Lipids | 9. Fats | 14. **A** | 19. c | 24. b |
| 5. Waxes | 10. hormone | 15. **A** | 20. b | 25. c |

## Solutions to In-Chapter Problems

**19.1**   Recall from Example 19.1 that lipids contain many nonpolar C–C and C–H bonds and few polar bonds.

a.   $HO_2CCH_2-\overset{\overset{\displaystyle CH_3}{|}}{\underset{\underset{\displaystyle OH}{|}}{C}}-CH_2CH_2OH$

two polar OH groups
and a COOH
six carbons
**not a lipid**

b.

one polar OH group
10 carbons
**lipid**

c.

two polar OH groups
18 carbons
**lipid**

**19.2** Since lipids contain many nonpolar C–C bonds, they are soluble in nonpolar and weakly polar organic solvents. Therefore, lipids are likely to be soluble in (a) $CH_2Cl_2$ and (c) $CH_3CH_2CH_2CH_2CH_3$. The 5% aqueous NaCl solution (b) is not a solution in which lipids are soluble since it is a polar solvent.

**19.3** Answer the questions as in Example 19.2.
- Skeletal structures have a carbon at the intersection of two lines and at the end of every line. The double bond must have the cis arrangement in an unsaturated fatty acid.
- The nonpolar C–C and C–H bonds comprise the hydrophobic portion of a molecule and the polar bonds comprise the hydrophilic portion.
- For the same number of carbons, increasing the number of double bonds decreases the melting point of a fatty acid.

a. and b.

$$CH_3(CH_2)_{16}COOH \quad = $$
**A**

hydrophobic portion      hydrophilic portion

$$CH_3(CH_2)_4CH=CHCH_2CH=CH(CH_2)_7COOH \quad = $$
**B**

hydrophobic portion      hydrophilic portion

c. **A** will have the higher melting point because the molecules can pack together better.

**19.4** In omega-*n* acids, *n* is the carbon at which the first double bond occurs in the carbon chain, beginning at the end of the chain that contains the $CH_3$ group.

a.

1

3

COOH

b.    first C=C at C3 ⟶ an **omega-3 acid**

**19.5** In omega-*n* acids, *n* is the carbon at which the first double bond occurs in the carbon chain, beginning at the end of the chain that contains the $CH_3$ group.

first C=C at C9 ⟶ an **omega-9 acid**

a.    $CH_3CH_2CH_2CH_2CH_2CH_2CH_2CH_2CH=CHCH_2CH_2CH_2CH_2CH_2CH_2CH_2COOH$

        1   2   3   4   5   6   7   8   9

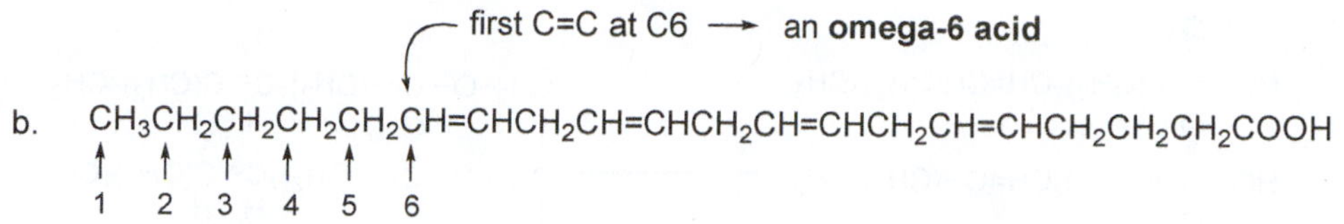

first C=C at C6 ⟶ an **omega-6 acid**

b.  CH₃CH₂CH₂CH₂CH₂CH=CHCH₂CH=CHCH₂CH=CHCH₂CH=CHCH₂CH₂CH₂COOH

    1  2  3  4  5  6

**19.6**  Use the steps in Example 19.3 to draw each wax.

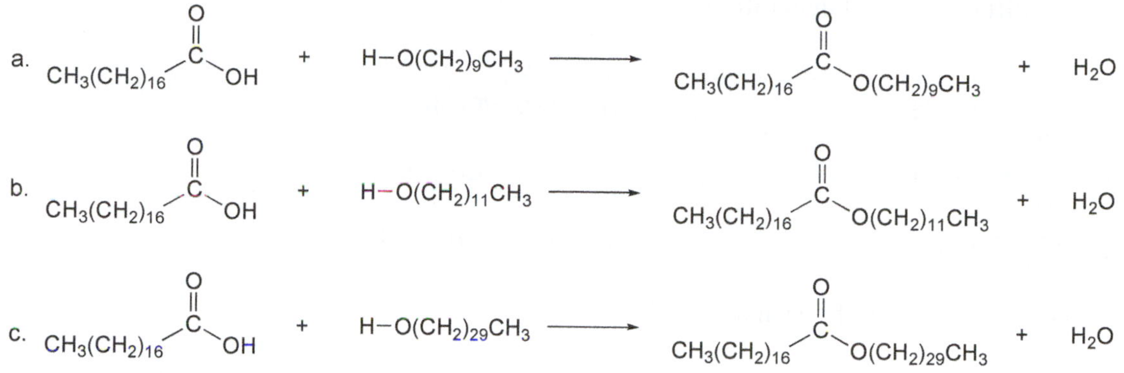

a.  $CH_3(CH_2)_{16}$ C(=O)OH  +  H–O(CH₂)₉CH₃  ⟶  $CH_3(CH_2)_{16}$ C(=O)O(CH₂)₉CH₃  +  $H_2O$

b.  $CH_3(CH_2)_{16}$ C(=O)OH  +  H–O(CH₂)₁₁CH₃  ⟶  $CH_3(CH_2)_{16}$ C(=O)O(CH₂)₁₁CH₃  +  $H_2O$

c.  $CH_3(CH_2)_{16}$ C(=O)OH  +  H–O(CH₂)₂₉CH₃  ⟶  $CH_3(CH_2)_{16}$ C(=O)O(CH₂)₂₉CH₃  +  $H_2O$

**19.7**  Draw the structure of the wax formed from eicosenoic acid.

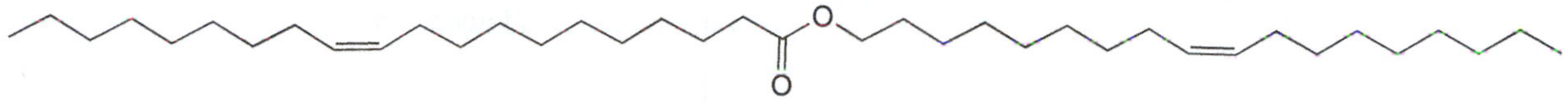

**19.8**  Beeswax is hydrophobic since it contains very long hydrocarbon chains.  It will therefore be insoluble in a very polar solvent like water, only slightly soluble in the less polar solvent ethanol, and very soluble in the weakly polar solvent chloroform.

**19.9**  Draw the products of the hydrolysis of cetyl myristate.

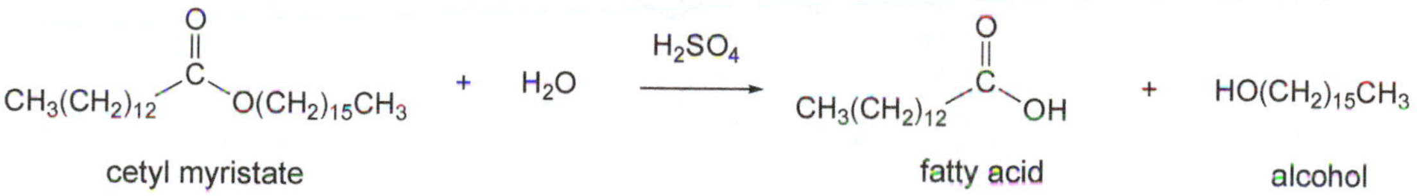

$CH_3(CH_2)_{12}$ C(=O)O(CH₂)₁₅CH₃  +  $H_2O$  $\xrightarrow{H_2SO_4}$  $CH_3(CH_2)_{12}$ C(=O)OH  +  HO(CH₂)₁₅CH₃

cetyl myristate                                          fatty acid            alcohol

**19.10**  Draw the structure of each triacylglycerol as in Example 19.4.

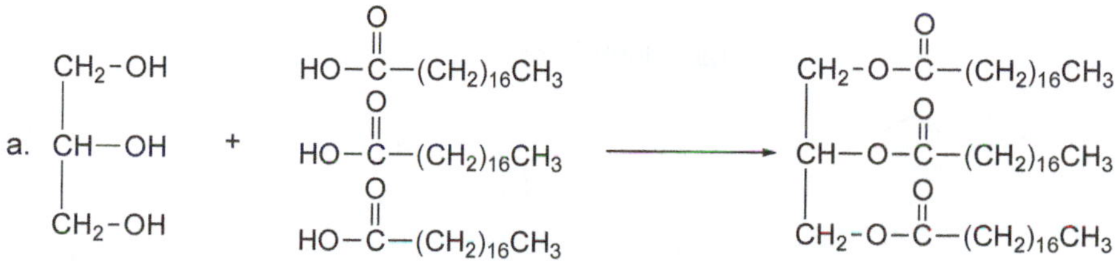

a.  CH₂–OH
    |
    CH—OH  +
    |
    CH₂–OH

    HO–C(=O)–(CH₂)₁₆CH₃
    HO–C(=O)–(CH₂)₁₆CH₃
    HO–C(=O)–(CH₂)₁₆CH₃

    ⟶

    CH₂–O–C(=O)–(CH₂)₁₆CH₃
    |
    CH–O–C(=O)–(CH₂)₁₆CH₃
    |
    CH₂–O–C(=O)–(CH₂)₁₆CH₃

b.

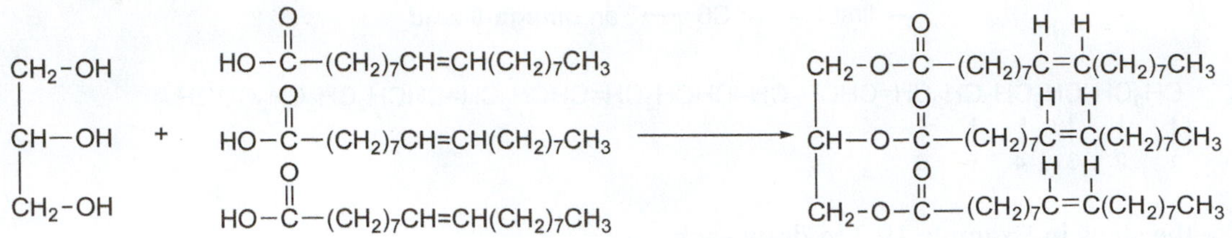

$$CH_2-OH$$
$$CH-OH$$
$$CH_2-OH$$

+

$$HO-\overset{O}{\underset{\|}{C}}-(CH_2)_7CH=CH(CH_2)_7CH_3$$
$$HO-\overset{O}{\underset{\|}{C}}-(CH_2)_7CH=CH(CH_2)_7CH_3$$
$$HO-\overset{O}{\underset{\|}{C}}-(CH_2)_7CH=CH(CH_2)_7CH_3$$

$\longrightarrow$

$$CH_2-O-\overset{O}{\underset{\|}{C}}-(CH_2)_7\overset{H}{\underset{}{C}}=\overset{H}{\underset{}{C}}(CH_2)_7CH_3$$
$$CH-O-\overset{O}{\underset{\|}{C}}-(CH_2)_7\overset{H}{\underset{}{C}}=\overset{H}{\underset{}{C}}(CH_2)_7CH_3$$
$$CH_2-O-\overset{O}{\underset{\|}{C}}-(CH_2)_7\overset{H}{\underset{}{C}}=\overset{H}{\underset{}{C}}(CH_2)_7CH_3$$

**19.11** Draw two different triacylglycerols.

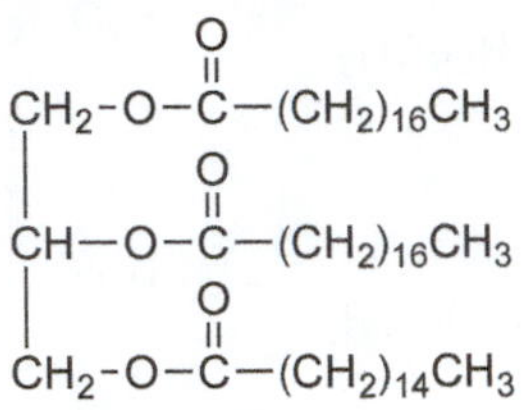

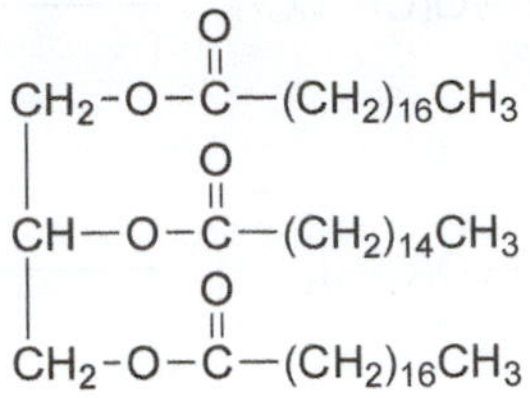

$$CH_2-O-\overset{O}{\underset{\|}{C}}-(CH_2)_{16}CH_3$$
$$CH-O-\overset{O}{\underset{\|}{C}}-(CH_2)_{16}CH_3$$
$$CH_2-O-\overset{O}{\underset{\|}{C}}-(CH_2)_{14}CH_3$$

$$CH_2-O-\overset{O}{\underset{\|}{C}}-(CH_2)_{16}CH_3$$
$$CH-O-\overset{O}{\underset{\|}{C}}-(CH_2)_{14}CH_3$$
$$CH_2-O-\overset{O}{\underset{\|}{C}}-(CH_2)_{16}CH_3$$

**19.12** Draw a triacylglycerol to fit each description.

a.

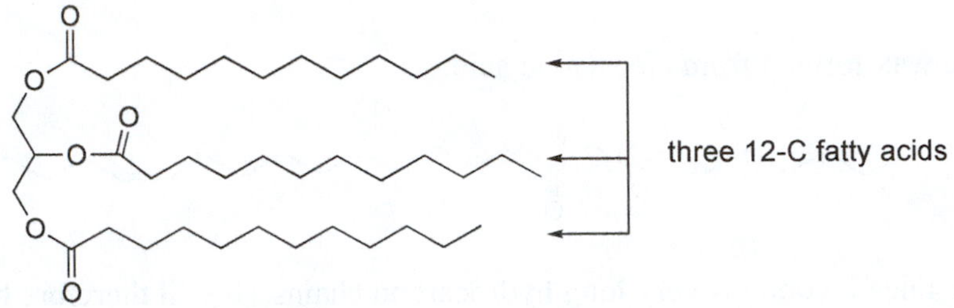

three 12-C fatty acids

b.

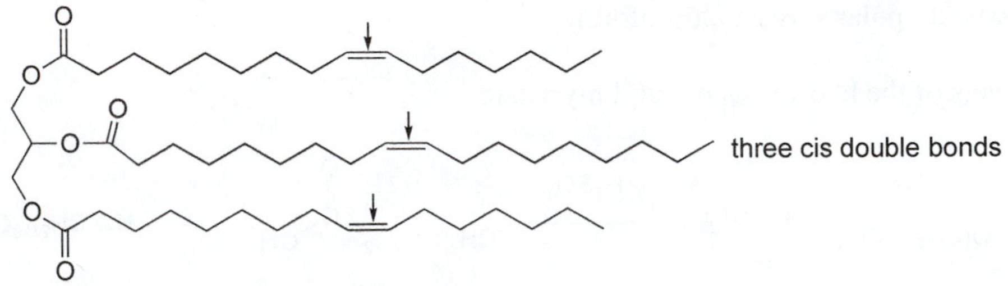

three cis double bonds

c.

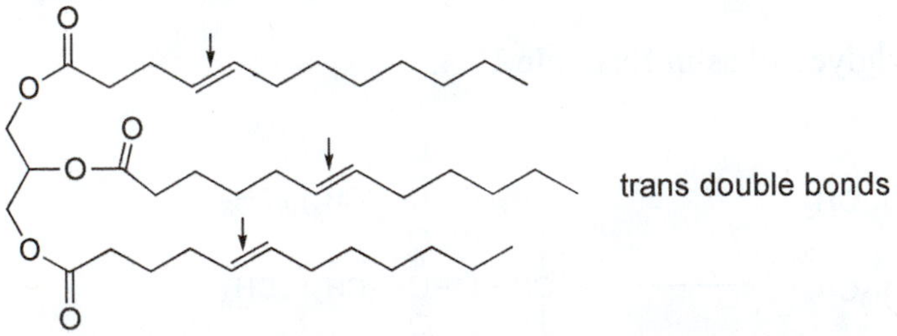

trans double bonds

**19.13** Draw the products of the hydrolysis of each triacylglycerol as in Example 19.5.

a.

$$CH_2-O-\overset{\overset{\displaystyle O}{\|}}{C}-(CH_2)_{12}CH_3$$
$$CH-O-\overset{\overset{\displaystyle O}{\|}}{C}-(CH_2)_{12}CH_3$$
$$CH_2-O-\overset{\overset{\displaystyle O}{\|}}{C}-(CH_2)_{12}CH_3$$

$\xrightarrow[H_2SO_4]{3\ H-OH}$

$$CH_2-OH$$
$$CH-OH$$
$$CH_2-OH$$

$+\ 3\ HO-\overset{\overset{\displaystyle O}{\|}}{C}-(CH_2)_{12}CH_3$

b.

$$CH_2-O-\overset{\overset{\displaystyle O}{\|}}{C}-(CH_2)_{12}CH_3$$
$$CH-O-\overset{\overset{\displaystyle O}{\|}}{C}-(CH_2)_7CH=CH(CH_2)_5CH_3$$
$$CH_2-O-\overset{\overset{\displaystyle O}{\|}}{C}-(CH_2)_7CH=CH(CH_2)_7CH_3$$

$\xrightarrow[H_2SO_4]{3\ H-OH}$

$$CH_2-OH$$
$$CH-OH$$
$$CH_2-OH$$

$+$

$$HO-\overset{\overset{\displaystyle O}{\|}}{C}-(CH_2)_{12}CH_3$$
$$HO-\overset{\overset{\displaystyle O}{\|}}{C}-(CH_2)_7CH=CH(CH_2)_5CH_3$$
$$HO-\overset{\overset{\displaystyle O}{\|}}{C}-(CH_2)_7CH=CH(CH_2)_7CH_3$$

**19.14** Balance the equation for the combustion of tristearin.

$2$
$$CH_2-O-\overset{\overset{\displaystyle O}{\|}}{C}-(CH_2)_{16}CH_3$$
$$CH-O-\overset{\overset{\displaystyle O}{\|}}{C}-(CH_2)_{16}CH_3$$
$$CH_2-O-\overset{\overset{\displaystyle O}{\|}}{C}-(CH_2)_{16}CH_3$$

tristearin
$C_{57}H_{110}O_6$

$+\ 163\ O_2 \xrightarrow{\text{enzymes}} 114\ CO_2\ +\ 110\ H_2O$

total of 338 O atoms on each side
220 H atoms
114 C atoms

**19.15** Draw the soap prepared by saponification of each triacylglycerol.

a.

$$CH_2-O-\overset{\overset{\displaystyle O}{\|}}{C}-(CH_2)_{12}CH_3$$
$$CH-O-\overset{\overset{\displaystyle O}{\|}}{C}-(CH_2)_{12}CH_3$$
$$CH_2-O-\overset{\overset{\displaystyle O}{\|}}{C}-(CH_2)_{12}CH_3$$

$\xrightarrow[H_2O]{NaOH}$

$3\ Na^+\ {}^-O-\overset{\overset{\displaystyle O}{\|}}{C}-(CH_2)_{12}CH_3$

b.

$$CH_2-O-\overset{\overset{\displaystyle O}{\|}}{C}-(CH_2)_{12}CH_3$$
$$CH-O-\overset{\overset{\displaystyle O}{\|}}{C}-(CH_2)_7CH=CH(CH_2)_5CH_3$$
$$CH_2-O-\overset{\overset{\displaystyle O}{\|}}{C}-(CH_2)_7CH=CH(CH_2)_7CH_3$$

$\xrightarrow[H_2O]{NaOH}$

$$Na^+\ {}^-O-\overset{\overset{\displaystyle O}{\|}}{C}-(CH_2)_{12}CH_3$$
$$Na^+\ {}^-O-\overset{\overset{\displaystyle O}{\|}}{C}-(CH_2)_7CH=CH(CH_2)_5CH_3$$
$$Na^+\ {}^-O-\overset{\overset{\displaystyle O}{\|}}{C}-(CH_2)_7CH=CH(CH_2)_7CH_3$$

**19.16** Draw the structure of the two cephalins as in Example 19.6.

$$CH_2-O-\overset{\displaystyle O}{\overset{\|}{C}}-(CH_2)_{14}CH_3 \longleftarrow \text{from palmitic acid}$$

$$\underset{\displaystyle |}{CH}-O-\overset{\displaystyle O}{\overset{\|}{C}}-(CH_2)_7\overset{H}{\underset{}{C}}=\overset{H}{\underset{}{C}}(CH_2)_7CH_3 \longleftarrow \text{from oleic acid}$$

$$CH_2-O-\overset{\displaystyle O}{\overset{\|}{P}}-O-CH_2CH_2\overset{+}{N}H_3$$
$$\underset{\displaystyle O^-}{|}$$

$$CH_2-O-\overset{\displaystyle O}{\overset{\|}{C}}-(CH_2)_7\overset{H}{\underset{}{C}}=\overset{H}{\underset{}{C}}(CH_2)_7CH_3 \longleftarrow \text{from oleic acid}$$

$$CH-O-\overset{\displaystyle O}{\overset{\|}{C}}-(CH_2)_{14}CH_3 \longleftarrow \text{from palmitic acid}$$

$$CH_2-O-\overset{\displaystyle O}{\overset{\|}{P}}-O-CH_2CH_2\overset{+}{N}H_3$$
$$\underset{\displaystyle O^-}{|}$$

**19.17** A **triacylglycerol** has glycerol as the backbone, and three nonpolar side chains formed from esters with fatty acids.

A **phosphoacylglycerol** has glycerol as the backbone, a phosphodiester on a terminal carbon, and two nonpolar side chains formed from esters with fatty acids.

A **sphingomyelin** has sphingosine as the backbone, one amide, and a phosphodiester on a terminal carbon.

a. phosphoacylglycerol, cephalin

c. sphingomyelin

b. triacylglycerol

**19.18** Phospholipids are present in cell membranes because they have an ionic polar head and two nonpolar tails, and can form a lipid bilayer needed for cell membrane function. Triacylglycerols are basically nonpolar compounds, so they have no polar head to attract water on the outside of a membrane.

**19.19** Membrane **A** is formed from the phospholipids linoleic and oleic acids, and will be more fluid or pliable because it contains unsaturated fatty acids. Membrane **B** is formed from the saturated fatty acids stearic and palmitic acids, which have no double bonds and therefore pack very tightly, making it more rigid.

**19.20** Ions don't diffuse readily through the interior of the cell membrane because it is hydrophobic and ions are insoluble in a nonpolar medium.

**19.21** Cholesterol is a lipid since it contains many C–C and C–H bonds and it is insoluble in water.

**19.22** Answer each question about cholesterol.

OH at C3     double bond between C5 and C6

d. Cholesterol contains one polar C–O bond and one polar O–H bond from the polar OH group. The large hydrocarbon skeleton with nonpolar C–C and C–H bonds makes it water insoluble.

**19.23** Triacylglycerols would be found in the interior of a lipoprotein particle, since this is the hydrophobic region of lipoproteins.

**19.24** Label the functional groups.

**19.25** a. Estrone has a phenol (a benzene ring with a hydroxyl group) and progesterone has a ketone and C=C in ring A. Progesterone also has a methyl group bonded to C10.
b. Estrone has a ketone at C17 and progesterone has a C–C bond, which is attached to a ketone.

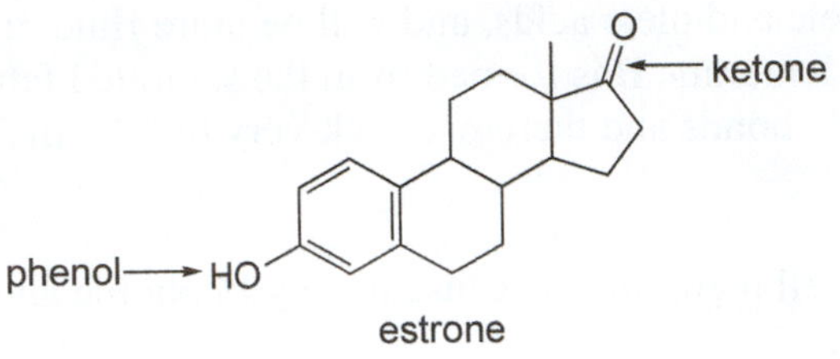

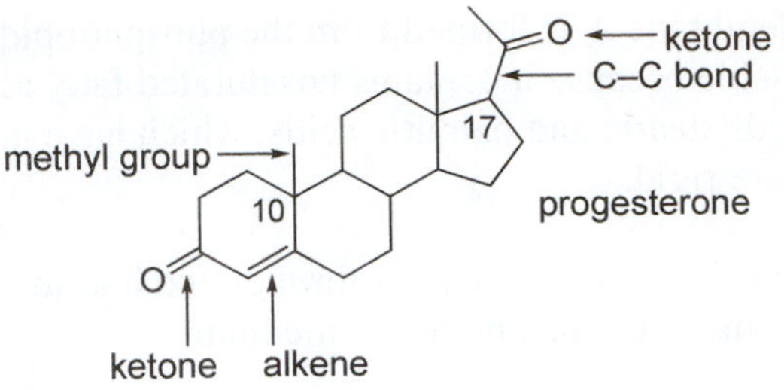

estrone — phenol → HO, ketone

progesterone — ketone, C–C bond, methyl group, 10, 17, ketone, alkene

**19.26** Label the functional groups in aldosterone.

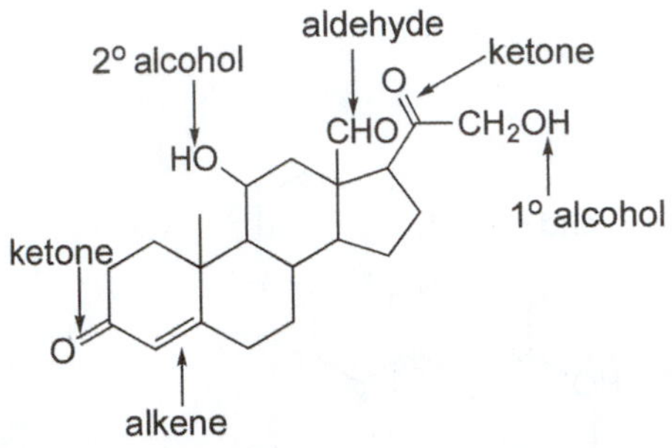

2° alcohol, aldehyde, ketone, CHO, CH₂OH, HO, 1° alcohol, ketone, O, alkene

**19.27** Testosterone has a methyl group at C10 that nandrolone lacks.

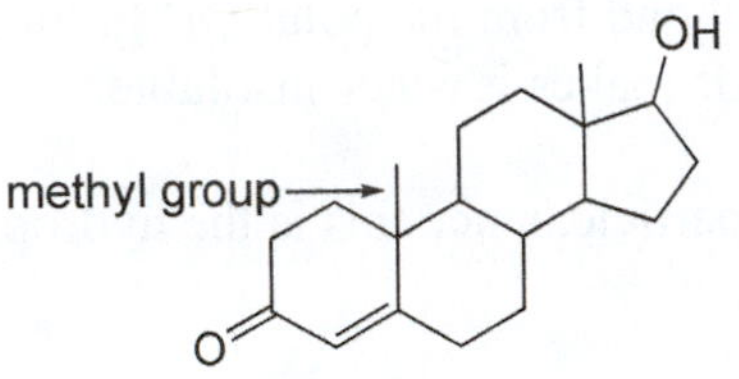

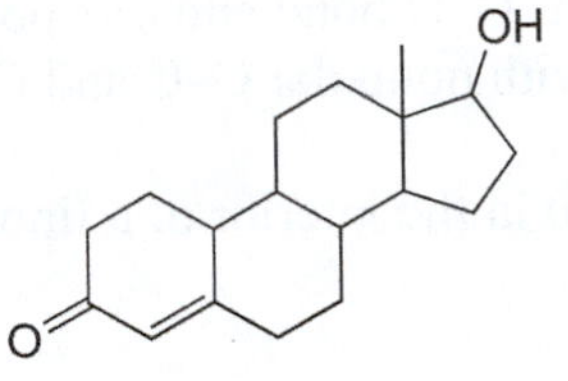

methyl group →

testosterone          nandrolone

**19.28** Water-soluble vitamins are excreted in the urine whereas fat-soluble vitamins are stored in the body. When a person ingests a large quantity of a water-soluble vitamin, much of it is excreted in the urine. When a person ingests a large quantity of a fat-soluble vitamin, it can be retained in the fat in the body, potentially building up to toxic levels.

**19.29** Label the functional groups and draw the skeletal structure for PGE$_2$.

a. ketone, carboxylic acid, $CH_2CH=CH(CH_2)_3COOH$, alkene, $CH=CHCH(CH_2)_4CH_3$, HO, OH, alcohol

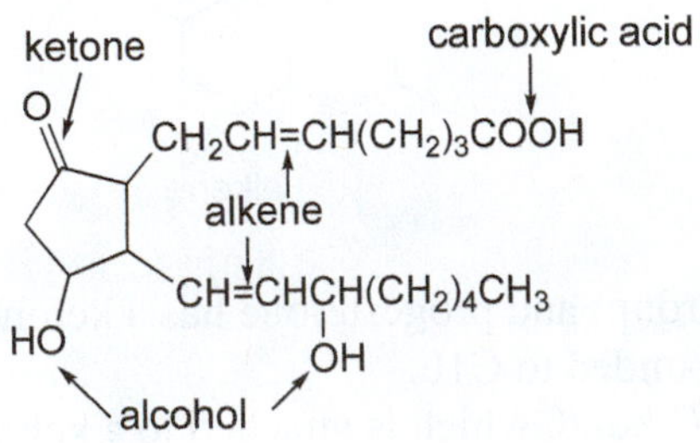

b. cis, COOH, HO, trans OH

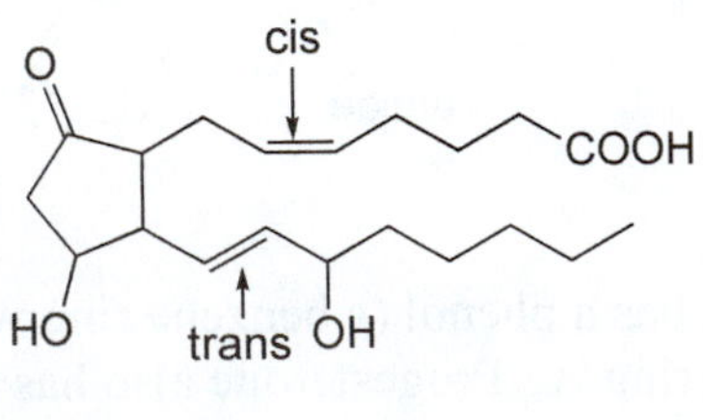

**19.30** Label the functional groups and the double bonds as cis or trans in LTC₄.

a. and b.

*(structure with labels: trans alkene, cis alkene, alcohol, amide, carboxylic acid, cis alkene, amide, amine)*

H–C=C–C=C–CHCH(OH)(CH₂)₃COOH

S–CH₂
|
CHCONHCH₂COOH
|
NHCOCH₂CH₂CHCOOH
|
NH₂

## Solutions to Odd-Numbered End-of-Chapter Problems

**19.31**   **Hydrolyzable lipids** include waxes, triacylglycerols, and phospholipids.
**Nonhydrolyzable lipids** include steroids, fat-soluble vitamins, and eicosanoids.

a. prostaglandin—nonhydrolyzable
b. triacylglycerol—hydrolyzable
c. leukotriene—nonhydrolyzable
d. vitamin A—nonhydrolyzable

e. phosphoacylglycerol—hydrolyzable
f. lecithin—hydrolyzable
g. cholesterol—nonhydrolyzable

**19.33**   A wax is hydrophobic. As a result, it is soluble in (b) $CH_2Cl_2$ and (c) $CH_3CH_2OCH_2CH_3$, both organic solvents, but insoluble in (a) water, which is polar.

**19.35**   For the same number of carbons, increasing the number of double bonds decreases the melting point of a fatty acid.

a. $CH_3(CH_2)_3CH=CH(CH_2)_7COOH$, $CH_3(CH_2)_{12}COOH$, $CH_3(CH_2)_{14}COOH$
b. $CH_3(CH_2)_5CH=CH(CH_2)_7COOH$, $CH_3(CH_2)_7CH=CH(CH_2)_7COOH$, $CH_3(CH_2)_{16}COOH$

**19.37**   a. Increasing the number of carbon atoms increases the melting point of fatty acids.
b. Increasing the number of double bonds decreases the melting point of fatty acids.

**19.39**   Answer each question about 7,10,13,16,19-docosapentaenoic acid.
a. and b.

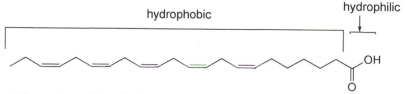

c. The melting point of the cis isomer would be lower than the melting point of the trans isomer.
d. This fatty acid will be a liquid at room temperature.
e. 7,10,13,16,19-Docosapentaenoic acid is an omega-3 fatty acid because there is a double bond at the third C from the left (numbered from the CH₃ group).

**19.41**

a. 
(cis) ...CO$_2$H

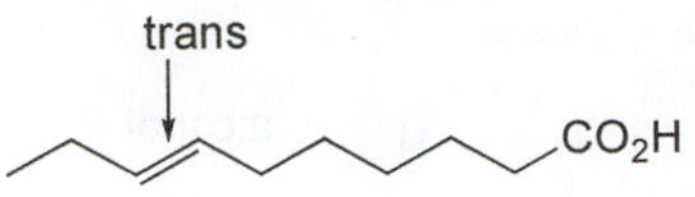

b. omega-3 acid

c.
(trans) ...CO$_2$H

d.

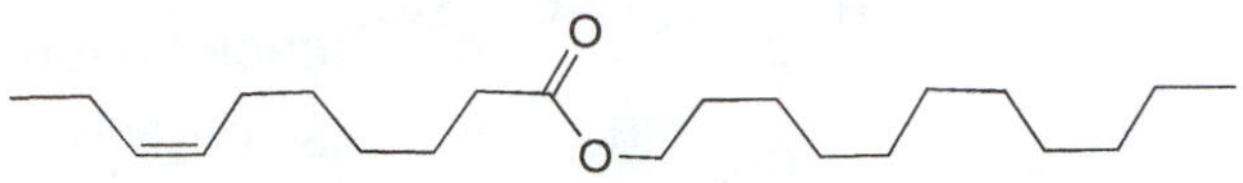

**19.43**   Draw the structure of each wax.

a.  $CH_3(CH_2)_{14}$—C(=O)—OH  +  H—O(CH$_2$)$_{21}$CH$_3$  $\longrightarrow$  $CH_3(CH_2)_{14}$—C(=O)—O(CH$_2$)$_{21}$CH$_3$  +  H$_2$O

b.  $CH_3(CH_2)_{14}$—C(=O)—OH  +  H—O(CH$_2$)$_{11}$CH$_3$  $\longrightarrow$  $CH_3(CH_2)_{14}$—C(=O)—O(CH$_2$)$_{11}$CH$_3$  +  H$_2$O

c.  $CH_3(CH_2)_{14}$—C(=O)—OH  +  H—O(CH$_2$)$_9$CH$_3$  $\longrightarrow$  $CH_3(CH_2)_{14}$—C(=O)—O(CH$_2$)$_9$CH$_3$  +  H$_2$O

**19.45**   Draw the products of the hydrolysis of each wax.

a.  $CH_3(CH_2)_{16}$—C(=O)—O(CH$_2$)$_{17}$CH$_3$  +  H$_2$O  $\xrightarrow{\text{H}_2\text{SO}_4}$  $CH_3(CH_2)_{16}$—C(=O)—OH  +  HO(CH$_2$)$_{17}$CH$_3$

b.  $CH_3(CH_2)_{12}$—C(=O)—O(CH$_2$)$_{25}$CH$_3$  +  H$_2$O  $\xrightarrow{\text{H}_2\text{SO}_4}$  $CH_3(CH_2)_{12}$—C(=O)—OH  +  HO(CH$_2$)$_{25}$CH$_3$

c.  $CH_3(CH_2)_{14}$—C(=O)—O(CH$_2$)$_{27}$CH$_3$  +  H$_2$O  $\xrightarrow{\text{H}_2\text{SO}_4}$  $CH_3(CH_2)_{14}$—C(=O)—OH  +  HO(CH$_2$)$_{27}$CH$_3$

d.  $CH_3(CH_2)_{22}$—C(=O)—O(CH$_2$)$_{13}$CH$_3$  +  H$_2$O  $\xrightarrow{\text{H}_2\text{SO}_4}$  $CH_3(CH_2)_{22}$—C(=O)—OH  +  HO(CH$_2$)$_{13}$CH$_3$

**19.47** Draw a triacylglycerol that fits each description.

a.

$$CH_2-O-\overset{O}{\overset{\|}{C}}-(CH_2)_{10}CH_3 \longleftarrow \text{lauric acid}$$

$$CH-O-\overset{O}{\overset{\|}{C}}-(CH_2)_{12}CH_3 \longleftarrow \text{myristic acid}$$

$$CH_2-O-\overset{O}{\overset{\|}{C}}-(CH_2)_7CH=CHCH_2CH=CH(CH_2)_4CH_3$$

triacylglycerol            linoleic acid

c.

$$CH_2-O-\overset{O}{\overset{\|}{C}}-(CH_2)_{12}CH_3$$

$$CH-O-\overset{O}{\overset{\|}{C}}-(CH_2)_{12}CH_3$$

$$CH_2-O-\overset{O}{\overset{\|}{C}}-(CH_2)_{12}CH_3$$

saturated triacylglycerol

b.

$$CH_2-O-\overset{O}{\overset{\|}{C}}$$

$$CH-O-\overset{O}{\overset{\|}{C}}$$

two cis double bonds

$$CH_2-O-\overset{O}{\overset{\|}{C}}-(CH_2)_7C=CCH_2C=C(CH_2)_4CH_3$$
$$\quad\quad H\ H\quad\quad H\ H$$

unsaturated triacylglycerol

d.

$$CH_2-O-\overset{O}{\overset{\|}{C}}-(CH_2)_{12}CH_3$$

$$CH-O-\overset{O}{\overset{\|}{C}}-(CH_2)_{12}CH_3$$

one double bond

$$CH_2-O-\overset{O}{\overset{\|}{C}}-(CH_2)_7C=C(CH_2)_7CH_3$$
$$\quad\quad H\ H$$

monounsaturated triacylglycerol

**19.49** Draw the triacylglycerol that fits the description.

$$CH_2-O-\overset{O}{\overset{\|}{C}}-(CH_2)_3CH=CH(CH_2)_3CH_3$$

$$CH-O-\overset{O}{\overset{\|}{C}}-(CH_2)_3CH=CH(CH_2)_3CH_3 \quad \xrightarrow[\text{in base}]{\text{hydrolysis}} \quad 3\ \ Na^+\ {}^-O-\overset{O}{\overset{\|}{C}}-(CH_2)_3CH=CH(CH_2)_3CH_3$$

$$CH_2-O-\overset{O}{\overset{\|}{C}}-(CH_2)_3CH=CH(CH_2)_3CH_3$$

**19.51** Answer each question.

| Compound | a. General structure | b. Example | c. Water soluble (Y/N) | d. Hexane soluble (Y/N) |
|---|---|---|---|---|
| [1] Fatty acid | RCOOH | ~~~~~COOH | N | Y |
| [2] Soap | RCOO⁻ Na⁺ | ~~~~~COO⁻ Na⁺ | Y | N |
| [3] Wax | RCOOR' | ~~~COO~~~ | N | Y |
| [4] Triacylglycerol | $CH_2-O-\overset{O}{\overset{\|}{C}}-R$ $CH-O-\overset{O}{\overset{\|}{C}}-R'$ $CH_2-O-\overset{O}{\overset{\|}{C}}-R''$ | $CH_2-O-\overset{O}{\overset{\|}{C}}-(CH_2)_{12}CH_3$ $CH-O-\overset{O}{\overset{\|}{C}}-(CH_2)_{12}CH_3$ $CH_2-O-\overset{O}{\overset{\|}{C}}-(CH_2)_{12}CH_3$ | N | Y |

**19.53** Answer each question about the triacylglycerol.

a.

CH$_2$-O-C(=O)-(CH$_2$)$_{18}$CH$_3$ ← arachidic acid

CH-O-C(=O)-(CH$_2$)$_{16}$CH$_3$ ← stearic acid

CH$_2$-O-C(=O)-(CH$_2$)$_{10}$CH$_3$ ← lauric acid

e.

CH$_2$-OH      HOOC(CH$_2$)$_{18}$CH$_3$

CH-OH    +    HOOC(CH$_2$)$_{16}$CH$_3$

CH$_2$-OH      HOOC(CH$_2$)$_{10}$CH$_3$

b. It would be a solid at room temperature since it is formed from saturated fatty acids.
c. The long hydrocarbon chains are hydrophobic.
d. The ester linkages are hydrophilic.

**19.55** Draw the products of triacylglycerol hydrolysis.

a.

CH$_2$-O-C(=O)-(CH$_2$)$_{14}$CH$_3$

CH-O-C(=O)-(CH$_2$)$_{14}$CH$_3$

CH$_2$-O-C(=O)-(CH$_2$)$_{16}$CH$_3$

$\xrightarrow[\text{H}_2\text{SO}_4]{\text{H}_2\text{O}}$

CH$_2$-OH

CH-OH    +

CH$_2$-OH

2 HOC(=O)-(CH$_2$)$_{14}$CH$_3$

HOC(=O)-(CH$_2$)$_{16}$CH$_3$

CH$_2$-O-C(=O)-(CH$_2$)$_{14}$CH$_3$

CH-O-C(=O)-(CH$_2$)$_{14}$CH$_3$

CH$_2$-O-C(=O)-(CH$_2$)$_{16}$CH$_3$

$\xrightarrow[\text{NaOH}]{\text{H}_2\text{O}}$

CH$_2$-OH

CH-OH    +

CH$_2$-OH

2 Na$^+$ $^-$OC(=O)-(CH$_2$)$_{14}$CH$_3$

Na$^+$ $^-$OC(=O)-(CH$_2$)$_{16}$CH$_3$

b.

CH$_2$-O-C(=O)-(CH$_2$)$_{14}$CH$_3$

CH-O-C(=O)-(CH$_2$)$_7$CH=CH(CH$_2$)$_7$CH$_3$

CH$_2$-O-C(=O)-(CH$_2$)$_7$CH=CH(CH$_2$)$_5$CH$_3$

$\xrightarrow[\text{H}_2\text{SO}_4]{\text{H}_2\text{O}}$

CH$_2$-OH

CH-OH    +

CH$_2$-OH

HOC(=O)-(CH$_2$)$_{14}$CH$_3$

HOC(=O)-(CH$_2$)$_7$CH=CH(CH$_2$)$_7$CH$_3$

HOC(=O)-(CH$_2$)$_7$CH=CH(CH$_2$)$_5$CH$_3$

CH$_2$-O-C(=O)-(CH$_2$)$_{14}$CH$_3$

CH-O-C(=O)-(CH$_2$)$_7$CH=CH(CH$_2$)$_7$CH$_3$

CH$_2$-O-C(=O)-(CH$_2$)$_7$CH=CH(CH$_2$)$_5$CH$_3$

$\xrightarrow[\text{NaOH}]{\text{H}_2\text{O}}$

CH$_2$-OH

CH-OH    +

CH$_2$-OH

Na$^+$ $^-$OC(=O)-(CH$_2$)$_{14}$CH$_3$

Na$^+$ $^-$OC(=O)-(CH$_2$)$_7$CH=CH(CH$_2$)$_7$CH$_3$

Na$^+$ $^-$OC(=O)-(CH$_2$)$_7$CH=CH(CH$_2$)$_5$CH$_3$

**19.57** Phospholipids are lipids that contain a phosphorus atom. Sphingomyelins (c) contain phosphorus. Triacylglycerols (a), leukotrienes (b), and fatty acids (d) do not.

**19.59** Draw a phosphoacylglycerol that fits each description.

a.

$$CH_2-O-\overset{\overset{\displaystyle O}{\|}}{C}-(CH_2)_7CH=CH(CH_2)_5CH_3 \leftarrow \text{from palmitoleic acid}$$
$$CH-O-\overset{\overset{\displaystyle O}{\|}}{C}-(CH_2)_7CH=CH(CH_2)_5CH_3 \leftarrow \text{from palmitoleic acid}$$
$$CH_2-O-\overset{\overset{\displaystyle O}{\|}}{\underset{\underset{\displaystyle O^-}{|}}{P}}-O-CH_2CH_2\overset{+}{N}H_3$$

cephalin

b.

$$CH_2-O-\overset{\overset{\displaystyle O}{\|}}{C}-(CH_2)_{10}CH_3 \leftarrow \text{from lauric acid}$$
$$CH-O-\overset{\overset{\displaystyle O}{\|}}{C}-(CH_2)_{10}CH_3 \leftarrow \text{from lauric acid}$$
$$CH_2-O-\overset{\overset{\displaystyle O}{\|}}{\underset{\underset{\displaystyle O^-}{|}}{P}}-O-CH_2CH_2\overset{+}{N}(CH_3)_3$$

phosphatidylcholine

c.

$$HO-CH-CH=CH(CH_2)_{12}CH_3$$
$$CH-NH-\overset{\overset{\displaystyle O}{\|}}{C}-(CH_2)_{16}CH_3 \leftarrow \text{from stearic acid}$$
$$CH_2-O-\overset{\overset{\displaystyle O}{\|}}{\underset{\underset{\displaystyle O^-}{|}}{P}}-O-CH_2CH_2\overset{+}{N}H_3 \leftarrow \text{from ethanolamine}$$

sphingomyelin

**19.61** Triacylglycerols do not have a strongly hydrophilic region contained in a polar head, so they cannot form a lipid bilayer.

**19.63** Diffusion is the movement of small molecules through a membrane along a concentration gradient. Facilitated transport is the transport of molecules through channels in the cell membrane. $O_2$ and $CO_2$ move by diffusion, whereas glucose and Cl$^-$ move by facilitated transport.

**19.65** Draw the anabolic steroid 4-androstene-3,17-dione.

methyl group at C13    carbonyl at C17
methyl group at C10
double bond between C4 and C5
carbonyl at C3

**19.67** Cholesterol is insoluble in the aqueous medium of the bloodstream. By being bound to a lipoprotein particle, however, it can be transported in the aqueous solution of the blood.

**19.69** Low-density lipoproteins (LDLs) transport cholesterol from the liver to the tissues where it is incorporated in cell membranes. High-density lipoproteins (HDLs) transport cholesterol from the tissues back to the liver. When LDLs supply more cholesterol than is needed, LDLs deposit cholesterol on the walls of arteries, forming plaque. Atherosclerosis is a disease that results from the buildup of these fatty deposits, restricting the flow of blood, increasing blood pressure, and increasing the likelihood of a heart attack or stroke. As a result, LDL cholesterol is often called "bad" cholesterol.  Since HDL cholesterol transports cholesterol back to the liver and removes it from the bloodstream, high levels of HDLs reduce the risk of heart disease and stroke, and it is called "good" cholesterol.

**19.71** Answer each question about estrone and testosterone.

a.

estrone          testosterone

b. The estrogen (left) and androgen (right) both contain the four rings of the steroid skeleton.  Both contain a methyl group bonded to C13.

c. The estrogen has an aromatic A ring and a hydroxyl group on this ring.  The androgen has a carbonyl on the A ring but does not contain an aromatic ring. The androgen also contains a C=C in the A ring and an additional $CH_3$ group at C10.  The D rings are also different. The estrogen contains a carbonyl at C17, whereas the androgen has an OH group.

d. Estrogens, synthesized in the ovaries, control the menstrual cycle and secondary sexual characteristics of females.  Androgens, synthesized in the testes, control the development of male secondary sexual characteristics.

**19.73** Prostaglandins and leukotrienes are two types of eicosanoids, a group of biologically active compounds containing 20 carbon atoms derived from the fatty acid arachidonic acid. Prostaglandins are a group of carboxylic acids that contain a five-membered ring.  Leukotrienes do not contain a ring.  They both mediate biological activity at the site where they are formed. Prostaglandins mediate inflammation and uterine contractions. Leukotrienes stimulate smooth muscle contraction in the lungs, leading to the narrowing of airways in individuals with asthma.

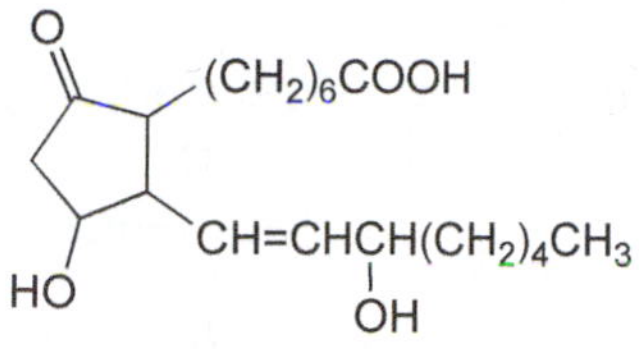

a prostaglandin
PGE$_1$
five-membered ring

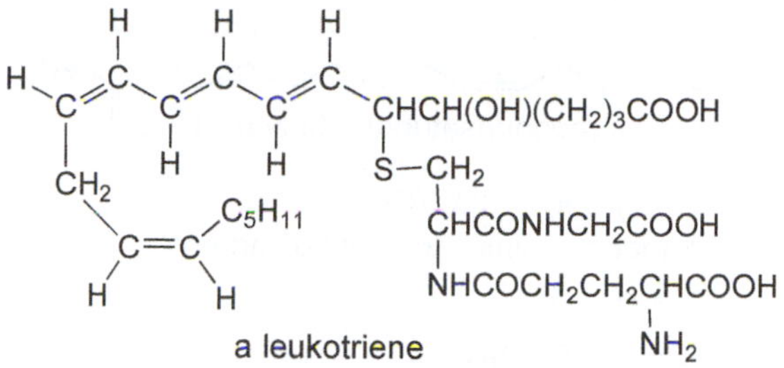

a leukotriene
LTC$_4$
no ring

**19.75** All prostaglandins contain a five-membered ring and a carboxyl group (COOH).

**19.77** Aspirin and celecoxib are both anti-inflammatory medicines. Aspirin inhibits the activity of both the COX-1 and COX-2 enzymes, whereas celecoxib inhibits the activity of COX-2 only.

**19.79** Vitamins are organic compounds required in small quantities for normal metabolism and must be obtained in the diet since our cells cannot synthesize these compounds.

**19.81** Answer each question about vitamins A and D.

| | Vitamin A | Vitamin D |
|---|---|---|
| a. | 10 Tetrahedral carbons | 21 Tetrahedral carbons |
| b. | 10 Trigonal planar carbons | Six trigonal planar carbons |
| e. | Required for normal vision | Regulates calcium and phosphorus metabolism |
| f. | Deficiency causes night blindness. | Deficiency causes rickets and skeletal deformities. |
| g. | Found in liver, kidney, oily fish, and dairy | Found in milk and breakfast cereals |

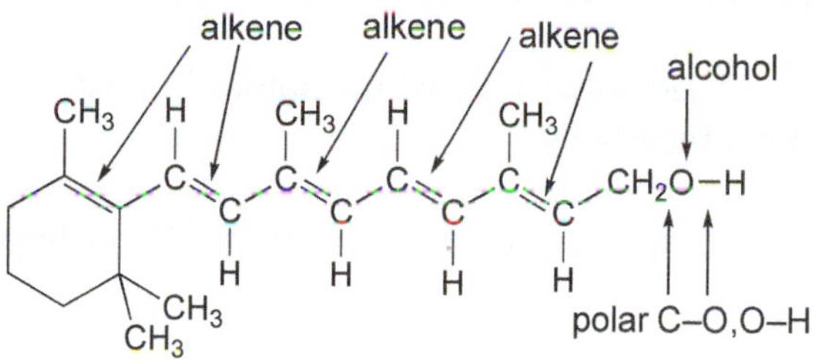

c. and d.

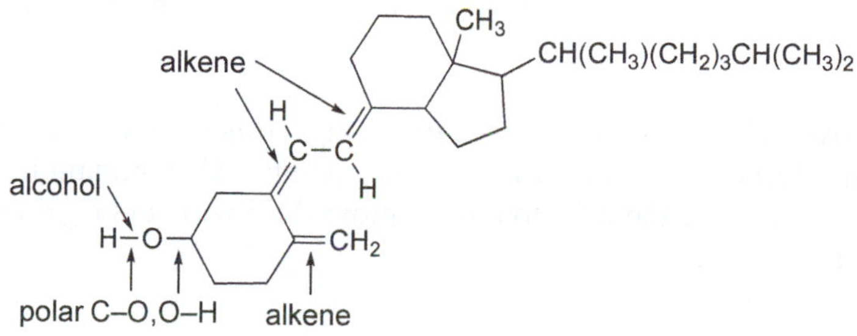

**19.83** Draw an example of each type of lipid.

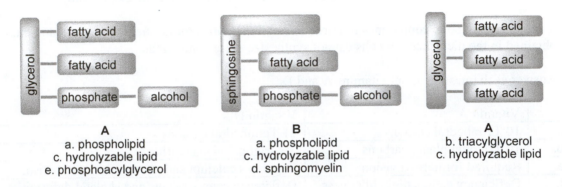

a.

a monounsaturated fatty acid

b. $CH_3(CH_2)_{12}COO(CH_2)_{15}CH_3$
a wax that contains a total of 30 carbons

c.
$$CH_2-O-\overset{\overset{O}{\|}}{C}-(CH_2)_{12}CH_3$$
$$CH-O-\overset{\overset{O}{\|}}{C}-(CH_2)_{12}CH_3 \quad \text{a saturated triacylglycerol}$$
$$CH_2-O-\overset{\overset{O}{\|}}{C}-(CH_2)_{12}CH_3$$

d.   $HO-CH-CH=CH(CH_2)_{12}CH_3$
$$CH-NH-\overset{\overset{O}{\|}}{C}-(CH_2)_{12}CH_3$$
$$CH_2-O-\overset{\overset{O}{\|}}{\underset{\underset{O^-}{|}}{P}}-O-CH_2CH_2\overset{+}{N}H_3$$

a sphingomyelin derived
from ethanolamine

**19.85** Decide which terms describe each block diagram.

**A**
a. phospholipid
c. hydrolyzable lipid
e. phosphoacylglycerol

**B**
a. phospholipid
c. hydrolyzable lipid
d. sphingomyelin

**A**
b. triacylglycerol
c. hydrolyzable lipid

**19.87** Phosphoacylglycerols contain an ionic head, making them more polar than triacylglycerols.

**19.89** Coconut oil is a liquid at room temperature because the hydrocarbon chains of lauric acid have only 12 carbons in them, making them short enough that the triacylglycerol remains a liquid.

**19.91** Vegetable oils are composed of triacylglycerols, while motor oil, derived from petroleum, is mostly alkanes and other long chain hydrocarbons.

**19.93** Humans cannot survive on a completely fat-free diet. Certain fatty acids and fat-soluble vitamins are required in the diet.

**19.95** Saturated fats should be avoided in the diet because they are more likely to lead to atherosclerosis and heart disease.

**19.97** Animals that live in colder climates have triacylglycerols with more unsaturated fatty acid side chains because the unsaturated fats have lower melting points. They remain liquid at the lower temperatures of their climate, allowing the cells to remain more fluid with less rigid cell membranes than saturated triacylglycerols would allow for.

**19.99**  Recall that increasing the number of C–O bonds is an oxidation, and increasing the number of C–H bonds is a reduction.

a.

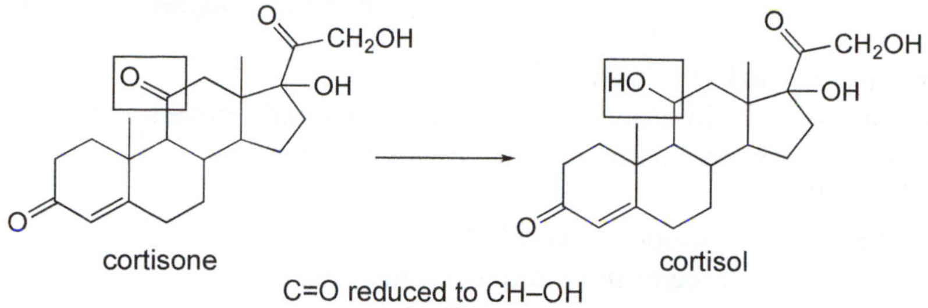

cortisone → cortisol

C=O reduced to CH–OH
**reduction**

b.

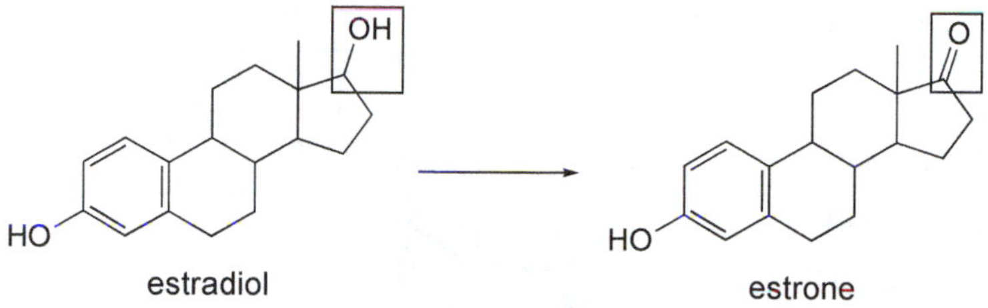

estradiol → estrone

CH–OH oxidized to C=O
**oxidation**

c.

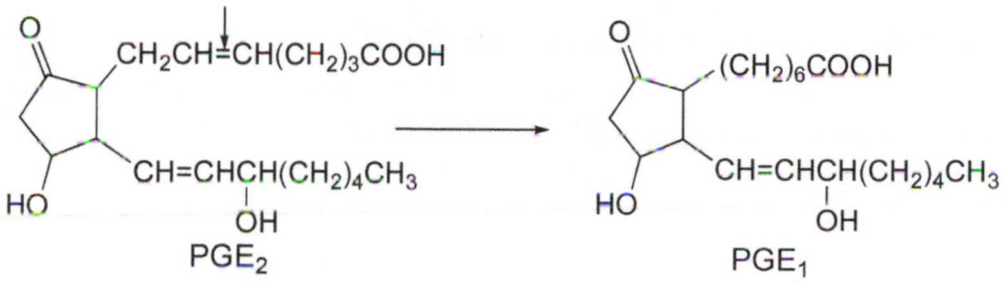

PGE$_2$ → PGE$_1$

C=C reduced to CH$_2$CH$_2$
**reduction**

# Chapter 20 Carbohydrates

## Chapter Review

**[1] Identify the three major types of carbohydrates. (20.1)**

- Monosaccharides, which cannot be hydrolyzed to simpler compounds, have three to six carbons with a carbonyl group at either the terminal carbon or the carbon adjacent to it. Generally, all other carbons have OH groups bonded to them.
- Disaccharides are composed of two monosaccharides.
- Polysaccharides are composed of three or more monosaccharides.

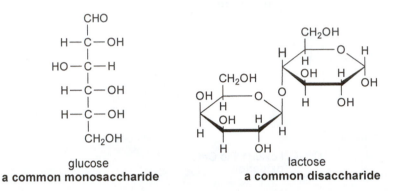

glucose
**a common monosaccharide**

lactose
**a common disaccharide**

**[2] What are the major structural features of monosaccharides? (20.2)**

- Monosaccharides with a carbonyl group at C1 are called aldoses and those with a carbonyl at C2 are called ketoses. Generally, OH groups are bonded to all other carbons. The terms triose, tetrose, and so forth are used to indicate the number of carbons in the chain.

- The acyclic form of monosaccharides is drawn with Fischer projection formulas. A D sugar has the OH group of the chirality center farthest from the carbonyl on the right side. An L sugar has the OH group of the chirality center farthest from the carbonyl on the left side.

**Two enantiomers of glyceraldehyde**

**[3] How are the cyclic forms of monosaccharides drawn? (20.3)**

- In aldohexoses, the OH group on C5 reacts with the aldehyde carbonyl to give two cyclic hemiacetals called anomers. The acetal carbon is called the anomeric carbon. The $\alpha$ anomer has the OH group drawn down for a D sugar and the $\beta$ anomer has the OH group drawn up.

**[4] What reduction and oxidation products are formed from monosaccharides? (20.4)**

- Monosaccharides are reduced to alditols with $H_2$ and Pd.

- Monosaccharides are oxidized to aldonic acids with Benedict's reagent. Sugars that are oxidized with Benedict's reagent are called reducing sugars.

**[5] What are the major structural features of disaccharides? (20.5)**

- Disaccharides contain two monosaccharides joined by an acetal C–O bond called a glycosidic linkage. An $\alpha$ glycoside has the glycosidic linkage oriented down and a $\beta$ glycoside has the glycosidic linkage oriented up.
- Disaccharides are hydrolyzed to two monosaccharides by the cleavage of the glycosidic C–O bond.

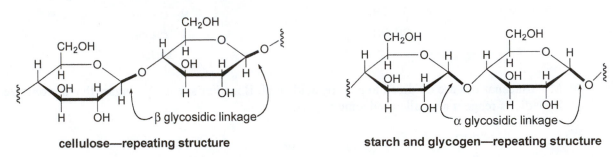

- Lactose, the principal disaccharide in milk, and sucrose (table sugar) are common disaccharides.

## [6] What are the differences in the polysaccharides cellulose, starch, and glycogen? (20.6)

- Cellulose, starch, and glycogen are all polymers of the monosaccharide glucose.
- Cellulose is an unbranched polymer composed of repeating glucose units joined in 1→4-β-glycosidic linkages. Cellulose forms long chains that stack in three-dimensional sheets. The human digestive system does not contain the needed enzyme to metabolize cellulose.
- There are two forms of starch—amylose, which is an unbranched polymer, and amylopectin, which is a branched polysaccharide polymer. Both forms contain 1→4-α-glycosidic linkages, and the polymer winds in a helical arrangement. Starch is digestible since the human digestive system has the needed amylase enzyme to catalyze hydrolysis.
- Glycogen resembles amylopectin but is more extensively branched. Glycogen is the major form in which polysaccharides are stored in animals.

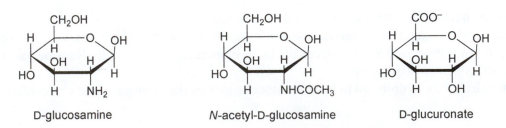

cellulose—repeating structure          starch and glycogen—repeating structure

## [7] Give examples of some carbohydrate derivatives that contain amino groups, amides, or carboxylate anions. (20.7)

- Glycosaminoglycans are a group of unbranched carbohydrates derived from amino sugars (such as D-glucosamine) and glucuronate units. Examples include hyaluronate, which forms a gel-like matrix in joints and the vitreous humor of the eye; chondroitin, which is a component of cartilage and tendons; and heparin, an anticoagulant.

D-glucosamine          N-acetyl-D-glucosamine          D-glucuronate

- Chitin, a polymer of N-acetyl-D-glucosamine, forms the hard exoskeleton of crabs, lobsters, and shrimp.

**[8] What role do carbohydrates play in determining blood type? (20.8)**

- Human blood type—A, B, AB, or O—is determined by three or four monosaccharides attached to a membrane protein on the surface of red blood cells. There are three different carbohydrate sequences, one for each of the A, B, and O blood types. Blood type AB contains the sequences for both blood type A and blood type B. Since the blood of an individual may contain antibodies to another blood type, blood type must be known before giving a transfusion.

## Problem Solving

### [1] Monosaccharides (20.2)

**Example 20.1** Classify each monosaccharide by the type of carbonyl group and the number of carbons in the chain.

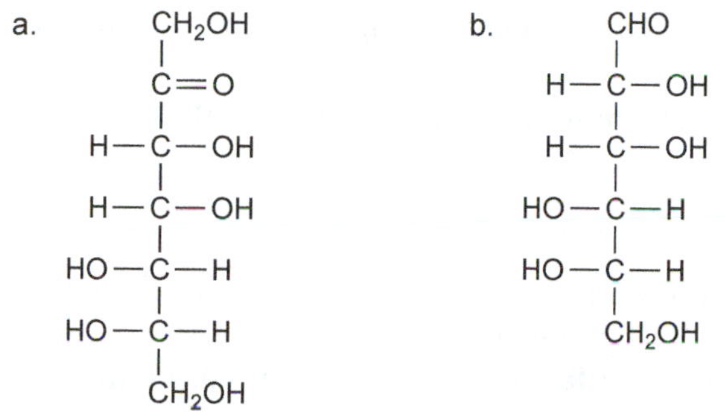


**Analysis**

Identify the type of carbonyl group to label the monosaccharide as an aldose or ketose. An aldose has the C=O at C1 so that a hydrogen atom is bonded to the carbonyl carbon. A ketose has two carbons bonded to the carbonyl carbon. Count the number of carbons in the chain to determine the suffix, -*triose*, -*tetrose*, and so forth.

**Solution**

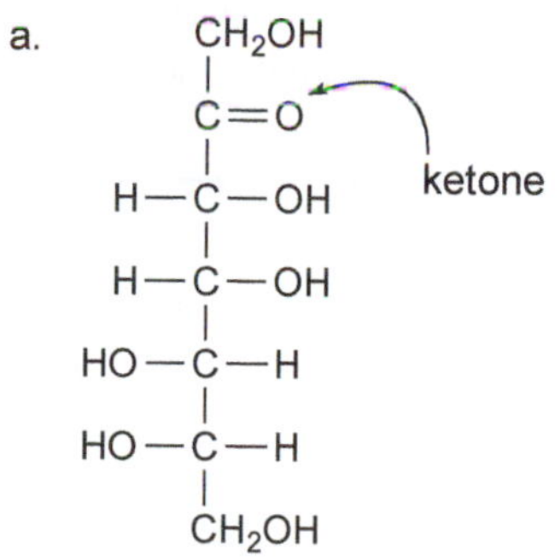


7 C's in the chain

**Answer:** ketoheptose

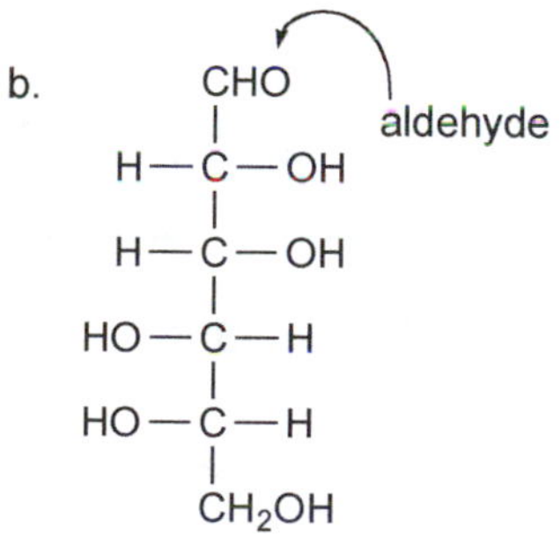


6 C's in the chain

**Answer:** aldohexose

**Example 20.2** Consider the given ketohexose. (a) Label all chirality centers. (b) Classify the ketohexose as a D or L monosaccharide. (c) Draw the enantiomer.

```
        CH₂OH
         |
        C=O
         |
    H — C — OH
         |
   HO — C — H
         |
   HO — C — H
         |
        CH₂OH
```

**Analysis**

- A chirality center has four different groups around a carbon atom.
- The labels D and L are determined by the position of the OH group on the chirality center farthest from the carbonyl group: a D sugar has the OH group on the right and an L sugar has the OH group on the left.
- To draw an enantiomer, draw the mirror image so that each group is a reflection of the group in the original compound.

**Solution**

a. The three carbons that contain a H and OH group in the ketohexose are chirality centers.

```
        CH₂OH
         |
        C=O
         |
    H — C*— OH
         |
   HO — C*— H
         |
   HO — C*— H
         |
        CH₂OH

   * = chirality center
```

b. The ketohexose is an L sugar since the OH group on the chirality center farthest from the carbonyl is on the left.

```
        CH₂OH
         |
        C=O
         |
    H — C — OH
         |
   HO — C — H
         |
   HO — C — H
         |
        CH₂OH

      L sugar
```

c. The enantiomer of the L-ketohexose has all three OH groups on the chirality centers on the opposite side of the carbon chain.

```
        CH₂OH
         |
        C=O
         |
   HO — C — H
         |
    H — C — OH
         |
    H — C — OH
         |
        CH₂OH

     enantiomer
```

---

## [2] The Cyclic Forms of Monosaccharides (20.3)

**Example 20.3** Draw the β anomer of the D monosaccharide.

```
        CHO
         |
    H — C — OH
         |
    H — C — OH
         |
   HO — C — H
         |
    H — C — OH
         |
        CH₂OH
```

**Analysis and Solution**

**[1] Draw a hexagon with an O atom in the upper right corner. Add the CH₂OH above the ring on the first carbon to the left of the O atom.**

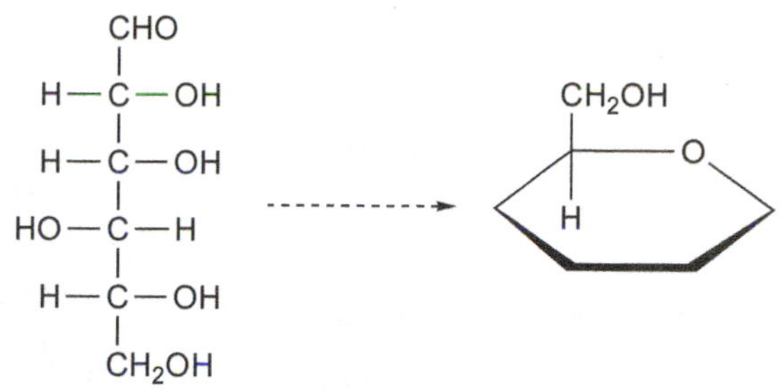

---

**[2] Draw the anomeric carbon on the first carbon clockwise from the O atom.**

- The β anomer has the OH group drawn up.

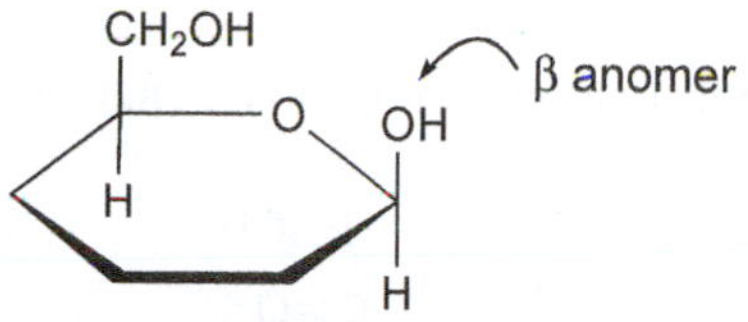

---

**[3] Add the OH groups and H atoms to the three remaining carbons (C2–C4).**

- Groups on the *right* side in the acyclic form are drawn *down*, below the six-membered ring, and groups on the *left* side in the acyclic form are drawn *up*, above the six-membered ring.

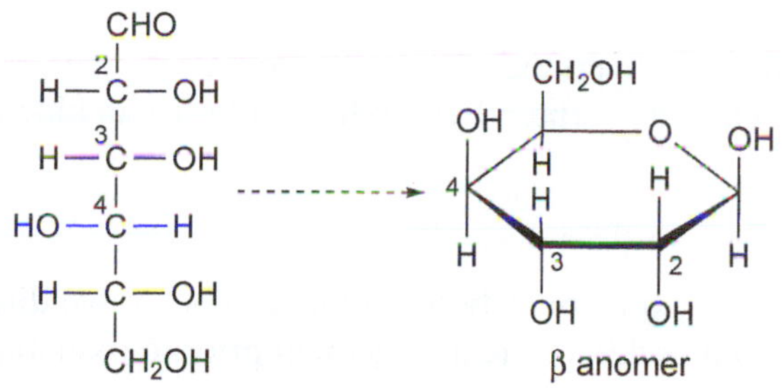

---

# [3] Reduction and Oxidation of Monosaccharides (20.4)

**Example 20.4** Draw the structure of the compound formed when the following monosaccharide is treated with H₂ in the presence of a Pd catalyst?

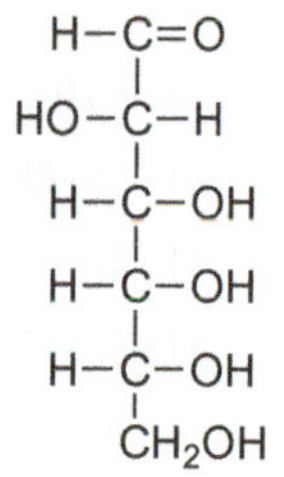

## Analysis
- Locate the C=O and mentally break one bond in the double bond.
- Mentally break the H–H bond of the reagent and add one H atom to each atom of the C=O, forming new C–H and O–H bonds.

## Solution

---

**Example 20.5** Draw the product formed when each monosaccharide is oxidized with Benedict's reagent.

## Analysis
To draw the oxidation product of an aldose, convert the CHO group to COOH. To oxidize a ketose, first rearrange the ketose to an aldose by moving the carbonyl group to C1 to give an aldehyde. Then convert the CHO group to COOH.

## Solution
The monosaccharide in part (a) is an aldose, so its oxidation product can be drawn directly. The monosaccharide in part (b) is a ketose, so it undergoes rearrangement prior to oxidation.

## [4] Disaccharides (20.5)

**Example 20.6** (a) Locate the glycosidic linkage in the disaccharide. (b) Number the carbon atoms in both rings. (c) Classify the glycosidic linkage as α or β, and use numbers to designate its location.

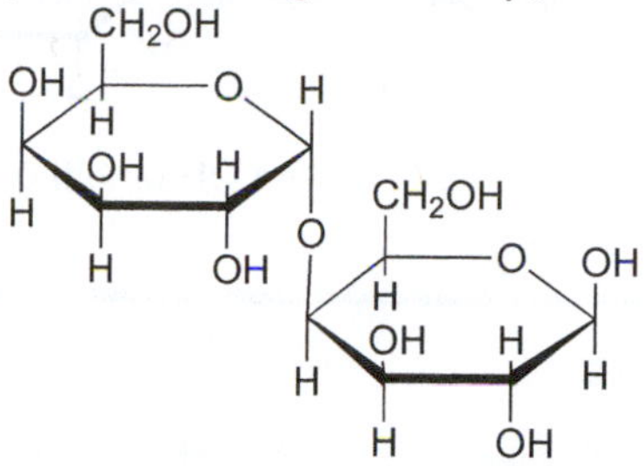

**Analysis and Solution**

a and b. The glycosidic linkage is the acetal C–O bond that joins the two monosaccharides. Each ring is numbered beginning at the anomeric carbon, the carbon bonded to two oxygen atoms.

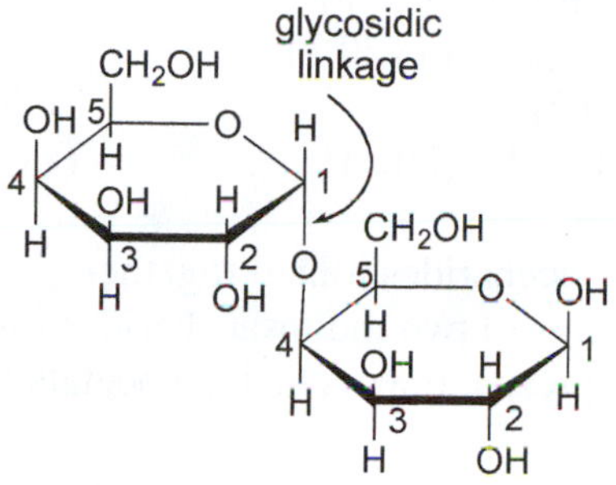

c. The disaccharide has an α glycosidic linkage since the C–O bond is drawn down. The glycosidic linkage joins C1 of one ring to C4 of the other ring, so it is called a 1→4-α-glycosidic linkage.

---

**Example 20.7** Draw the products formed when the following disaccharide is hydrolyzed with water.

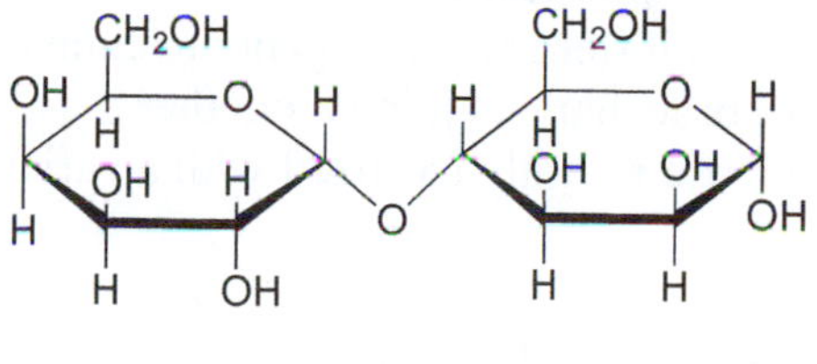

**Analysis**

Locate the glycosidic linkage, the acetal C–O bond that joins the two monosaccharides. Cleave the C–O bond by adding the elements of $H_2O$ across the bond.

**Solution**

Two monosaccharides are formed by cleaving the glycosidic linkage.

glycosidic linkage

Add OH onto the ring.

Add H to the O atom.

## Self-Test

**[1] Fill in the blank with one of the terms listed below.**

Aldonic acid (20.4)
Anomeric carbon (20.3)
Carbohydrates (20.1)
Disaccharides (20.1)

Glycosidic linkage (20.5)
Haworth projections (20.3)
Hexose (20.2)
Monosaccharides (20.1)

Nonreducing sugars (20.4)
Polysaccharides (20.1)
Reducing sugars (20.4)
Triose (20.2)

1. _____ are composed of two monosaccharides joined together.
2. The new C–O bond that joins the rings of two monosaccharides together is called a _____.
3. Flat, six-membered rings used to represent the cyclic hemiacetals of glucose and other sugars are called _____.
4. _____, commonly referred to as sugars and starches, are polyhydroxy aldehydes and ketones, or compounds that can be hydrolyzed to them.
5. A _____ is a monosaccharide that has six carbons.
6. In the cyclic form of monosaccharides, the carbon atom that is part of the hemiacetal is a new chirality center called the _____.
7. A _____ is a monosaccharide that has three carbons.
8. Carbohydrates that are oxidized with Benedict's reagent are called _____.
9. _____ or simple sugars are the simplest carbohydrates.
10. Carbohydrates that do not react with Benedict's reagent are called _____.
11. _____ have three or more monosaccharides joined together.
12. The aldehyde carbonyl of an aldose is easily oxidized with a variety of reagents to form a carboxyl group, yielding an _____.

**[2] Label each monosaccharide as an aldose or a ketose.**

a. aldose

b. ketose

13.

$$\begin{array}{c} CHO \\ | \\ H-C-OH \\ | \\ H-C-OH \\ | \\ CH_2OH \end{array}$$

14.

$$\begin{array}{c} CH_2OH \\ | \\ C=O \\ | \\ H-C-OH \\ | \\ HO-C-H \\ | \\ CH_2OH \end{array}$$

15.

$$\begin{array}{c} CHO \\ | \\ H-C-OH \\ | \\ HO-C-H \\ | \\ HO-C-H \\ | \\ H-C-OH \\ | \\ CH_2OH \end{array}$$

**[3] Label each monosaccharide as an L or D sugar.**

a. L sugar           b. D sugar

16.
```
        CHO
         |
   H —— C —— OH
         |
  HO —— C —— H
         |
       CH2OH
```

17.
```
       CH2OH
         |
        C = O
         |
   H —— C —— OH
         |
  HO —— C —— H
         |
       CH2OH
```

18.
```
        CHO
         |
   H —— C —— OH
         |
  HO —— C —— H
         |
  HO —— C —— H
         |
   H —— C —— OH
         |
       CH2OH
```

**[4] Label each cyclic monosaccharide as an α or β anomer.**

a. α anomer           b. β anomer

19.

20.

21.

**[5] Pick the product of each reaction.**

a.
```
        COOH
         |
   H —— C —— OH
         |
  HO —— C —— H
         |
   H —— C —— OH
         |
       CH2OH
```

b.
```
       CH2OH
         |
  HO —— C —— H
         |
  HO —— C —— H
         |
   H —— C —— OH
         |
   H —— C —— OH
         |
       CH2OH
```

c.
```
       CH2OH
         |
   H —— C —— OH
         |
  HO —— C —— H
         |
   H —— C —— OH
         |
       CH2OH
```

d.
```
        COOH
         |
  HO —— C —— H
         |
  HO —— C —— H
         |
   H —— C —— OH
         |
   H —— C —— OH
         |
       CH2OH
```

22.
```
        CHO
         |
  HO —— C —— H
         |
  HO —— C —— H
         |
   H —— C —— OH
         |
   H —— C —— OH
         |
       CH2OH
```
$\xrightarrow[\text{OH}^-]{2\ Cu^{2+}}$

23.
```
       CH2OH
         |
        C = O
         |
  HO —— C —— H
         |
   H —— C —— OH
         |
       CH2OH
```
$\xrightarrow[\text{OH}^-]{2\ Cu^{2+}}$

24.
```
        CHO
         |
  HO —— C —— H
         |
  HO —— C —— H
         |
   H —— C —— OH
         |
   H —— C —— OH
         |
       CH2OH
```
$\xrightarrow[\text{Pd}]{\text{H}_2}$

25.
```
   H —— C = O
         |
   H —— C —— OH
         |
  HO —— C —— H
         |
   H —— C —— OH
         |
       CH2OH
```
$\xrightarrow[\text{Pd}]{\text{H}_2}$

## Answers to Self-Test

| | | | | |
|---|---|---|---|---|
| 1. Disaccharides | 6. anomeric carbon | 11. Polysaccharides | 16. a | 21. b |
| 2. glycosidic linkage | 7. triose | 12. aldonic acid | 17. a | 22. d |
| 3. Haworth projections | 8. reducing sugars | 13. a | 18. b | 23. a |
| 4. Carbohydrates | 9. Monosaccharides | 14. b | 19. a | 24. b |
| 5. hexose | 10. nonreducing sugars | 15. a | 20. a | 25. c |

## Solutions to In-Chapter Problems

**20.1** Draw the Lewis structure for glucose with all bonds and lone pairs drawn in.

**20.2** Label the hemiacetal carbon and the hydroxyl groups in lactose. Recall that a 1° hydroxyl group is bonded to a carbon bonded to one other carbon. A 2° hydroxyl group is bonded to a carbon bonded to two other carbons.

**20.3** Classify each monosaccharide by the type of carbonyl group and the number of carbons in the chain as in Example 20.1.

**20.4** Draw the structure of each monosaccharide.

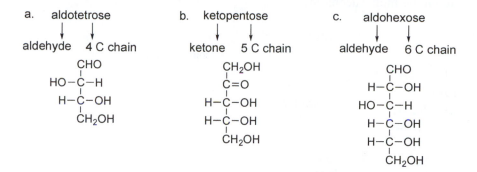

a. aldotetrose

aldehyde   4 C chain

CHO
HO—C—H
H—C—OH
CH₂OH

b. ketopentose

ketone   5 C chain

CH₂OH
C=O
H—C—OH
H—C—OH
CH₂OH

c. aldohexose

aldehyde   6 C chain

CHO
H—C—OH
HO—C—H
H—C—OH
H—C—OH
CH₂OH

**20.5** Water solubility increases with an increased number of polar groups. Hexane is a nonpolar hydrocarbon, insoluble in water. 1-Decanol is a long-chain alcohol and is slightly soluble in water due to the polar OH group. Glucose, with multiple hydroxyl groups, is water soluble.

hexane  <  1-decanol  <  glucose

→ Increasing water solubility

**20.6** Recall that in a Fischer projection:
- A carbon atom is located at the intersection of the two lines of the cross.
- The horizontal bonds come forward, on wedges.
- The vertical bonds go back, on dashed lines.

CHO
HO—C—H
CH₂OH

L-glyceraldehyde

- - - - - - - →

CHO
HO———H
CH₂OH

Fischer projection formula

**20.7** Use the definitions in Example 20.2 to [1] label all chirality centers; [2] classify the monosaccharide as D or L; [3] draw the enantiomer.

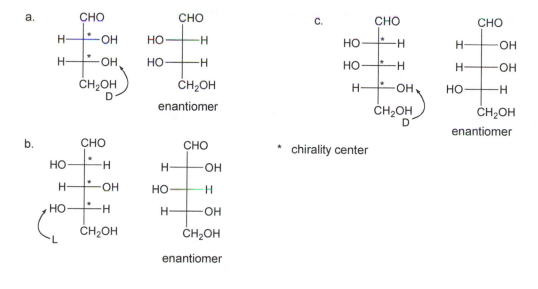

a.

CHO
H—*—OH
H—*—OH
CH₂OH
D

CHO
HO———H
HO———H
CH₂OH

enantiomer

b.

CHO
HO—*—H
H—*—OH
HO—*—H
CH₂OH
L

CHO
H———OH
HO———H
H———OH
CH₂OH

enantiomer

c.

CHO
HO—*—H
HO—*—H
H—*—OH
CH₂OH
D

CHO
H———OH
H———OH
HO———H
CH₂OH

enantiomer

* chirality center

**20.8** Answer each question about the monosaccharide.

a.

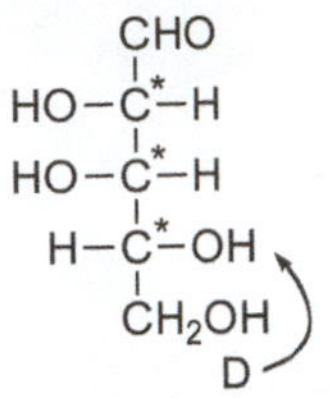

CHO
HO–C*–H
HO–C*–H
H–C*–OH
CH₂OH
D

b.

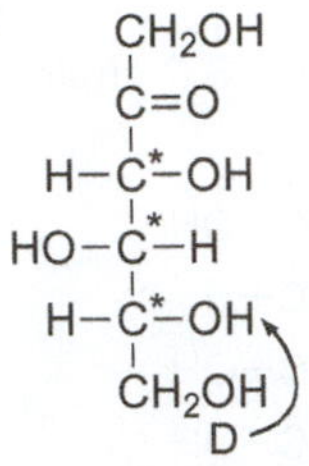

CH₂OH
C=O
H–C*–OH
HO–C*–H
H–C*–OH
CH₂OH
D

c.

CH₂OH
C=O
H–C*–OH
CH₂OH
D

* chirality center

5 C's
aldehyde
aldopentose

6 C's
ketone
ketohexose

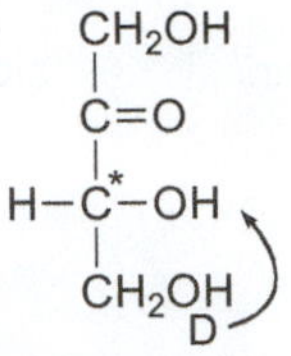

4 C's
ketone
ketotetrose

**20.9** D-Glucose and D-fructose are constitutional isomers (same molecular formula, different connectivity) because glucose has an aldehyde group and fructose has a ketone group. D-Galactose and D-fructose are also constitutional isomers because galactose has an aldehyde group and fructose has a ketone group.

**20.10** Draw the Fischer projection for D-galactose and its enantiomer.

a.

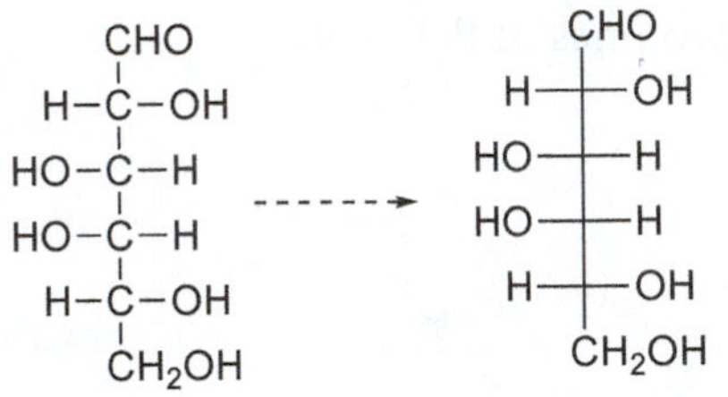

CHO
H–C–OH
HO–C–H
HO–C–H
H–C–OH
CH₂OH

D-galactose

CHO
H——OH
HO——H
HO——H
H——OH
CH₂OH

Fischer
projection

b.

CHO
HO——H
H——OH
H——OH
HO——H
CH₂OH

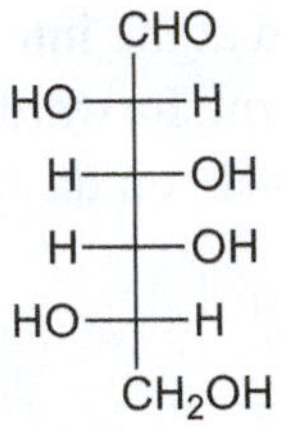

enantiomer
All of the H and OH groups are
on the opposite side.

**20.11** The hemiacetal carbon is the carbon bonded to an OH and an OR group.
- The α anomer has the OH group drawn down, below the ring.
- The β anomer has the OH group drawn up, above the ring.

a.

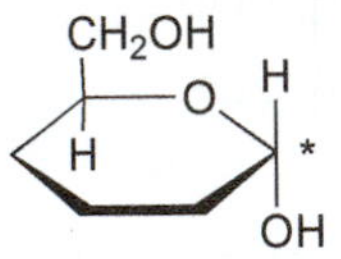

CH₂OH
H
O
H
OH
*

α anomer

b.

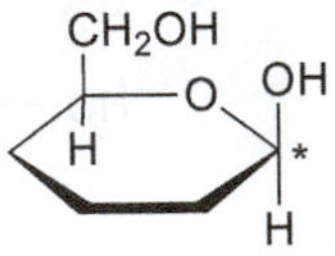

CH₂OH
OH
H
H
*

β anomer

c.

CH₂OH
H
O
OH
H
HO
H
*

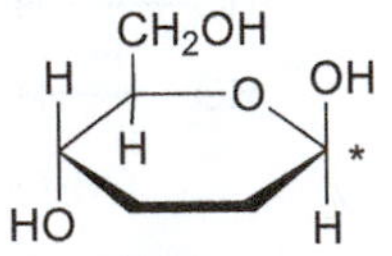

β anomer

* hemiacetal C

**20.12** Convert each monosaccharide to the indicated anomer as in Example 20.3.

a.

```
      CHO
  H ——— OH
  H ——— OH
  H ——— OH
  H ——— OH
      CH₂OH

   α anomer
```

- - - - →  (α anomer ring structure, CH₂OH)  - - - - →  (CH₂OH ring structure, positions 1, 2, 3, 4)

b.

```
      CHO
 HO ——— H
 HO ——— H
 HO ——— H
  H ——— OH
      CH₂OH

   β anomer
```

- - - - →  (β anomer ring structure, CH₂OH)  - - - - →  (CH₂OH ring structure, positions 1, 2, 3, 4)

c.

```
      CHO
 HO ——— H
  H ——— OH
  H ——— OH
  H ——— OH
      CH₂OH

   α anomer
```

- - - - →  (α anomer ring structure, CH₂OH)  - - - - →  (CH₂OH ring structure, positions 1, 2, 3, 4)

**20.13** The anomeric carbon is the carbon bonded to an OH and an OR group.
- The α anomer has the OH group drawn down, below the ring.
- The β anomer has the OH group drawn up, above the ring.

a.  (ring structure)  α anomer

b.  (ring structure)  α anomer

c.  (ring structure)  β anomer

\* anomeric C

**20.14** Draw the products of each reduction reaction as in Example 20.4.

a.

```
      CHO                      CH₂OH
  H—C—OH          H₂      H—C—OH
  H—C—OH       ——————→   H—C—OH
  H—C—OH          Pd      H—C—OH
      CH₂OH                    CH₂OH
```

b.

```
      CHO                      CH₂OH
  H—C—OH                   H—C—OH
 HO—C—H           H₂      HO—C—H
 HO—C—H        ——————→    HO—C—H
  H—C—OH           Pd      H—C—OH
      CH₂OH                    CH₂OH
```

c.

```
      CHO                          CH₂OH
   H—C—OH      H₂             H—C—OH
   H—C—OH   ────────→         H—C—OH
      CH₂OH      Pd              CH₂OH
```

**20.15** Draw the products of the oxidation of each monosaccharide as in Example 20.5.

a.

```
      CHO                          COOH
   H—C—OH                       H—C—OH
   H—C—OH     2 Cu²⁺            H—C—OH
   H—C—OH   ────────→           H—C—OH
      CH₂OH     OH⁻               CH₂OH
```

c.

```
     CH₂OH                         COOH
      C=O                       H—C—OH
   HO—C—H      2 Cu²⁺          HO—C—H
   HO—C—H    ────────→         HO—C—H
   H—C—OH       OH⁻            H—C—OH
      CH₂OH                       CH₂OH
```

b.

```
      CHO                          COOH
   H—C—OH     2 Cu²⁺           H—C—OH
   H—C—OH   ────────→          H—C—OH
      CH₂OH      OH⁻              CH₂OH
```

**20.16** Draw the products of each reaction.

a.

```
      CHO                         CH₂OH
   HO—C—H      H₂            HO—C—H
   H—C—OH    ────────→        H—C—OH
   H—C—OH       Pd            H—C—OH
      CH₂OH                      CH₂OH
```

b.

```
      CHO                         COOH
   HO—C—H      2 Cu²⁺        HO—C—H
   H—C—OH    ────────→        H—C—OH
   H—C—OH       OH⁻           H—C—OH
      CH₂OH                      CH₂OH
```

**20.17** Answer each question about the disaccharide as in Example 20.6.

glycosidic linkage between C1 and C4
C—O bond of the glycosidic linkage drawn up = β
1→ 4-β-disaccharide

glycosidic linkage

**20.18** Draw the products of the hydrolysis of cellobiose as in Example 20.7.

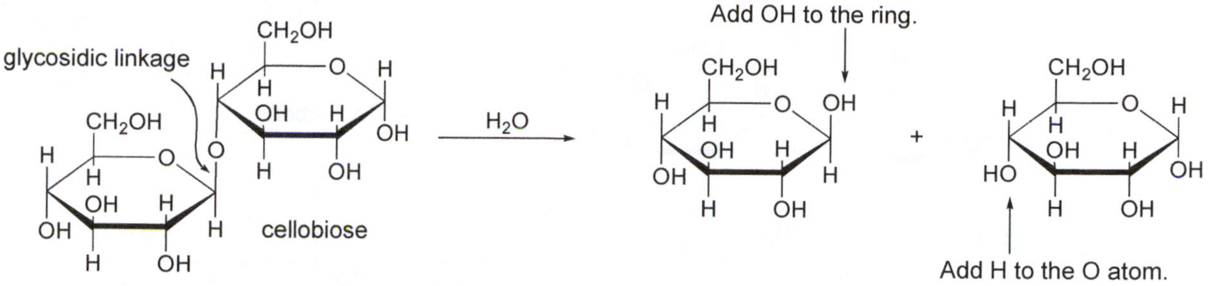

Add OH to the ring.

glycosidic linkage

cellobiose

$H_2O$

Add H to the O atom.

**20.19** Label the acetal (C bonded to two OR groups) and hemiacetal (C bonded to an OH and an OR) groups in lactose.

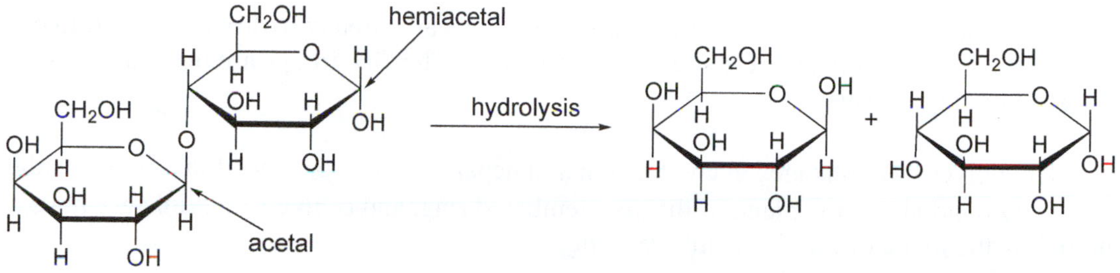

hemiacetal

acetal

hydrolysis

**20.20** Locate the three acetals in rebaudioside A.

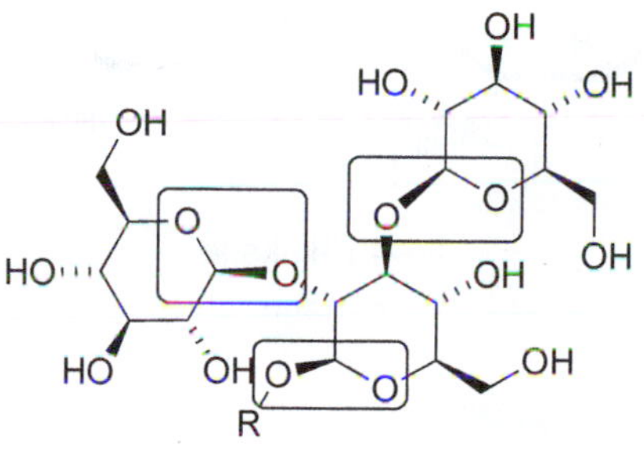

**20.21** Label all the acetal carbons in sucralose, and classify the alkyl halides as 1°, 2°, or 3°.

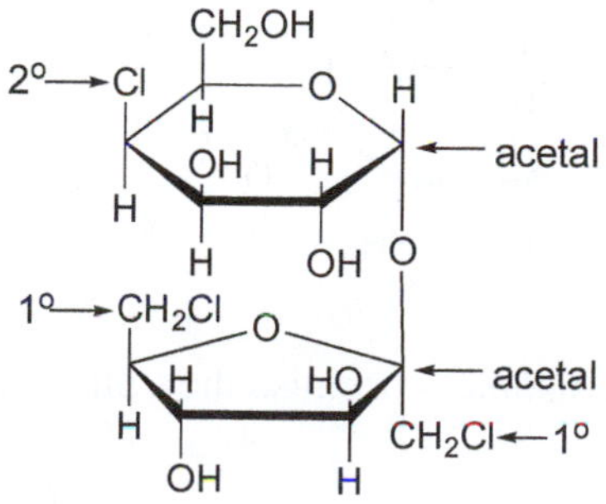

2° → Cl

acetal

1° → $CH_2Cl$

acetal

$CH_2Cl$ ← 1°

**20.22** Label all the acetal carbons in amylopectin.

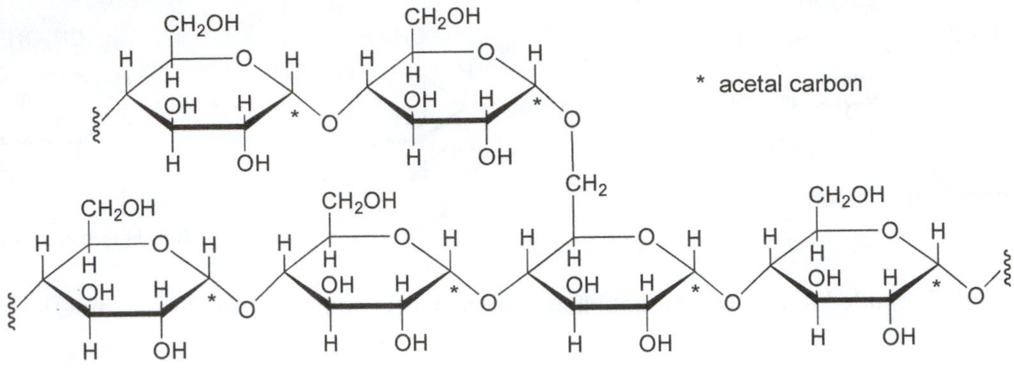

\* acetal carbon

**20.23** Cellulose is water insoluble because the many OH groups are already hydrogen bonded to other OH groups in the interior of the three-dimensional structure. The OH groups are therefore less available for hydrogen bonding to water.

**20.24** Classify the glycosidic linkages in chondroitin and heparin as α or β. Recall that β glycosidic linkages have a C–O bond above the plane of the six-membered ring, and α glycosidic linkages have a C–O bond below the plane of the six-membered ring.

1→3-β-linkage

chondroitin

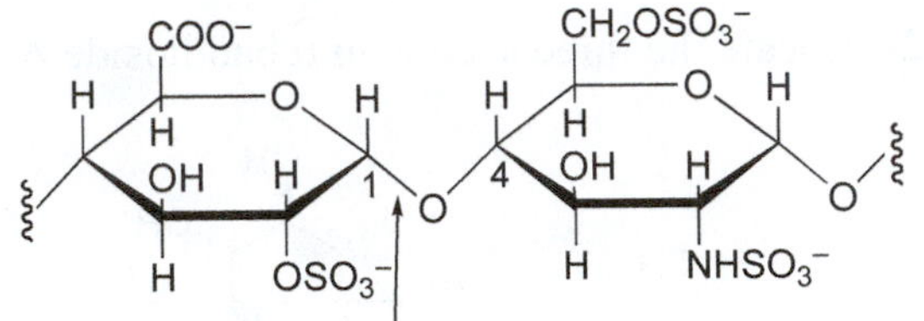

1→4-α-linkage

heparin

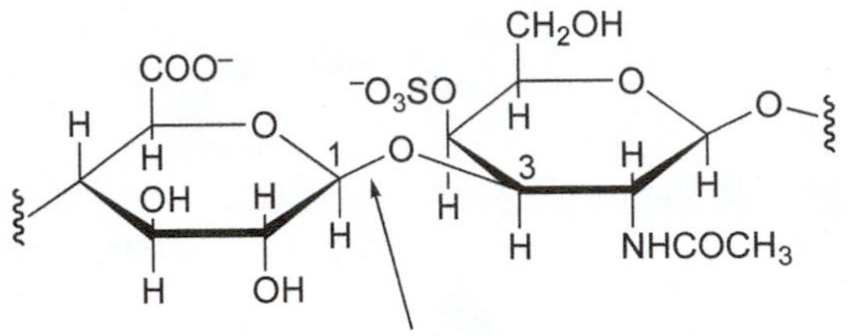

**20.25** Change the β glycosidic linkages to α glycosidic linkages in chitin.

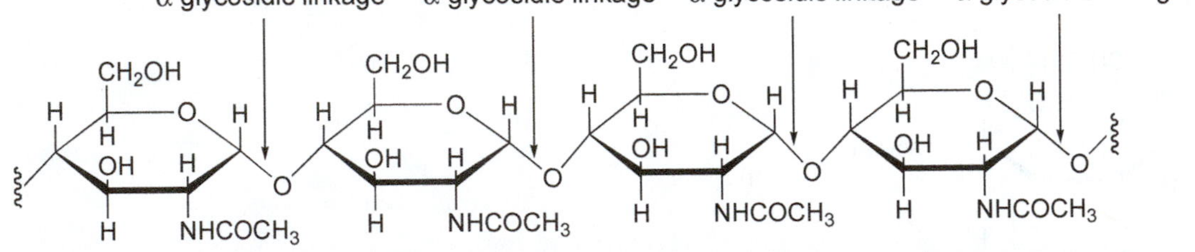

α glycosidic linkage    α glycosidic linkage    α glycosidic linkage    α glycosidic linkage

**20.26** *N*-Acetyl-D-glucosamine and *N*-acetyl-D-galactosamine are stereoisomers because they differ only in the configuration at C4.

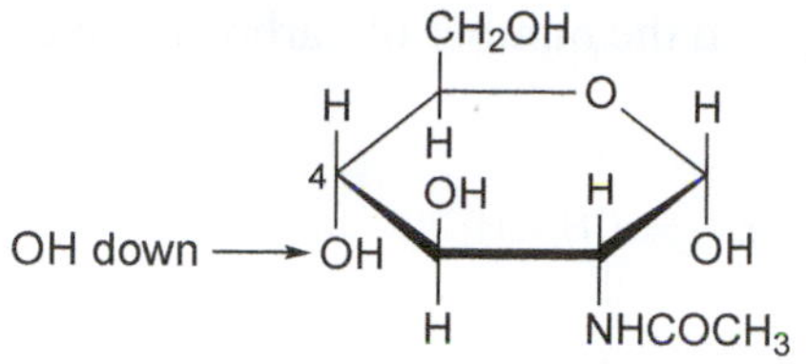

CH$_2$OH
H
H
4
OH
H
OH down → OH
H
O
H
OH
H
NHCOCH$_3$

*N*-acetyl-D-glucosamine

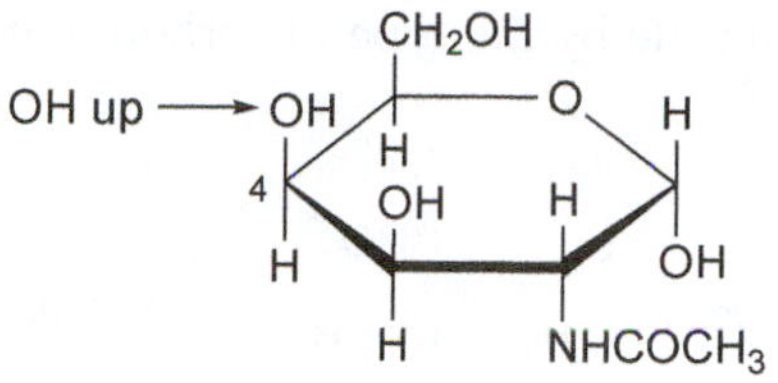

OH up → OH
CH$_2$OH
H
H
4
OH
H
H
O
H
OH
H
NHCOCH$_3$

*N*-acetyl-D-galactosamine

## Solutions to Odd-Numbered End-of-Chapter Problems

**20.27** Aldoses are monosaccharides with a carbonyl group at C1, forming an aldehyde, and ketoses are monosaccharides with a carbonyl group at C2, forming a ketone.

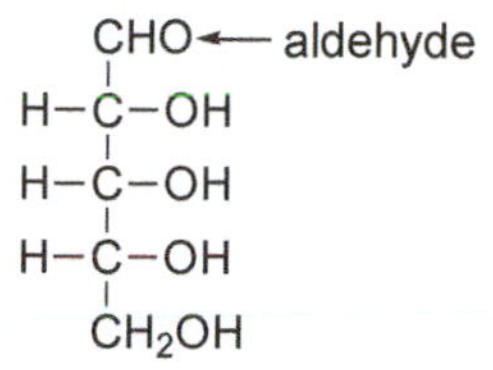

CHO ← aldehyde
H–C–OH
H–C–OH
H–C–OH
CH$_2$OH

ribose, an aldose

CH$_2$OH
C=O ← ketone
HO–C–H
H–C–OH
H–C–OH
CH$_2$OH

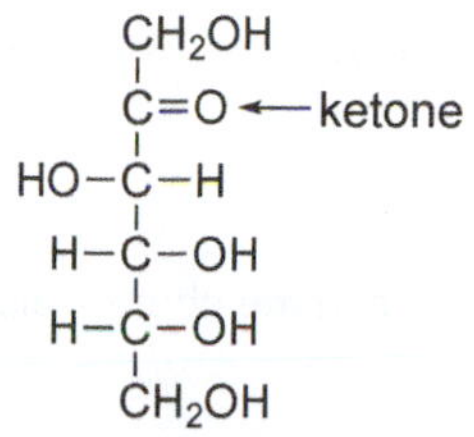

fructose, a ketose

**20.29** Draw the structure of each monosaccharide.

a. an L-aldopentose

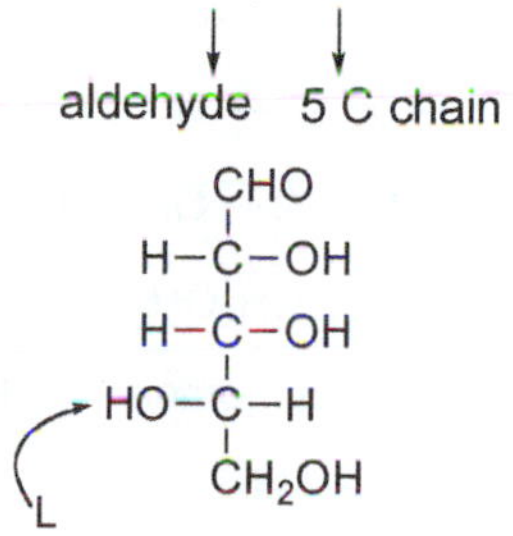

aldehyde    5 C chain

CHO
H–C–OH
H–C–OH
→ HO–C–H
CH$_2$OH
L

b. a D-aldotetrose

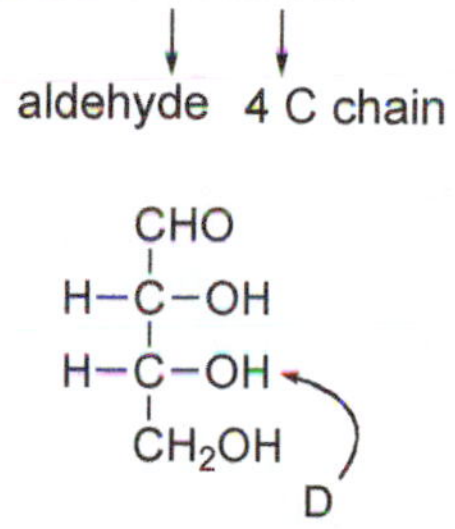

aldehyde   4 C chain

CHO
H–C–OH
H–C–OH ←
CH$_2$OH
D

c. a five-carbon alditol

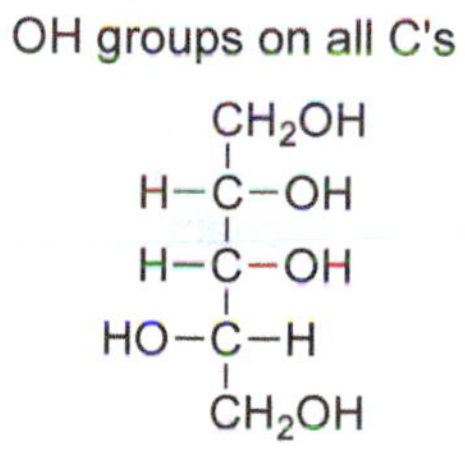

OH groups on all C's

CH$_2$OH
H–C–OH
H–C–OH
HO–C–H
CH$_2$OH

**20.31** α-D-Glucose and β-D-glucose are not enantiomers because they differ in the orientation of only one OH at C1. Enantiomers are mirror images of each other, and therefore differ at all chirality centers.

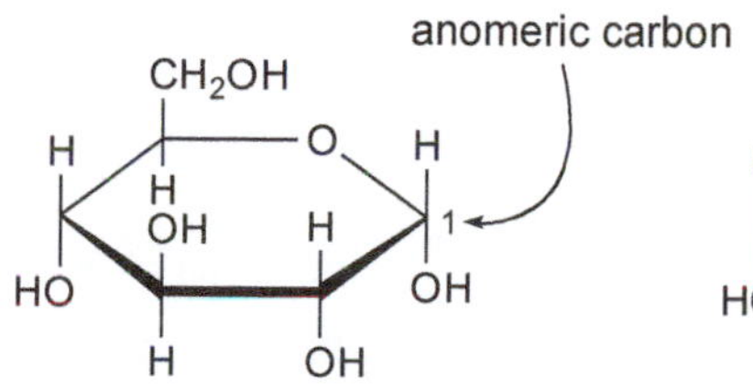

anomeric carbon

CH$_2$OH
H
H
OH
H
O
H
HO
1
OH
H
OH

α-D-glucose

anomeric carbon

CH$_2$OH
H
H
OH
H
O
OH
HO
1
H
H
OH

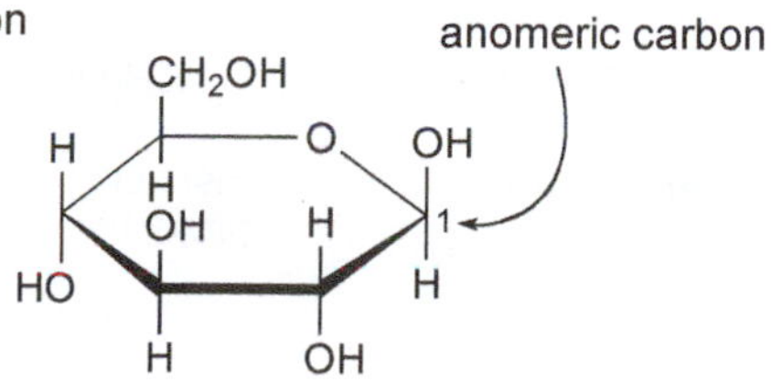

β-D-glucose

**20.33** Classify each monosaccharide by the type of carbonyl group and the number of carbons in the chain as in Example 20.1.

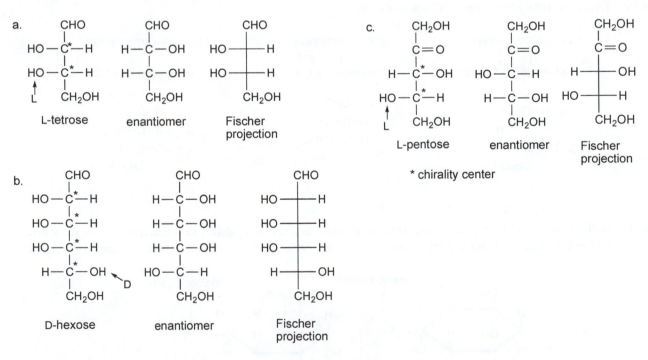

a.
CHO
aldehyde
HO—C—H
HO—C—H
CH₂OH

4 C's in the chain
**aldotetrose**

b.
CHO
aldehyde
HO—C—H
HO—C—H
HO—C—H
H—C—OH
CH₂OH

6 C's in the chain
**aldohexose**

c.
CH₂OH
C=O ketone
H—C—OH
HO—C—H
CH₂OH

5 C's in the chain
**ketopentose**

**20.35** Draw the Fischer projection and then classify the monosaccharide as in Example 20.1.

a.
CHO aldehyde
H——OH
CH₂OH

b. 3 C's in the chain = aldotriose

**20.37** Answer each question about the monosaccharides.

a.
CHO
HO—C*—H
HO—C*—H
L  CH₂OH

**L-tetrose**

CHO
H—C—OH
H—C—OH
CH₂OH

**enantiomer**

CHO
HO——H
HO——H
CH₂OH

**Fischer projection**

c.
CH₂OH
C=O
H—C*—OH
HO—C*—H
L  CH₂OH

**L-pentose**

CH₂OH
C=O
HO—C—H
H—C—OH
CH₂OH

**enantiomer**

CH₂OH
C=O
H——OH
HO——H
CH₂OH

**Fischer projection**

* chirality center

b.
CHO
HO—C*—H
HO—C*—H
HO—C*—H
H—C—OH
CH₂OH  D

**D-hexose**

CHO
H—C—OH
H—C—OH
H—C—OH
HO—C—H
CH₂OH

**enantiomer**

CHO
HO——H
HO——H
HO——H
H——OH
CH₂OH

**Fischer projection**

**20.39** Answer each question about monosaccharides **A, B,** and **C.**

a. **A** and **B** are stereoisomers with a different three-dimensional arrangement at the indicated chirality centers.

```
      CHO                    CHO
  H—C—OH                 H—C—OH
  H—C—OH                HO—C—H
    CH2OH                  CH2OH
      A                      B
```

b. **B** and **C** (or **A** and **C**) are constitutional isomers.

```
      CHO          CH2OH              CHO          CH2OH
  H—C—OH           C=O            H—C—OH           C=O
 HO—C—H         H—C—OH           H—C—OH         H—C—OH
    CH2OH          CH2OH             CH2OH          CH2OH
      B              C                 A              C
   aldehyde        ketone          aldehyde        ketone
```

c. Enantiomer of **B**:

```
      CHO                    CHO
  H—C—OH                HO—C—H
 HO—C—H      ------->    H—C—OH
    CH2OH                  CH2OH
      B
```

d. Fischer projection of **A**:

```
      CHO                    CHO
  H—C—OH                H——OH
  H—C—OH      ------->   H——OH
    CH2OH                  CH2OH
      A
```

**20.41** Convert the monosaccharide to both anomers as in Example 20.3.

```
      CHO
 HO——H
  H——OH
 HO——H
  H——OH
    CH2OH
```

α anomer   β anomer

**20.43** Answer each question about the cyclic monosaccharide, drawn in a skeletal structure.

1° HO—
2° HO—        —OH ← OH on the anomeric C
hemiacetal C
HO   OH
2°   2°

**20.45** Answer each question about the cyclic monosaccharide.

a. and b.

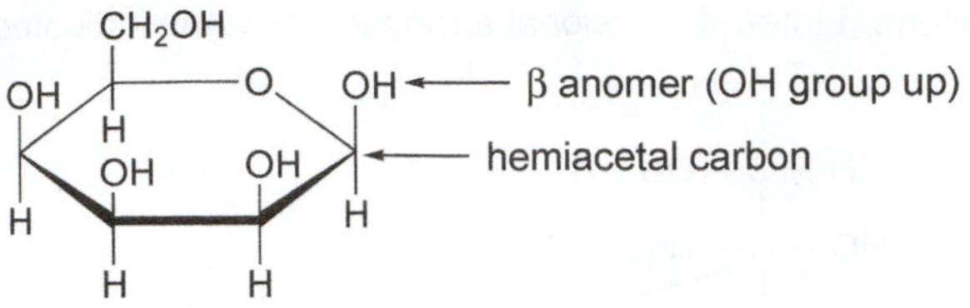

β anomer (OH group up)
hemiacetal carbon

c.

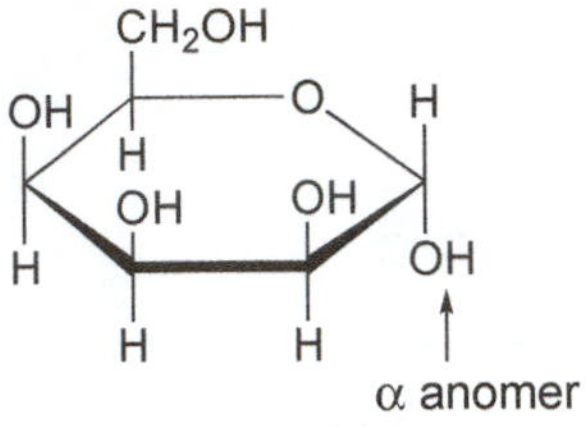

α anomer

d. stereoisomer:

different arrangement

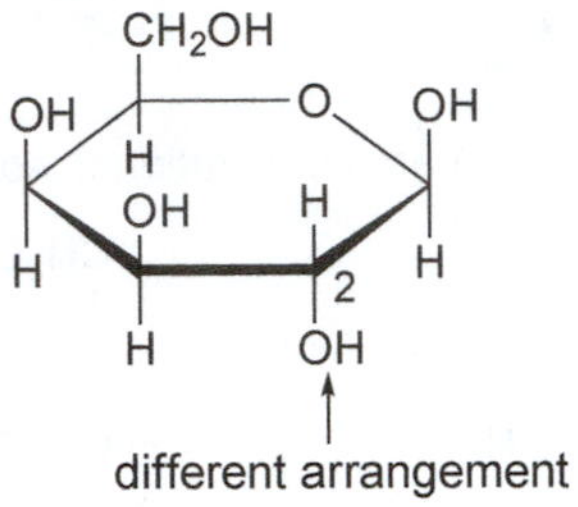

**20.47** Answer the questions for each cyclic monosaccharide.

a.

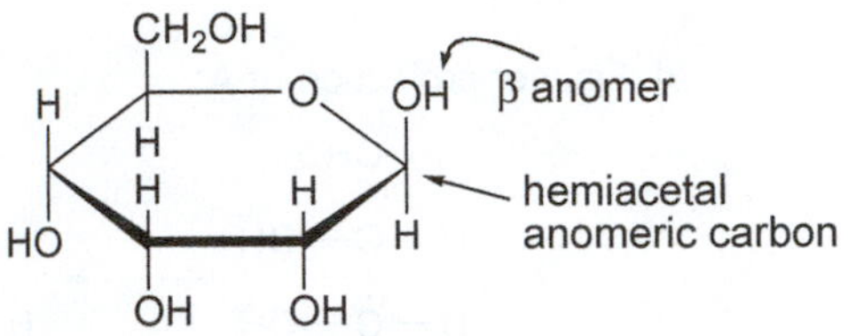

β anomer
hemiacetal
anomeric carbon

b.

hemiacetal
anomeric carbon
α anomer

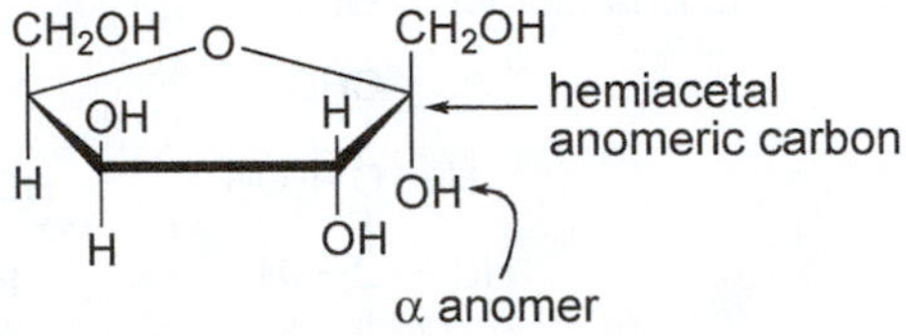

**20.49** A ketotriose has no chirality centers and no stereoisomers, so the D or L classification is not needed. The only ketotriose is dihydroxyacetone.

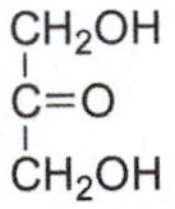

$$CH_2OH$$
$$C=O$$
$$CH_2OH$$

dihydroxyacetone
ketotriose

**20.51** The hemiacetal carbon is the carbon bonded to an OH and an OR group.
- The α anomer has the OH group drawn down, below the ring.
- The β anomer has the OH group drawn up, above the ring.

a.

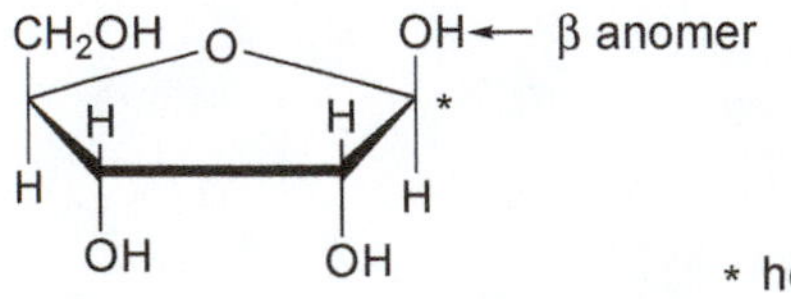

β anomer

* hemiacetal C

b.

α anomer

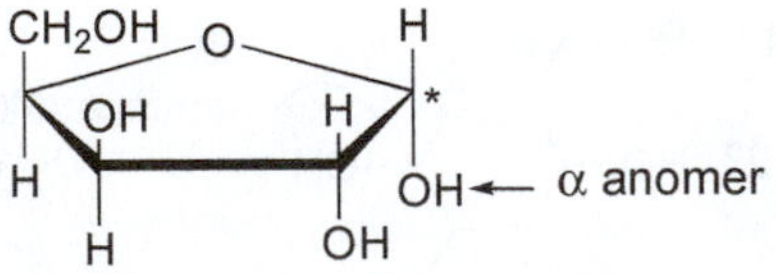

**20.53** Draw the structures of an alditol and an aldonic acid that contain four carbons.

$$
\begin{array}{cc}
\text{CH}_2\text{OH} & \text{COOH} \\
\text{H}-\text{C}-\text{OH} & \text{H}-\text{C}-\text{OH} \\
\text{HO}-\text{C}-\text{H} & \text{HO}-\text{C}-\text{H} \\
\text{CH}_2\text{OH} & \text{CH}_2\text{OH} \\
\text{alditol} & \text{aldonic acid}
\end{array}
$$

**20.55** Draw the products of each reaction.

a.

$$
\begin{array}{cccc}
\text{CHO} & & \text{CH}_2\text{OH} & \\
\text{HO}-\text{C}-\text{H} & \xrightarrow[\text{Pd}]{\text{H}_2} & \text{HO}-\text{C}-\text{H} & \\
\text{H}-\text{C}-\text{OH} & & \text{H}-\text{C}-\text{OH} & \\
\text{CH}_2\text{OH} & & \text{CH}_2\text{OH} &
\end{array}
$$

$$
\begin{array}{cccc}
\text{CHO} & & \text{COOH} & \\
\text{HO}-\text{C}-\text{H} & \xrightarrow[\text{OH}^-]{2\ \text{Cu}^{2+}} & \text{HO}-\text{C}-\text{H} & \\
\text{H}-\text{C}-\text{OH} & & \text{H}-\text{C}-\text{OH} & \\
\text{CH}_2\text{OH} & & \text{CH}_2\text{OH} &
\end{array}
$$

b.

$$
\begin{array}{cccc}
\text{CHO} & & \text{CH}_2\text{OH} & \\
\text{H}-\text{C}-\text{OH} & \xrightarrow[\text{Pd}]{\text{H}_2} & \text{H}-\text{C}-\text{OH} & \\
\text{HO}-\text{C}-\text{H} & & \text{HO}-\text{C}-\text{H} & \\
\text{HO}-\text{C}-\text{H} & & \text{HO}-\text{C}-\text{H} & \\
\text{CH}_2\text{OH} & & \text{CH}_2\text{OH} &
\end{array}
$$

$$
\begin{array}{cccc}
\text{CHO} & & \text{COOH} & \\
\text{H}-\text{C}-\text{OH} & \xrightarrow[\text{OH}^-]{2\ \text{Cu}^{2+}} & \text{H}-\text{C}-\text{OH} & \\
\text{HO}-\text{C}-\text{H} & & \text{HO}-\text{C}-\text{H} & \\
\text{HO}-\text{C}-\text{H} & & \text{HO}-\text{C}-\text{H} & \\
\text{CH}_2\text{OH} & & \text{CH}_2\text{OH} &
\end{array}
$$

c.

$$
\begin{array}{cccc}
\text{CHO} & & \text{CH}_2\text{OH} & \\
\text{H}-\text{C}-\text{OH} & & \text{H}-\text{C}-\text{OH} & \\
\text{H}-\text{C}-\text{OH} & \xrightarrow[\text{Pd}]{\text{H}_2} & \text{H}-\text{C}-\text{OH} & \\
\text{HO}-\text{C}-\text{H} & & \text{HO}-\text{C}-\text{H} & \\
\text{HO}-\text{C}-\text{H} & & \text{HO}-\text{C}-\text{H} & \\
\text{CH}_2\text{OH} & & \text{CH}_2\text{OH} &
\end{array}
$$

$$
\begin{array}{cccc}
\text{CHO} & & \text{COOH} & \\
\text{H}-\text{C}-\text{OH} & & \text{H}-\text{C}-\text{OH} & \\
\text{H}-\text{C}-\text{OH} & \xrightarrow[\text{OH}^-]{2\ \text{Cu}^{2+}} & \text{H}-\text{C}-\text{OH} & \\
\text{HO}-\text{C}-\text{H} & & \text{HO}-\text{C}-\text{H} & \\
\text{HO}-\text{C}-\text{H} & & \text{HO}-\text{C}-\text{H} & \\
\text{CH}_2\text{OH} & & \text{CH}_2\text{OH} &
\end{array}
$$

**20.57** Draw the products of the oxidation of each monosaccharide as in Example 20.5.

a.

$$
\begin{array}{cccc}
\text{CHO} & & \text{COOH} & \\
\text{H}-\text{C}-\text{OH} & & \text{H}-\text{C}-\text{OH} & \\
\text{H}-\text{C}-\text{OH} & \xrightarrow[\text{OH}^-]{2\ \text{Cu}^{2+}} & \text{H}-\text{C}-\text{OH} & \\
\text{HO}-\text{C}-\text{H} & & \text{HO}-\text{C}-\text{H} & \\
\text{CH}_2\text{OH} & & \text{CH}_2\text{OH} &
\end{array}
$$

b.

$$
\begin{array}{cccc}
\text{CH}_2\text{OH} & & \text{COOH} & \\
\text{C}=\text{O} & & \text{H}-\text{C}-\text{OH} & \\
\text{H}-\text{C}-\text{OH} & \xrightarrow[\text{OH}^-]{2\ \text{Cu}^{2+}} & \text{H}-\text{C}-\text{OH} & \\
\text{H}-\text{C}-\text{OH} & & \text{H}-\text{C}-\text{OH} & \\
\text{CH}_2\text{OH} & & \text{CH}_2\text{OH} &
\end{array}
$$

**20.59** Reducing sugars are carbohydrates that are oxidized with Benedict's reagent ($Cu^{2+}$). Glucose, an aldohexose, is a reducing sugar.

**20.61** Draw the products of the hydrolysis of the disaccharides as in Example 20.7.

a.

b.

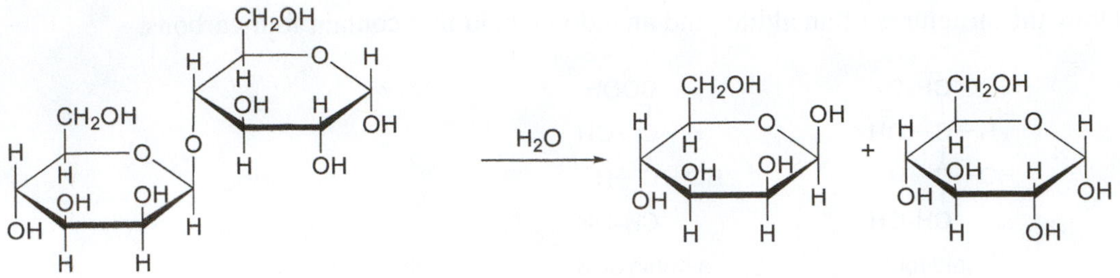

**20.63** An α anomer has a hydroxyl at C1, the hemiacetal C, in the down position. An α glycoside has the glycosidic linkage in the down position.

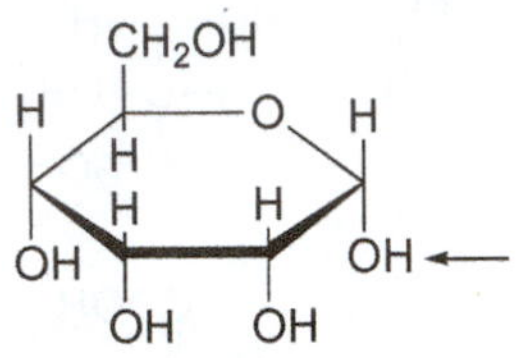

α anomer

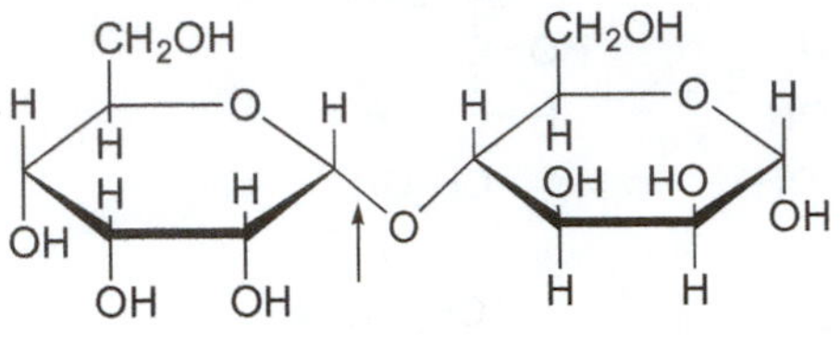

α glycosidic linkage

**20.65** Draw the disaccharide with an α glycosidic linkage.

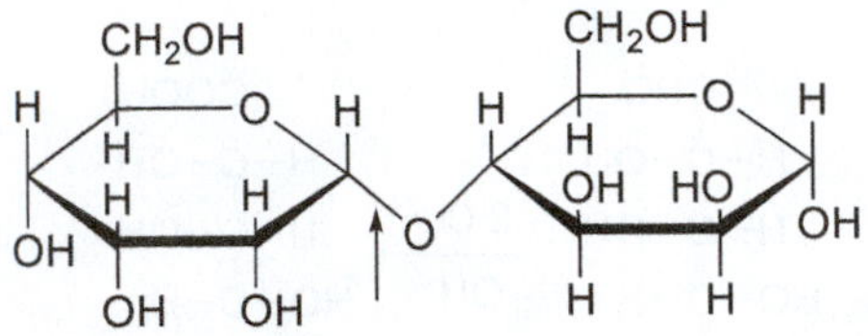

α glycosidic linkage

**20.67** Answer each question about isomaltose.

a. and b.

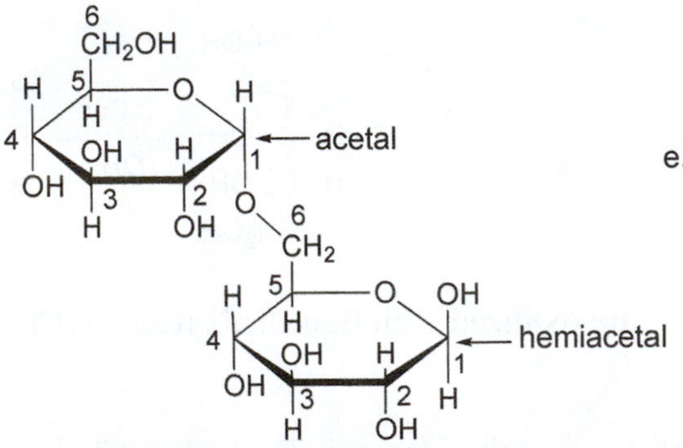

acetal

hemiacetal

e.

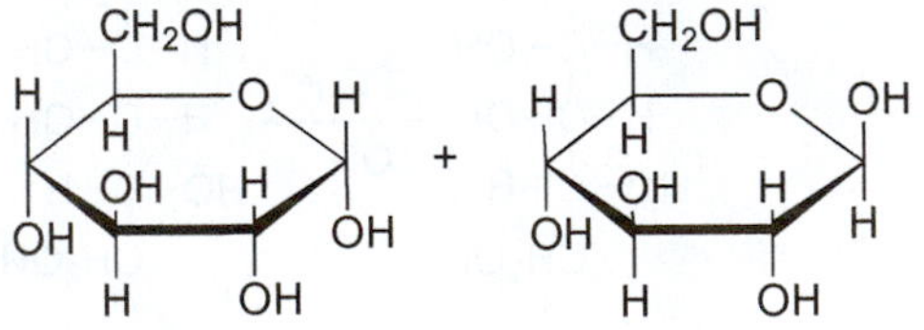

α anomer

β anomer

c. 1→6-α-glycosidic linkage

d. β anomer (hemiacetal OH group up)

**20.69** Draw the disaccharide formed from two galactose units joined by a 1→4-β-glycosidic linkage.

1→4-β-glycosidic linkage

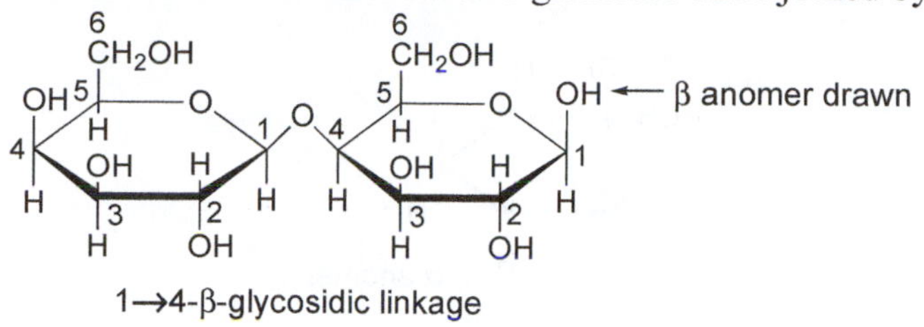

**20.71** Cellulose and amylose are both composed of repeating glucose units. In cellulose the glucose units are joined by a 1→4-β-glycosidic linkage, but in amylose they are joined by a 1→4-α-glycosidic linkage. This leads to very different three-dimensional shapes, with cellulose forming sheets and amylose forming helices.

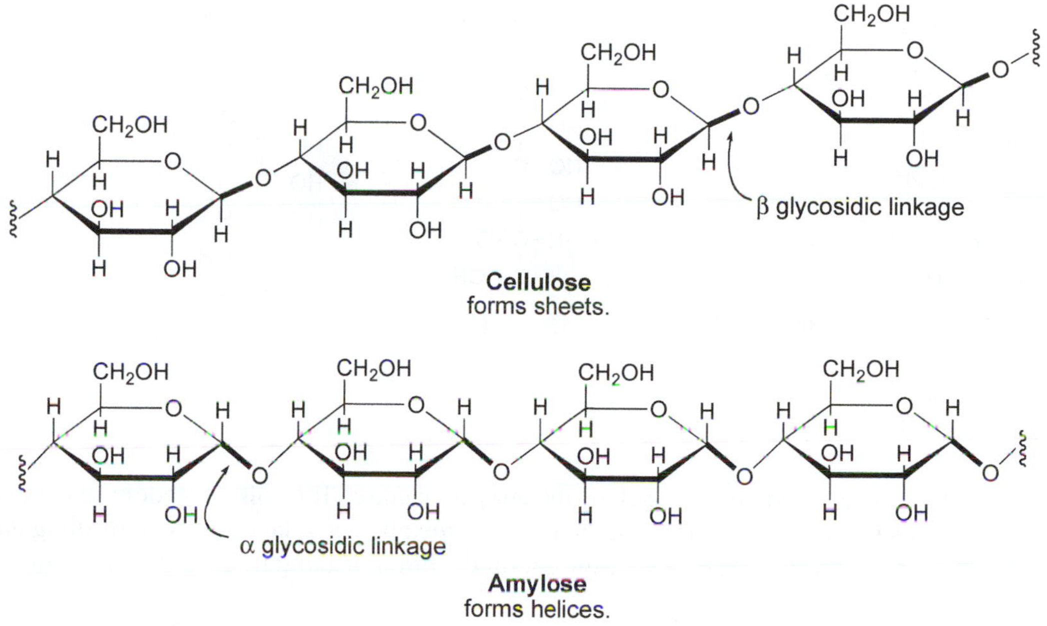

**Cellulose**
forms sheets.

**Amylose**
forms helices.

**20.73** Draw a short segment of a polysaccharide that contains three galactose units joined together in 1→4-α-glycosidic linkages.

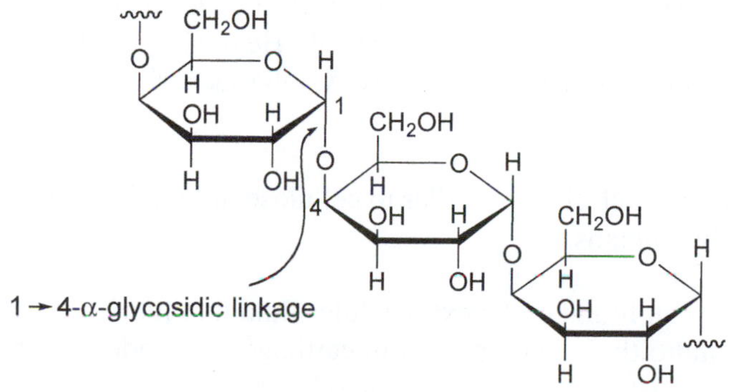

1→4-α-glycosidic linkage

**20.75** Answer each question about the monosaccharide.

a. and b.

$$CHO$$
$$HO-C-H$$
$$HO-C-H$$
$$HO-C-H$$
$$H-C-OH$$
$$CH_2OH$$

D sugar

6 C's
aldehyde
**aldohexose**

c.

$$CHO$$
$$H-C-OH$$
$$H-C-OH$$
$$H-C-OH$$
$$HO-C-H$$
$$CH_2OH$$

enantiomer

d.

$$CHO$$
$$HO-\overset{*}{C}-H$$
$$HO-\overset{*}{C}-H$$
$$HO-\overset{*}{C}-H$$
$$H-\overset{*}{C}-OH$$
$$CH_2OH$$

* chirality center

e.

$$CH_2OH$$
OH ─── O ─ H
H
OH  HO
H        OH
H      H

α anomer

f.

$$CHO$$
$$HO-C-H$$
$$HO-C-H$$
$$HO-C-H$$
$$H-C-OH$$
$$CH_2OH$$

$\xrightarrow[OH^-]{2\ Cu^{2+}}$

$$COOH$$
$$HO-C-H$$
$$HO-C-H$$
$$HO-C-H$$
$$H-C-OH$$
$$CH_2OH$$

g.

$$CHO$$
$$HO-C-H$$
$$HO-C-H$$
$$HO-C-H$$
$$H-C-OH$$
$$CH_2OH$$

$\xrightarrow[Pd]{H_2}$

$$CH_2OH$$
$$HO-C-H$$
$$HO-C-H$$
$$HO-C-H$$
$$H-C-OH$$
$$CH_2OH$$

h. a reducing sugar

**20.77** Lactose intolerance results from a lack of the enzyme lactase. It results in abdominal cramping and diarrhea. Galactosemia results from the inability to metabolize galactose. As a result, galactose accumulates in the liver, causing cirrhosis, and in the brain, leading to mental retardation.

**20.79** Fructose is a naturally occurring sugar with more perceived sweetness per gram than sucrose, and so fructose provides the same amount of sweetness in fewer grams and fewer calories. Sucralose is a synthetic sweetener; that is, it is not naturally occurring and is therefore labeled artificial.

**20.81** An individual with type A blood can receive only blood types A and O, because he or she will produce antibodies and an immune response to B or AB blood. He or she can donate to individuals with either type A or AB blood, however, because the type A polysaccharides are common to both and no immune response will be generated.

**20.83** The long sheets of polysaccharides in chitin are similar to cellulose in that they have β glycosidic linkages and are not digestible by humans.

**20.85** Hyaluronate is a glycosaminoglycan found in the extracellular fluid that lubricates joints and the vitreous humor of the eye. Chondroitin is a component of cartilage and tendons. Heparin is a glycosaminoglycan stored in the mast cells of the liver and other organs and prevents blood clotting.

**20.87**   Answer the questions about fructose.

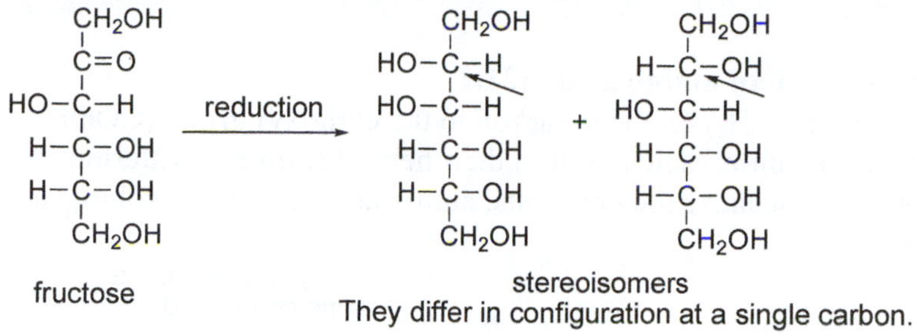

$$CH_2OH$$
$$C=O$$
$$HO-C-H$$
$$H-C-OH$$
$$H-C-OH$$
$$CH_2OH$$

reduction ⟶

$$CH_2OH$$
$$HO-C-H$$
$$HO-C-H$$
$$H-C-OH$$
$$H-C-OH$$
$$CH_2OH$$

+

$$CH_2OH$$
$$H-C-OH$$
$$HO-C-H$$
$$H-C-OH$$
$$H-C-OH$$
$$CH_2OH$$

fructose

stereoisomers
They differ in configuration at a single carbon.

# Chapter 21 Amino Acids, Proteins, and Enzymes

## Chapter Review

**[1] What are the main structural features of an amino acid? (21.2)**
- Amino acids contain an amino group ($NH_2$) on the $\alpha$ carbon to the carboxyl group (COOH). Because they contain both an acid and a base, amino acids exist in their neutral form as zwitterions having the general structure $^+H_3NCH(R)COO^-$. Because they are salts, amino acids are water soluble and have high melting points.

- All amino acids except glycine (R = H) have a chirality center on the $\alpha$ carbon. The L-amino acids are naturally occurring.
- Amino acids are subclassified as neutral, acidic, or basic by the functional groups present in the R group, as shown in Table 21.2.

**[2] Describe the acid–base properties of amino acids. (21.3)**
- Neutral, uncharged amino acids exist as zwitterions containing an ammonium cation ($-NH_3^+$) and a carboxylate anion ($-COO^-$).
- When strong acid is added, the carboxylate anion gains a proton and the amino acid has a net +1 charge. When strong base is added, the ammonium cation loses a proton and the amino acid has a net −1 charge.

**[3] What are the main structural features of peptides? (21.4)**
- Peptides contain amino acids, called amino acid residues, joined together by amide (peptide) bonds. The amino acid that contains the free $-NH_3^+$ group on the $\alpha$ carbon is called the N-terminal amino acid, and the amino acid that contains the free $-COO^-$ group on the $\alpha$ carbon is called the C-terminal amino acid.
- Peptides are written from left to right, from the N-terminal to C-terminal end, using the one- or three-letter abbreviations for the amino acids listed in Table 21.2.

$$H_3\overset{+}{N}-\underset{\underset{\text{CH}_3}{|}}{\text{CH}}-\overset{\overset{\text{O}}{\|}}{\text{C}}-\underset{\underset{\text{H}}{|}}{\text{N}}-\underset{\underset{\text{CH}_2\text{OH}}{|}}{\text{CH}}-\overset{\overset{\text{O}}{\|}}{\text{C}}-\text{O}^-$$

N-terminal        C-terminal
amino acid      amino acid

alanylserine
Ala–Ser

**[4] Give examples of simple biologically active peptides. (21.5)**

- Enkephalins are pentapeptides that act as sedatives and pain killers by binding to pain receptors.
- Oxytocin is a nonapeptide hormone that stimulates the contraction of uterine muscles and initiates the flow of milk in nursing mothers.
- Vasopressin is a nonapeptide hormone that serves as an antidiuretic; that is, vasopressin causes the kidneys to retain water.

**[5] What are the general characteristics of the primary, secondary, tertiary, and quaternary structures of proteins? (21.6)**

- The primary structure of a protein is the particular sequence of amino acids joined together by amide bonds.
- The two most common types of secondary structure are the α-helix and the β-pleated sheet. Both structures are stabilized by hydrogen bonds between the N–H and C=O groups.
- The tertiary structure is the three-dimensional shape adopted by the entire peptide chain. London dispersion forces stabilize hydrophobic interactions between nonpolar amino acids. Hydrogen bonding and ionic interactions occur between polar or charged amino acid residues. Disulfide bonds are covalent sulfur–sulfur bonds that occur between cysteine residues in different parts of the peptide chain.
- When a protein contains more than one polypeptide chain, the quaternary structure describes the shape of the protein complex formed by two or more chains.

**[6] What are the basic features of fibrous proteins like α-keratin and collagen? (21.7)**

- Fibrous proteins are composed of long linear polypeptide chains that serve structural roles and are water insoluble.
- α-Keratin in hair is a fibrous protein composed almost exclusively of α-helix units that wind together to form a superhelix. Disulfide bonds between chains make the resulting bundles of protein chains strong.
- Collagen, found in connective tissue, is composed of a superhelix formed from three elongated left-handed helices.

**[7] What are the basic features of globular proteins like hemoglobin and myoglobin? (21.7)**

- Globular proteins have compact shapes and are folded to place polar amino acids on the outside to make them water soluble. Hemoglobin and myoglobin are both conjugated proteins composed of a protein unit and a heme molecule. The $Fe^{2+}$ ion of the heme binds oxygen. While myoglobin has a single polypeptide chain, hemoglobin contains four peptide chains that form a single protein molecule.

heme

## [8] What products are formed when a protein is hydrolyzed? (21.8)

- Hydrolysis breaks up the primary structure of a protein to form the amino acids that compose it. All the amide bonds are broken by the addition of water, forming a carboxylate anion ($-COO^-$) in one amino acid and an ammonium cation ($-NH_3^+$) in the other.

Break the peptide bonds.

## [9] What is denaturation? (21.8)

- Denaturation is a process that alters the shape of a protein by disrupting the secondary, tertiary, or quaternary structure. High temperature, acid, base, and agitation can denature a protein. Compact water-soluble proteins uncoil and become less water soluble.

## [10] What are the main structural features of enzymes? (21.9)

- Enzymes are biological catalysts that greatly increase the rate of biological reactions and are highly specific for a substrate or a type of substrate. An enzyme binds a substrate at its active site, forming an enzyme–substrate complex by either the lock-and-key model or the induced-fit model.
- Enzyme inhibitors cause an enzyme to lose activity. Irreversible inhibition occurs when an inhibitor covalently binds the enzyme and permanently destroys its activity. Competitive reversible inhibition occurs when the inhibitor is structurally similar to the substrate and competes with it for occupation of the active site. Noncompetitive reversible inhibition occurs when an inhibitor binds to a location other than the active site, altering the shape of the active site.

## [11] How are enzymes used in medicine? (21.10)

- Measuring blood enzyme levels is used to diagnose heart attacks and diseases that cause higher-than-normal concentrations of certain enzymes to enter the blood.

- Drugs that inhibit the action of an enzyme can be used to kill bacteria. ACE inhibitors are used to treat high blood pressure. HIV protease inhibitors are used to treat HIV by binding to an enzyme needed by the virus to replicate itself.

## Problem Solving

## [1] Amino Acids (21.2)

**Example 21.1** Draw the Fischer projection for each amino acid.

a. L-serine    b. D-methionine

**Analysis**
To draw an amino acid in a Fischer projection, place the –COO⁻ group at the top and the R group at the bottom. The L isomer has the $-NH_3^+$ on the left side and the D isomer has the $-NH_3^+$ on the right side.

**Solution**

a. For serine, R = $CH_2OH$

$$COO^-$$
$$H_3\overset{+}{N} —\!\!\!|—\!\!\! H$$
$$CH_2OH$$

$\overset{+}{N}H_3$ on left
L isomer

b. For methionine, R = $CH_2CH_2SCH_3$

$$COO^-$$
$$H —\!\!\!|—\!\!\! \overset{+}{N}H_3$$
$$CH_2CH_2SCH_3$$

$\overset{+}{N}H_3$ on right
D isomer

## [2] Acid–Base Behavior of Amino Acids (21.3)

**Example 21.2** Draw the structure of the amino acid leucine at each pH: (a) 6; (b) 2; (c) 11.

**Analysis**
A neutral amino acid exists in its zwitterionic form (no net charge) at its isoelectric point, which is pH ≈ 6. The zwitterionic forms of neutral amino acids appear in Table 21.2. At low pH ($\leq 2$), the carboxylate anion is protonated and the amino acid has a net positive (+1) charge. At high pH ($\geq 10$), the ammonium cation loses a proton and the amino acid has a net negative (–1) charge.

**Solution**

a. At pH = 6, the neutral, zwitterionic form of leucine predominates.

$$H$$
$$H_3\overset{+}{N} — \overset{|}{C} — COO^-$$
$$CH_2CH(CH_3)_2$$
neutral
pH = 6

b. At pH = 2, leucine has a net +1 charge.

$$H$$
$$H_3\overset{+}{N} — \overset{|}{C} — COOH$$
$$CH_2CH(CH_3)_2$$
+1 charge
pH = 2

c. At pH = 11, leucine has a net –1 charge.

$$H$$
$$H_2N — \overset{|}{C} — COO^-$$
$$CH_2CH(CH_3)_2$$
–1 charge
pH = 11

## [3] Peptides (21.4)

**Example 21.3** Identify the individual amino acids used to form the following tripeptide. What is the name of the tripeptide?

$$H_3\overset{+}{N}-\underset{\underset{CH_2OH}{|}}{CH}-\overset{\overset{O}{||}}{C}-\underset{\underset{H}{|}}{N}-\underset{\underset{H}{|}}{CH}-\overset{\overset{O}{||}}{C}-\underset{\underset{H}{|}}{N}-\underset{\underset{\underset{C=O}{|}}{\underset{|}{CH_2}}}{CH}-\overset{\overset{O}{||}}{C}-O^-$$

with $NH_2$ at the bottom.

### Analysis

- Locate the amide bonds in the peptide. To draw the structures of the individual amino acids, break the amide bonds by adding water. Add $O^-$ to the carbonyl carbon to form a carboxylate anion ($-COO^-$) and 2 H's to the N atom to form an ammonium cation ($-NH_3^+$).
- Identify the amino acids by comparing with Table 21.2.
- To name the peptide: [1] Name the C-terminal amino acid. [2] Name the other amino acids as substituents by changing the -*ine* (or -*ic acid*) ending to the suffix -*yl*. Place the names of the substituent amino acids in order from left to right.

### Solution

To determine the amino acids that form the peptide, work backwards. Break the amide bonds (in bold) that join the amino acids together. This forms serine, glycine, and asparagine.

serine          glycine          asparagine

The tripeptide is named as a derivative of the C-terminal amino acid, asparagine, with serine and glycine as substituents; thus, the tripeptide is named **serylglycylasparagine.**

## [4] Protein Hydrolysis and Denaturation (21.8)

**Example 21.4** Draw the structures of the amino acids formed by the hydrolysis of the peptide below.

### Analysis

Locate each amide bond in the protein or peptide backbone. To draw the hydrolysis products, break each amide bond by adding the elements of $H_2O$ to form a carboxylate anion (–COO⁻) in one amino acid and an ammonium cation (–NH₃⁺) in the other.

### Solution

[1] Locate the amide bonds in the peptide backbone.
[2] Break each bond by adding $H_2O$.

---

**[1] Fill in the blank with one of the terms listed below.**

| | | |
|---|---|---|
| Cofactor (21.9) | Inhibitor (21.9) | Primary structure (21.6) |
| C-Terminal amino acid (21.4) | Noncompetitive inhibitor (21.9) | Secondary structure (21.6) |
| Dipeptide (21.4) | N-Terminal amino acid (21.4) | Tertiary structure (21.6) |
| Enzymes (21.9) | Peptide bonds (21.4) | Tripeptide (21.4) |
| α-Helix (21.6) | β-Pleated sheet (21.6) | Zwitterion (21.2) |

1.   The _____ of a protein is the particular sequence of amino acids that is joined together by peptide bonds.
2.   The amide bonds in peptides and proteins are called _____.
3.   A _____ contains both a positive and a negative charge.
4.   A _____ is a metal ion or a nonprotein organic molecule needed for an enzyme-catalyzed reaction to occur.
5.   _____ are proteins that serve as biological catalysts for reactions in all living organisms.
6.   The amino acid in a peptide with the free $-NH_3^+$ group on the $\alpha$ carbon is called the _____.
7.   The three-dimensional shape adopted by the entire peptide chain is called its _____.
8.   The _____ forms when two or more peptide chains, called strands, line up side-by-side.
9.   A _____ has two amino acids joined together by one amide bond.
10.  An _____ is a molecule that causes an enzyme to lose activity.
11.  The _____ forms when a peptide chain twists into a right-handed or clockwise spiral.
12.  The amino acid in a peptide with the free $-COO^-$ group on the $\alpha$ carbon is called the _____.
13.  A _____ has three amino acids joined together by two amide bonds.
14.  The three-dimensional arrangement of localized regions of a protein is called its _____.
15.  A _____ binds to an enzyme but does not bind at the active site.

**[2] Choose the term for each label in the figures below.**

a. enzyme–substrate complex     b. substrate     c. active site     d. products

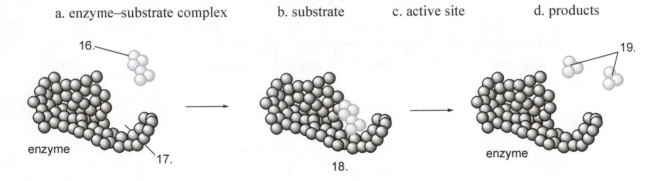

16. enzyme    17.

18.

enzyme    19.

**[3] Pick the appropriate abbreviation for each amino acid.**

a. Gln     b. Val     c. Phe     d. Pro

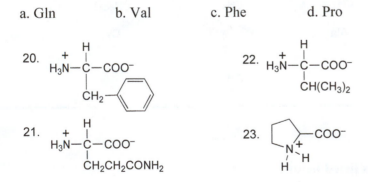

20.

21.

22.

23.

**[4] Label each amino acid as an L or D isomer.**

a. L isomer

b. D isomer

24.
$$
\begin{array}{c}
\text{COO}^- \\
H_3\overset{+}{N}\!-\!\!\vert\!\!-\!H \\
\text{CH}_2\text{CH}_2\text{CONH}_2
\end{array}
$$

25.
$$
\begin{array}{c}
\text{COO}^- \\
H\!-\!\!\vert\!\!-\!\overset{+}{N}H_3 \\
\text{CH(CH}_3)_2
\end{array}
$$

## Answers to Self-Test

1. primary structure
2. peptide bonds
3. zwitterion
4. cofactor
5. Enzymes

6. N-terminal amino acid
7. tertiary structure
8. β-pleated sheet
9. dipeptide
10. inhibitor

11. α-helix
12. C-terminal amino acid
13. tripeptide
14. secondary structure
15. noncompetitive inhibitor

16. b
17. c
18. a
19. d
20. c

21. a
22. b
23. d
24. a
25. b

## Solutions to In-Chapter Problems

**21.1** Identify the other functional groups in each amino acid.

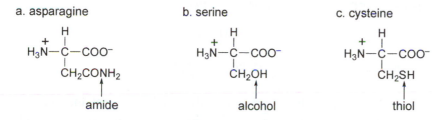

a. asparagine — amide

b. serine — alcohol

c. cysteine — thiol

**21.2** Compare the OH groups in each amino acid.

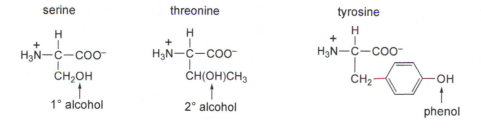

serine — 1° alcohol

threonine — 2° alcohol

tyrosine — phenol

**21.3** Draw the enantiomers as in Example 21.1.

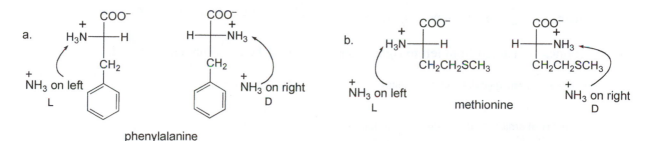

a. $H_3\overset{+}{N}$ on left — L; $\overset{+}{N}H_3$ on right — D; phenylalanine

b. $\overset{+}{N}H_3$ on left — L; $\overset{+}{N}H_3$ on right — D; methionine

**21.4**  Refer to Table 21.2 to determine which amino acids are naturally occurring. Naturally occurring amino acids are L-amino acids.

a.
$$\begin{array}{c} COO^- \\ | \\ H_3\overset{+}{N}-\!\!\!\overset{|}{\phantom{C}}\!\!\!-H \\ | \\ CH_2OH \end{array}$$

L-serine, naturally occurring

b.
$$\begin{array}{c} COO^- \\ | \\ H-\!\!\!\overset{|}{\phantom{C}}\!\!\!-\overset{+}{N}H_3 \\ | \\ CH_2CH_2COO^- \end{array}$$

D-glutamic acid

c.
$$\begin{array}{c} COO^- \\ | \\ H_3\overset{+}{N}-\!\!\!\overset{|}{\phantom{C}}\!\!\!-H \\ | \\ CH_2COO^- \end{array}$$

L-aspartic acid, naturally occurring

**21.5**  Draw the structure of valine at each pH as in Example 21.2.

a.
$$\begin{array}{c} H \\ | \\ H_3\overset{+}{N}-C-COO^- \\ | \\ CH(CH_3)_2 \end{array}$$

pH = 6
predominant at p*I*

b.
$$\begin{array}{c} H \\ | \\ H_3\overset{+}{N}-C-COOH \\ | \\ CH(CH_3)_2 \end{array}$$

pH = 2

c.
$$\begin{array}{c} H \\ | \\ H_2N-C-COO^- \\ | \\ CH(CH_3)_2 \end{array}$$

pH = 11

**21.6**  Draw the different forms of phenylalanine.

$$\begin{array}{c} H \\ | \\ H_3\overset{+}{N}-C-COOH \\ | \\ CH_2- \end{array}$$ (benzene ring)

predominant at pH 1

$$\begin{array}{c} H \\ | \\ H_3\overset{+}{N}-C-COO^- \\ | \\ CH_2- \end{array}$$ (benzene ring)

neutral

$$\begin{array}{c} H \\ | \\ H_2N-C-COO^- \\ | \\ CH_2- \end{array}$$ (benzene ring)

predominant at pH 11

**21.7**  The amino acid with the free $-NH_3^+$ group on the α carbon is called the N-terminal amino acid. The amino acid with the free $-COO^-$ group on the α carbon is called the C-terminal amino acid.

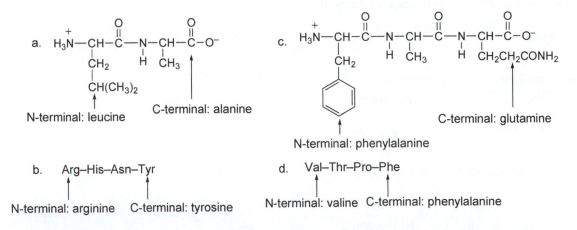

a. N-terminal: leucine   C-terminal: alanine

c. N-terminal: phenylalanine   C-terminal: glutamine

b.   Arg–His–Asn–Tyr
N-terminal: arginine   C-terminal: tyrosine

d.   Val–Thr–Pro–Phe
N-terminal: valine  C-terminal: phenylalanine

**21.8**  Peptides are named from left to right as substituents of the C-terminal amino acid.

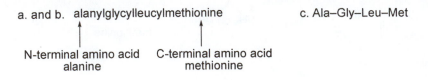

a. and b.  alanylglycylleucylmethionine

N-terminal amino acid    C-terminal amino acid
alanine                  methionine

c. Ala–Gly–Leu–Met

**21.9** Draw the structure of each dipeptide by joining the –COO⁻ and –NH₃⁺ groups of the two different amino acids.

a. Gly–Phe

$$H_3\overset{+}{N}-\underset{\underset{H}{|}}{\overset{\overset{H}{|}}{C}}-\overset{\overset{O}{||}}{C}-\underset{\underset{CH_2-\bigcirc}{|}}{N}-\overset{\overset{H}{|}}{C}-COO^-$$

b. Gln–Ile

$$H_3\overset{+}{N}-\underset{\underset{CH_2CONH_2}{|}}{\overset{\overset{H}{|}}{\underset{CH_2}{|}}{C}}-\overset{\overset{O}{||}}{C}-\underset{\underset{H}{|}}{N}-\underset{CH(CH_3)CH_2CH_3}{\overset{\overset{H}{|}}{C}}-COO^-$$

c. Leu–Cys

$$H_3\overset{+}{N}-\underset{\underset{CH(CH_3)_2}{|}}{\overset{\overset{H}{|}}{\underset{CH_2}{|}}{C}}-\overset{\overset{O}{||}}{C}-\underset{\underset{H}{|}}{N}-\underset{CH_2SH}{\overset{\overset{H}{|}}{C}}-COO^-$$

**21.10** Draw the two dipeptides formed from leucine and asparagine.

a.

$$H_3\overset{+}{N}-\underset{\underset{CH_2CH(CH_3)_2}{|}}{\overset{\overset{H}{|}}{C}}-\overset{\overset{O}{||}}{C}-\underset{\underset{H}{|}}{N}-\underset{CH_2CONH_2}{\overset{\overset{H}{|}}{C}}-COO^-$$

$$H_3\overset{+}{N}-\underset{\underset{CH_2CONH_2}{|}}{\overset{\overset{H}{|}}{C}}-\overset{\overset{O}{||}}{C}-\underset{\underset{H}{|}}{N}-\underset{CH_2CH(CH_3)_2}{\overset{\overset{H}{|}}{C}}-COO^-$$

b. N-terminal: leucine
   C-terminal: asparagine

   N-terminal: asparagine
   C-terminal: leucine

c. Leu–Asn

   Asn–Leu

**21.11** Identify the amino acids in the dipeptides as in Example 21.3. The amide bonds joining the two amino acids are drawn in **bold**.

a.

$$H_3\overset{+}{N}-\underset{\underset{CH_3}{|}}{CH}-\overset{\overset{O}{||}}{\mathbf{C}}\mathbf{-}\underset{\underset{H}{|}}{\mathbf{N}}-\underset{CH(CH_3)CH_2CH_3}{CH}-\overset{\overset{O}{||}}{C}-O^-$$

$$H_3\overset{+}{N}-\underset{\underset{CH_3}{|}}{CH}-\overset{\overset{O}{||}}{C}-O^- \qquad H_3\overset{+}{N}-\underset{\underset{CH(CH_3)CH_2CH_3}{|}}{CH}-\overset{\overset{O}{||}}{C}-O^-$$

alanine and isoleucine
Ala–Ile

b.

$$H_3\overset{+}{N}-\underset{\underset{CH_2}{|}}{CH}-\overset{\overset{O}{||}}{\mathbf{C}}\mathbf{-}\underset{\underset{H}{|}}{\mathbf{N}}-\underset{CH(CH_3)_2}{CH}-\overset{\overset{O}{||}}{C}-O^-$$

(benzene ring with OH)

$$H_3\overset{+}{N}-\underset{\underset{CH_2}{|}}{CH}-\overset{\overset{O}{||}}{C}-O^- \qquad H_3\overset{+}{N}-\underset{\underset{CH(CH_3)_2}{|}}{CH}-\overset{\overset{O}{||}}{C}-O^-$$

(benzene ring with OH)

tyrosine and valine
Tyr–Val

**21.12** Name the dipeptide and indicate the peptide bond as in Example 21.3.

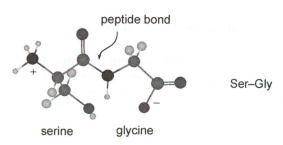

peptide bond

Ser–Gly

serine     glycine

**21.13** Answer each question about met-enkephalin.

a.

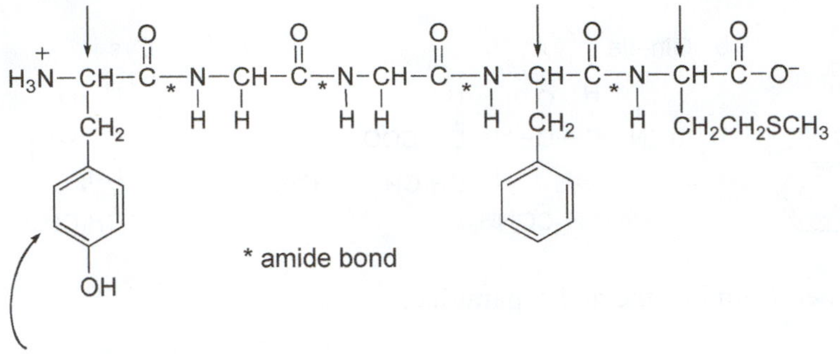

\* amide bond

b. N-terminal amino acid: tyrosine
c. three chirality centers (labeled with arrows)

**21.14** When cysteine is oxidized, a disulfide bond forms.

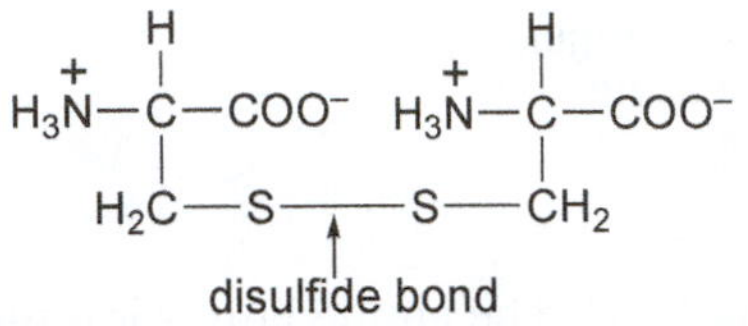

disulfide bond

**21.15** Yes, two different proteins can be composed of the same amino acids since the amino acids can be ordered differently. For example: Ala–Gly–Ile–Trp and Gly–Trp–Ala–Ile.

**21.16** Draw each pair of amino acids.

a.

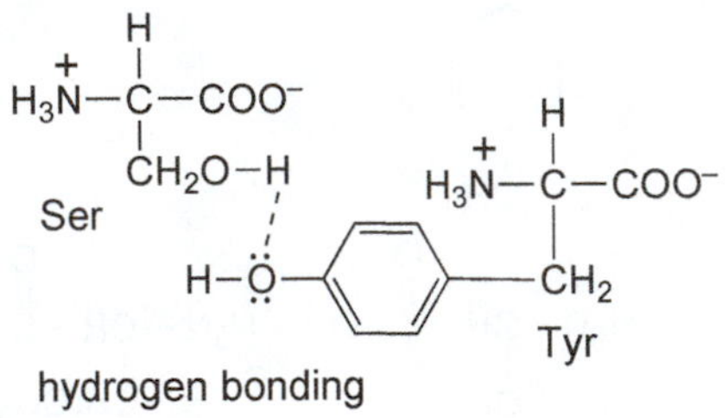

Ser

Tyr

hydrogen bonding

b.

Val

Leu

London dispersion forces between
the nonpolar side chains

c.

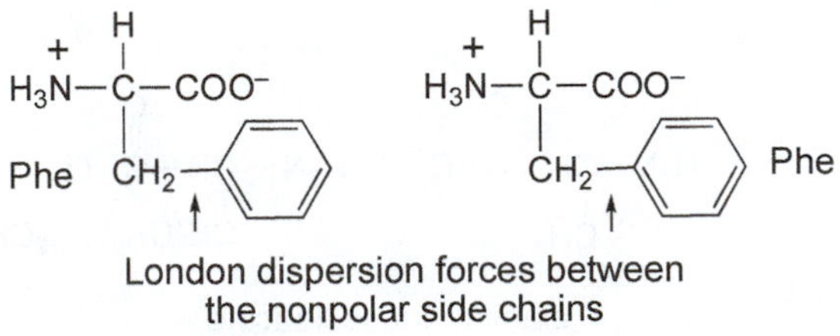

Phe

Phe

London dispersion forces between
the nonpolar side chains

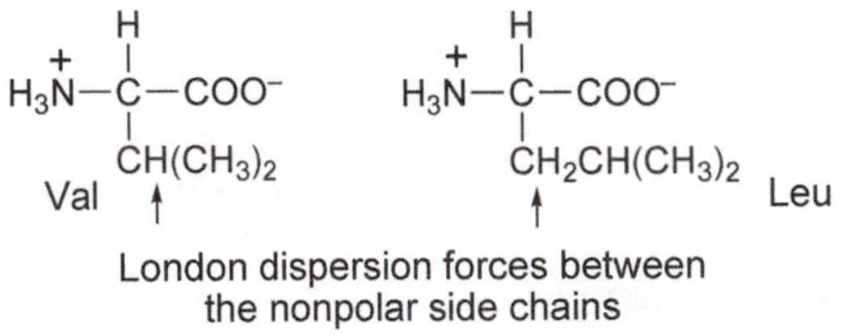

**21.17** Glycine has no large side chain and this allows for the β sheets to stack well together.

**21.18** Hemoglobin is more water soluble than keratin because it is a globular protein that folds with its polar groups on its exterior, thus allowing for hydrogen bonding. Keratin is water insoluble because it has more nonpolar groups on its exterior, and this does *not* allow for hydrogen bonding.

**21.19** Draw the products formed by the hydrolysis of each tripeptide as in Example 21.4. The products of hydrolysis are the individual amino acids.

a. Ala–Leu–Gly

| Ala | Leu | Gly |

$$H_3\overset{+}{N}-\overset{\overset{\displaystyle H}{|}}{\underset{\underset{\displaystyle CH_3}{|}}{C}}-COO^-$$

Ala

$$H_3\overset{+}{N}-\overset{\overset{\displaystyle H}{|}}{\underset{\underset{\displaystyle CH_2CH(CH_3)_2}{|}}{C}}-COO^-$$

Leu

$$H_3\overset{+}{N}-\overset{\overset{\displaystyle H}{|}}{\underset{\underset{\displaystyle H}{|}}{C}}-COO^-$$

Gly

b. Ser–Thr–Phe

$$H_3\overset{+}{N}-\overset{\overset{\displaystyle H}{|}}{\underset{\underset{\displaystyle CH_2OH}{|}}{C}}-COO^-$$

Ser

$$H_3\overset{+}{N}-\overset{\overset{\displaystyle H}{|}}{\underset{\underset{\displaystyle CH(OH)CH_3}{|}}{C}}-COO^-$$

Thr

$$H_3\overset{+}{N}-\overset{\overset{\displaystyle H}{|}}{\underset{\underset{\displaystyle CH_2}{|}}{C}}-COO^-$$

Phe

c. Leu–Tyr–Asn

$$H_3\overset{+}{N}-\overset{\overset{\displaystyle H}{|}}{\underset{\underset{\displaystyle CH_2CH(CH_3)_2}{|}}{C}}-COO^-$$

Leu

$$H_3\overset{+}{N}-\overset{\overset{\displaystyle H}{|}}{\underset{\underset{\displaystyle CH_2}{|}}{C}}-COO^-$$

Tyr

$$H_3\overset{+}{N}-\overset{\overset{\displaystyle H}{|}}{\underset{\underset{\displaystyle CH_2CONH_2}{|}}{C}}-COO^-$$

Asn

**21.20** Heating collagen disrupts the hydrogen bonding and other intermolecular forces. The superhelix unwinds, making it less ordered, turning it into the jelly-like substance gelatin.

**21.21** The names of most enzymes end in the suffix *-ase*. Sucrase (b), lactase (d), and phosphofructokinase (e) are all enzymes.

**21.22** Answer each question about fumarase.

a.

fumarate →(fumarase)→ malate    OH H ← H₂O added

$$^-OOC-\overset{\overset{\displaystyle OH}{|}}{\underset{\underset{\displaystyle H}{|}}{C}}-\overset{\overset{\displaystyle H}{|}}{\underset{\underset{\displaystyle H}{|}}{C}}-COO^-$$

b. The enzyme is a hydratase because it adds the elements of water to a double bond; that is, the enzyme is needed to *hydrate* the double bond.

**21.23** A competitive inhibitor has a shape and structure similar to the substrate, so it competes with the substrate for binding to the active site.

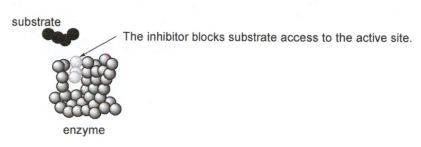

substrate

The inhibitor blocks substrate access to the active site.

enzyme

**21.24** Sarin is an irreversible inhibitor since it forms a covalent bond at the enzyme's active site.

**21.25** Fibrin and thrombin circulate as inactive zymogens (fibrinogen and prothrombin) so that the blood does not clot unnecessarily. They are activated as required at a bleeding point to form a clot.

**21.26** Both captopril and enalapril are derived from the amino acid proline and contain an amide and a carboxylate anion. Captopril also contains a thiol (SH), while enalapril contains three additional functional groups—an amine, a carboxylate anion, and a benzene ring.

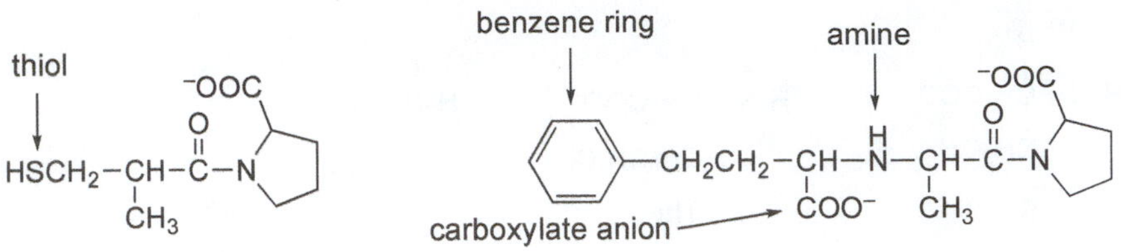

Generic name: captopril
Trade name: Capoten

Generic name: enalapril
Trade name: Vasotec

## Solutions to Odd-Numbered End-of-Chapter Problems

**21.27** The L designation refers to the configuration at the chirality center. With a vertical carbon chain in the Fischer projection, the L isomer has the $-NH_3^+$ drawn on the left side. The α-amino acid designation indicates that the amino group is bonded to the carbon adjacent to the carbonyl group.

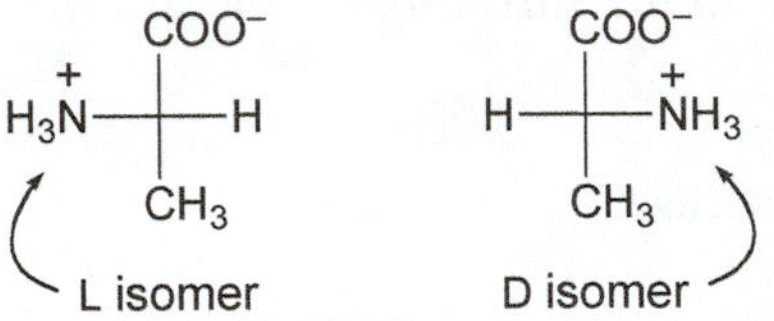

L isomer    D isomer

**21.29** Alanine is an ionic salt with extremely strong electrostatic forces, leading to its high melting point, and making it a solid at room temperature. Pyruvic acid is a neutral polar molecule with weaker intermolecular forces, so it is a liquid at room temperature.

**21.31** Draw an amino acid to fit each requirement.

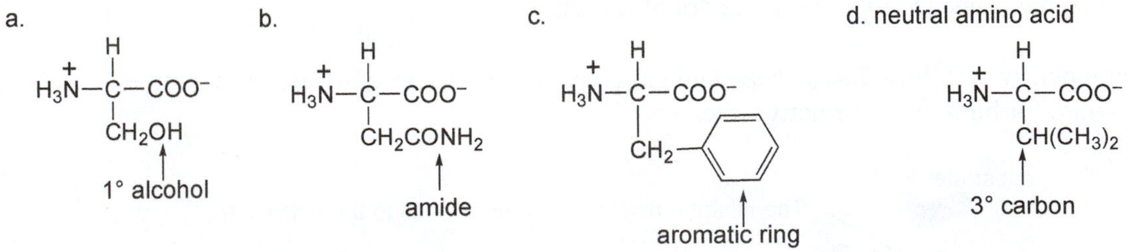

a.

1° alcohol

b.

amide

c.

aromatic ring

d. neutral amino acid

3° carbon

**21.33** Both isoleucine and threonine contain two chirality centers.

isoleucine    chirality center    threonine

$$H_3\overset{+}{N}-\underset{\underset{\displaystyle CH(CH_3)CH_2CH_3}{|}}{\overset{\overset{\displaystyle H}{|}}{C}}-COO^-$$

$$H_3\overset{+}{N}-\underset{\underset{\displaystyle CH(OH)CH_3}{|}}{\overset{\overset{\displaystyle H}{|}}{C}}-COO^-$$

chirality center    chirality center

**21.35** Answer each question about the amino acids.

| [1] L Enantiomer | [2] Classification | [3] Three-letter Symbol | [4] One-letter Symbol |
|---|---|---|---|
| a. leucine | neutral | Leu | L |
| b. tryptophan | neutral | Trp | W |
| c. lysine | basic | Lys | K |
| d. aspartic acid | acidic | Asp | D |

**21.37** Draw and label the enantiomers of each amino acid.

a. methionine      b. asparagine

L      D      L      D

**21.39** Answer each question about the amino acids.

| [1] Amino Acid | [2] Three-letter Symbol | [3] One-letter Symbol | [4] Classification |
|---|---|---|---|
| a. glutamine | Gln | Q | neutral |
| b. tyrosine | Tyr | Y | neutral |

a. $COO^-$, $H_3\overset{+}{N}$, $H$, $CH_2CH_2CONH_2$ — glutamine

b. $COO^-$, $H_3\overset{+}{N}$, $H$, $CH_2$—〈 〉—OH — tyrosine

**21.41** Draw the structure of leucine at each pH as in Example 21.2.

a. $H_3\overset{+}{N}-\overset{\overset{H}{|}}{C}-COO^-$
$CH_2CH(CH_3)_2$
pH = 6
predominant form at p$I$

b. $H_2N-\overset{\overset{H}{|}}{C}-COO^-$
$CH_2CH(CH_3)_2$
pH = 10

c. $H_3\overset{+}{N}-\overset{\overset{H}{|}}{C}-COOH$
$CH_2CH(CH_3)_2$
pH = 2

**21.43** Draw the structure of tyrosine at each pH as in Example 21.2.

$H_3\overset{+}{N}-\overset{\overset{H}{|}}{C}-COO^-$
$CH_2$—〈 〉—OH
neutral
predominant form at p$I$

$H_3\overset{+}{N}-\overset{\overset{H}{|}}{C}-COOH$
$CH_2$—〈 〉—OH
positively charged
pH = 1

$H_2N-\overset{\overset{H}{|}}{C}-COO^-$
$CH_2$—〈 〉—OH
negatively charged
pH = 11

**21.45** Locate the peptide bond and name the dipeptide as in Example 21.3.

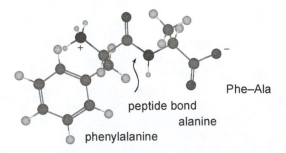

Phe–Ala

peptide bond
alanine

phenylalanine

**21.47** The amino acid with the free $-NH_3^+$ group on the α carbon is called the N-terminal amino acid. The amino acid with the free $-COO^-$ group on the α carbon is called the C-terminal amino acid.

N-terminal amino acid        N-terminal amino acid

a, b, c.    $H_3\overset{+}{N}-\overset{\overset{H}{|}}{C}-\overset{\overset{O}{||}}{C}-\overset{\overset{H}{|}}{N}-\overset{\overset{H}{|}}{C}-COO^-$

CH(CH$_3$)$_2$    CH$_2$

Val–Phe

$H_3\overset{+}{N}-\overset{\overset{H}{|}}{C}-\overset{\overset{O}{||}}{C}-\overset{\overset{H}{|}}{N}-\overset{\overset{H}{|}}{C}-COO^-$

CH$_2$    CH(CH$_3$)$_2$

Phe–Val

C-terminal amino acid        C-terminal amino acid

**21.49**    Answer each question about the tripeptides.

a. leucylvalyltryptophan

[1]    $H_3\overset{+}{N}-\overset{\overset{H}{|}}{C}-\overset{\overset{O}{||}}{\overset{*}{C}}-\overset{\overset{H}{|}}{N}-\overset{\overset{H}{|}}{C}-\overset{\overset{O}{||}}{\overset{*}{C}}-\overset{\overset{H}{|}}{N}-\overset{\overset{H}{|}}{C}-COO^-$

CH$_2$CH(CH$_3$)$_2$    CH(CH$_3$)$_2$    CH$_2$

[2]    * amide bond labeled

[3]    N-terminal: leucine
       C-terminal: tryptophan

[4]    Leu–Val–Trp

b. phenylalanylserylthreonine

[1]    $H_3\overset{+}{N}-\overset{\overset{H}{|}}{C}-\overset{\overset{O}{||}}{\overset{*}{C}}-\overset{\overset{H}{|}}{N}-\overset{\overset{H}{|}}{C}-\overset{\overset{O}{||}}{\overset{*}{C}}-\overset{\overset{H}{|}}{N}-\overset{\overset{H}{|}}{C}-COO^-$

CH$_2$    CH$_2$OH    CH(OH)CH$_3$

[2]    * amide bond labeled

[3]    N-terminal: phenylalanine
       C-terminal: threonine

[4]    Phe–Ser–Thr

**21.51**    Answer each question about the tripeptide.

a.    $H_3\overset{+}{N}-\overset{\overset{|}{|}}{CH}-\overset{\overset{O}{||}}{C}-\overset{\overset{H}{|}}{N}-\overset{\overset{|}{|}}{CH}-\overset{\overset{O}{||}}{C}-\overset{\overset{H}{|}}{N}-\overset{\overset{|}{|}}{CH}-\overset{\overset{O}{||}}{C}-O^-$

CH(CH$_3$)$_2$    H    CH$_2$

valine            glycine        phenylalanine
N-terminal                        C-terminal
amino acid                        amino acid

Val–Gly–Phe

b.    $H_3\overset{+}{N}-\overset{\overset{|}{|}}{CH}-\overset{\overset{O}{||}}{C}-\overset{\overset{H}{|}}{N}-\overset{\overset{|}{|}}{CH}-\overset{\overset{O}{||}}{C}-\overset{\overset{H}{|}}{N}-\overset{\overset{|}{|}}{CH}-\overset{\overset{O}{||}}{C}-O^-$

CH$_2$    CH$_2$    CH$_2$CH$_2$SCH$_3$
CH(CH$_3$)$_2$

leucine        OH        methionine
N-terminal                C-terminal
amino acid    tyrosine    amino acid

Leu–Tyr–Met

**21.53**  Draw the structure of the three tripeptides.

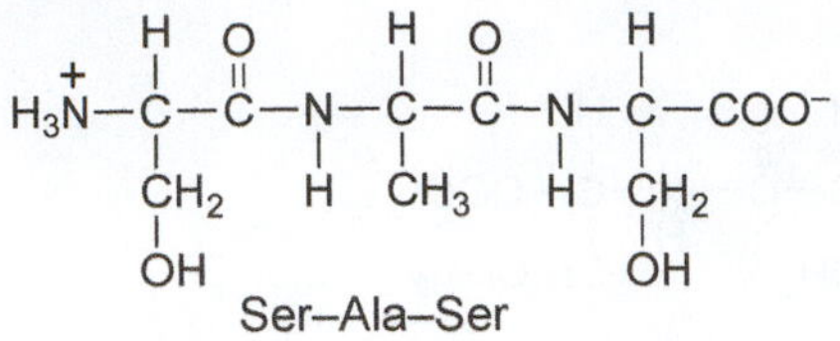

Ser–Ala–Ser

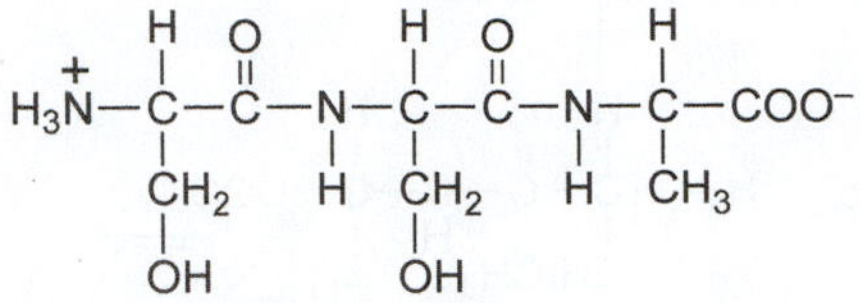

Ser–Ser–Ala

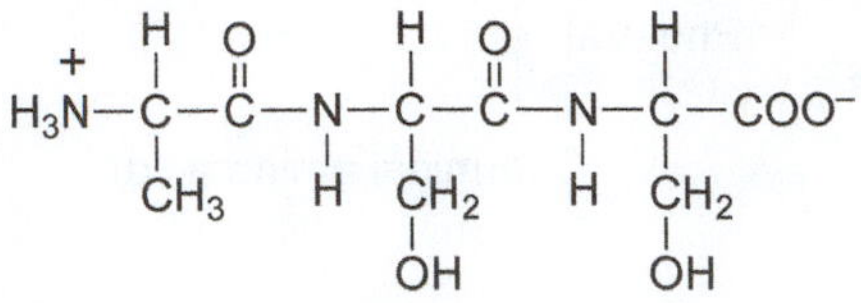

Ala–Ser–Ser

**21.55**  Draw the products formed by the hydrolysis of the tripeptide as in Example 21.4.

hydrolysis

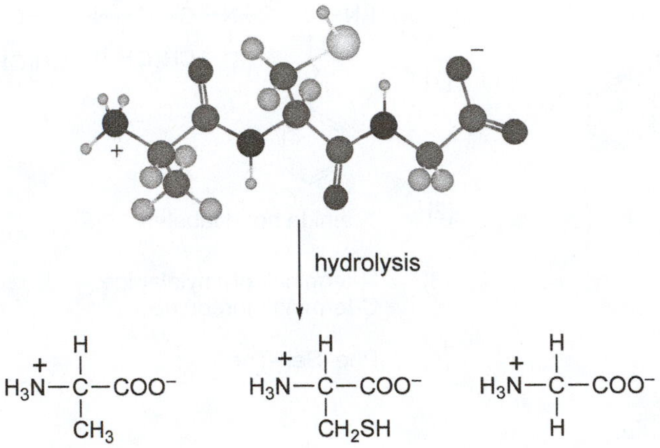

alanine        cysteine        glycine

**21.57**  Draw the products formed by the hydrolysis of each tripeptide as in Example 21.4.

a.

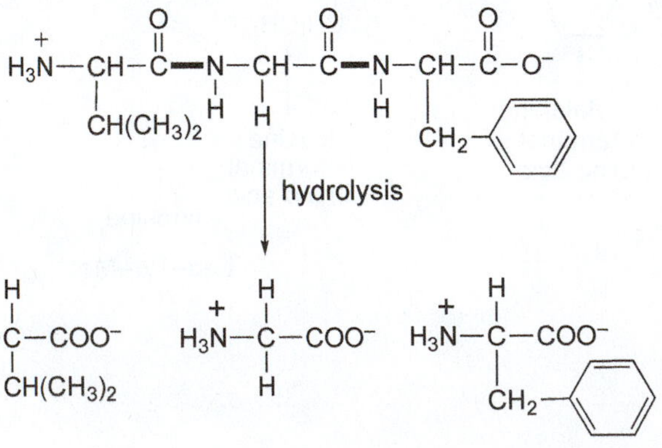

hydrolysis

b.

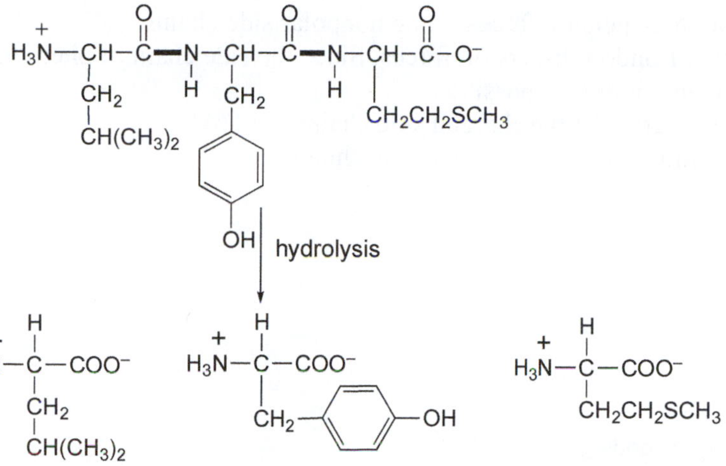

**21.59**  Draw the products of the hydrolysis of bradykinin.

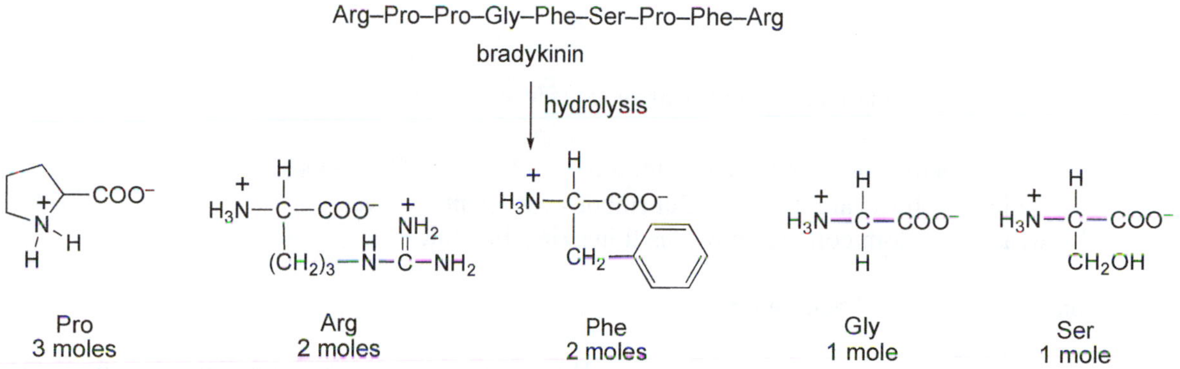

Arg–Pro–Pro–Gly–Phe–Ser–Pro–Phe–Arg

bradykinin

hydrolysis

| Pro | Arg | Phe | Gly | Ser |
|-----|-----|-----|-----|-----|
| 3 moles | 2 moles | 2 moles | 1 mole | 1 mole |

**21.61**  Draw the products of the hydrolysis of the peptide with chymotrypsin.

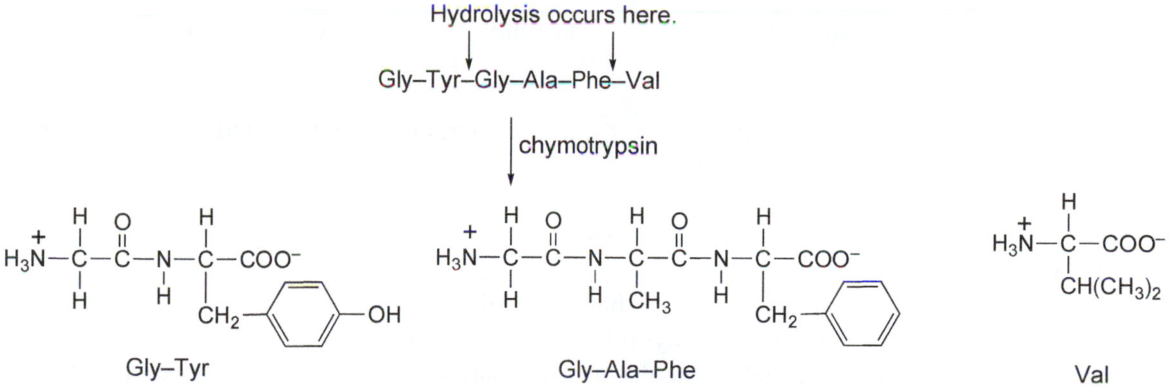

Hydrolysis occurs here.

Gly–Tyr–Gly–Ala–Phe–Val

chymotrypsin

Gly–Tyr        Gly–Ala–Phe        Val

**21.63**  The primary structure of a protein is the order of its amino acids. The secondary structure refers to the three-dimensional arrangement of regions within the protein.

**21.65**  Use Figure 21.10 to determine which types of intermolecular forces occur between each amino acid pair.

a. isoleucine and valine: London dispersion forces of the nonpolar side chains
b. threonine and phenylalanine: London dispersion forces. Since the side chain of phenylalanine has no O or N atom, no hydrogen bonding is possible.
c. Lys and Glu: electrostatic attraction of the charged side chains
d. Arg and Asp: electrostatic attraction of the charged side chains

**21.67**  Draw the structure.

**21.69**  Use Figure 21.6 to label the regions of secondary structure.

The α-helix is the corkscrew-shaped region.
The β-pleated sheets are drawn as flat ribbons with gentle curves.
The areas of random coil are shown as thin string-like lines.

**21.71**  Compare keratin and hemoglobin.

|  | a. Secondary Structure | b. H$_2$O Solubility | c. Function | d. Location |
|---|---|---|---|---|
| Hemoglobin | globular with much α-helix | soluble | carries oxygen to tissues | blood |
| Keratin | α-helix | insoluble | gives strength to tissues | nails, hair |

**21.73**  When a protein is heated and denatured, the primary structure is unaffected. The 2°, 3°, and 4° structures may be altered by unfolding.

**21.75**  a. Insulin is a hormone that controls glucose levels.
b. Myoglobin stores oxygen in muscle.
c. α-Keratin forms hard tissues such as hair and nails.
d. Chymotrypsin is a protease that hydrolyzes peptide bonds.
e. Oxytocin is a hormone that stimulates uterine contractions and induces the release of breast milk.

**21.77**  Reversible enzyme inhibition occurs when an enzyme's activity is restored when the inhibitor is released. Irreversible inhibition permanently renders the enzyme incapable of further activity.

**21.79**  Captopril inhibits the angiotensin-converting enzyme, blocking the conversion of angiotensinogen to angiotensin. This reduces the concentration of angiotensin, which in turn lowers blood pressure.

**21.81**   A noncompetitive inhibitor binds to the enzyme but does not bind at the active site.

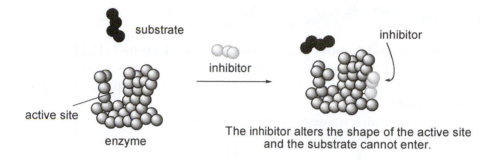

The inhibitor alters the shape of the active site and the substrate cannot enter.

**21.83**   The α-keratin in nails has more cysteine residues to form disulfide bonds.  The larger the number of disulfide bonds, the harder the substance.  Therefore, fingernails are harder than skin.

**21.85**   Humans cannot synthesize the amino acids methionine and lysine, and they therefore must be obtained in the diet. Diets that include animal products readily supply all of the needed amino acids, but no one plant source has sufficient amounts of all the essential amino acids.  Grains—wheat, rice, and corn—are low in lysine, and legumes—beans, peas, and peanuts—are low in methionine, but a combination of these foods provides all of the needed amino acids.

**21.87**   Cauterization denatures the proteins in a wound.

**21.89**   Sickle cell disease results from abnormal hemoglobin.  In sickle hemoglobin there is a substitution of a single amino acid, valine for glutamic acid, which changes the shape of the hemoglobin.

**21.91**   Penicillin inhibits the formation of the bacterial cell wall by irreversibly binding to an enzyme needed for its construction.  It does not affect humans since human cells have a cell membrane, instead of a rigid cell wall.  Sulfanilamide inhibits the production of folic acid and therefore reproduction in bacteria, but humans do not synthesize folic acid (they must ingest it instead), so it does not affect humans.

**21.93**   Both aspartic acid and glutamic acid have two carboxylic acid groups.  At low pH they have a +1 charge with both acid groups protonated, but at a high pH both acid groups are ionized, leading to a net charge of −2.

$$H_3\overset{+}{N}-\underset{\underset{CH_2COOH}{|}}{\overset{\overset{H}{|}}{C}}-COOH$$

Asp at low pH
+1 charge

$$H_2N-\underset{\underset{CH_2COO^-}{|}}{\overset{\overset{H}{|}}{C}}-COO^-$$

Asp at high pH
−2 charge

# Chapter 22 Nucleic Acids and Protein Synthesis

## Chapter Review

**[1] What are the main structural features of nucleosides and nucleotides? (22.1)**
- A nucleoside contains a monosaccharide joined to a base at the anomeric carbon.
- A nucleotide contains a monosaccharide joined to a base, and a phosphate bonded to the 5'-OH group of the monosaccharide.
- The monosaccharide is either ribose or 2-deoxyribose, and the bases are abbreviated as A, G, C, T, and U.

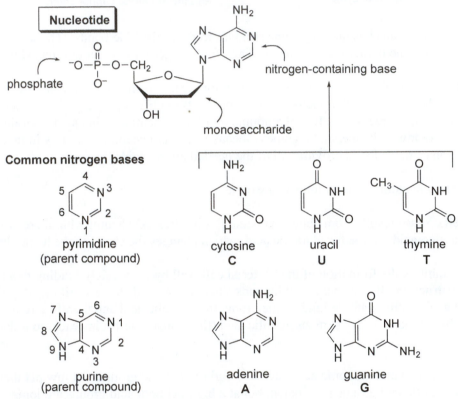

**[2] How do the nucleic acids DNA and RNA differ in structure? (22.2)**
- DNA is a polymer of deoxyribonucleotides where the sugar is 2-deoxyribose and the bases are A, G, C, and T. DNA is double stranded.
- RNA is a polymer of ribonucleotides where the sugar is ribose and the bases are A, G, C, and U. RNA is single stranded.

**[3] Describe the basic features of the DNA double helix. (22.3)**
- DNA consists of two polynucleotide strands that wind into a right-handed double helix. The sugar–phosphate backbone lies on the outside of the helix and the bases lie on the inside. The two strands run in opposite directions; that is, one strand runs from the 5' end to the 3' end and the other runs from the 3' end to the 5' end. The double helix is stabilized by hydrogen bonding between complementary base pairs; A pairs with T, and C pairs with G.

**[4] Outline the main steps in the replication of DNA. (22.4)**
- Replication of DNA is semiconservative; an original DNA molecule forms two DNA molecules, each of which has one strand from the parent DNA and one new strand.

- In replication, DNA unwinds and the enzyme DNA polymerase catalyzes replication on both strands. The identity of the bases on the template strand determines the order of the bases on the new strand, with A pairing with T and C pairing with G. Replication occurs from the 3' end to the 5' end of each strand. One strand, the leading strand, grows continuously, while the other strand, the lagging strand, is synthesized in fragments and then joined together with a DNA ligase enzyme.

**[5] List the three types of RNA molecules and describe their functions. (22.5)**
- Ribosomal RNA (rRNA), which consists of one large subunit and one small subunit, provides the site where proteins are assembled.
- Messenger RNA (mRNA) contains the sequence of nucleotides that determines the amino acid sequence in a protein. mRNA is transcribed from DNA such that each DNA gene corresponds to a different mRNA molecule.
- Transfer RNA (tRNA) contains an anticodon that identifies the amino acid that it carries on its acceptor stem and delivers that amino acid to a growing polypeptide.

**[6] What is transcription? (22.6)**
- Transcription is the synthesis of mRNA from DNA. The DNA helix unwinds and RNA polymerase catalyzes RNA synthesis from the 3' to 5' end of the template strand, forming mRNA with complementary bases.

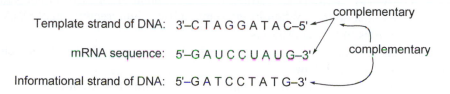

**[7] What are the main features of the genetic code? (22.7)**
- mRNAs contain sequences of three bases called codons that code for individual amino acids. There are 61 codons that correspond to the 20 amino acids, as well as three stop codons that signal the end of protein synthesis.

**[8] How are proteins synthesized by the process of translation? (22.8)**
- Translation begins with initiation, the binding of the ribosomal subunits to mRNA and the arrival of the first tRNA with an amino acid. During elongation, tRNAs bring individual amino acids to the ribosome one after another, and new peptide bonds are formed. Termination occurs when a stop codon is reached.

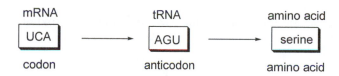

**[9] What is a mutation and how are mutations related to genetic diseases? (22.9)**

- Mutations are changes in the nucleotide sequence in a DNA molecule. A mutation that causes an inherited condition may result in a genetic disease.
  - A point mutation results in the substitution of one nucleotide for another.

- Deletion and insertion mutations result in the loss or addition of nucleotides, respectively.

**[10] What are the principal features of three techniques that use DNA in the laboratory—recombinant DNA, the polymerase chain reaction (PCR), and DNA fingerprinting? (22.10)**

- Recombinant DNA is synthetic DNA formed when a segment of DNA from one source is inserted into the DNA of another source. When inserted into a bacterium, recombinant DNA can be used to prepare large quantities of useful proteins.
- The polymerase chain reaction (PCR) is used to amplify a portion of a DNA molecule to produce millions of copies of a single gene.
- DNA fingerprinting is used to identify an individual by cutting DNA with restriction endonucleases to give a unique set of fragments.

**[11] What are the main characteristics of viruses? (22.11)**

- A virus is an infectious agent that contains either DNA or RNA within a protein coat. When the virus invades a host cell, it uses the biochemical machinery of the host to replicate. A retrovirus contains RNA and a reverse transcriptase that allow the RNA to synthesize viral DNA, which then transcribes RNA that directs the synthesis of viral proteins.

```
viral RNA  ──────────────▶  viral DNA  ──────────────▶  viral RNA  ──────────▶  viral proteins
            reverse                       transcription
            transcriptase
```

## Problem Solving

## [1] Nucleosides and Nucleotides (22.1)

**Example 22.1** Identify the base and monosaccharide used to form the following nucleoside, and then name it.

**Analysis**

- The sugar portion of the nucleoside contains the five-membered ring. If there is an OH group at C2', the sugar is ribose, but if there is no OH group at C2', the sugar is deoxyribose.
- The base is joined to the five-membered ring as an *N*-glycoside. A pyrimidine base has one ring, and is derived from cytosine, uracil, or thymine. A purine base has two rings and is derived from either adenine or guanine.
- Nucleosides derived from pyrimidines end in the suffix *-idine*. Nucleosides derived from purines end in the suffix *-osine*.

**Solution**

The sugar has an OH at C2', so it is derived from ribose. The base is uracil. To name the ribonucleoside, change the suffix of the base to *-idine*; thus, uracil → uridine.

---

**Example 22.2** Draw the structure of the nucleotide dTMP.

**Analysis**

Translate the abbreviation to a name; dTMP is deoxythymidine 5'-monophosphate. First, draw the sugar. Since there is a *deoxy* prefix in the name, dTMP is a deoxyribonucleotide and the sugar is deoxyribose. Then draw the base, in this case thymine, bonded to C1' of the sugar ring. Finally, add the phosphate. dTMP has one phosphate group bonded to the 5'-OH of the nucleoside.

**Solution**

---

## [2] Nucleic Acids (22.2)

**Example 22.3** (a) Draw the structure of a dinucleotide formed by joining the 3'-OH group of GMP to the 5'-phosphate in CMP. (b) Label the 5' and 3' ends. (c) Name the dinucleotide.

**Analysis**

Draw the structure of each nucleotide, including the sugar, the phosphate bonded to C5', and the base at C1'. In this case the sugar is ribose since the names of the mononucleotides do not contain the prefix *deoxy*. Bond the 3'-OH group to the 5'-phosphate to form the phosphodiester bond. The name of the dinucleotide begins with the nucleotide that contains the free phosphate at the 5' end.

**Solution**
a. and b.

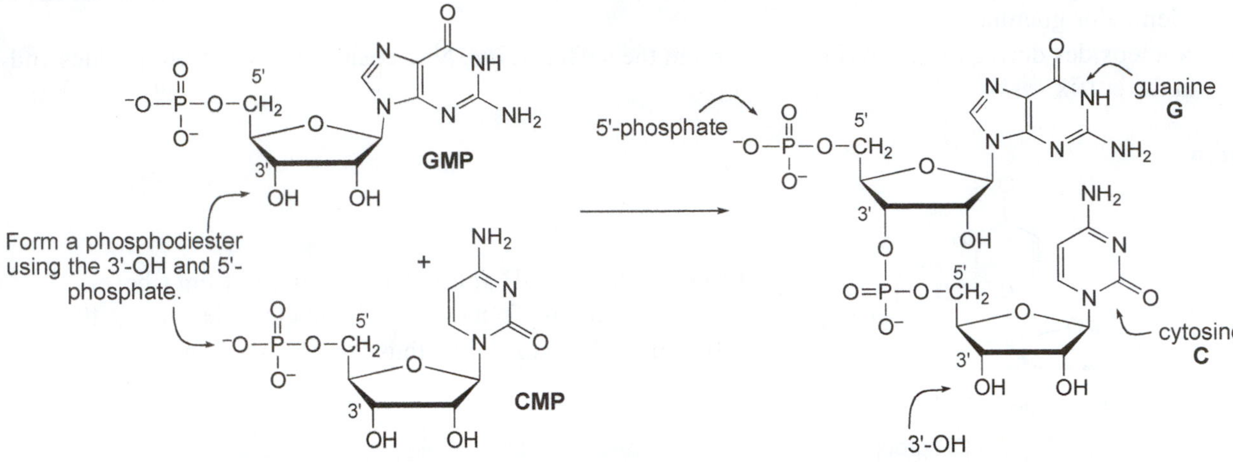

c. Since polynucleotides are named beginning at the 5' end, this dinucleotide is named GC.

---

## [3] The DNA Double Helix (22.3)

**Example 22.4** Write the sequence of the complementary strand of the following portion of a DNA molecule:
5'–AGGTATC–3'.

**Analysis**
The complementary strand runs in the opposite direction, from the 3' to the 5' end. Use base pairing to determine the corresponding sequence on the complementary strand: A pairs with T and C pairs with G.

**Solution**

Original strand:    5'–A G G T A T C–3'
                         ↓ ↓ ↓ ↓ ↓ ↓ ↓
Complementary strand:    3'–T C C A T A G–5'

---

## [4] Replication (22.4)

**Example 22.5** What is the sequence of a newly synthesized DNA segment if the template strand has the sequence 3'–CACGTC–5'?

**Analysis**
The newly synthesized strand runs in the opposite direction, from the 5' end to the 3' end in this example. Use base pairing to determine the corresponding sequence on the new strand: A pairs with T and C pairs with G.

**Solution**

Template strand:  3'–C A C G T C–5'

New strand:  5'–G T G C A G–3'

---

## [5] Transcription (22.6)

**Example 22.6** If a portion of the template strand of a DNA molecule has the sequence 3'–TACACATTC–5', what is the sequence of the mRNA molecule produced from this template? What is the sequence of the informational strand of this segment of the DNA molecule?

**Analysis**

mRNA has a base sequence that is complementary to the template from which it is prepared. mRNA has a base sequence that is identical to the informational strand of DNA, except that it contains the base U instead of T.

**Solution**

Template strand of DNA:  3'–T A C A C A T T C–5'          complementary

mRNA sequence:  5'–A U G U G U A A G–3'          complementary

Informational strand of DNA:  5'–A T G T G T A A G–3'

---

## [6] The Genetic Code (22.7)

**Example 22.7** Derive the amino acid sequence that is coded for by the following mRNA sequence.

5' UUG  CAG  AGA  GCA  CCG  GAG 3'

**Analysis**

Use Table 22.3 to identify the codons that correspond to each amino acid. Codons are written from the 5' to 3' end of an mRNA molecule and correspond to a peptide written from the N-terminal to C-terminal end.

**Solution**

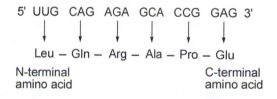

5' UUG  CAG  AGA  GCA  CCG  GAG 3'

Leu – Gln – Arg – Ala – Pro – Glu

N-terminal          C-terminal
amino acid          amino acid

---

## [7] Translation and Protein Synthesis (22.8)

**Example 22.8** What sequence of amino acids would be formed from the following mRNA sequence:
5' ACA UAU CAG ACC UGG 3'? List the anticodons contained in each of the needed tRNA molecules.

**Analysis**
Use Table 22.3 to determine the amino acid that is coded for by each codon. The anticodons contain complementary bases to the codons: A pairs with U, and C pairs with G.

**Solution**

```
Amino acid sequence      Thr – Tyr– Gln– Thr –Trp
                          ↑    ↑    ↑    ↑    ↑

        mRNA codons    5' ACA UAU CAG ACC UGG 3'
                          ↓    ↓    ↓    ↓    ↓

       tRNA anticodons    UGU AUA GUC UGG ACC
```

---

**Example 22.9** What polypeptide would be synthesized from the following template strand of DNA:
3' GCG AGT GCT ACG 5'?

**Analysis**
To determine what polypeptide is synthesized from a DNA template, two steps are needed. First, use the DNA sequence to determine the transcribed mRNA sequence: C pairs with G, T pairs with A, and A (on DNA) pairs with U (on mRNA). Then, use the codons in Table 22.3 to determine what amino acids are coded for by a given codon in mRNA.

**Solution**

```
DNA template strand ⟶  3' GCG AGT GCT ACG 5'

           mRNA ⟶  5' CGC UCA CGA UGC 3'

       Polypeptide ⟶     Arg– Ser– Arg–Cys
```

---

## [8] Mutations and Genetic Diseases (22.9)

**Example 22.10** (a) What dipeptide is produced from the following segment of DNA: CAGATG?
(b) What happens to the dipeptide when a point mutation occurs and the DNA segment contains the sequence CTGATG instead?

**Analysis**
Transcribe the DNA sequence to an mRNA sequence with complementary base pairs. Then use Table 22.3 to determine what amino acids are coded for by each codon.

**Solution**
a. Since GUC codes for valine and UAC codes for tyrosine, the dipeptide Val–Tyr results.

b. Since GAC codes for aspartic acid, the point mutation results in the synthesis of the dipeptide Asp–Tyr.

## Self-Test

**[1] Fill in the blank with one of the terms listed below.**

Codon (22.7)                           Messenger RNA (22.5)                    Ribosomal RNA (22.5)
Deoxyribonucleic acid (22.1, 22.3)     Nucleic acids (22.1, 22.2)             Transcription (22.6)
Gene (22.1)                            Nucleoside (22.1)                      Transfer RNA (22.5)
Lagging strand (22.4)                  Nucleotides (22.1)                     Virus (22.11)
Leading strand (22.4)                  Ribonucleic acids (22.1, 22.5)

1.   _____ stores the genetic information of an organism and transmits that information from one generation to another.
2.   During DNA replication, the _____ grows continuously.
3.   _____ are polymers of nucleotides.
4.   Nucleic acids are unbranched polymers composed of repeating monomers called _____.
5.   _____ interprets the genetic information in mRNA and brings specific amino acids to the site of protein synthesis in the ribosome.
6.   A _____ is an infectious agent consisting of a DNA or RNA molecule that is contained within a protein coating.
7.   _____ is the carrier of information from DNA (in the cell nucleus) to the ribosomes (in the cytoplasm).
8.   A sequence of three nucleotides (a triplet) codes for a specific amino acid. Each triplet is called a _____.
9.   _____ is the synthesis of messenger RNA from DNA.
10.  During DNA replication, the _____ is synthesized in small fragments, which are then joined together by a DNA ligase enzyme.
11.  A _____ is formed by joining the anomeric carbon of a monosaccharide with a nitrogen atom of a base.
12.  A _____ is a portion of a DNA molecule responsible for the synthesis of a single protein.
13.  _____ translate the genetic information contained in DNA into proteins needed for all cellular functions.
14.  _____ is found in the ribosomes in the cytoplasm of the cell.

**[2] Decide if the statement is describing RNA or DNA.**

        a. RNA                        b. DNA

15. A polymer of deoxyribonucleotides      17. The bases are A, G, C, and U.
16. The monosaccharide is ribose.           18. The bases are A, G, C, and T.

**[3] What amino acid is coded for by each codon?**

19. AAC                        21. UGG
20. GCU                        22. CGU

**[4] Label each structure as a purine or pyrimidine.**

        a. purine                     b. pyrimidine

23.                  24.                  25.

## Answers to Self-Test

| | | | | |
|---|---|---|---|---|
| 1. Deoxyribonucleic acid | 6. virus | 11. nucleoside | 16. a | 21. Trp |
| 2. leading strand | 7. Messenger RNA | 12. gene | 17. a | 22. Arg |
| 3. Nucleic acids | 8. codon | 13. Ribonucleic acids | 18. b | 23. b |
| 4. nucleotides | 9. Transcription | 14. Ribosomal RNA | 19. Asn | 24. a |
| 5. Transfer RNA | 10. lagging strand | 15. b | 20. Ala | 25. b |

## Solutions to In-Chapter Problems

**22.1**  Name each nucleoside as in Example 22.1.

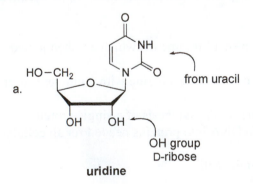

a.

from uracil

OH group
D-ribose

**uridine**

b.

from guanine

no OH group
D-2-deoxyribose

**deoxyguanosine**

**22.2** Draw the structure of guanosine.

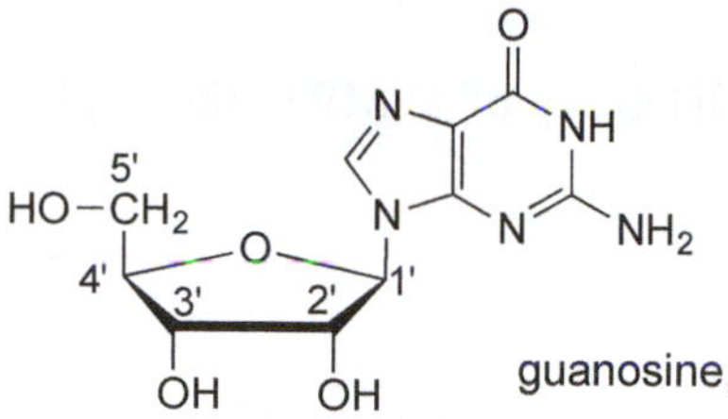

guanosine, a ribonucleoside

**22.3** Draw the structure of each nucleotide.

a.

from uracil

ribose

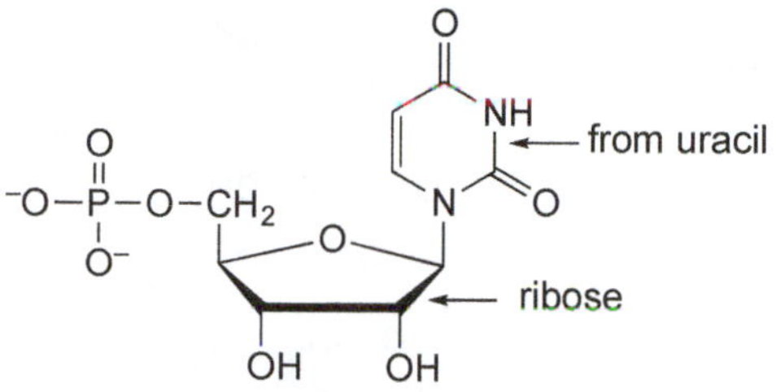

uridine 5'-monophosphate, UMP

b.

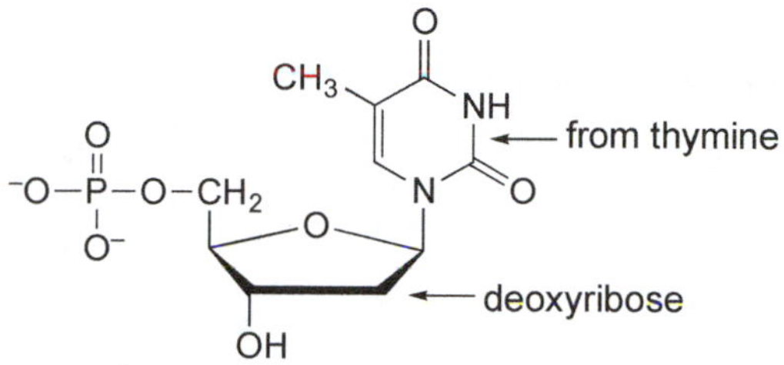

from thymine

deoxyribose

deoxythymidine 5'-monophosphate, dTMP

**22.4** Answer each question about the nucleic acids.

a. the sugar ribose: RNA
b. the sugar deoxyribose: DNA
c. the base T: DNA

d. the base U: RNA
e. the nucleotide GMP: RNA
f. the nucleotide dCMP: DNA

**22.5** Draw the structure of each nucleotide as in Example 22.2.

a.

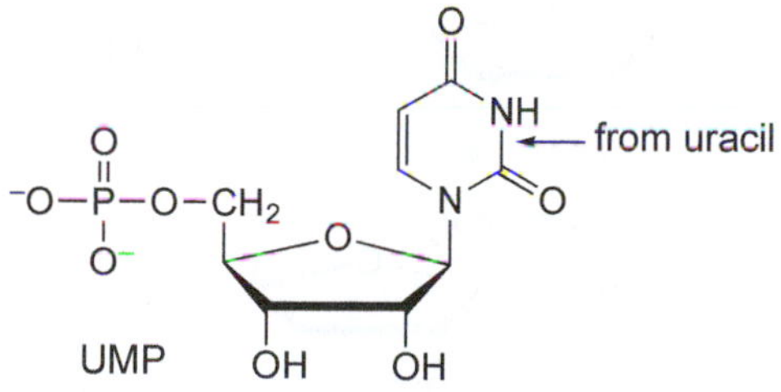

from uracil

UMP

c.

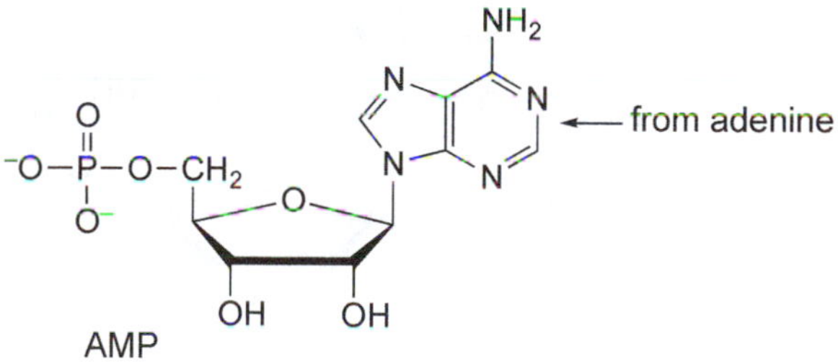

from adenine

AMP

b.

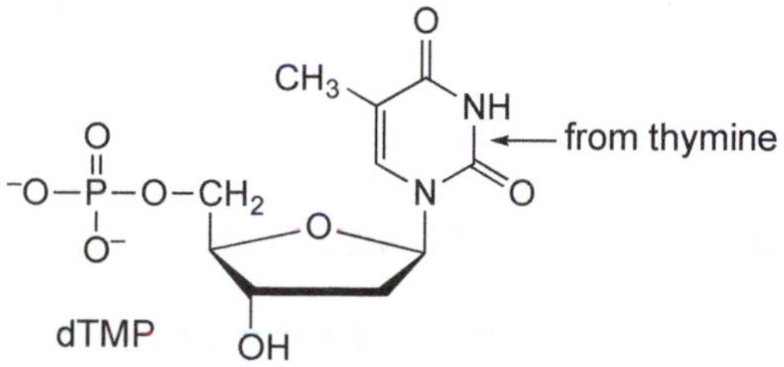

from thymine

dTMP

**22.6** Write out the name for each abbreviation using Table 22.1. Recall that the first capital letter indicates the base, and the second indicates the number of phosphate groups (mono-, di-, or tri- = M, D, or T, respectively).

a. GTP: guanosine 5'-triphosphate
b. dCDP: deoxycytidine 5'-diphosphate

c. dTTP: deoxythymidine 5'-triphosphate
d. UDP: uridine 5'-diphosphate

**22.7** Draw the structure of a dinucleotide formed by joining the 3'-OH group of dTMP to the 5'-phosphate in dGMP as in Example 22.3.

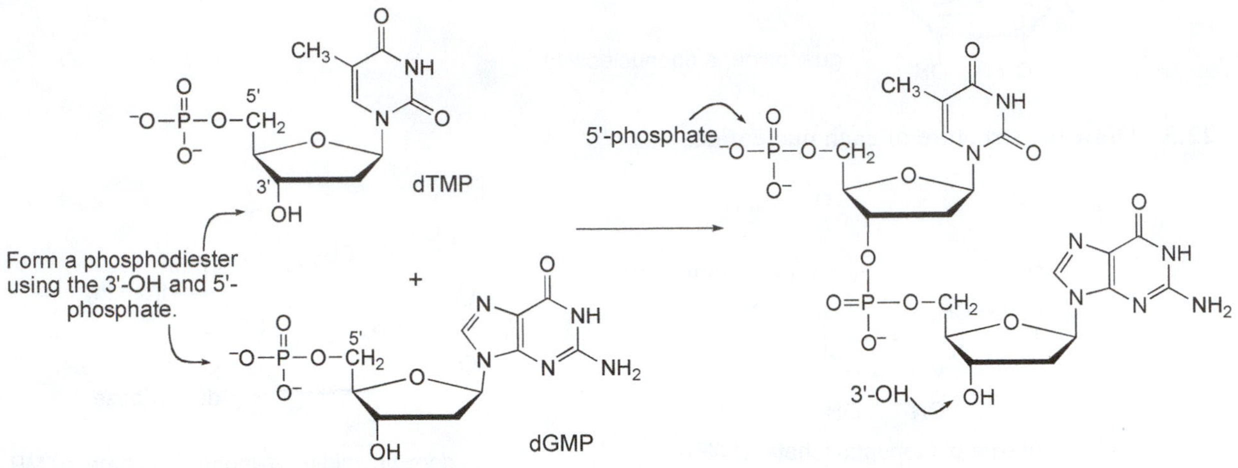

**22.8** Draw the structure of each polynucleotide. In (a), the monosaccharide must be ribose since the dinucleotide contains the base U. In (b), the monosaccharide must be deoxyribose since the trinucleotide contains the base T.

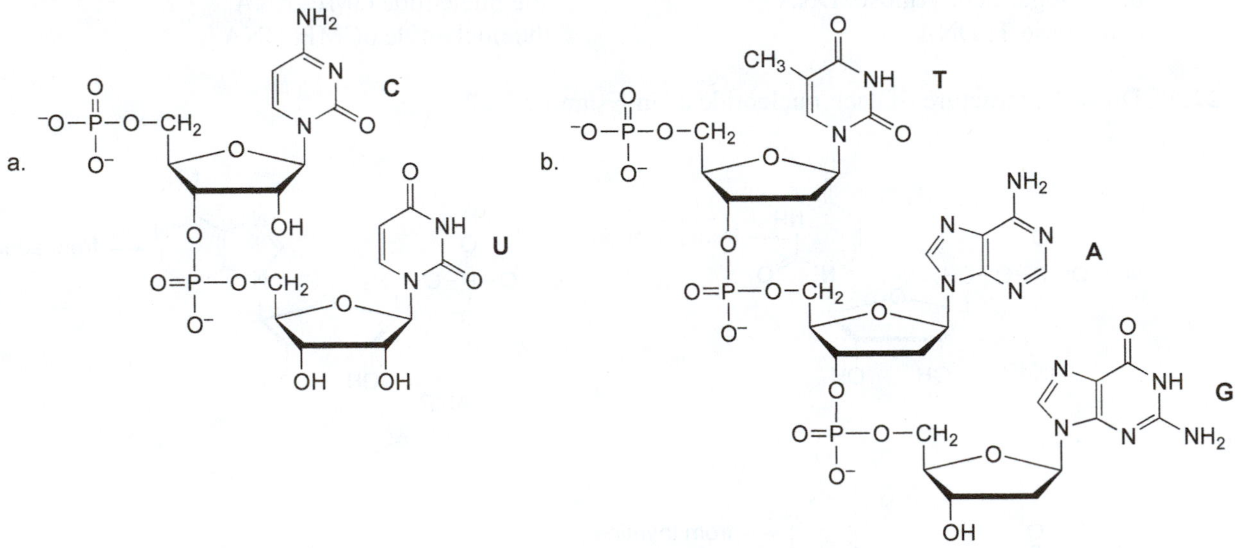

**22.9** Write the complementary strand for each of the following strands of DNA as in Example 22.4.

a. 5'–A A A C G T C C–3'

Complementary
strand:  3'–T T T G C A G G–5'

b. 5'–T A T A C G C C–3'

Complementary
strand:  3'–A T A T G C G G–5'

c. 5'–A T T G C A C C C G C–3'

Complementary strand:  3'–T A A C G T G G G C G–5'

d. 5'–C A C T T G A T C G G–3'

Complementary strand:  3'–G T G A A C T A G C C–5'

**22.10** Draw the sequence of a newly synthesized DNA segment as in Example 22.5.

a. 3'–A G A G T C T C–5'

New strand: 5'–T C T C A G A G–3'

c. 3'–A T C C T G T A C–5'

New strand: 5'–T A G G A C A T G–3'

b. 5'–A T T G C T C–3'

New strand: 3'–T A A C G A G–5'

d. 5'–G G C C A T A C T C–3'

New strand: 3'–C C G G T A T G A G–5'

**22.11** Draw the mRNA and the informational strand of DNA as in Example 22.6.

a. 3'–TGCCTAACG–5'  Template strand of DNA

[1] 5'–ACGGAUUGC–3'  mRNA sequence

[2] 5'–ACGGATTGC–3'  Informational strand

c. 3'–TTAACGCGA–5'  Template strand of DNA

[1] 5'–AAUUGCGCU–3'  mRNA sequence

[2] 5'–AATTGCGCT–3'  Informational strand

b. 3'–GACTCC–5'  Template strand of DNA

[1] 5'–CUGAGG–3'  mRNA sequence

[2] 5'–CTGAGG–3'  Informational strand

d. 3'–CAGTGACCGTAC–5'  Template strand of DNA

[1] 5'–GUCACUGGCAUG–3'  mRNA sequence

[2] 5'–GTCACTGGCATG–3'  Informational strand

**22.12** Work backwards to draw the template strand of DNA from which each mRNA was synthesized.

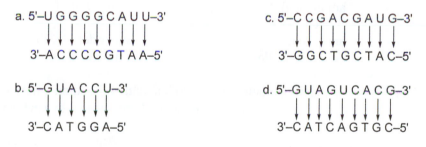

a. 5'–U G G G G C A U U–3'

3'–A C C C C G T A A–5'

c. 5'–C C G A C G A U G–3'

3'–G G C T G C T A C–5'

b. 5'–G U A C C U–3'

3'–C A T G G A–5'

d. 5'–G U A G U C A C G–3'

3'–C A T C A G T G C–5'

**22.13** Use Table 22.3 to determine what amino acid is coded for by each codon.

a. GCC: Ala
b. AAU: Asn

c. CUA: Leu
d. AGC: Ser

e. CAA: Gln
f. AAA: Lys

**22.14** Use Table 22.3 to determine what codons code for each amino acid.

a. glycine: GGU, GGC, GGA, GGG
b. isoleucine: AUU, AUC, AUA

c. lysine: AAA, AAG
d. glutamic acid: GAA, GAG

**22.15** Derive the amino acid sequence that is coded for by each mRNA sequence as in Example 22.7.

a. 5' CAA  GAG  GUA  UCC  UAC  AGA 3'
　　↓　　↓　　↓　　↓　　↓　　↓
　　Gln – Glu – Val – Ser – Tyr – Arg

c. 5' CUA  UGC  AGU  AGG  ACA  CCC 3'
　　↓　　↓　　↓　　↓　　↓　　↓
　　Leu – Cys – Ser – Arg – Thr – Pro

b. 5' GUC  AUC  UGG  AGG  GGC  AUU 3'
　　↓　　↓　　↓　　↓　　↓　　↓
　　Val – Ile – Trp – Arg – Gly – Ile

**22.16** Use Table 22.3 to determine possible codons for each amino acid sequence.

a. Met–Arg–His–Phe
　　5'–AUG CGC CAU UUU–3'
b. Gly–Ala–Glu–Gln
　　5'–GGC GCC GAA CAA–3'

c. Gln–Asn–Gly–Ile–Val
　　5'–CAA AAU GGA AUU GUG–3'
d. Thr–His–Asp–Cys–Trp
　　5'–ACU CAU GAU UGC UGG–3'

**22.17** Answer each question about the mRNA sequence.

5' GAG CCC GUA UAC GCC ACG 3'
｜｜｜｜｜｜｜｜｜｜｜｜｜｜｜｜｜｜
↓↓↓　↓↓↓　↓↓↓　↓↓↓　↓↓↓　↓↓↓
a. 3' CTC GGG CAT ATG CGG TGC 5'

　　DNA template strand

5' GAG CCC GUA UAC GCC ACG 3'
　↓　　↓　　↓　　↓　　↓　　↓
b. Glu – Pro – Val – Tyr – Ala – Thr

　　peptide

**22.18** The anticodon has three nucleotides that are complementary to the codon, and is found on the tRNA. Use Table 22.3 to determine the amino acid represented by each codon.

a. CGG
　anticodon: GCC
　amino acid: Arg
b. GGG
　anticodon: CCC
　amino acid: Gly

c. UCC
　anticodon: AGG
　amino acid: Ser
d. AUA
　anticodon: UAU
　amino acid: Ile

e. CCU
　anticodon: GGA
　amino acid: Pro
f. GCC
　anticodon: CGG
　amino acid: Ala

**22.19** The anticodon has three nucleotides that are complementary to the codon, and is found on a tRNA molecule, as in Example 22.8. Use Table 22.3 to determine the amino acid represented by each codon.

Pro – Pro – Ala–Asn–Glu–Ala　amino acid sequence
　↑　　↑　　↑　↑　↑　↑
a. 5' CCA CCG GCA AAC GAA GCA 3'　mRNA codons
　↓　　↓　　↓　↓　↓　↓
　GGU GGC CGU UUG CUU CGU　tRNA anticodons

Ala–Pro–Leu–Arg–Asp　amino acid sequence
　↑　↑　↑　↑　↑
b. 5' GCA CCA CUA AGA GAC 3'　mRNA codons
　↓　↓　↓　↓　↓
　CGU GGU GAU UCU CUG　tRNA anticodons

**22.20** Draw the polypeptide synthesized from each template DNA, as in Example 22.8.

a. 3' TCT CAT CGT AAT GAT TCG 5'  ⟵— DNA template strand —⟶  b. 3' GCT CCT AAA TAA CAC TTA 5'

5' AGA GUA GCA UUA CUA AGC 3'  ⟵  mRNA  ⟶  5' CGA GGA UUU AUU GUG AAU 3'

Arg – Val – Ala – Leu – Leu – Ser  ⟵  Polypeptide  ⟶  Arg – Gly – Phe – Ile – Val – Asn

**22.21** Draw the dipeptide and explain the effect of the mutations as in Example 22.10.

a.  AAC TGA ————⟶ UUG ACU ————⟶ Leu – Thr
    DNA                    mRNA                    dipeptide

b.  [1] AAC GGA ————⟶ UUG CCU ————⟶ [1] Leu – Pro

    [2] ATC TGA ————⟶ UAG ACU ————⟶ [2] Stop codon, no dipeptide is formed.

    [3] AAT TGA ————⟶ UUA ACU ————⟶ [3] Leu – Thr
        DNA                    mRNA                    dipeptide

**22.22** Compare the peptides made from the amino acids.

DNA:  TAT GCA CTT  ——deletion——⟶  TAA CTT

mRNA: AUA CGU GAA                  AUU GAA

Ile – Arg – Glu  ——loss of Arg——⟶  Ile – Glu

**22.23** Draw the cut segment of DNA.

5' ⌇⌇⌇⌇ C C A ⌐———┐  sticky ends  A G C T T G G A T T ⌇⌇⌇⌇ 3'
3' ⌇⌇⌇⌇ G G T T C G A                 └———┘ A C C T A A ⌇⌇⌇⌇ 5'

**22.24** The antibiotics sulfanilamide and penicillin act upon enzymes found only in bacteria; therefore, they affect bacterial growth and reproduction. Viruses use the starting materials and biochemical processes of the host to reproduce, so antibiotics are ineffective in treating viral infections.

## Solutions to Odd-Numbered End-of-Chapter Problems

**22.25**  A ribonucleoside contains ribose (a monosaccharide) and a nitrogen-containing base (A, G, C or U), and a ribonucleotide has these components plus a phosphate group attached to the 5' position of the ribose.

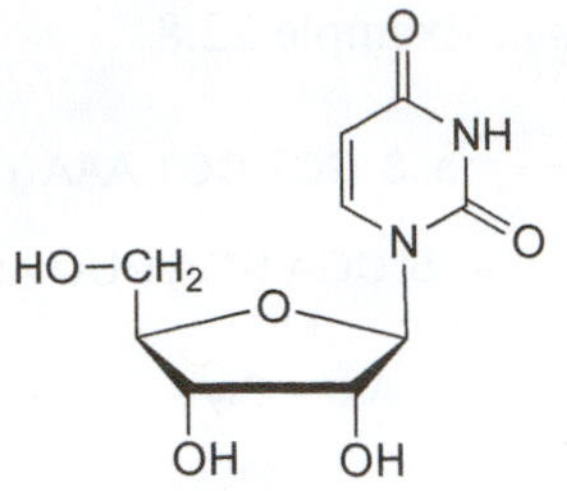
HO–CH2
NH
N
O
O
OH OH

uridine
a ribonucleoside

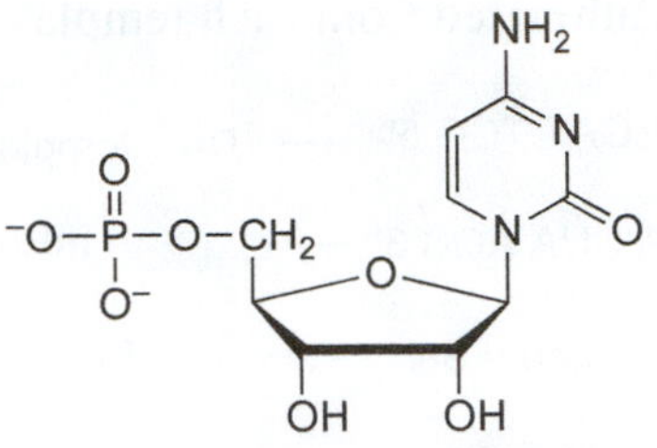
NH2
N
N
O
O
⁻O–P–O–CH2
O⁻
O
OH OH

cytidine 5'-monophosphate
a ribonucleotide

**22.27** A gene is a portion of the DNA molecule responsible for the synthesis of a single protein. Many genes form each chromosome.

**22.29** a. The polynucleotide is double stranded: DNA.
b. The polynucleotide may contain adenine: DNA and RNA.
c. The polynucleotide may contain dGMP: DNA.
d. The polynucleotide is a polymer of ribonucleotides: RNA.

**22.31** Name each nucleoside as in Example 22.1.

a.

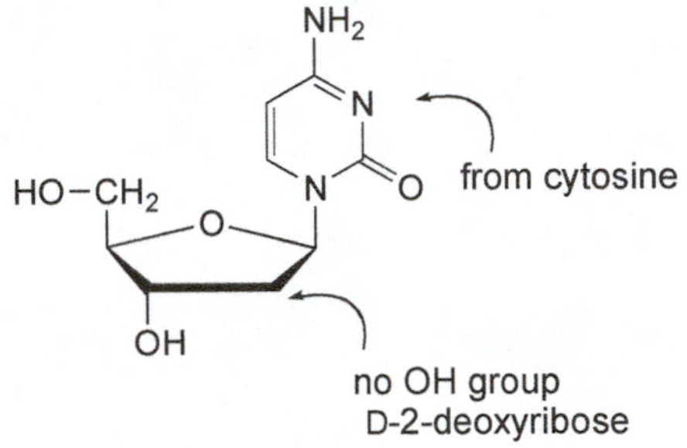
NH2
from cytosine
HO–CH2
OH
no OH group
D-2-deoxyribose

**deoxycytidine**

b.

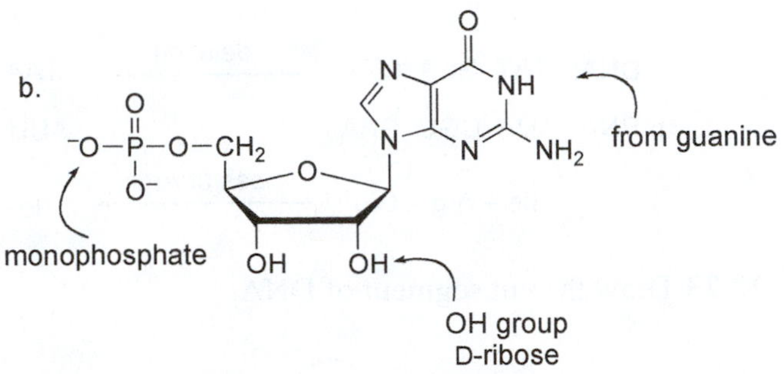
from guanine
⁻O–P–O–CH2
NH
NH2
monophosphate
OH OH
OH group
D-ribose

**guanosine 5'-monophosphate**

**22.33** Draw each structure.

a.

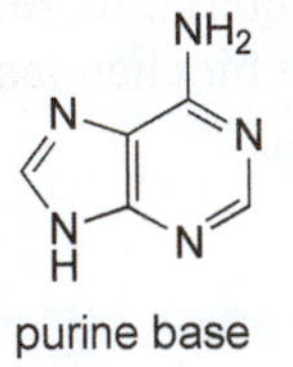
NH2
purine base

c.

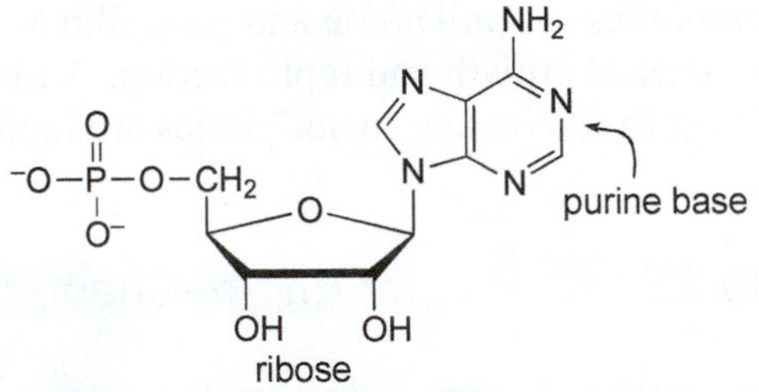
NH2
⁻O–P–O–CH2
purine base
OH OH
ribose

b.

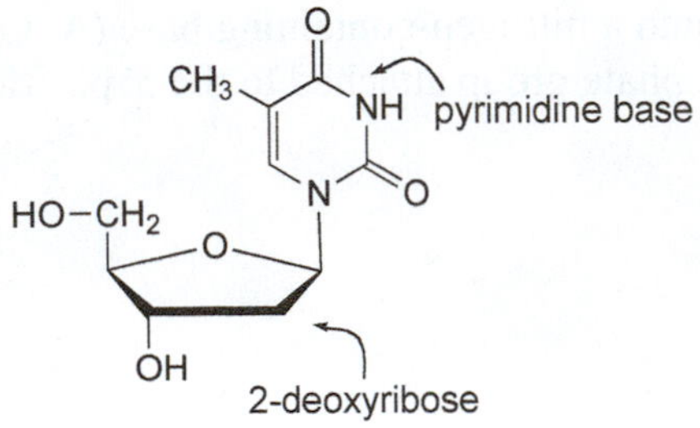
CH3
NH
pyrimidine base
HO–CH2
OH
2-deoxyribose

d.

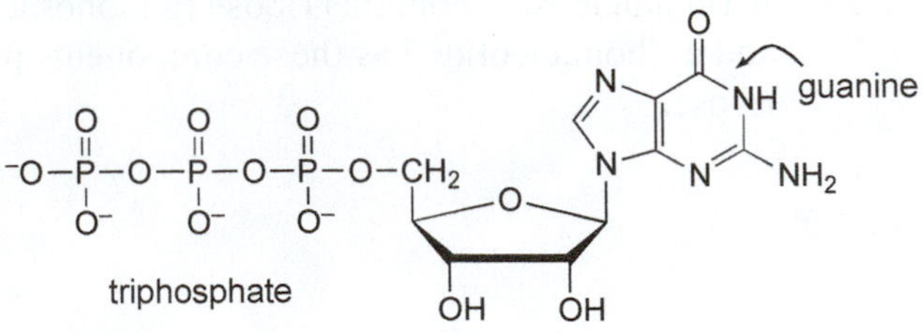
guanine
⁻O–P–O–P–O–P–O–CH2
NH2
triphosphate
OH OH

**22.35** Draw the structure of each nucleoside or nucleotide.

a. adenosine

b. deoxyguanosine

c. GDP

d. dTDP

**22.37**

a., b., and c.

uracil, UMP

adenine, AMP

5'

3'

d. This dinucleotide is a ribonucleotide because the sugar rings contain an OH group on C2'.

e. UA

**22.39** Draw the two possible dinucleotides.

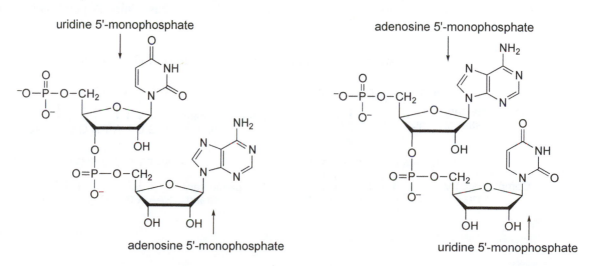

uridine 5'-monophosphate

adenosine 5'-monophosphate

adenosine 5'-monophosphate

uridine 5'-monophosphate

**22.41** Draw the structure of each dinucleotide.

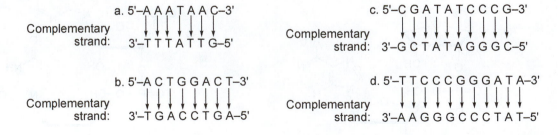

**22.43** Draw the deoxyribonucleotide.

**22.45** a. The DNA double helix has 2-deoxyribose as the only sugar. The sugar–phosphate groups are on the outside of the helix.
   b. The 5' end has a phosphate and the 3' end has an OH group.
   c. Hydrogen bonding occurs in the interior of the helix between base pairs: A pairs with T and G pairs with C.

**22.47** Write the complementary strand for each of the following strands of DNA as in Example 22.4.

a. 5'–A A A T A A C–3'
Complementary strand:   3'–T T T A T T G–5'

b. 5'–A C T G G A C T–3'
Complementary strand:   3'–T G A C C T G A–5'

c. 5'–C G A T A T C C C G–3'
Complementary strand:   3'–G C T A T A G G G C–5'

d. 5'–T T C C C G G G A T A–3'
Complementary strand:   3'–A A G G G C C C T A T–5'

**22.49** If 27% of the bases in DNA are A, the same percentage must be T because they are complementary base pairs. The remaining 46% are equally divided between G and C (23% each).

**22.51** Complementary base pairing between G and C occurs between strands of DNA, two sites on RNA, and between DNA and RNA, whereas A pairs with either T or U. In DNA, A–T base pairs occur, whereas in RNA, A–U base pairs occur. When DNA pairs with RNA, A in DNA pairs with U in RNA, while T in DNA pairs with A in RNA.

**22.53** Draw the sequence of a newly synthesized DNA segment as in Example 22.5.

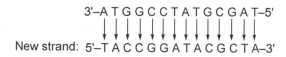

**22.55** Replication occurs in only one direction, from the 3' end to the 5' end of the template strand, but the strands of DNA run in opposite directions. The leading strand can be formed continuously from the 3' end as it is unwound. The lagging strand runs in the opposite direction, so it must be formed in segments that are then joined by DNA ligase.

**22.57** Messenger RNA carries the specific sequence of the DNA code from the cell nucleus to the ribosomes in the cytoplasm to make a protein. Each transfer RNA brings a specific amino acid to the growing protein chain on the ribosome according to the sequence specified by the mRNA.

**22.59** Draw the mRNA sequence transcribed from each sequence of DNA as in Example 22.6.
- a. DNA: 5'–AAATAAC–3'
  mRNA: 3'–UUUAUUG–5'
- b. DNA: 5'–ACTGGACT–3'
  mRNA: 3'–UGACCUGA–5'
- c. DNA: 5'–CGATATCCCG–3'
  mRNA: 3'–GCUAUAGGGC–5'
- d. DNA: 5'–TTCCCGGGATA–3'
  mRNA: 3'–AAGGGCCCUAU–5'

**22.61** Draw the mRNA and the informational strand of DNA as in Example 22.6.
- a. 3'–ATGGCTTA–5'  Template strand of DNA
  [1] 5'–UACCGAAU–3'  mRNA sequence
  [2] 5'–TACCGAAT–3'  Informational strand of DNA
- b. 3'–CGGCGCTTA–5'  Template strand of DNA
  [1] 5'–GCCGCGAAU–3'  mRNA sequence
  [2] 5'–GCCGCGAAT–3'  Informational strand of DNA
- c. 3'–GGTATACCG–5'  Template strand of DNA
  [1] 5'–CCAUAUGGC–3'  mRNA sequence
  [2] 5'–CCATATGGC–3'  Informational strand of DNA
- d. 3'–TAGGCCGTA–5'  Template strand of DNA
  [1] 5'–AUCCGGCAU–3'  mRNA sequence
  [2] 5'–ATCCGGCAT–3'  Informational strand of DNA

**22.63** The anticodon has the three nucleotides that are complementary to the codon, and is found on the tRNA. Use Table 22.3 to determine the amino acid represented by each codon.
- a. CUG
  anticodon: GAC
  amino acid: Leu
- b. UUU
  anticodon: AAA
  amino acid: Phe
- c. AAG
  anticodon: UUC
  amino acid: Lys
- d. GCA
  anticodon: CGU
  amino acid: Ala

**22.65**  Fill in the missing information.

mRNA:       G C G
anticodon:   C G C
amino acid:  alanine

**22.67**  Derive the amino acid sequence that is coded for by each mRNA sequence as in Example 22.7.

a. 5' CCA  ACC  UGG  GUA  GAA 3'

Pro – Thr – Trp – Val – Glu

c. 5' GUC  GAC  GAA  CCG  CAA 3'

Val – Asp – Glu – Pro – Gln

b. 5' AUG  UUU  UUA  UGG  UGG 3'

Met – Phe – Leu – Trp – Trp

**22.69**  Work backwards to derive the mRNA sequence from the amino acid sequence.

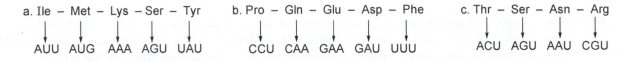

a. Ile – Met – Lys – Ser – Tyr

AUU  AUG  AAA  AGU  UAU

b. Pro – Gln – Glu – Asp – Phe

CCU  CAA  GAA  GAU  UUU

c. Thr – Ser – Asn – Arg

ACU  AGU  AAU  CGU

**22.71**  Work backwards to write the sequence of the DNA template strand from which the mRNA was synthesized, and then give the peptide synthesized by the mRNA.

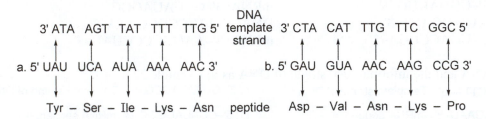

3' ATA  AGT  TAT  TTT  TTG 5'

DNA
template
strand

3' CTA  CAT  TTG  TTC  GGC 5'

a. 5' UAU  UCA  AUA  AAA  AAC 3'

Tyr – Ser – Ile – Lys – Asn

b. 5' GAU  GUA  AAC  AAG  CCG 3'

peptide   Asp – Val – Asn – Lys – Pro

**22.73**  A point mutation results in the substitution of one nucleotide for another in a DNA molecule.  A silent mutation is a point mutation in DNA that results in no change in an amino acid sequence.  A point mutation can be a silent mutation if the new codon results in the same amino acid residue.

**22.75**  Answer each question about the mRNA sequence.

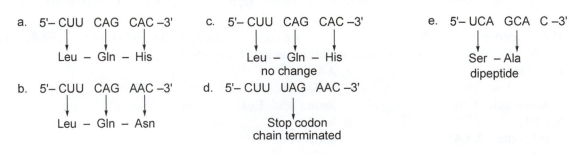

a.  5'– CUU  CAG  CAC –3'

Leu – Gln – His

c.  5'– CUU  CAG  CAC –3'

Leu – Gln – His
no change

e.  5'– UCA  GCA  C –3'

Ser – Ala
dipeptide

b.  5'– CUU  CAG  AAC –3'

Leu – Gln – Asn

d.  5'– CUU  UAG  AAC –3'

Stop codon
chain terminated

**22.77** Answer each question about the DNA sequence.

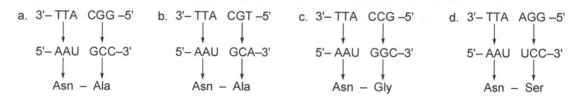

a. 3'– TTA   CGG –5'       b. 3'– TTA   CGT –5'       c. 3'– TTA   CCG –5'       d. 3'– TTA   AGG –5'

5'– AAU   GCC–3'           5'– AAU   GCA–3'           5'– AAU   GGC–3'           5'– AAU   UCC–3'

Asn – Ala                 Asn – Ala                 Asn – Gly                 Asn – Ser

**22.79** A restriction endonuclease is an enzyme that cleaves plasmid DNA at a specific site, based on the sequence of base pairs.

**22.81** Draw the cut segment of DNA.

5' ∿∿∿ C C G G T T G ⌐———⌐          G A T C C T T ∿∿∿ 3'
3' ∿∿∿ G G C C A A C C T A G    sticky ends    ⌐———G A A ∿∿∿ 5'

**22.83** Circular plasmid DNA is isolated from bacteria and cleaved at a specific site with a restriction endonuclease. This forms "sticky ends" of DNA that are then combined with another DNA source, perhaps human, with complementary "sticky ends" to the plasmid DNA. When these segments of DNA come together in a presence of a DNA ligase enzyme, circular DNA is re-formed. This recombinant DNA can then be re-inserted into bacteria.

**22.85** The DNA fragments can help determine genetic relationships. A line is drawn through each fragment found in the parents. Ovals are draw around the fragments in the children that are not found in either parent.

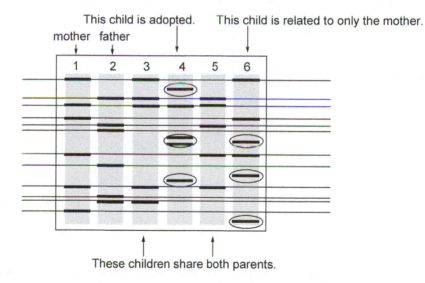

This child is adopted.   This child is related to only the mother.
mother father
1   2   3   4   5   6

These children share both parents.

a. Lanes 3 and 5 represent DNA of children that share both parents because they have DNA fragments from both parents (lines drawn through each fragment).

b. Lane 4 represents DNA from an adopted child because the DNA fragments have little relationship to the parental DNA fragments (four of five fragments are not found in either parent).

**22.87**  a.  Lamivudine resembles the nucleoside cytidine.

b.  Lamivudine inhibits reverse transcription because it is a nucleoside analogue that gets incorporated into a DNA chain, but since it does not contain a 3' hydroxyl group, synthesis is terminated.

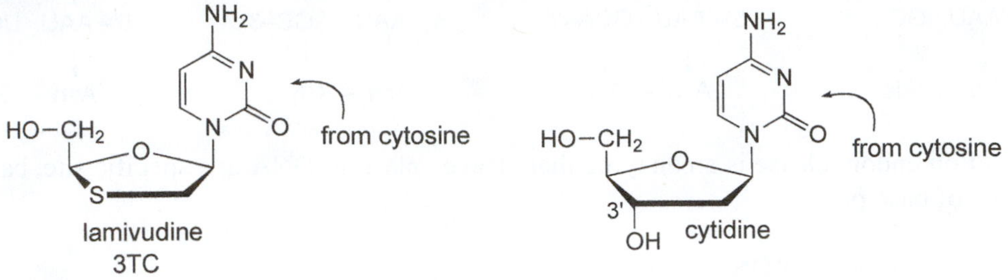

**22.89**  Answer each question about the DNA sequence.

a.  5' GTT ACA TAA AAA CGA 3'  Informational strand

3' CAA TGT ATT TTT GCT 5'  Template strand

b.  5' GUU ACA UAA AAA CGA 3'  mRNA

c.  Val – Thr  stop

**22.91**

| DNA informational strand: | 5' end | AAC | TAT | CCA | ACG | AAG | ATG | 3' end |
|---|---|---|---|---|---|---|---|---|
| DNA template strand: | 3' end | TTG | ATA | GGT | TGC | TTC | TAC | 5' end |
| mRNA codons: | 5' end | AAC | UAU | CCA | ACG | AAG | AUG | 3' end |
| tRNA anticodons: | | UUG | AUA | GGU | UGC | UUC | UAC | |
| Polypeptide: | | Asn | Tyr | Pro | Thr | Lys | Met | |

**22.93**

| DNA informational strand: | 5' end | AAC | GTA | TCA | ACT | CAC | ATG | 3' end |
|---|---|---|---|---|---|---|---|---|
| DNA template strand: | 3' end | TTG | CAT | AGT | TGA | GTG | TAC | 5' end |
| mRNA codons: | 5' end | AAC | GUA | UCA | ACU | CAC | AUG | 3' end |
| tRNA anticodons: | | UUG | CAU | AGU | UGA | GUG | UAC | |
| Polypeptide: | | Asn | Val | Ser | Thr | His | Met | |

**22.95**  To determine the number of bases in the gene, multiply the number of amino acids by three, since each amino acid is coded for by a codon of three bases.

$$325 \times 3 = 975 \text{ bases}$$

**22.97** Work backwards from the amino acid sequence to determine a base pair sequence that could code for met-enkephalin.

ATA  CCA  CCA  AAA  TAC  DNA

↑　　↑　　↑　　↑　　↑

UAU  GGU  GGU  UUU  AUG  mRNA

↑　　↑　　↑　　↑　　↑

Tyr – Gly – Gly – Phe – Met　　peptide

**22.99** Identify each amino acid in the peptide, and then use Table 22.3 to determine the DNA sequence.

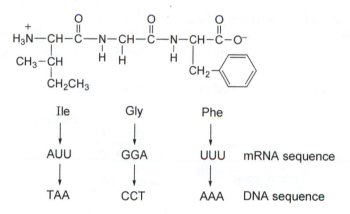

Ile　　　　Gly　　　　Phe

↓　　　　↓　　　　↓

AUU　　　GGA　　　UUU　　mRNA sequence

↓　　　　↓　　　　↓

TAA　　　CCT　　　AAA　　DNA sequence

# Chapter 23 Metabolism and Energy Production

## Chapter Review

**[1] What is metabolism and where is energy produced in cells? (23.1)**
- Metabolism is the sum of all the chemical reactions that take place in an organism. Catabolic reactions break down large molecules and release energy, while anabolic reactions synthesize larger molecules and require energy.
- Energy is produced in the mitochondria, sausage-shaped organelles that contain an outer and inner cell membrane. Energy is produced in the matrix, the area surrounded by the inner membrane.

**[2] What are the four stages of metabolism? (23.2)**
- Metabolism begins with digestion in stage [1], in which large molecules—polysaccharides, proteins, and triacylglycerols—are hydrolyzed to smaller molecules—monosaccharides, amino acids, fatty acids, and glycerol.
- In stage [2], biomolecules are degraded into two-carbon acetyl units.

- The citric acid cycle comprises stage [3]. The citric acid cycle converts two carbon atoms to two molecules of $CO_2$, and forms reduced coenzymes NADH and $FADH_2$ that carry electrons to the electron transport chain.
- In stage [4], the electron transport chain and oxidative phosphorylation produce ATP, and oxygen is converted to water.

**[3] What is ATP and how do coupled reactions with ATP drive energetically unfavorable reactions? (23.3)**
- ATP is the primary energy-carrying molecule in metabolic pathways. The hydrolysis of ATP cleaves one phosphate group and releases 7.3 kcal/mol of energy.
- The hydrolysis of ATP provides the energy to drive a reaction that requires energy. A pair of reactions of this sort is said to be coupled.

**[4] List the main coenzymes in metabolism and describe their roles. (23.4)**
- Nicotinamide adenine dinucleotide ($NAD^+$) is a biological oxidizing agent that accepts electrons and protons, thus generating its reduced form, NADH. NADH is a reducing agent that donates electrons and protons, thus re-forming $NAD^+$.

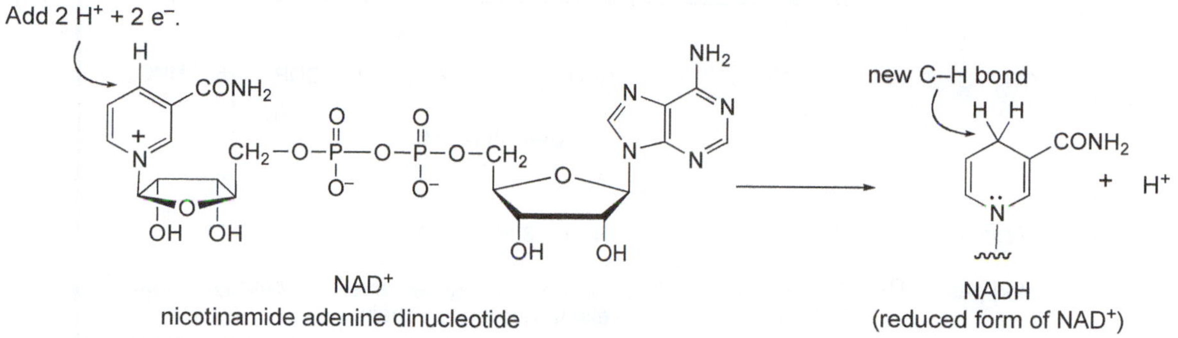

Add 2 H$^+$ + 2 e$^-$.

new C–H bond

NAD$^+$

nicotinamide adenine dinucleotide

NADH

(reduced form of NAD$^+$)

- Flavin adenine dinucleotide (FAD) is a biological oxidizing agent that accepts electrons and protons, thus yielding its reduced form, FADH$_2$. FADH$_2$ is a reducing agent that donates electrons and protons, thus re-forming FAD.

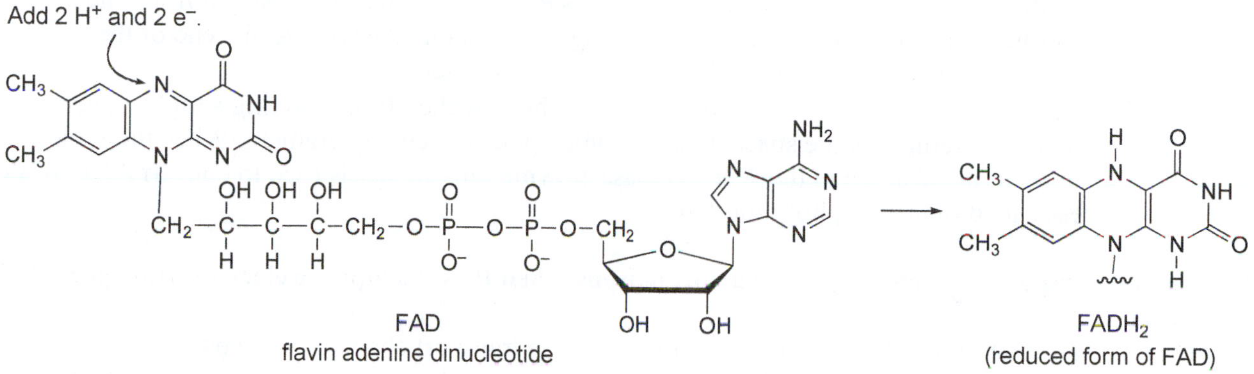

Add 2 H$^+$ and 2 e$^-$.

FAD

flavin adenine dinucleotide

FADH$_2$

(reduced form of FAD)

- Coenzyme A reacts with acetyl groups (CH$_3$CO–) to form a high-energy thioester that delivers two-carbon acetyl groups to other substrates.

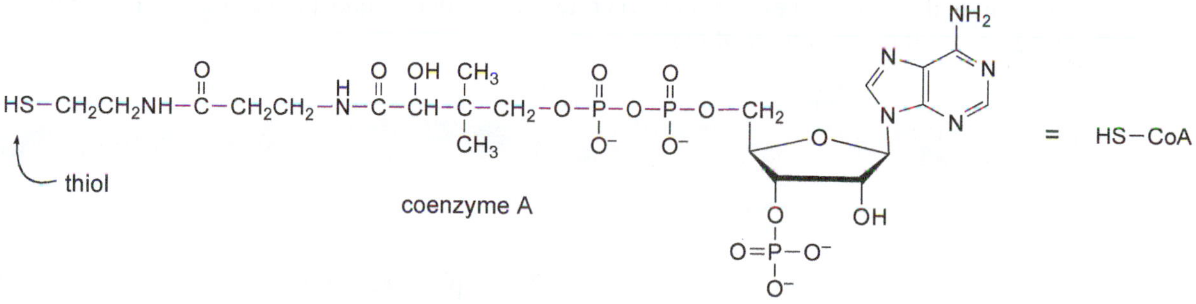

coenzyme A

= HS–CoA

### [5] What are the main features of the citric acid cycle? (23.5)

- The citric acid cycle is an eight-step cyclic pathway that begins with the addition of acetyl CoA to a four-carbon substrate. In the citric acid cycle, two carbons are converted to $CO_2$ and four molecules of reduced coenzymes (3 NADH + FADH$_2$) are formed. One molecule of GTP, a high-energy nucleoside triphosphate, is also formed.

$$CH_3-\overset{\overset{\displaystyle O}{\|}}{C}-SCoA \quad + \quad 2\,H_2O \quad + \quad 3\,NAD^+ \quad + \quad FAD \quad + \quad GDP \quad + \quad HPO_4{}^{2-}$$

**overall reaction**

$$2\,CO_2 \quad + \quad HSCoA \quad + \quad \underbrace{3\,NADH \; + \; 3\,H^+ \; + \; FADH_2} \quad + \quad GTP$$

exhaled gas     The coenzyme re-       The reduced coenzymes enter     energy source
enters the cycle.       the electron transport chain.

**[6] What are the main components of the electron transport chain and oxidative phosphorylation? (23.6)**

- The electron transport chain is a multistep process that takes place in the inner membrane of mitochondria. Electrons from reduced coenzymes enter the chain and are passed from one molecule to another in a series of redox reactions, releasing energy along the way. At the end of the chain, electrons and protons react with inhaled oxygen to form water.
- $H^+$ ions are pumped across the inner membrane of the mitochondrion, forming a high concentration of $H^+$ ions in the intermembrane space, thus creating a potential energy gradient. When the $H^+$ ions travel through the channel in the ATP synthase enzyme, this energy is used to convert ADP to ATP—a process called oxidative phosphorylation.

**[7] Why do compounds such as cyanide act as poisons when they disrupt the electron transport chain? (23.7)**

- Since all catabolic pathways converge at the electron transport chain, these steps are needed to produce energy for normal cellular processes. Compounds that disrupt a single step can halt ATP synthesis, so that an organism cannot survive.
- Cyanide from HCN irreversibly binds to the $Fe^{3+}$ ion of the enzyme cytochrome oxidase and, as a result, $Fe^{3+}$ cannot be reduced to $Fe^{2+}$ and water cannot be formed from oxygen. Since the electron transport chain is disrupted, energy is not generated for oxidative phosphorylation, ATP is not synthesized, and cell death often follows.

## Problem Solving

## [1] ATP and Energy Production (23.3)

**Example 23.1** The phosphorylation of glycerol to glycerol 3-phosphate requires 2.2 kcal/mol of energy. This unfavorable reaction can be driven by the hydrolysis of ATP to ADP. (a) Write the equation for the coupled reaction. (b) How much energy is released in the coupled reaction?

$$glycerol \quad + \quad HPO_4{}^{2-} \quad \longrightarrow \quad glycerol\ 3\text{-phosphate} \quad + \quad H_2O$$

**Analysis**
Write the equation for the hydrolysis of ATP to ADP, which releases 7.3 kcal/mol of energy (Section 23.3A). Add together the substances in this equation and the given equation to give the net equation—that is, the equation for the coupled reaction. To determine the overall energy change, add together the energy changes for each step.

**Solution**

a. The coupled equation shows the overall reaction that combines the phosphorylation of glycerol and the hydrolysis of ATP.

Cross out compounds that appear on both sides of the reaction arrows.

Energy change

Phosphorylation: [1]  glycerol  +  H~~PO~~$_4{}^{2-}$  ⟶  glycerol 3-phosphate  +  H~~2~~O  +2.2 kcal/mol

Hydrolysis: [2]  ATP  +  H~~2~~O  ⟶  ADP  +  H~~PO~~$_4{}^{2-}$  −7.3 kcal/mol

**Coupled reaction:** glycerol  +  ATP  ⟶  glycerol 3-phosphate  +  ADP

b. The overall energy is the sum of the energy changes for each step:
+2.2 kcal/mol + (−7.3) kcal/mol = −5.1 kcal/mol.

## [2] Coenzymes in Metabolism (23.4)

**Example 23.2** Label the reaction as an oxidation or reduction, and give the reagent, $NAD^+$ or NADH, that would be used to carry out the reaction.

ascorbic acid          dehydroascorbic acid

**Analysis**

Count the number of C–O bonds in the starting material and product. Oxidation increases the number of C–O bonds and reduction decreases the number of C–O bonds. $NAD^+$ is a coenzyme used for an oxidation, and NADH is a coenzyme used for a reduction.

**Solution**

The conversion of ascorbic acid to dehydroascorbic acid is an oxidation, since the product has more C–O bonds than the reactant. To carry out the oxidation, the oxidizing agent $NAD^+$ could be used.

ascorbic acid          dehydroascorbic acid

## [3] The Citric Acid Cycle (23.5)

**Example 23.3** (a) Using curved arrow symbolism, write out the reaction of isocitrate with $NAD^+$ to form α-ketoglutarate and carbon dioxide. (b) Classify the reaction as an oxidation, reduction, or decarboxylation.

## Analysis

Use Figure 23.7 to draw the structures for isocitrate and $\alpha$-ketoglutarate. Draw the organic reactant and product on the horizontal arrow and the oxidizing reagent $NAD^+$, which is converted to NADH, on the curved arrow. Oxidation reactions result in a loss of electrons, a loss of hydrogen, or a gain of oxygen. Reduction reactions result in a gain of electrons, a gain of hydrogen, or a loss of oxygen. A decarboxylation results in the loss of $CO_2$.

## Solution

a. Equation:

b. Two processes occur in this reaction. An alcohol with one C–O bond is converted to a carbonyl group with two C–O bonds. This reaction is an oxidation since the number of C–O bonds increases. Also, $CO_2$ is removed from the starting material, so decarboxylation occurs as well.

## Self-Test

**[1] Fill in the blank with one of the terms listed below.**

Acetyl CoA (23.4)
Adenosine 5'-diphosphate (ADP, 23.3)
Adenosine 5'-triphosphate (ATP, 23.3)
Anabolism (23.1)

Catabolism (23.1)
Citric acid cycle (23.5)
Coenzyme A (23.4)
Coupled reactions (23.3)

Electron transport chain (23.6)
Metabolism (23.1)
Oxidative phosphorylation (23.6)

1. _____ is formed by adding three phosphates to the 5'-OH group of a nucleoside composed of the sugar ribose and the base adenine.
2. The _____ is a multistep process that relies on four enzyme systems, called complexes I, II, III, and IV, as well as mobile electron carriers.
3. _____ differs from other coenzymes in this chapter because it is not an oxidizing or reducing agent.
4. The _____ is a cyclic metabolic pathway that begins with the addition of acetyl CoA to a four-carbon substrate and ends when the same four-carbon compound is formed as a product eight steps later.
5. _____ is formed by adding two phosphates to the 5'-OH group of a nucleoside composed of the sugar ribose and the base adenine.
6. _____ are pairs of reactions that occur together. The energy released by one reaction provides the energy to drive the other reaction.
7. _____ is a process in which energy from the oxidation of reduced coenzymes is used to transfer a phosphate group.
8. When an acetyl group is bonded to coenzyme A, the product is called _____.
9. _____ is the breakdown of large molecules into smaller ones with the release of energy.
10. _____ is the sum of all the chemical reactions that take place in an organism.

11. _____ is the synthesis of large molecules from smaller ones, and energy is generally absorbed in the process.

**[2] Fill in the blank with one of the terms listed below.**

a. organelles          c. mitochondria          e. intermembrane space
b. matrix              d. cytoplasm

12. The _____ is the region of the cell between the cell membrane and the nucleus.
13. _____ are specialized structures, each of which has a specific function.
14. _____ are small, sausage-shaped organelles in which energy production takes place.
15. The area between the outer membrane and inner membrane of a mitochondrion is called the _____.
16. Energy production occurs within the _____, the area surrounded by the inner membrane of the mitochondrion.

**[3] Classify each molecule as an oxidizing agent, a reducing agent, or neither.**

a. oxidizing agent          b. reducing agent          c. neither

17. NADH
18. FAD
19. ADP
20. A coenzyme that gains hydrogen atoms
21. A coenzyme that loses hydrogen atoms

**[4] Fill in the table with the terms listed.**

a. monosaccharides          c. triacylglycerols          e. protein
b. glycolysis               d. amino acid catabolism     f. fat

| Food | Basic Unit | Components | Catabolic Pathway |
|------|-----------|-----------|-------------------|
| Olive oil | 22. | Fatty acids and glycerol | Fatty acid oxidation |
| Pasta | Carbohydrates | 23. | 24. |
| Chicken breast | Protein | Amino acids | 25. |

## Answers to Self-Test

1. Adenosine 5'-triphosphate (ATP)
2. electron transport chain
3. Coenzyme A
4. citric acid cycle
5. Adenosine 5'-diphosphate (ADP)
6. Coupled reactions
7. Oxidative phosphorylation
8. acetyl CoA
9. Catabolism
10. Metabolism
11. Anabolism
12. d
13. a
14. c
15. e
16. b
17. b
18. a
19. c
20. a
21. b
22. c
23. a
24. b
25. d

## Solutions to In-Chapter Problems

**23.1** By funneling all catabolic pathways into a single common pathway, it is possible to use the same reactions and the same enzyme systems to metabolize all types of biomolecules.

**23.2** Hydrolysis of GTP to GDP is similar to the hydrolysis of ATP to ADP shown in Figure 23.4.

$$GTP \ + \ H_2O \ \longrightarrow \ GDP \ + \ HPO_4{}^{2-}$$

**23.3** Write the equation for the coupled reaction and determine how much energy is released as in Example 23.1.

Cross out compounds that appear on both sides of the reaction arrows.

a. [1] glucose + H̶P̶O̶₄²⁻ ⟶ glucose 1-phosphate + H̶₂̶O̶

[2] ATP + H̶₂̶O̶ ⟶ ADP + H̶P̶O̶₄²⁻

**Coupled reaction:** glucose + ATP ⟶ glucose 1-phosphate + ADP

b. Overall energy change: +5.0 kcal/mol + (−7.3) kcal/mol = −2.3 kcal/mol
(Reaction [1])  (Reaction [2])

**23.4** Write out each reaction using curved arrow symbolism.

a. fructose ⟶ fructose 6-phosphate (ATP → ADP)

b. glucose ⟶ glucose 6-phosphate (ATP → ADP)

**23.5** Calculate the energy change in the reaction.

−10.3 kcal/mol of energy released

creatine phosphate + ADP ⟶ creatine + ATP

+7.3 kcal/mol of energy absorbed

Overall energy change: −10.3 kcal/mol + 7.3 kcal/mol = −3.0 kcal/mol of energy released

**23.6** Label each reaction as an oxidation or reduction and give the reagent that would be used to carry out the reaction as in Example 23.2.

a. $H_2C{=}O$ ⟶ $CH_3OH$ (NADH + H⁺ → NAD⁺)

one fewer C–O bond
**reduction**

b. $CH_3{-}\underset{H}{\overset{OH}{C}}{-}COO^-$ ⟶ $CH_3{-}\overset{O}{C}{-}COO^-$ (NAD⁺ → NADH + H⁺)

one more C–O bond
**oxidation**

**23.7** Riboflavin has numerous polar groups (OH and NH) that can hydrogen bond to water, making it water soluble.

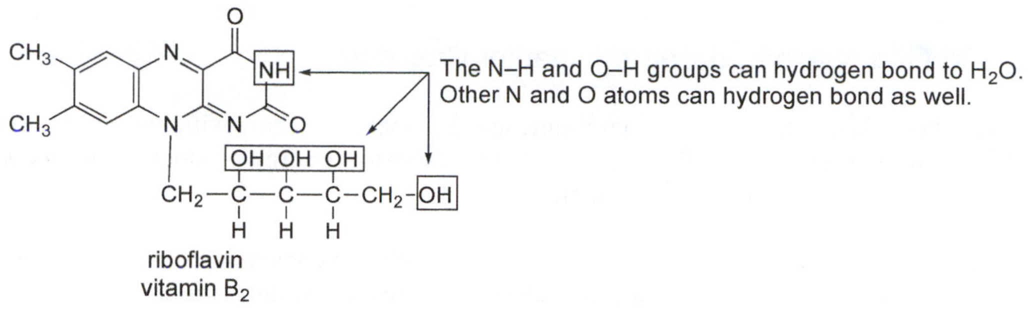

riboflavin
vitamin B₂

The N–H and O–H groups can hydrogen bond to $H_2O$.
Other N and O atoms can hydrogen bond as well.

**23.8** The many polar groups in vitamin $B_5$ make it water soluble.

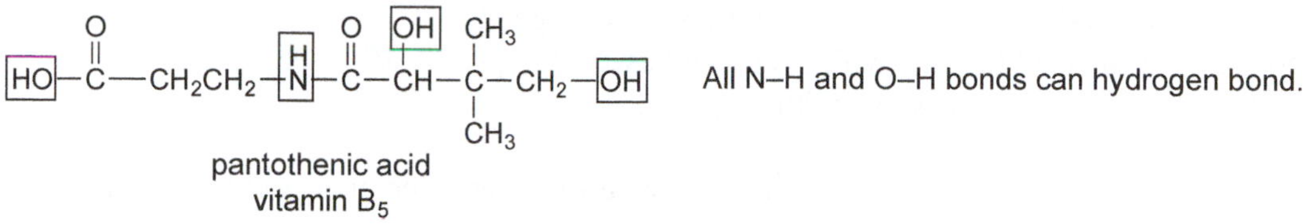

pantothenic acid
vitamin B₅

All N–H and O–H bonds can hydrogen bond.

**23.9** Draw the products of hydrolysis of the thioester $CH_3CH_2CH_2COSCoA$.

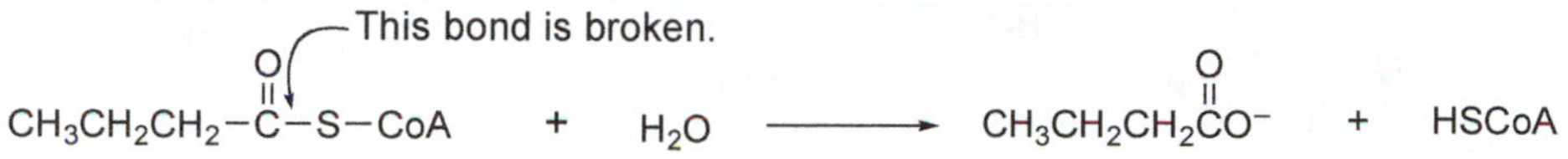

**23.10** Write out the reaction with curved arrow symbolism, and classify the reaction as an oxidation, reduction, or decarboxylation as in Example 23.3.

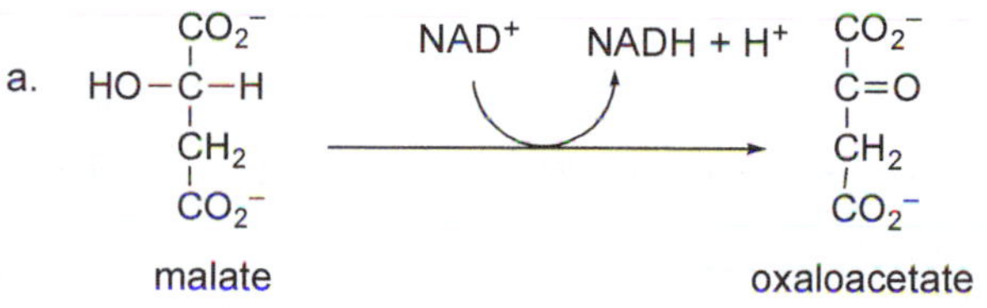

a.

malate                    oxaloacetate

b. This reaction is an oxidation, since an alcohol with one C–O bond is converted to a carbonyl group with two C–O bonds.

**23.11** Succinate dehydrogenase is so named because the elements of hydrogen ($H_2$) are removed from succinate, and dehydrogenase enzymes remove hydrogen atoms.

**23.12** If NADH and $FADH_2$ were not oxidized in the electron transport chain, the citric acid cycle and aerobic metabolism would cease, since $NAD^+$ and FAD are required in the reactions of the citric acid cycle.

**23.13**  a. The conversion of $Fe^{3+}$ to $Fe^{2+}$ is a reduction reaction because $Fe^{3+}$ gains an electron.
 b. $Fe^{3+}$ is an oxidizing agent because it gains electrons, causing another species to be oxidized.

**23.14** The pH would be lower in the intermembrane space since the $H^+$ concentration is higher in the intermembrane space than in the matrix.

## Solutions to Odd-Numbered End-of-Chapter Problems

**23.15**   Mitochondria contain an outer membrane and an inner membrane with many folds. The area between these two membranes is called the intermembrane space. Energy production occurs within the matrix, the area surrounded by the inner membrane.

**23.17**   The steps of catabolism in order: hydrolysis of starch, the conversion of glucose to acetyl CoA, the citric acid cycle, the electron transport chain, oxidative phosphorylation.

**23.19**   a. cleavage of a protein with chymotrypsin: stage [1]
b. oxidation of a fatty acid to acetyl CoA: stage [2]
c. oxidation of malate to oxaloacetate with $NAD^+$: stage [3]
d. conversion of ADP to ATP with ATP synthase: stage [4]
e. hydrolysis of starch to glucose with amylase: stage [1]

**23.21**   Coupled reactions are pairs of reactions that occur together.  The energy released by one reaction provides the energy to drive the other reaction.

**23.23**

$$CH_3-\overset{\overset{O}{\|}}{C}-O-\overset{\overset{O}{\|}}{\underset{\underset{O^-}{|}}{P}}-O^- \xrightarrow{\;H_2O\;} CH_3\overset{\overset{O}{\|}}{C}O^- \;+\; HPO_4^{2-}$$

This bond is broken.

acetyl phosphate

**23.25**   Write the equation for the coupled reaction and determine how much energy is released as in Example 23.1.

|  |  |  | Energy change |
|---|---|---|---|
| a. | succinate + HSCoA $\longrightarrow$ succinyl CoA + $\cancel{H_2O}$ | | +9.4 kcal/mol |
|  | ATP + $\cancel{H_2O}$ $\longrightarrow$ ADP + $HPO_4^{2-}$ | | −7.3 kcal/mol |

**Coupled reaction:**  ATP + succinate + HSCoA $\longrightarrow$ succinyl CoA + ADP + $HPO_4^{2-}$

b.  The reaction is not energetically favored because an input of energy is required (+2.1 kcal/mol).

Overall energy change:  +9.4 kcal/mol + (−7.3) kcal/mol = +2.1 kcal/mol

**23.27**   Answer each question about the phosphorylation of glucose.

a. glucose + $HPO_4^{2-}$ $\longrightarrow$ glucose 1-phosphate + $H_2O$

b. Coupled reaction: glucose + ATP $\longrightarrow$ glucose 1-phosphate + ADP

c. Overall energy change:  +5.0 kcal/mol + (−7.3) kcal/mol = −2.3 kcal/mol
   phosphorylation   ATP hydrolysis

**23.29**  Answer each question about the phosphorylation of fructose 6-phosphate.

a.  The energy change for the reverse reaction is –3.9 kcal/mol.

Energy change

b.

[1]  fructose 6-phosphate  +  $HPO_4^{2-}$  ⟶  fructose 1,6-bisphosphate  +  H₂O   +3.9 kcal/mol

[2]      ATP  +  H₂O    ⟶    ADP  +  $HPO_4^{2-}$      –7.3 kcal/mol

Coupled reaction:     fructose 6-phosphate  +  ATP  ⟶  fructose 1,6-bisphosphate + ADP

c. Overall energy change:  +3.9 kcal/mol  +  (–7.3) kcal/mol  =  –3.4 kcal/mol
          (Reaction [1])     (Reaction [2])

**23.31**  Answer each question about the reaction.

a.  Coupled reaction:  1,3-bisphosphoglycerate  +  ADP  ⟶  3-phosphoglycerate  +  ATP

b.  –11.8 kcal/mol + 7.3 kcal/mol = –4.5 kcal/mol

c.  Yes, this is an energetically favorable coupled reaction that could be used to synthesize ATP from ADP.

**23.33**  Draw the structure of GTP and the hydrolysis product.

a. 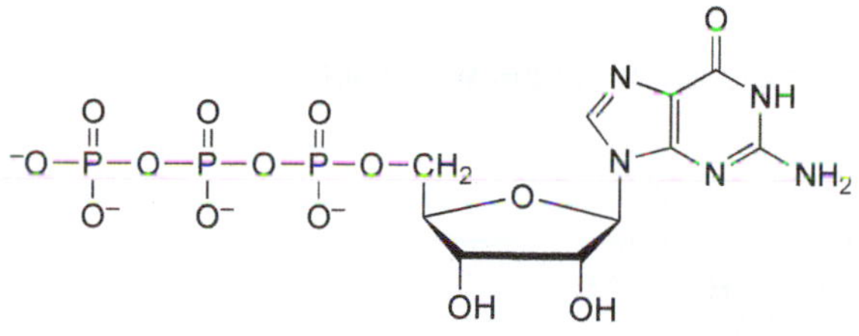 GTP

b. 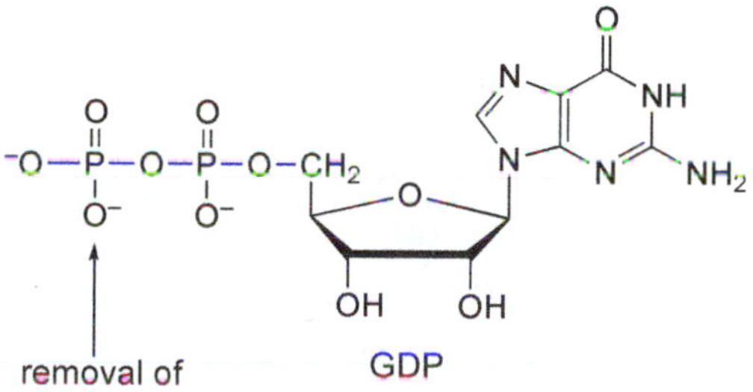 removal of one phosphate group    GDP

**23.35**  An oxidizing agent causes an oxidation reaction to occur, so the oxidizing agent is reduced. A reducing agent causes a reduction reaction to occur, so the reducing agent is oxidized.

a. $FADH_2$: reducing agent        b. ATP: neither        c. $NAD^+$: oxidizing agent

**23.37**  When a substrate is oxidized, $NAD^+$ is reduced to NADH because it gains a hydrogen.  $NAD^+$ is an oxidizing agent.

**23.39** Label each reaction as an oxidation or reduction and give the reagent that would be used to carry out the reaction as in Example 23.2.

a.

$NAD^+$ $NADH + H^+$

$-OH \longrightarrow =O$

one more C–O bond
**oxidation**

b.

$NADH + H^+$ $NAD^+$

CHO $\longrightarrow$ $CH_2OH$

one more C–H bond
**reduction**

**23.41** Find the step in the citric acid cycle with each characteristic. Use Figure 23.7 as a guide.

a. The reaction generates NADH: steps [3], [4], and [8].
b. $CO_2$ is removed: steps [3] and [4].
c. The reaction utilizes FAD: step [6].
d. The reaction forms a new carbon–carbon single bond: step [1].

**23.43** Draw the structural formula of **A**. Then find the structure in the citric acid cycle to answer each question.

a.

$\begin{array}{cc} H & CO_2^- \\ & C=C \\ -O_2C & H \end{array}$

**A**
fumarate

b. immediate precursor to **A**:
   succinate

c. formed from **A**:
   malate

**23.45** Look at Figure 23.7 to find intermediates in the citric acid cycle with each characteristic.

a.

$\begin{array}{c} CO_2^- \\ CH_2 \\ H-C-CO_2^- \\ {}^* \\ HO-C-H \\ {}^* \\ CO_2^- \end{array}$

isocitrate

* = chirality center

b.

$\begin{array}{c} CO_2^- \\ HO-C-H \\ CH_2 \\ CO_2^- \end{array}$ $\begin{array}{c} CO_2^- \\ CH_2 \\ H-C-CO_2^- \\ HO-C-H \\ CO_2^- \end{array}$

malate     isocitrate

2° alcohols

**23.47** Answer each question about the conversion of isocitrate to α-ketoglutarate.

$\begin{array}{c} CO_2^- \\ CH_2 \\ H-C-CO_2^- \\ HO-C-H \\ CO_2^- \end{array}$

isocitrate

$\xrightarrow[\substack{\text{a. oxidation} \\ \text{c. } NAD^+ \text{ here}}]{[3a]}$

$\begin{array}{c} CO_2^- \\ CH_2 \\ H-C-CO_2^- \\ C=O \\ CO_2^- \end{array}$

oxalosuccinate

$\xrightarrow[\text{b. decarboxylation}]{[3b]}$

$\begin{array}{c} CO_2^- \\ CH_2 \\ CH_2 \\ C=O \\ CO_2^- \end{array}$

α-ketoglutarate

d. The elements of $H_2$ (hydrogen) are removed in converting isocitrate to oxalosuccinate, so the enzyme that catalyzes this step is called isocitrate dehydrogenase.

**23.49**  A hydration reaction (addition of $H_2O$) occurs in step [7] of the citric acid cycle.

$$\underset{\text{fumarate}}{\underset{\displaystyle CO_2^-}{\overset{\displaystyle CO_2^-}{\underset{|}{\overset{|}{\underset{CO_2^-}{\overset{CH}{\underset{||}{HC}}}}}}}} \quad + \quad H_2O \quad \longrightarrow \quad \underset{\text{malate}}{\underset{\displaystyle CO_2^-}{\overset{\displaystyle CO_2^-}{\underset{|}{\overset{|}{\underset{CO_2^-}{\overset{HO-C-H}{\underset{|}{CH_2}}}}}}}}$$

**23.51**  Aerobic respiration means that a reaction requires oxygen. $O_2$ is converted to $H_2O$ in the final stage of electron transport.

**23.53**  a.  $FADH_2$ donates electrons to the electron transport chain.
       b.  ADP is a substrate for the formation of ATP.
       c.  ATP synthase catalyzes the formation of ATP from ADP.
       d.  The inner mitochondrial membrane contains the four complexes for the electron transport chain. ATP synthase is also embedded in the membrane and contains the $H^+$ ion channel that allows $H^+$ to return to the matrix.

**23.55**  The final products of the electron transport chain are $NAD^+$, FAD, and $H_2O$.

**23.57**  $FADH_2$ enters the electron transport chain at complex II, whereas NADH enters at complex I, so they result in different amounts of ATP formation.

**23.59**  a.  Pepsin catalyzes the hydrolysis of proteins to amino acids in the stomach.
       b.  Succinate dehydrogenase catalyzes the conversion of succinate to fumarate in step [6] of the citric acid cycle.
       c.  ATP synthase catalyzes the oxidative phosphorylation of ADP to ATP.

**23.61**  Each two-carbon fragment of acetyl CoA that enters the citric acid cycle produces three molecules of NADH and one molecule of $FADH_2$. These then enter the electron transport chain and serve as reducing agents.  For each NADH that enters the electron transport chain, there are 2.5 ATPs produced, and for each $FADH_2$, there are 1.5 ATPs produced.  In addition, there is one GTP (an ATP equivalent) produced in the citric acid cycle.

| In total: | | | |
|---|---|---|---|
| 3 NADH × 2.5 ATP/NADH | = | 7.5 | ATPs |
| 1 $FADH_2$ × 1.5 ATP/$FADH_2$ | = | 1.5 | ATPs |
| 1 GTP (ATP equivalent) | = | 1 | ATP |
| | | **10** | **ATPs** |

**23.63**  ATP has three phosphate groups and ADP has two.

**23.65**  Answer each question about glucose 1-phosphate.

a.

glucose 1-phosphate

b.

glucose 1-phosphate

**23.67** The citric acid cycle generates the reducing agents NADH and $FADH_2$ that enter the electron transport chain. In the process, $NAD^+$ and FAD, coenzymes needed for oxidation reactions in the citric acid cycle, are also produced.

**23.69** There would be more mitochondria in the heart because it has far more metabolic needs than bone.

**23.71** Creatine phosphate is a high-energy phosphate that generates energy upon hydrolysis.

**23.71** Use conversion factors to solve the problem.

$$500. \text{Cal} \times \frac{1 \text{ kcal}}{1 \text{ Cal}} \times \frac{1 \text{ mol ATP}}{7.3 \text{ kcal}} = \textbf{68 moles}$$

$$68 \text{ mol} \times \frac{6.02 \times 10^{23} \text{ molecules}}{1 \text{ mol}} = \textbf{4.1} \times \textbf{10}^{25} \textbf{ molecules of ATP}$$

# Chapter 24 Carbohydrate, Lipid, and Protein Metabolism

## Chapter Review

**[1] What are the main elements that provide clues to the outcome of a biochemical reaction? (24.2)**
- To understand the course of a biochemical reaction, examine the functional groups that are added or removed, the reagents (coenzymes or other materials), and the enzyme. The name of an enzyme is often a clue as to the type of reaction.

**[2] What are the main aspects of glycolysis? (24.3)**
- Glycolysis is a linear, 10-step pathway that converts glucose to two three-carbon pyruvate molecules. In the energy-investment phase, steps [1]–[5], the energy from two ATP molecules is used for phosphorylation and two three-carbon products are formed. In the energy-generating phase, steps [6]–[10], the following species are generated: two pyruvate molecules ($CH_3COCO_2^-$), 2 NADHs, and 4 ATPs.
- The net result of glycolysis, considering both phases, is 2 $CH_3COCO_2^-$, 2 NADHs, and 2 ATPs.

**[3] What are the major pathways for pyruvate metabolism? (24.4)**
- When oxygen is plentiful, pyruvate is converted to acetyl CoA, which can enter the citric acid cycle.
- When the oxygen level is low, the anaerobic metabolism of pyruvate forms lactate and $NAD^+$.
- In yeast and other microorganisms, pyruvate is converted to ethanol and $CO_2$ by fermentation.

**[4] How much ATP is formed by the complete catabolism of glucose? (24.5)**
- To calculate the amount of ATP formed in the catabolism of glucose, we must take into account the ATP yield from glycolysis, the oxidation of two molecules of pyruvate to two molecules of acetyl CoA, the citric acid cycle, and oxidative phosphorylation.
- As shown in Figure 24.6, the complete catabolism of glucose forms six $CO_2$ molecules and 32 molecules of ATP.

**[5] What are the main features of gluconeogenesis? (24.6)**
- Gluconeogenesis is the synthesis of glucose from noncarbohydrate sources—lactate, amino acids, or glycerol. Gluconeogenesis converts two molecules of pyruvate to glucose. Conceptually, gluconeogenesis is the reverse of glycolysis, but three steps in gluconeogenesis require different enzymes.

$$2 \ CH_3-\overset{\overset{\displaystyle O}{\|}}{C}-CO_2^- \quad \xrightarrow{\text{gluconeogenesis}} \quad C_6H_{12}O_6$$

pyruvate                               glucose

- Gluconeogenesis occurs when the body has depleted its supplies of glucose and stored glycogen, and takes place during sustained physical exercise and fasting.

## [6] Describe the main features of the β-oxidation of fatty acids. (24.7)
- β-Oxidation is a spiral metabolic pathway that sequentially cleaves two-carbon acetyl CoA units from an acyl CoA derived from a fatty acid. Each cycle of β-oxidation consists of a four-step sequence that forms one molecule each of acetyl CoA, NADH, and $FADH_2$.

$$CH_3(CH_2)_{16}-\overset{\overset{\displaystyle O}{\|}}{C}-OH \quad \longrightarrow \quad CH_3(CH_2)_{16}-\overset{\overset{\displaystyle O}{\|}}{C}-SCoA \quad \longrightarrow \quad 9 \ CH_3-\overset{\overset{\displaystyle O}{\|}}{C}-SCoA$$

fatty acid                         thioester                       acetyl CoA

## [7] How much ATP is formed from complete fatty acid oxidation? (24.7)
- To determine the ATP yield from the complete catabolism of a fatty acid, we must consider the ATP used up in the synthesis of the acyl CoA, the ATP generated from coenzymes produced during β-oxidation, and the ATP that results from the catabolism of each acetyl CoA.
- As an example, the complete catabolism of stearic acid, $C_{18}H_{36}O_2$, yields 120 ATPs.

## [8] What are ketone bodies and how do they play a role in metabolism? (24.8)
- Ketone bodies—acetoacetate, β-hydroxybutyrate, and acetone—are formed when acetyl CoA levels exceed the capacity of the citric acid cycle. Ketone bodies can be re-converted to acetyl CoA and metabolized for energy. When the level of ketone bodies is high, the pH of the blood can be lowered, causing ketoacidosis.

$$2 \ CH_3-\overset{\overset{\displaystyle O}{\|}}{C}-SCoA \xrightarrow[\text{several steps}]{} CH_3-\overset{\overset{\displaystyle O}{\|}}{C}-CH_2-\overset{\overset{\displaystyle O}{\|}}{C}-O^- \xrightarrow{\curvearrowright} CH_3-\overset{\overset{\displaystyle OH}{|}}{\underset{\underset{\displaystyle H}{|}}{C}}-CH_2-\overset{\overset{\displaystyle O}{\|}}{C}-O^-$$

NADH + H⁺     NAD⁺

acetoacetate                            β-hydroxybutyrate

$$\searrow CO_2$$

$$CH_3-\overset{\overset{\displaystyle O}{\|}}{C}-CH_3$$

acetone

## [9] What are the main features of amino acid catabolism? (24.9)
- The catabolism of amino acids involves two parts. First, the amino group is removed by transamination followed by oxidative deamination. The $NH_4^+$ ion formed enters the urea cycle where it is converted to urea and eliminated in urine. The carbon skeletons of the amino acids are catabolized by a variety of pathways to yield pyruvate, acetyl CoA, or an intermediate in the citric acid cycle.

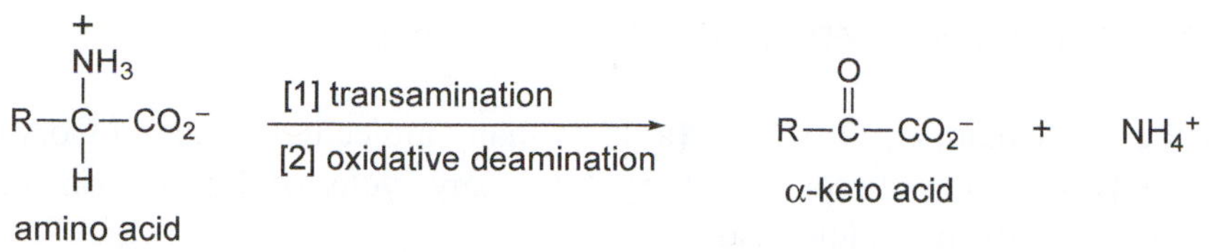

$$
\underset{\text{amino acid}}{R-\overset{\overset{\displaystyle +}{\overset{\displaystyle NH_3}{|}}}{\underset{|}{\underset{H}{C}}}-CO_2^-}
\xrightarrow[\text{[2] oxidative deamination}]{\text{[1] transamination}}
\underset{\alpha\text{-keto acid}}{R-\overset{\overset{\displaystyle O}{||}}{C}-CO_2^-} \quad + \quad NH_4^+
$$

## Problem Solving

## [1] Understanding Biochemical Reactions (24.2)

**Example 24.1** Analyze the following reaction by considering the functional groups that change and the name of the enzyme.

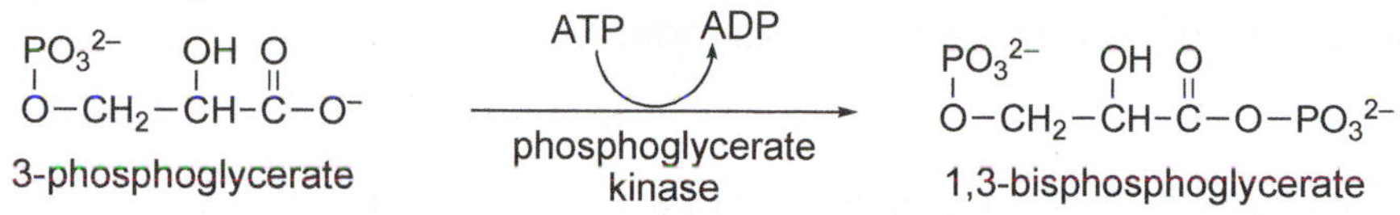

$$
\underset{\text{3-phosphoglycerate}}{\overset{PO_3^{2-}\quad OH\ \ O}{O-CH_2-CH-C-O^-}}
\xrightarrow[\substack{\text{phosphoglycerate}\\\text{kinase}}]{\overset{ATP\qquad ADP}{\frown}}
\underset{\text{1,3-bisphosphoglycerate}}{\overset{PO_3^{2-}\quad OH\ \ O}{O-CH_2-CH-C-O-PO_3^{2-}}}
$$

### Analysis

- Consider the functional groups that are added or removed as a clue to the type of reaction.
- Classify the reagent (when one is given) as to the type of reaction it undergoes.
- Use the name of the enzyme as a clue to the reaction type.

### Solution

A phosphate group is added to 3-phosphoglycerate in the reaction, meaning that a **phosphorylation** has occurred. Since kinase enzymes catalyze the addition or removal of phosphates, the enzyme used for the reaction is phosphoglycerate kinase. One phosphate bond is broken in ATP, giving ADP as one of the by-products.

---

**Example 24.2** Analyze the following reaction by considering the functional groups that change, the coenzyme utilized, and the name of the enzyme.

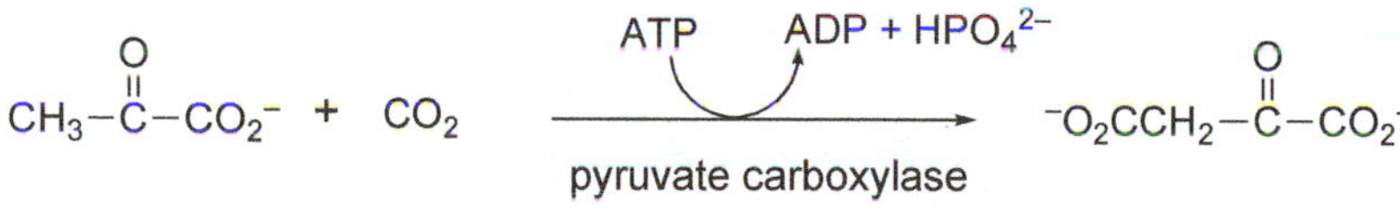

$$
\underset{}{CH_3-\overset{\overset{\displaystyle O}{||}}{C}-CO_2^-} \ + \ CO_2
\xrightarrow[\text{pyruvate carboxylase}]{\overset{ATP\qquad ADP+HPO_4^{2-}}{\frown}}
{}^-O_2CCH_2-\overset{\overset{\displaystyle O}{||}}{C}-CO_2^-
$$

### Analysis

- Consider the functional groups that are added or removed as a clue to the type of reaction.
- Classify the reagent—that is, any coenzyme or other reactant—as to the type of reaction it undergoes.
- Use the name of the enzyme as a clue to the reaction type.

### Solution

In this reaction, a carboxylate (–COO$^-$) is added to the reactant pyruvate from $CO_2$. Since carboxylase enzymes catalyze the addition of a carboxylate, the enzyme is called pyruvate carboxylase. ATP is the coenzyme used, and hydrolysis of one phosphate group provides the energy for the reaction, forming ADP and $HPO_4^{2-}$.

## [2] The Catabolism of Triacylglycerols (24.7)

**Example 24.3** Consider oleic acid, $C_{17}H_{33}CO_2H$. (a) How many molecules of acetyl CoA are formed from the complete β-oxidation of oleic acid? (b) How many cycles of β-oxidation are needed for the complete catabolism of oleic acid?

**Analysis**

The number of carbons in the fatty acid determines the number of molecules of acetyl CoA formed and the number of times β-oxidation occurs.

- The number of molecules of acetyl CoA equals one-half the number of carbons in the original fatty acid.
- Because the final turn of the cycle forms *two* molecules of acetyl CoA, the number of cycles is one fewer than the number of acetyl CoA molecules formed.

**Solution**

Since oleic acid has 18 carbons, it forms nine molecules of acetyl CoA from eight cycles of β-oxidation.

## [3] Amino Acid Metabolism (24.9)

**Example 24.4** What products are formed in the following transamination reaction?

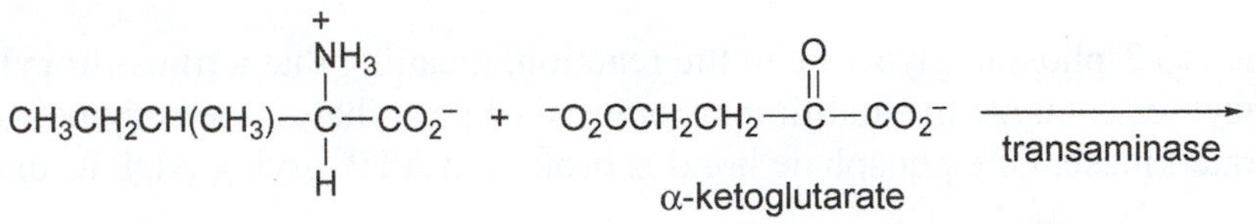

**Analysis**

To draw the products of transamination, convert the C–H and C–NH$_3^+$ groups of the amino acid to a C=O, forming an α-keto acid.

**Solution**

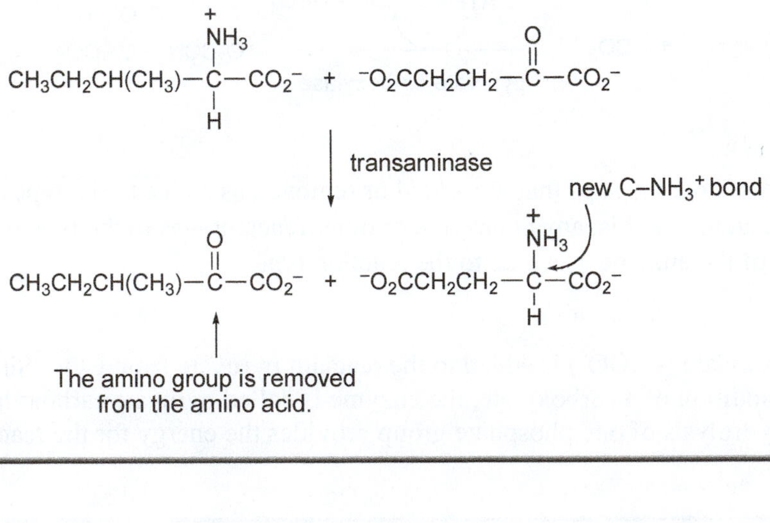

**Example 24.5** What final products are formed when the following amino acid is subjected to transamination followed by oxidative deamination?

$$\overset{\overset{+}{N}H_3}{\underset{\underset{H}{|}}{(CH_3)_2CH-C-CO_2^-}}$$

**Analysis**

To draw the organic product, replace the C–H and C–$NH_3^+$ on the $\alpha$ carbon of the amino acid by C=O. $NH_4^+$ is formed from the amino group.

**Solution**

$\alpha$ carbon $\overset{\overset{+}{N}H_3}{\underset{\underset{H}{|}}{(CH_3)_2CH-C-CO_2^-}}$ $\xrightarrow[\text{[2] oxidative deamination}]{\text{[1] transamination}}$ $\overset{\overset{O}{||}}{(CH_3)_2CH-C-CO_2^-}$ + $NH_4^+$

---

## Self-Test

### [1] Fill in the blank with one of the terms listed below.

Cori cycle (24.6)　　　　　　Isomerase (24.2)　　　　　　Kinase (24.2)
Fermentation (24.4)　　　　　Ketogenesis (24.8)　　　　　$\beta$-Oxidation (24.7)
Glucogenic amino acids (24.9)　Ketogenic amino acids (24.9)　Oxidative deamination (24.9)
Gluconeogenesis (24.6)　　　　Ketone bodies (24.8)　　　　Transamination (24.9)
Glycolysis (24.3)　　　　　　　Ketosis (24.8)

1. _____ is an anabolic process that synthesizes glucose from pyruvate.
2. A _____ catalyzes the transfer of a phosphate group from one substrate to another.
3. Cycling compounds from the muscle to the liver and back to the muscle is called the _____.
4. _____ is a process in which two-carbon acetyl CoA units are sequentially cleaved from a fatty acid.
5. Under some circumstances ketone bodies accumulate, a condition called _____.
6. _____ are catabolized to pyruvate or an intermediate in the citric acid cycle, so these compounds can be used to synthesize glucose.
7. _____ is the anaerobic conversion of glucose to ethanol and $CO_2$.
8. _____ is the synthesis of ketone bodies from acetyl CoA.
9. An _____ catalyzes the conversion of one isomer to another.
10. _____ is the transfer of an amino group from an amino acid to an $\alpha$-keto acid, usually $\alpha$-ketoglutarate.
11. In _____, the C–H and C–$NH_3^+$ bonds on the $\alpha$ carbon of glutamate are converted to C=O and an ammonium ion, $NH_4^+$.
12. _____ is a catabolic process that converts glucose to pyruvate.
13. Acetoacetate, $\beta$-hydroxybutyrate, and acetone are all _____.

14. _____ are converted to acetyl CoA, or the related thioester acetoacetyl CoA, $CH_3COCH_2COSCoA$. These catabolic products cannot be used to synthesize glucose, but they can be converted to ketone bodies and yield energy by this pathway.

**[2] Match the type of enzyme with the action.**

   a. carboxylase                b. decarboxylase                c. dehydrogenase

15. Addition of a carboxylate ($-COO^-$)
16. Removal of two hydrogen atoms
17. Removal of carbon dioxide ($CO_2$)

**[3] Pick the reaction conditions under which pyruvate is converted to each of the following:**

   a. acetyl CoA          b. lactate          c. ethanol

18. fermentation
19. aerobic conditions
20. anaerobic conditions

**[4] Match the enzyme with the reaction it catalyzes.**

   a. hexokinase                c. pyruvate decarboxylase          e. phosphohexose isomerase
   b. alcohol dehydrogenase     d. phosphofructokinase

21. fructose → fructose 6-phosphate
22. pyruvate → acetaldehyde
23. glucose → glucose 6-phosphate
24. acetaldehyde → ethanol
25. glucose 6-phosphate → fructose 6-phosphate

## Answers to Self-Test

| | | | | |
|---|---|---|---|---|
| 1. Gluconeogenesis | 6. Glucogenic amino acids | 11. oxidative deamination | 16. c | 21. d |
| 2. kinase | 7. Fermentation | 12. Glycolysis | 17. b | 22. c |
| 3. Cori cycle | 8. Ketogenesis | 13. ketone bodies | 18. c | 23. a |
| 4. β-Oxidation | 9. isomerase | 14. Ketogenic amino acids | 19. a | 24. b |
| 5. ketosis | 10. Transamination | 15. a | 20. b | 25. e |

## Solutions to In-Chapter Problems

**24.1** Analyze the reaction as in Example 24.1.

dihydroxyacetone phosphate → (triose phosphate isomerase) → glyceraldehyde 3-phosphate

The reactant and product are constitutional isomers that differ in the location of a carbonyl and hydroxyl group. Thus, an isomerase enzyme converts one isomer to another.

**24.2** Analyze the reaction as in Example 24.2.

The aldehyde (CHO) in the reactant is oxidized to a carboxylate anion (COO⁻). $NAD^+$ is the oxidizing agent, which gets reduced to NADH. A dehydrogenase enzyme removes hydrogen from the reactant in the oxidation.

**24.3** Compare glucose 6-phosphate and fructose 6-phosphate.

**24.4** The energy of the coupled reaction (–3.4 kcal/mol) represents the sum of the energies of the phosphorylation of fructose 6-phosphate and the hydrolysis of ATP (–7.3 kcal/mol).

phosphorylation + (–7.3 kcal/mol) = –3.4 kcal/mol
phosphorylation = +3.9 kcal/mol

Thus, the phosphorylation of fructose 6-phosphate requires 3.9 kcal/mol.

**24.5** Use Figures 24.2, 24.3, and 24.4 to determine which steps utilize kinases, and write the species that is phosphorylated.

step [1]: glucose; step [3]: fructose 6-phosphate; step [7]: ADP; step [10]: ADP

**24.6** Use Figures 24.2, 24.3, and 24.4 to determine which steps involve the conversion of one constitutional isomer to another.

step [2]: phosphohexose isomerase
step [5]: triose phosphate isomerase
step [8]: phosphoglycerate mutase

**24.7** Citrate is one intermediate in the citric acid cycle, so the rate of glycolysis decreases when there are high levels of citrate.

**24.8** Compare galactose and mannose.

a.

same molecular formula
same connectivity of atoms
different arrangement
**stereoisomers**

galactose

mannose

b.

galactose 1-phosphate

mannose 6-phosphate

c. same molecular formula
different connectivity of atoms
**constitutional isomers**

**24.9** NADH serves as the reducing agent for the conversion of pyruvate to both lactate and ethanol. The $NAD^+$ formed in the process can then be used as the oxidant in step [6] of glycolysis, making the process anaerobic.

**24.10** a. The conversion of pyruvate to acetyl CoA and the conversion of pyruvate to ethanol both create a two-carbon compound from the three-carbon pyruvate with loss of $CO_2$.
  b. The conversions are different in that the production of acetyl CoA generates NADH, whereas the production of ethanol consumes NADH.

**24.11** a. The conversions of pyruvate to lactate and pyruvate to ethanol both use NADH to generate a reduced compound.
  b. The conversions are different in that ethanol is a two-carbon product formed by decarboxylation, whereas lactate is a three-carbon compound formed by the addition of hydrogen.

**24.12** Use Figure 24.6 to determine how much ATP is produced in each transformation.

a. glucose $\rightarrow$ 2 pyruvate: 7 ATP
b. pyruvate $\rightarrow$ acetyl CoA: 2.5 ATP per pyruvate

c. glucose $\rightarrow$ 2 acetyl CoA: 12 ATP
d. 2 acetyl CoA $\rightarrow$ 4 $CO_2$: 20 ATP

**24.13** The conversion of pyruvate to acetyl CoA and steps [3] and [4] of the citric acid cycle form $CO_2$.

**24.14** Aerobic oxidation of glucose to $CO_2$ is much more efficient than anaerobic conversion of glucose to lactate. Thirty-two (32) ATPs are generated with aerobic metabolism and only two ATPs are generated with anaerobic metabolism to lactate.

**24.15** Since gluconeogenesis is formally the reverse of glycolysis, use Figures 24.2, 24.3, and 24.4 to determine the starting material and product for each step in gluconeogenesis.

|  | Starting Material | Product |
|---|---|---|
| a. | pyruvate | phosphoenolpyruvate |
| b. | 2-phosphoglycerate | 3-phosphoglycerate |
| c. | glucose 6-phosphate | glucose |

**24.16** The reaction involves the transfer of a phosphate group from ATP to glycerol by glycerol kinase. No oxidation or reduction occurs.

**24.17** For each fatty acid, determine the number of molecules of acetyl CoA formed from the complete catabolism and the number of cycles of β-oxidation needed for complete oxidation as in Example 24.3.

a. arachidic acid ($C_{20}H_{40}O_2$): 20 carbons, [1] 10 molecules of acetyl CoA, [2] 9 cycles
b. palmitoleic acid ($C_{16}H_{30}O_2$): 16 carbons, [1] 8 molecules of acetyl CoA, [2] 7 cycles

**24.18** Calculate the number of molecules of ATP released on the catabolism of palmitic acid.
[1] Determine the amount of ATP required to synthesize the acyl CoA from the fatty acid.
[2] Add up the ATP generated from the coenzymes produced during β-oxidation.
[3] Determine the amount of ATP that results from each acetyl CoA, and add the results for steps [1]–[3].

[1] 2 ATPs are required for the conversion of $C_{16}H_{32}O_2$ to $C_{15}H_{31}COSCoA$ (–2 ATPs).
[2] 7 cycles of β-oxidation:

| 7 NADH | × | 2.5 ATP/NADH | = | 17.5 ATPs |
|---|---|---|---|---|
| 7 FADH$_2$ | × | 1.5 ATP/FADH$_2$ | = | 10.5 ATPs |
|  |  | From reduced coenzymes: |  | 28 ATPs |

[3] From acetyl CoA:

| 8 acetyl CoA | × | 10 ATP/acetyl CoA | = | 80 ATPs |
|---|---|---|---|---|

Total steps [1]–[3]:

| (–2) + 28 + 80 | = | 106 ATP molecules from palmitic acid |
|---|---|---|
|  |  | **Answer** |

**24.19** Calculate the number of molecules of ATP released on the catabolism of arachidic acid using the steps in Answer 24.18.

[1] 2 ATPs are required for the conversion of $C_{20}H_{40}O_2$ to $C_{19}H_{39}COSCoA$ (–2 ATPs).

[2] 9 cycles of β-oxidation:

| | | | |
|---|---|---|---|
| 9 NADH | × | 2.5 ATP/NADH | = | 22.5 ATPs |
| 9 FADH$_2$ | × | 1.5 ATP/FADH$_2$ | = | 13.5 ATPs |
| | | From reduced coenzymes: | | 36 ATPs |

[3] From acetyl CoA:

10 acetyl CoA  ×  10 ATP/acetyl CoA  =  100 ATPs

Total steps [1]–[3]:

(–2) + 36 + 100   =   134 ATP molecules from arachidic acid

**Answer**

**24.20** Calculate the number of moles of ATP formed per gram of palmitic acid.

$$\frac{106 \text{ ATP}}{\text{mol palmitic acid}} \times \frac{1 \text{ mol}}{256 \text{ g}} = \frac{0.414 \text{ mol ATP}}{\text{g palmitic acid}}$$

**24.21** Compare the number of ATPs generated per carbon atom.

a.  $\dfrac{32 \text{ ATP}}{6 \text{ C's in glucose}}$  =  5.3 ATP/C in glucose

b.  $\dfrac{120 \text{ ATP}}{18 \text{ C's in stearic acid}}$  =  6.7 ATP/C in stearic acid

more energy per carbon

c. Since stearic acid produces more ATP per carbon atom than glucose, this supports the fact that lipids are more effective energy-storing molecules than carbohydrates.

**24.22** All three compounds contain at least one carbonyl group.

acetoacetate          acetone          β-hydroxybutyrate

**24.23** β-Hydroxybutyrate contains an alcohol and a carboxylate anion, but no ketone.

β-hydroxybutyrate

**24.24** When acetoacetate is converted to acetone, a $CO_2$ group is removed. Therefore, the enzyme is a decarboxylase.

acetoacetate                    acetone

**24.25** An increased concentration of acetyl CoA increases the production of ketone bodies, since acetyl CoA is the reactant that starts ketogenesis.

**24.26** Draw the products when each amino acid undergoes transamination as in Example 24.4.

a.
$$HOCH_2-\overset{\overset{+}{N}H_3}{\underset{H}{C}}-CO_2^- \;+\; {}^-O_2CCH_2CH_2-\overset{O}{C}-CO_2^- \longrightarrow HOCH_2-\overset{O}{C}-CO_2^- \;+\; {}^-O_2CCH_2CH_2-\overset{\overset{+}{N}H_3}{\underset{H}{C}}-CO_2^-$$

serine

b.
$$CH_3SCH_2CH_2-\overset{\overset{+}{N}H_3}{\underset{H}{C}}-CO_2^- \;+\; {}^-O_2CCH_2CH_2-\overset{O}{C}-CO_2^- \longrightarrow CH_3SCH_2CH_2-\overset{O}{C}-CO_2^- \;+\; {}^-O_2CCH_2CH_2-\overset{\overset{+}{N}H_3}{\underset{H}{C}}-CO_2^-$$

methionine

c.
$$HSCH_2-\overset{\overset{+}{N}H_3}{\underset{H}{C}}-CO_2^- \;+\; {}^-O_2CCH_2CH_2-\overset{O}{C}-CO_2^- \longrightarrow HSCH_2-\overset{O}{C}-CO_2^- \;+\; {}^-O_2CCH_2CH_2-\overset{\overset{+}{N}H_3}{\underset{H}{C}}-CO_2^-$$

cysteine

**24.27** Draw the products of transamination and oxidative deamination as in Example 24.5.

a.
$$CH_3CH(OH)-\overset{\overset{+}{N}H_3}{\underset{H}{C}}-CO_2^- \longrightarrow CH_3CH(OH)-\overset{O}{C}-CO_2^- \;+\; NH_4^+$$

b.
$$H-\overset{\overset{+}{N}H_3}{\underset{H}{C}}-CO_2^- \longrightarrow H-\overset{O}{C}-CO_2^- \;+\; NH_4^+$$

c.
$$CH_3CH_2CH(CH_3)-\overset{\overset{+}{N}H_3}{\underset{H}{C}}-CO_2^- \longrightarrow CH_3CH_2CH(CH_3)-\overset{O}{C}-CO_2^- \;+\; NH_4^+$$

**24.28** Use Figure 24.10 to determine what metabolic intermediates are produced from each amino acid.

a. cysteine: pyruvate  
b. aspartic acid: oxaloacetate  

c. valine: succinyl CoA  
d. threonine: acetyl CoA, pyruvate, and succinyl CoA

## Solutions to Odd-Numbered End-of-Chapter Problems

**24.29**   Analyze the reactions as in Examples 24.1 and 24.2.

a.   $CH_3-\overset{\overset{\displaystyle OH}{|}}{CH}-CH_2-\overset{\overset{\displaystyle O}{||}}{C}-SCoA$   $\xrightarrow[\substack{\text{β-hydroxyacyl CoA}\\\text{dehydrogenase}}]{\overset{\displaystyle NAD^+ \quad NADH + H^+}{\curvearrowright}}$   $CH_3-\overset{\overset{\displaystyle O}{||}}{C}-CH_2-\overset{\overset{\displaystyle O}{||}}{C}-SCoA$

Oxidation occurs with loss of hydrogen by a dehydrogenase enzyme.  The coenzyme $NAD^+$ serves as an oxidant.

b.   $^-O_2CCH_2-\overset{\overset{\displaystyle O}{||}}{C}-CO_2^-$   $\xrightarrow[\substack{\text{phosphoenolpyruvate}\\\text{carboxykinase}}]{\overset{\displaystyle GTP \quad GDP}{\curvearrowright}}$   $CH_2=\overset{\overset{\displaystyle PO_3^{2-}}{\overset{\displaystyle |}{\overset{\displaystyle O}{|}}}}{C}-\overset{\overset{\displaystyle O}{||}}{C}-O^-$   $+$   $CO_2$

Transfer of phosphate from GTP to the reactant occurs, forming GDP.  Since a kinase enzyme catalyzes the transfer of a phosphate group, the enzyme used here is phosphoenolpyruvate carboxykinase. One carboxylate ($-COO^-$) is also lost, breaking a C–C bond and forming $CO_2$.

**24.31**   This reaction converts one isomer to another because there is a change in position of the phosphate; thus, the enzyme is called a "mutase" (phosphoglucomutase).

**24.33**   Convert **A** to a structural formula, and then answer each question about glycolysis.

a.   $^-O-\overset{\overset{\displaystyle O^-}{|}}{\underset{\underset{\displaystyle O}{|}}{P}}=O$

$CH_2=\overset{\overset{\displaystyle O}{|}}{C}-\overset{\overset{\displaystyle O}{||}}{C}-O^-$

**A**
phosphoenolpyruvate

b. immediate precursor to **A**: 2-phosphoglycerate
c. formed from **A**: pyruvate

**24.35**   Compare the energy-investment phase and the energy-generating phase of glycolysis.

|  | Energy-Investment Phase: | Energy-Generating Phase: |
|---|---|---|
| a. reactant/final product | glucose, glyceraldehyde 3-phosphate | glyceraldehyde 3-phosphate, pyruvate |
| b. ATP used/formed | 2 ATP used | 4 ATP/glucose produced |
| c. coenzymes used or formed | no coenzymes used or formed | 2 $NAD^+$ used and 2 NADH produced per glucose |

**24.37**

a. steps that form ATP: 7, 10
b. steps that use ATP: 1, 3

c. step that forms a reduced coenzyme: 6
d. step that breaks a C–C bond: 4

**24.39** Six molecules of $CO_2$ are formed during the catabolism of glucose. The conversion of pyruvate to acetyl CoA forms one $CO_2$ for each pyruvate. The conversions of isocitrate to α-ketoglutarate and α-ketoglutarate to succinyl CoA in the citric acid cycle form one $CO_2$ for each acetyl CoA.

| | |
|---|---|
| 2 pyruvate to 2 acetyl CoA | 2 $CO_2$ released |
| 2 isocitrate to 2 α-ketoglutarate | 2 $CO_2$ released |
| 2 α-ketoglutarate to 2 succinyl CoA | 2 $CO_2$ released |
| | 6 $CO_2$ total |

**24.41**

a.   $C_6H_{12}O_6$ + 2 NAD$^+$ $\xrightarrow{\text{2 ADP} \quad \text{2 ATP}}$ 2 $CH_3COCO_2^-$ + 2 NADH + 2 H$^+$
glucose                                                      pyruvate

b.   $C_6H_{12}O_6$ $\xrightarrow{\text{2 ADP} \quad \text{2 ATP}}$ 2 $CH_3CH_2OH$ + 2 $CO_2$
glucose                                      ethanol

c.   pyruvate $\xrightarrow{\text{NADH + H}^+ \quad \text{NAD}^+}$ lactate

**24.43**

   a. Under aerobic conditions, pyruvate carbons become $CO_2$.
      Under anaerobic conditions, pyruvate carbons become lactate.
   b. Under aerobic conditions, NAD$^+$ and FAD are used and NADH and FADH$_2$ are formed.
      Under anaerobic conditions, NADH is used in converting pyruvate to lactate and NAD$^+$ is formed.

**24.45** Glycolysis takes glucose and breaks it into smaller units, producing pyruvate. Gluconeogenesis is the synthesis of glucose from lactate, amino acids, and glycerol.

**24.47** The rate of glycolysis is affected by each of the following conditions.

   a. low ATP concentration: increased         c. high carbohydrate diet: increased
   b. low ADP concentration: decreased         d. low carbohydrate diet: decreased

**24.49** In the Cori cycle, glucose in muscle is catabolized to pyruvate and then converted to lactate, which is transported to the liver. In the liver, lactate is converted to glucose by gluconeogenesis and then transported back to the muscle.

**24.51** The following metabolic products are formed from pyruvate under each condition.

   a. anaerobic conditions in the body: lactate         c. aerobic conditions: $CO_2$
   b. anaerobic conditions in yeast: ethanol

**24.53**    Each molecule of glucose produces 32 molecules of ATP during catabolism.

During glycolysis:
          2 ATP and 2 NADH generated
          1 NADH → 2.5 ATPs, so 2 NADH → 5 ATPs          7 ATPs from glycolysis
Conversion of 2 pyruvate to 2 acetyl CoA:
          2 NADH → 5 ATPs          5 ATPs from pyruvate
In the citric acid cycle, 20 additional ATPs are formed from
2 $CH_3COSCoA$:
          2 GTP (2 ATP), 6 NADH → 15 ATPs, 2 $FADH_2$ → 3 ATPs   <u>20 ATPs</u>
                         =   32 ATPs

**24.55**    In fermentation, the six carbons in glucose become two carbons in two molecules of $CO_2$ and four carbons in two molecules of $CH_3CH_2OH$.

**24.57**    When a fatty acid is converted to an acyl CoA, two phosphate bonds of ATP are hydrolyzed, forming AMP.  Therefore, 2 ATP equivalents are used.

**24.59**    For myristic acid, determine the number of molecules of acetyl CoA formed and the number of cycles of β-oxidation needed as in Example 24.3.

myristic acid, $C_{13}H_{27}CO_2H$: 14 carbons, [1] 7 molecules of acetyl CoA, [2] 6 cycles

**24.61**    For hexanoic acid, the compound depicted in the ball-and-stick model, determine the number of molecules of acetyl CoA formed and the number of cycles of β-oxidation needed as in Example 24.3.

hexanoic acid, $CH_3(CH_2)_4CO_2H$: 6 carbons: (a) 3 molecules of acetyl CoA; (b) 2 cycles

**24.63**    Calculate the number of molecules of ATP formed from myristic acid using the steps in Answer 24.18.

[1] 2 ATPs are required for the conversion of $CH_3(CH_2)_{12}CO_2H$ to $CH_3(CH_2)_{12}COSCoA$
    (−2 ATPs).
[2] 6 cycles of β-oxidation:
              6 NADH    ×    2.5 ATP/NADH    =    15 ATPs
              6 $FADH_2$    ×    1.5 ATP/$FADH_2$    =    <u>9 ATPs</u>
                        From reduced coenzymes:    24 ATPs

[3] From acetyl CoA:
              7 acetyl CoA   ×   10 ATP/acetyl CoA   =   70 ATPs

Total steps [1]–[3]:
            (−2) + 24 + 70    =   92 ATP molecules from myristic acid
                                    **Answer**

**24.65**

$$C_{15}H_{31}CO_2H$$

$$CH_3(CH_2)_{12}-CH_2-CH_2-\overset{\overset{O}{\|}}{C}-SCoA$$

$C_{16}$ acyl CoA

FAD   FADH$_2$

acyl CoA
dehydrogenase

**[1]
Oxidation**

$$CH_3(CH_2)_{12}-CH=CH-\overset{\overset{O}{\|}}{C}-SCoA$$

enoyl CoA hydratase

H$_2$O

**[2]
Hydration**

$$CH_3(CH_2)_{12}-\overset{OH}{\underset{|}{CH}}-\overset{H}{\underset{|}{CH}}-\overset{\overset{O}{\|}}{C}-SCoA$$

NAD$^+$

NADH + H$^+$

**[3]
Oxidation**

$$CH_3(CH_2)_{12}-\overset{\overset{O}{\|}}{C}-CH_2-\overset{\overset{O}{\|}}{C}-SCoA$$

acyl CoA
acyltransferase

HSCoA

**[4]
Cleavage**

$$CH_3(CH_2)_{12}-\overset{\overset{O}{\|}}{C}-SCoA \quad + \quad CH_3-\overset{\overset{O}{\|}}{C}-SCoA$$

C$_{14}$ acyl CoA          2 C's

**24.67**   Answer each question about decanoic acid.

a.   $$CH_3(CH_2)_6-\underset{\beta}{CH_2}-\underset{\alpha}{CH_2}-\overset{\overset{O}{\|}}{C}-OH$$

b.   $$CH_3(CH_2)_6-\underset{\beta}{CH_2}-\underset{\alpha}{CH_2}-\overset{\overset{O}{\|}}{C}-SCoA$$

c.   Five acetyl CoA molecules are formed, since there are 10 carbons.

d.   Four cycles of β-oxidation are needed for complete oxidation.

e.   [1] –2 ATPs for conversion to acyl CoA
[2] 4 cycles of β-oxidation:
     4 NADH x 2.5 ATP/NADH  = 10 ATPs
     4 FADH$_2$ x 1.5 ATP/FADH$_2$ = 6 ATPs
        From reduced coenzymes:  16 ATPs
[3] 5 acetyl CoA x 10 ATP/acetyl CoA = 50 ATPs

Total steps [1]–[3]
   (–2) + 16 + 50 = **64 molecules of ATP**

**24.69**   Calculate the number of moles of ATP formed per gram of decanoic acid.

$$\frac{64\ ATP}{mol\ decanoic\ acid} \quad \times \quad \frac{1\ mol}{172\ g} \quad = \quad \frac{0.37\ mol\ ATP}{g\ decanoic\ acid}$$

**24.71**   Ketosis is the condition under which ketone bodies accumulate.  Ketogenesis is the synthesis of ketone bodies from acetyl CoA.

**24.73** In poorly controlled diabetes, glucose cannot be metabolized due to a lack of insulin. Therefore, fatty acids are used for metabolism and ketone bodies are formed to a greater extent.

**24.75** Ketogenic amino acids are converted to acetyl CoA or a related thioester and can be converted to ketone bodies. Glucogenic amino acids are catabolized to pyruvate or another intermediate in the citric acid cycle and can be converted to glucose.

**24.77** Draw the products when each amino acid undergoes transamination as in Example 24.4.

a.
$$\underset{\overset{|}{\underset{H}{}}}{\overset{\overset{+}{N}H_3}{\underset{|}{H-C-CO_2^-}}} \longrightarrow H-\overset{O}{\overset{||}{C}}-CO_2^-$$

b.
$$\text{C}_6\text{H}_5-\text{CH}_2-\underset{\overset{|}{H}}{\overset{\overset{+}{N}H_3}{\underset{|}{C}}}-CO_2^- \longrightarrow \text{C}_6\text{H}_5-\text{CH}_2-\overset{O}{\overset{||}{C}}-CO_2^-$$

**24.79** Draw the products of transamination as in Example 24.4.

a.
$$CH_3-\underset{\overset{|}{H}}{\overset{\overset{+}{N}H_3}{\underset{|}{C}}}-CO_2^- + {}^-O_2CCH_2-\overset{O}{\overset{||}{C}}-CO_2^- \longrightarrow CH_3-\overset{O}{\overset{||}{C}}-CO_2^- + {}^-O_2CCH_2-\underset{\overset{|}{H}}{\overset{\overset{+}{N}H_3}{\underset{|}{C}}}-CO_2^-$$

b.
$$(CH_3)_2CHCH_2-\underset{\overset{|}{H}}{\overset{\overset{+}{N}H_3}{\underset{|}{C}}}-CO_2^- + {}^-O_2CCH_2-\overset{O}{\overset{||}{C}}-CO_2^- \longrightarrow (CH_3)_2CHCH_2-\overset{O}{\overset{||}{C}}-CO_2^- + {}^-O_2CCH_2-\underset{\overset{|}{H}}{\overset{\overset{+}{N}H_3}{\underset{|}{C}}}-CO_2^-$$

**24.81** Draw the products of oxidative deamination.

a.
$$(CH_3)_2CHCH_2-\underset{\overset{|}{H}}{\overset{\overset{+}{N}H_3}{\underset{|}{C}}}-CO_2^- \longrightarrow (CH_3)_2CHCH_2-\overset{O}{\overset{||}{C}}-CO_2^- + NH_4^+$$

b.
$$\text{C}_6\text{H}_5-CH_2-\underset{\overset{|}{H}}{\overset{\overset{+}{N}H_3}{\underset{|}{C}}}-CO_2^- \longrightarrow \text{C}_6\text{H}_5-CH_2-\overset{O}{\overset{||}{C}}-CO_2^- + NH_4^+$$

**24.83** Use Figure 24.10 to determine what metabolic intermediates are produced from each amino acid.

a. phenylalanine: acetoacetyl CoA and fumarate
b. glutamic acid: α-ketoglutarate

c. asparagine: oxaloacetate
d. glycine: pyruvate

**24.85** Yes, amino acids can be both ketogenic and glucogenic. Phenylalanine, for example, can be degraded by multiple routes and is therefore both ketogenic and glucogenic.

**24.87** A spiral pathway involves the same set of reactions repeated on increasingly smaller substrates; example: β-oxidation of fatty acids. A cyclic pathway is a series of reactions that regenerates the first reactant; example: citric acid cycle.

24.89 Glucose is metabolized anaerobically by yeast to form $CO_2$ and ethanol. The $CO_2$ causes bread to rise.

24.91 Pyruvate is metabolized anaerobically to lactate in the cornea, because the blood supply (and therefore oxygen) is very limited.

24.93 The Atkins diet calls for the ingestion of only protein and fat. The diet has an absence of carbohydrates to be metabolized, so ketone bodies are formed.

24.95 Compare the reactant and product to determine what type of reaction has occurred.

a. $CH_3-\overset{O}{\overset{||}{C}}-CH_2OH$ $\xrightarrow{\text{kinase}}$ $CH_3-\overset{O}{\overset{||}{C}}-CH_2OPO_3{}^{2-}$
addition of phosphate

b. $CH_3-\overset{O}{\overset{||}{C}}-CH_2OH$ $\xrightarrow{\text{dehydrogenase}}$ $CH_3-\overset{O}{\overset{||}{C}}-CHO$
loss of $H_2$

c. $CH_3-\overset{O}{\overset{||}{C}}-CH_2OH$ $\xrightarrow{\text{carboxylase}}$ $HO_2CCH_2-\overset{O}{\overset{||}{C}}-CH_2OH$
addition of $CO_2$

24.97 Calculate the number of ATP molecules from the catabolism of one glycerol molecule.

glycerol $\xrightarrow{[1]}$ glycerol 3-phosphate $\xrightarrow{[2]}$ dihydroxyacetone phosphate $\xrightarrow{[3]}$ pyruvate $\xrightarrow{[4]}$ acetyl CoA $\xrightarrow{[5]}$ $CO_2$

[1] 1 ATP required for conversion of glycerol to glycerol 3-phosphate      −1 ATP
[2] glycerol 3-phosphate → dihydroxyacetone phosphate:
     1 NADH → 2.5 ATPs      2.5 ATPs
[3] dihydroxyacetone phosphate → pyruvate:
     2 ATPs + 1 NADH → 2.5 ATPs =      4.5 ATPs
[4] pyruvate → acetyl CoA:
     1 NADH → 2.5 ATPs      2.5 ATPs
[5] acetyl CoA → $CO_2$:
     1 acetyl CoA × 10 ATP/acetyl CoA =      10 ATPs

**Total steps [1]–[5]:**      **18.5 ATPs**

24.99 Calculate the amount of ATP generated during catabolism of a gram of glucose compared to a gram of arachidic acid as in Section 24.7.

$$\frac{32 \text{ mol ATP}}{1 \text{ mol glucose}} \times \frac{1 \text{ mol glucose}}{180.2 \text{ g glucose}} = \frac{0.18 \text{ mol ATP}}{1 \text{ g glucose}} \quad \text{(Section 24.7)}$$

Answer 24.19 $\longrightarrow$ $$\frac{134 \text{ mol ATP}}{1 \text{ mol arachidic acid}} \times \frac{1 \text{ mol arachidic acid}}{312.5 \text{ g arachidic acid}} = \frac{0.429 \text{ mol ATP}}{1 \text{ g arachidic acid}}$$

One gram of arachidic acid produces more ATP than one gram of glucose (0.429 mol/g compared to 0.18 mol/g). This result supports the fact that lipids are more efficient at storing energy than carbohydrates.

# NOTES